4th International Conference on Applied Mathematics, Modeling and Simulation (AMMS 2021)

AA007198

Journal of Physics: Conference Series Volume 2068

Guangzhou, China
17 – 18 September 2021

ISBN: 978-1-7138-5039-7
ISSN: 1742-6588

Printed from e-media with permission by:

Curran Associates, Inc.
57 Morehouse Lane
Red Hook, NY 12571

Some format issues inherent in the e-media version may also appear in this print version.

This work is licensed under a Creative Commons Attribution 3.0 International Licence.
Licence details: http://creativecommons.org/licenses/by/3.0/.

No changes have been made to the content of these proceedings. There may be changes to pagination and minor adjustments for aesthetics.

Printed with permission by Curran Associates, Inc. (2022)

For permission requests, please contact the Institute of Physics at the address below.

Institute of Physics
Dirac House, Temple Back
Bristol BS1 6BE UK

Phone: 44 1 17 929 7481
Fax: 44 1 17 920 0979

techtracking@iop.org

Additional copies of this publication are available from:

Curran Associates, Inc.
57 Morehouse Lane
Red Hook, NY 12571 USA
Phone: 845-758-0400
Fax: 845-758-2633
Email: curran@proceedings.com
Web: www.proceedings.com

TABLE OF CONTENTS

Journal of Physics: Conference Series **2068** (2021) 011001 doi:10.1088/1742-6596/2068/1/011001

Preface

Thanks for your participations in 2021 4th International Conference on Applied Mathematics, Modeling and Simulation (AMMS2021), which was re-scheduled to be held in virtual meeting via Tencent Meeting on September 17, 2021 due to the current severe situation of the epidemic worldwide. AMMS2021 virtual meeting aims to provide an excellent and valuable opportunity for experts to exchange ideas and results covering Mathematics, Optimization Methods, Computing Methods and Advanced Computing, and strengthen the collaboration among researchers, faculty, and doctors from both academia and industry.

With its strong emphasis on innovative approaches, AMMS2021 Online Meeting offered a chance for scientists and educators working in different areas and help them advance their own research and forge new professional relationships and collaborations. This virtual meeting including Keynote Speech, Oral presentations, Poster presentations, Q & A section started at 09:00 am, Beijing time on September 17, 2021. And our staff set up a test room before the formal meeting to ensure the stability of all participants including video, audio, screen sharing, network connection, etc.

Prof. Mijia Yang from North Dakota State University, USA make keynote speech on Holistic Structural Displacement Monitoring and Condition Assessment through a Computer Vision Approach on 09:20-10:00, September 17, 2021. Dr. Ranjit Singh Sarban Singh from Universiti Teknikal Malaysia Melaka (UTeM), Malaysia shared his viewpoints on Constructive of Multi-Renewable Energy Resources Controller Modelling and Simulation – Simulink MATLAB on 10:00-10:40, September 17, 2021. Prof. Alexey Kolesnikov from Siberian State University of Geosystems and Technologies, Russia exhibited his advanced findings in Modeling of Traffic Flows and Supply Chains Based on Geospatial Knowledge on 10:40-11:20, September 17, 2021. And Dr. M. Madhiarasan from Indian Institute of Technology, India shared keynote speech video about Artificial Intelligence: Trends, Opportunity, and Applications on 11:20-12:00, September 17, 2021 with his poor network connection.

Here are 132 papers eventually in AMMS2021 Call for Papers. Combined with professional reviewing of AMMS2021 Technical Program Committee member and Ithenticate software detecting, 56 papers were accepted and selected in AMMS2021 Proceedings. All these accepted papers have been reviewed by 2-3 reviewers and editors majoring in relevant research areas and be detected instances of overlapping and similar text in submitted manuscripts via iThenticate software before submitting for publication in Journal of Physics: Conference Series.

Based on the main topics and directions of accepted papers submissions and publication chairs' suggestions, here are 2 parts in AMMS2021 Conference Proceedings as below:

1. Applied Mathematics and Algorithm Analysis
2. Modeling, Simulation and Optimization

Sincerely thanks for IOP Publishing with efficient publication for AMMS2021 paper submissions, all outstanding authors with original research findings, all the members and reviewers in AMMS2021 Technical Program Committee with professional reviewing and checking. We truly hope you made benefitial academic sharing and exchanging in this AMMS2021 virtual meeting. Looking forward for the future cooperation with you.

AMMS2021 Organization Committees

Content from this work may be used under the terms of the Creative Commons Attribution 3.0 licence. Any further distribution of this work must maintain attribution to the author(s) and the title of the work, journal citation and DOI.
Published under licence by IOP Publishing Ltd

General Chairs
Prof. Elżbieta Macioszek, Silesian University of Technology, Poland
Prof. Giuseppe Failla, University of Reggio Calabria, Italy

General Co-Chairs
Prof. Smail Bachir, Poitiers University, France
Prof. Mehul Mahrishi, Swami Keshvanand Institute of Technology, India
Prof. Shrikant D Pawar, Yale University, USA

Publication Chairs
Prof. Ricardo Branco, University of Coimbra, Portugal
Prof. Marcelo V. Garcia, University of the Basque Country, Spain

Publication Co-Chairs
Dr. Hsiao-Chi Ling, Chiao Tung University, Taiwan
Dr. Saša Ćuković, ETH Zurich, Switzerland

International Scientific Committees
Dr. Ramakrishna Hegde, Vidyavardhaka College of Engineering, India
Dr. S.BalaKrishnan, Sri Krishna College of Engineering and Technology, India
Dr. Kim Huat Yeap, Westlake International School, Malaysia
Dr. Mahmood Mohassel Feghhi, University of Tabriz, Iran
Dr. Natanael Karjanto, Sungkyunkwan University, South Korea
Dr. K. Dinesh Kumar, VIT University, India
Dr. Isham A lzoubi, University of Tehran, Iran
Dr. Niusha Shafiabady, Azad University, Iran
Dr. Arun Kishor Johar, Malaviya National Institute of Technology Jaipur, India
Dr. Amirreza Ghadimi Avval, University of Arkansas, USA
Dr. Shyamalendu Kandar, Indian Institute of Engineering Science and Technology, India
Prof. Farhad Qaderi, Tarbiat Modares University, Iran
Prof. Konstantinos Giannoutakis, University of Macedonia, Greece
Dr. Jan Kubicek, VSB-Technical University of Ostrava, Czech Republic
Prof. Abdolhossein Hemmati-Sarapardeh, Shahid Bahonar University of Kerman, Iran
Dr. Mohamed Arezki Mellal, M'Hamed Bougara University, Algeria
Dr. Sandeep Patil, Nissan Technical Center, United States
Dr. Ertugrul Dogruluk, University of Minho, Portugal
Prof. Carla Osthoff, National Laboratory for Scientific Computing, Brazil
Dr. Md Nazirul Islam Sarker, Neijiang Normal University, China
Dr. Ayman El-Saleh, A'Sharqiyah University, Oman
Dr. Djamel Boutagouga, University Larbi Tébessi, Algeria
Dr. Catherine Nayer Tadros Tawdros, Alexandria University, Egypt
Prof. Omar Abu Arqub, Al-Balqa Applied University, Jordan
Dr. Shahram Babaie, Islamic Azad University, Iran
Dr. Anupama Kumari, Indian Institute of Technology, India
Dr. Hassan Abdalla, University of Udine, Italy
Dr. Cheng Siong Chin, Newcastle University, Singapore
Dr. S. Revathi, VIT University, India
Dr. Isha Suwalka, Department of Electronics and Communication, India
Prof. Prem Kumar Singh, GITAM University, India
Dr. Shadha Mohammed Al Amri, University of Technology and Applied Science, Oman
Dr. Manisha G Waje, G H Raisoni College of Engineering and Management, India
Dr. Imane Fathi, Higher Institute of Health Sciences, Morocco
Prof. Ioan-Lucian Popa, "1 Decembrie 1918" University of Alba Iulia, Romania
Dr. Sangeetha V, M.S.Ramaiah Institute of Technology, India
Dr. Chen Zhang, KTH Royal Institute of Technology, Sweden

Dr. Mohamed El-Amine Slimani, University of Science and Technology "USTHB", Algeria
Prof. Safa Bhar Layeb, University of Tunis El Manar, Tunisia
Dr. Miodrag Forcan, University of East Sarajevo, Bosnia and Herzegovina
Dr. Agnese Rapposelli, "G. d'Annunzio" University of Chieti-Pescara, Italy
Prof. Basil T. Wong, Swinburne University of Technology Sarawak Campus, Malaysia
Dr. Vaishali Gupta, IPS Academy, India
Dr. Gordana Jovanovic Dolecek, University of Sarajevo, Mexico
Dr. Chaoxian Wu, University of Liverpool, UK
Dr. Medhat Ahmed Rakha, Suez Canal University, Egypt
Dr. Nikolaos Dimitriou, Information Technologies Institute, Greece
Dr. Abel Zhou, University of Canberra, Australia
Prof. Sofyan A Taya, Islamic University, Gaza, Palestine
Dr. Ali Ghorbanian, Esfarayen University of Technology, Iran
Prof. Schehrazad SELMANE, University of Sciences and Technology Houari Boumediene (USTHB), Algeria
Dr. Haniyeh Fallah, Amirkabir University of Technology, Iran
Prof. Ali Jahani, Research Center of Environment and Sustainable Development, Iran
Prof. Damir Magdić, Josip Juraj Strossmayer University of Osijek, Croatia
Prof. Shweta Sharma, MIT- World Peace University, India
Dr. Maryam Saffariha, University of Tehran, Iran
Prof. José Raúl Castro, Universidad Técnica Particular de Loja, Ecuador
Prof. Bapan Ghosh, Indian Institute of Technology Indore, India
Prof. Hina Sadaf, DBS&H, CEME-NUST, Pakistan
Prof. Yilun Shang, Northumbria University, UK
Prof. Yuri Leonidovich Menshikov, Dnipro University, Ukraine
Dr. Bappa Acherjee, BIT Mesra, India
Dr. Jency Rubia J, Anna Universit, India
Prof. Arnulfo Luévanos Rojas, Institute of Multidisciplinary Researches, Coahuila
Prof. José Joaquim de Moura Ramos, University of A Coruña, Spain
Prof. Gheorghe Grigoras, "Gheorghe Asachi" Technical University of Iasi, Romania
Dr. Jan Kubicek, VSB-Technical University of Ostrava, Czech Republic
Prof. Alexei Shishkin, Moscow State University, Russia
Prof. Wei Xie, Institute of Disaster Prevention, China
Prof. Pavlo Maruschak, Ternopil Ivan Puluj National Technical University, Ukraine
Dr. Raghu N, Jain University, India
Prof. Mihaela Popescu, Institution: University of Craiova, Romania

Peer review declaration

All papers published in this volume of Journal of Physics: Conference Series have been peer reviewed through processes administered by the Editors. Reviews were conducted by expert referees to the professional and scientific standards expected of a proceedings journal published by IOP Publishing.

- **Type of peer review: Double-blind**
- **Conference submission management system: All the papers were submitted papers via the conference email: amms2021@163.com**
- **Number of submissions received: 132**
- **Number of submissions sent for review: 105**
- **Number of submissions accepted: 56**
- **Acceptance Rate (Number of Submissions Accepted / Number of Submissions Received X 100):42.4%**
- **Average number of reviews per paper: 2**
- **Total number of reviewers involved:81**
- **Any additional info on review process: AMMS2021 uses a plagiarism software to detect instances of overlapping and similar text in submitted manuscripts.**
- **Contact person for queries:**
  Ms. Yolanda Xu, Advanced Science and Industry Research Center, Hong Kong, amms2021@163.com

Content from this work may be used under the terms of the Creative Commons Attribution 3.0 licence. Any further distribution of this work must maintain attribution to the author(s) and the title of the work, journal citation and DOI.
Published under licence by IOP Publishing Ltd

Controlled Fractional Order Processes with Distributed Parameters

Mashrabjon Mamatov[1,*] and Xakimjon Alimov[2]

[1]National University of Uzbekistan named after M. Ulugbek, Tashkent, Uzbekistan
[2]Jizzakh State Pedagogical Institute, Jizzakh, Uzbekistan

[*]Corresponding author email: mamatovmsh@mail.ru

Abstract. This work is devoted to the study of the problem of pursuit in systems controlled with distributed parameters of fractional order. Derivatives with respect to spatial variables are ordinary, integer, and even of arbitrary order. Designed and studied sampling schemes. A numerical method is constructed for finding strategies in suitable classes and for constructing the corresponding control laws. Theorems are proved for the possibility of completing the pursuit by the finite difference method, and they can be used to solve finite-difference optimal control problems or numerical solutions of differential games of fractional order. Problems of the type under study are encountered in modeling the processes of economic growth and in problems of stabilizing dynamic systems.

1. Introduction

There are many works devoted to differential equations of fractional order, among them we mention research [1,2]. They mainly study fractional-order differential equations with lumped parameters. Boundary-value problems are considered in detail in [3,4], and non-self-ad joint integral operators accompanying these boundary-value problems are investigated by methods of perturbation theory. Many, especially interdisciplinary problems, can be modeled using fractional derivatives. For example, a nonlinear oscillation of an earthquake can be modeled with a fractional derivative [3], and a hydrodynamic model of motion with fractional derivatives [4] can eliminate the disadvantage arising from a continuous traffic flow.

Numerous studies are devoted to pursuit and evasion problems for various classes of differential and discrete games [5-11]. In [6], linear differential games are considered, the main model for which is the process of pursuit of one controlled object by another controlled object, and fundamental results of the theory of differential games described by systems with lumped parameters are obtained. In [7, 9], the pursuit problem in controlled systems with high-order distributed parameters is investigated. Sufficient conditions are obtained for the possibility of completing the pursuit by the finite difference method. And in the article [10] the problem of pursuit of differential games is discussed. In [18], the method of processing digital images as a game problem with distributed parameters was first investigated. This article is related to works [6-10].

2. Methods

The general research methodology is based on the fundamental results of the mathematical theory and methods of sequential systems, the theory of linear multistep games, Bellman's methods of discrete control problems, the theory of dynamic and integer programming. The work also partially uses some numerical methods for solving fractional differential equations. Such methods have been investigated

Content from this work may be used under the terms of the Creative Commons Attribution 3.0 licence. Any further distribution of this work must maintain attribution to the author(s) and the title of the work, journal citation and DOI.
Published under licence by IOP Publishing Ltd

by many authors. For example, the variation iterative method, the method of homotopy perturbations, spectral methods and other methods [1-5].
In this paper, we use Pontryagin methods developed by specialists for discrete game pursuit problems. The well-known methods of discrete games are inapplicable to the corresponding discrete games, which are obtained after applying the method of finite differences to an equation with distributed parameters of fractional order. Basically, discrete games were studied as an independent problem or the corresponding differential games were of a whole order. Consequently, it was necessary to develop new sufficient conditions that could be applied to solve differential games described by an equation with distributed parameters of fractional order. By modifying the known methods, we managed to obtain such a sufficient condition. Thus, the mathematical apparatus used in this work is new.
Other methods are known as well. Currently, only algorithms for solving matrix games have been developed. Let us point out the methods of Brown, Neumann and their numerous modifications. Some approaches to solving continuous games are given in [11], but the algorithms proposed in them turn out to be very cumbersome in the case, especially when variable multidimensional vectors.
The methods for solving discrete games proposed in this paper are new; it allows solving difference game problems with initial and boundary conditions for differential games of fractional order.

3. Results

We consider the following game problem described by the fractional-order equation

$$D_{0t}^{\alpha} z + (-1)^m \frac{\partial^{2m} z}{\partial x^{2m}} + u(x,t) = \upsilon(x,t), \qquad m = 1,2,..., \tag{1}$$

$$z(x,0) = f(x), \; 0 < x < 1; \tag{2}$$

$$\frac{\partial^{2p} z(0,t)}{\partial x^{2p}} = \frac{\partial^{2p} z(1,t)}{\partial x^{2p}} = 0, \quad p = 0,1,..., m-1; \;\; 0 \le t \le T, \tag{3}$$

$z \in R^1$, $u, \upsilon -$ control parameters, $u -$ pursuit department $u \in \overline{P}$, $\upsilon -$ escape control $\upsilon \in \overline{Q}$, $x \in \Omega = \{x : 0 \le x \le 1\}$, $(x,t) \in Q_T = \{(x,t) : x \in \Omega, \; 0 < t \le T\}$, $S_T = \{(x,t) | x \in \partial\Omega, \; t \in (0,T]\}$, $\partial\Omega$ – area boundary $\Omega, 0 < \alpha \le 1$.

The fractional derivative will be understood in the usual sense. $D_{0t}^{\alpha} z(x,t) = \frac{\partial}{\partial t} I_{0t}^{1-\alpha}$, where - $I_{0t}^{1-\alpha}$ partial fractional integrals [1].

$$I_{0t}^{\alpha} z(x,t) = \frac{1}{\Gamma(\alpha)} \int_0^t \frac{z(x,s)}{(t-s)^{1-\alpha}} ds$$

In R^1 selected non-empty terminal set $\overline{M}_1 \subset R^1$.

Definition 1. In problem (1) - (3), it is possible ε – completion $\varepsilon > 0$ pursuit from starting position $f(\cdot)$, if there are number $T = T(f(\cdot))$ and function $u(\upsilon, x, t) \in \overline{P}$, $\upsilon \in \overline{Q}$, $x \in \Omega$, $t \in [0,T]$, such that for an arbitrary function $\upsilon_0(x,t) \in \overline{Q}$, $x \in \Omega$, $t \in [0,T]$ solution $z_0(x,t)$ of problem (1) - (3), where $u = u(\upsilon_0(x,t), x, t)$, $\upsilon = \upsilon_0(x,t)$, falls into the set $\varepsilon I + \overline{M}1$, at some $(\tilde{x}, \tilde{t})$, $\tilde{x} \in \Omega$, $\tilde{t} \in [0,T]$: $z_0(\tilde{x}, \tilde{t}) \in \varepsilon I + \overline{M}1$, where $I = (-1,1)$.

On the segment $[t_k, t_{k+1}]$ get

$$D_{0t}^{\alpha} z(x,t)\Big|_{t_k} = \frac{1}{\Gamma(1-\alpha)} \left(\frac{z(x,t_k)}{(t_{k+1} - t_k)^{\alpha}} + \int_{t_k}^{t_{k+1}} \frac{z'(x,s)ds}{(t_{k+1} - s)^{\alpha}} \right)$$

Hence we have

$$D_{0t}^{\alpha}z(x,t)\Big|_{t_k}\approx\frac{1}{\Gamma(1-\alpha)}\left(\frac{z(x,t_k)}{(t_{k+1}-t_k)^{\alpha}}+\frac{z(x,t_{k+1})-z(x,t_k)}{l}\int_{t_k}^{t_{k+1}}\frac{ds}{(t_{k+1}-s)^{\alpha}}\right)=$$

$$=\frac{z(x,t_{k+1})-\alpha z(x,t_k)}{l^{\alpha}(1-\alpha)\Gamma(1-\alpha)}=\frac{z(x,t_{k+1})-\alpha z(x,t_k)}{l^{\alpha}\,\Gamma(2-\alpha)}.$$

Problems (1) - (3), are replaced by the following equations

$$\frac{z_{i,k+1}-\alpha z_{i,k}}{l^{\alpha}\Gamma(2-\alpha)}+\frac{(-1)^m}{h^{2m}}\sum_{\alpha=1}^{n}\left[a\Delta^{2m}z_{i-m,k+1}+(1-a)\Delta^{2m}z_{i-m,k}\right]=-u_{i,k}+\upsilon_{i,k},\tag{4}$$

where h and l – the values of the steps, respectively x and t, $i=1,2,\ldots,r-1$, $k=0,1,\ldots,\theta-1$,

$$z_{i,k}=z(ih,kl),\ z_{i,0}=f(ih),\ i=1,2,\ldots,r-1,\ z_{-i,k}=-z_{i,k},\ i=0,1,\ldots,m-1,\ k=0,1,\ldots,\theta,$$

$$z_{r+i,k}=-z_{r-i,k},\ i=0,1,\ldots,m-1,k=0,1,\ldots,\theta,$$

$$\Delta^{2m}z_{i-m,k}=\sum_{j=0}^{2m}(-1)^jC_{2m}^{j}z_{i+m-j,k},z_{i-m,k}=z((i-m)h,\,kl),$$

$$z_{i+m-j,k}=z((i+m-j)h,kl),\ \ C_{2m}^{j}=\frac{(2m)!}{j!(2m-j)!},\ \ 0\le a\le 1,\ u_{i,k}=u(ih,kl),\ \ \upsilon_{i,k}=\upsilon(ih,kl)$$

It is easy to verify [5] that the solution satisfies the following convergence rate estimate

$$\left\|(z)_{hl}-z_{i,k}\right\|_{\Phi_{hl}}\le K_1l+K_2h^2 \text{ at } a\ne\frac{1}{2},\ \left\|(z)_{hl}-z_{i,k}\right\|_{\Phi_{hl}}\le\overline{K}_1l^2+\overline{K}_2h^2 \text{ at } a=\frac{1}{2},\tag{5}$$

Φ_{hl} – space of grid functions, $\|\cdot\|_{\Phi_{hl}}$ – the norm of this space, $K_1,\overline{K_1},$, $K_2,\overline{K_2}$, – constants.
After some transformations and notation, we obtain the following equation

$$Az_{k+1}-Bz_k=-\Gamma(2-\alpha)\ l^{\alpha}u_k+\Gamma(2-\alpha)\ l^{\alpha}\upsilon_k,\ k=0,1,2,\ldots,\theta-1,\ \ z_0=\overline{f},\tag{6}$$

where z_k, u_k, υ_k – H - matrices, $H=r-1$, here

$$z_k=(z_{1,k},z_{2,k},\ldots,z_{r-1,k})^T,$$

$$A=E+a\frac{(-1)^m l^{\alpha}\Gamma(2-\alpha)}{h^{2m}}C^m;\ \ B=\alpha E+(1-a)\frac{(-1)^m l^{\alpha}\Gamma(2-\alpha)}{h^{2m}}C^m,$$

Where E - identity matrix, $C=(c_{ij})$ - a square matrix in which $c_{ii}=-2$ and $c_{i-1,i}=c_{i,i+1}=1$. It is clear that there is $A^{-1}\ne 0$. We multiply both sides of equation (6) by A^{-1}, and denoting

$$A^{-1}B=G,\ A^{-1}=F \text{ get}$$

$$z_{k+1}=Gz_k-F\Gamma(2-\alpha)\ l^{\alpha}u_k+F\Gamma(2-\alpha)\ l^{\alpha}\upsilon_k,\quad k=0,1,2,\ldots,\theta-1,\ z_0=\overline{f}.\tag{7}$$

Let M some set be selected in in R^H .

Definition 2. In Game (7) from $z_0=\overline{f}$ you can complete the chase in N steps, if in any sequence $\overline{\upsilon}_k,k=\overline{0,N-1}$ one can construct such a sequence $\overline{u}_k,k=\overline{0,N-1}$, what's the solution $z_0,\overline{z}_k,k=\overline{1,N}$ equations $z_{k+1}=Gz_k-F\Gamma(2-\alpha)\ l^{\alpha}\left(\overline{u}_k-\overline{\upsilon}_k\right)$, $k=\overline{0,N-1}$, for some $d\le N$ falls on M.

Let's pretend that $M=M_0+M_1$, where M_0 – $(H-\gamma)$ - dimensional linear subspace R^H. M_1 – subset of subspace L – orthogonal complement M_0 in R^H. Through Π denote the matrix of the orthogonal projection operator from R^H to L. $A+B$ and $A\overset{*}{-}B$ algebraic sum and geometric difference of sets A,B respectively. Let $M_1=\underbrace{\overline{M}_1\times\overline{M}_1\times\ldots\times\overline{M}_1}_{\gamma}$, $1\le\gamma\le H$,

$$\beta_m(\cdot)=\left\{\beta_0,\beta_1,\ldots,\beta_{m-1}:\beta_k\ge0,\ \sum_{k=0}^{m-1}\beta_k=1\right\}, \text{ and}$$

$$W(\beta_m(\cdot))=\sum_{k=0}^{m-1}\left[(\beta_k M_1+\Pi G^k F\Gamma(2-\alpha)\ l^\alpha P)\overset{*}{-}\Pi G^k F\Gamma(2-\alpha)\ l^\alpha Q\right],\ n=\overline{0,N}. \text{ We put}$$

$$W_1(0)=M_1,\ W_1(m)=\bigcup_{\beta_m(\cdot)} W(\beta_m(\cdot)),\ u_n,\upsilon_n. \tag{8}$$

Theorem 1. If the inclusion takes place $\Pi G^m z_0\in W_1(m)$, then in the game (7) the pursuit can be completed from the given initial position.

$$W(0)=W\{0\},\ W(m)=\sum_{k=0}^{m-1}\left[\Pi G^k F\Gamma(2-\alpha)\ l^\alpha P\overset{*}{-}\Pi G^k F\Gamma(2-\alpha)\ l^\alpha Q\right],\ W_2(m)=M_1+W(m),\ m=1,2,\ldots,\theta \tag{9}$$

Theorem 2. Let's pretend that $\Pi G^m z_0\in W_2(m)$. Then in game (7) from $z_0=\overline{f}$ you can complete the pursuit.

Let now $W_3(0)=M_1$, $W_3(1)=[W_3(0)+\Pi F\Gamma(2-\alpha)\ l^\alpha P]\overset{*}{-}\Pi F\Gamma(2-\alpha)\ l^\alpha Q, \ldots$,

$$W_3(m)=\left[W_3(m-1)+\Pi G^{m-1}F\Gamma(2-\alpha)\ l^\alpha P\right]\overset{*}{-}\Pi G^{m-1}F\Gamma(2-\alpha)\ l^\alpha Q.$$

Theorem 3. If $\Pi G^m z_0\in W_3(m)$, then in game (7) from $z_0=\overline{f}$ you can complete the pursuit for m steps.

Theorem 4. Let in inequality (5) $K_1 l+K_2 h^2<\varepsilon$, at $a\ne\frac{1}{2}$, and $\overline{K}_1 l^2+\overline{K}_2 h^2<\varepsilon$, at $a=\frac{1}{2}$, and in game (7) from point $z_0=\overline{f}$ the pursuit can be completed in the sense of Definition 2. Then the game can be completed in the sense of Definition 1.

4. Discussion

Proof of Theorem 1. Instead of including the condition of the theorem, consider the equivalent inclusion

$$\Pi G^m z_0\in\sum_{k=0}^{m-2}\left[(\overline{\beta}_k M_1+\Pi G^k F\Gamma(2-\alpha)\ l^\alpha P)\overset{*}{-}\Pi G^k F\Gamma(2-\alpha)\ l^\alpha Q\right]+$$
$$(\overline{\beta}_{m-1}M_1+\Pi G^{m-1}F\Gamma(2-\alpha)\ l^\alpha P)\overset{*}{-}\Pi G^{m-1}F\Gamma(2-\alpha)\ l^\alpha Q \tag{10}$$

Where (10) for $\overline{\upsilon}_0$ get

$$\Pi G^m z_0+\Pi G^{m-1}F\Gamma(2-\alpha)\ l^\alpha\overline{\upsilon}_0\in\sum_{k=0}^{m-2}\left[(\overline{\beta}_k M_1+\Pi G^k F\Gamma(2-\alpha)\ l^\alpha P)\overset{*}{-}\Pi G^k F\Gamma(2-\alpha)\ l^\alpha Q\right]+$$
$$\overline{\beta}_{m-1}M_1+\Pi G^{m-1}F\Gamma(2-\alpha)\ l^\alpha P \tag{11}$$

Control $\overline{u}_0\in P$ construct as a solution to the following equation

$$\Pi G^{m-1}F\Gamma(2-\alpha)\ l^\alpha\overline{\upsilon}_0-\Pi G^{m-1}F\Gamma(2-\alpha)\ l^\alpha\overline{u}_0=\overline{\beta}_{m-1}a_1,\quad a_1\in M_1.$$

Further, by virtue of (11), we have

$$\Pi G^{m-1}(Gz_0-F\Gamma(2-\alpha)\ l^\alpha\overline{u}_0+F\Gamma(2-\alpha)\ l^\alpha\overline{\upsilon}_0)\in$$

$$\sum_{k=0}^{m-2}\left[(\overline{\beta}_k M_1 + \Pi G^k F\Gamma(2-\alpha)\ l^\alpha P) \overset{*}{-} \Pi G^k F\Gamma(2-\alpha)\ l^\alpha Q\right] + \overline{\beta}_{m-1} a_1 .$$

Then (7) we find

$$\Pi G^{m-1} z_1 \in \sum_{k=0}^{m-2}\left[(\overline{\beta}_k M_1 + \Pi G^k F\Gamma(2-\alpha)\ l^\alpha P) \overset{*}{-} \Pi G^k F\Gamma(2-\alpha)\ l^\alpha Q\right] + \overline{\beta}_{m-1} a_1 .$$

In the same way, we obtain the inclusion

$$\Pi G^{m-2} z_2 \in \sum_{k=0}^{m-3}\left[(\overline{\beta}_k M_1 + \Pi G^k F\Gamma(2-\alpha)\ l^\alpha P) \overset{*}{-} \Pi G^k F\Gamma(2-\alpha)\ l^\alpha Q\right] + \overline{\beta}_{m-1} a_1 + \overline{\beta}_{m-2} a_2 ,$$

$$a_2 \in M_1 ,$$

etc. Thus, we have

$$\Pi z_m = \overline{\beta}_{m-1} a_1 + \overline{\beta}_{m-2} a_2 + \ldots + \overline{\beta}_0 a_m \in \overline{\beta}_{m-1} M_1 + \overline{\beta}_{m-2} M_1 + \ldots + \overline{\beta}_0 M_1 \in M_1 ,$$

hence we have

$$z_m \in M .$$

Theorem 2-4 is proved similarly.

5. Conclusion

In conclusion, it can be argued that the application of the finite difference method to solving fractional order differential pursuit games is a powerful and effective tool. The problem is investigated when the movements of the players are described by fractional order in time and high order in spatial variables. When solving the problem, using the finite difference method, we go over the game, which is described by the grid equation, after some transformations we went over to the discrete game. With the help of auxiliary theorems, sufficient conditions are obtained for solving the pursuit problem. The properties of the dynamics of change in the set of meeting points of players are studied, with the help of which the game is resolved for the case under consideration. In many works, the stability of the grid equation has been studied, which consists in finding sufficient conditions under which the error tends to zero, or at least remains bounded.

If the rounding errors generated during the counting process tend to decrease, or at least not increase, as they arise, then Eq. (4) is called stable. For the case under consideration, it is naturally assumed that the stability conditions are satisfied. The results obtained can be used to solve problems in the theory of dynamic games, in particular, in the development of the method of resolving functions.

References

[1] Kilbas, A.A., Srivastava, H.M. and Trujillo, J.J. 2006. *Theory and Applications of Fractional Differential Equations*. (Amsterdam: North-Holland/American Elsevier) p 500.

[2] Mainardi F 1997 *Fractional calculus: some basic problems in continuum and statistical mechanics, Fractals and fractional calculus in continuum mechanics* (Springer-Verlag, New York) pp 291-348.

[3] J. He, 1998 Nonlinear oscillation with fractional derivative and its applications, *International Conference on Vibrating Engineering, Dalian, China*, pp 288-291.

[4] J. He, 1999 Some applications of nonlinear fractional differential equations and their approximations, *Bull. Sci. Technol, Vol15* pp 86-90.

[5] Pontreagin L. 1980 Linear differential games of pursuit *(Mat. Sb. Vol 112)* pp 307-330.

[6] Alimov, K, Mamatov, M 2014 Solving a pursuit problem in high-order controlled distributed systems (*Siberian Advances in Mathematics, Vol 24*) pp 229–239.

[7] Mamatov M, Tashmanov E, Alimov H 2015 Zwquasi Linear Discrete Games of Pursuit Described by High Order Equation Systems *(Automatic Control and Computer Sciences, Vol 49)* pp 148-152.

[8] Mamatov M, Sobirov K 2020 Journal of Mathematical Sciences (US), **245(3)**, pp 332–340.

[9] Mamatov M, Tashmanov E, Alimov K 2015 Automatic Control and Computer Sciences, **49(3)**, pp 148-152.

[10] Tukhtasinov M, Mamatov M 2009 Differential Equations, **45(3)**, pp 439-444.
[11] Yevtushenko Y, Smolyakov E 1999 Equilibria in Differential Games and Problems of Proposal Acceptance (Math and Phys, Vol 39), pp 897-905.

Mathematical Modeling and Functional Order Pursuits with Separated Dynamics

Mashrabjon Mamatov[1,*] and Xakimjon Alimov[2]

[1]National University of Uzbekistan named after M. Ulugbek, Tashkent, Uzbekistan
[2]Jizzakh State Pedagogical Institute, Jizzakh, Uzbekistan

[*]Corresponding author email: mamatovmsh@mail.ru

Abstract. This work is devoted to the study of the pursuit problem in controlled systems described by a fractional-order equation with divided dynamics. For fixed player controls, representations of solutions are established in the form of analogs of the Cauchy formula using generalized matrix functions. Sufficient conditions are obtained for the possibility of completing the pursuit. Specific types of fractional differential equations and models of fractional dynamical systems are considered. The qualitative dynamics, issues of stability and controllability of such systems are discussed. Considered, try which, the motion of the equation is described with irrational orders. Problems of the type under study are encountered in modeling the processes of economic growth and in problems of stabilizing dynamic systems.

1. Introduction

The increased interest in differential equations of fractional order is due to their physical interpretation, which caused their widespread use in classical mechanics, hydrodynamics, diffusion, dynamics of a turbulent medium, the theory of phase transitions, dynamic chaos, geology, and so on [1-4]. In the future, this direction was developed in the works [5-7]. In addition to research in the field of modeling, fractional dynamical systems, in recent years, research on control problems, such as differential game [8-10]. This work is devoted to the study of the problem of pursuit, and adjacent works [9, 10].
Suppose that the movement of the pursuing object is described by the equation

$$D^{\alpha}x = Ax + u,\ x \in R^{m}, \qquad (1)$$

D^{α} – Fractional order differentiation operator α , $n_1 - 1 < \alpha < n_1$, the evader, by the equation

$$D^{\beta}y = By + \upsilon,\ y \in R^{m} \qquad (2)$$

where D^{β} – fractional order differentiation operator β , $n_2 - 1 < \beta < n_2$, $u-$ pursuing player control parameter, $u \in P \subset R^{m}$, $\upsilon-$ runaway player's control parameter, $\upsilon \in Q \subset R^{m}$.
Suppose that in game (1), (2) $M = M_0 + M_1$, where $M_0 -$ linear subspace R^{m} , $M_1 \subset L$, $M_0 \times L = R^{m}$. Further, through Π, denote the matrix of the operator orthogonal projections, respectively, from R^{m} to L. The game is considered over if the conditions are met $x - y \in M$.

2. Methods

There are various methods for solving differential games, first the methods of Krasovsky, Pontryagin and the method of resolving functions. In this paper, we use Pontryagin's methods and its modifications developed by specialists for differential pursuit games, with lumped parameters with no separated

Content from this work may be used under the terms of the Creative Commons Attribution 3.0 licence. Any further distribution of this work must maintain attribution to the author(s) and the title of the work, journal citation and DOI.
Published under licence by IOP Publishing Ltd

dynamics. Consequently, it was necessary to develop new sufficient conditions that could be used to solve differential games described by a fractional-order equation with divided dynamics. By modifying the known methods, we managed to obtain such a sufficient condition. Thus, the mathematical apparatus used in this work is new.

3. Results

Further, everywhere $\sum_{\lambda=0}^{\infty}\frac{M^{\lambda}}{\Gamma(\frac{\lambda}{\xi}+\varsigma)}=E_{\xi}(M;\varsigma)$ – the Mittag-Lefler matrix function [1]. In the dynamical system (1), (2) it is assumed that

$$x_k^0=x^{(k)}(0),\quad k=\overline{0,n_1-1},\ y_l^0=y^{(l)}(0),\quad l=\overline{0,n_2-1}. \tag{3}$$

Let be $\Pi r^{\alpha-1}E_{\frac{1}{\alpha}}(Ar^{\alpha};\alpha)P=\hat{u}(r)$, $\Pi r^{\beta-1}E_{\frac{1}{\beta}}(Br^{\beta};\beta)Q=\hat{\upsilon}(r)$. Let us denote by $\hat{w}$ lots of $[\hat{u}(r)-\frac{1}{\tau}M_1]\overset{*}{-}\hat{\upsilon}(r)$. We introduce the control function

$$W(\tau)=\int_0^{\tau}\hat{w}(r,\tau)dr. \tag{4}$$

Theorem. The (1), (2), (3) for τ_0, switching on

$$-\Pi h_x(x^0,\tau_0)+\Pi h_y(y^0,\tau_0)\in W(\tau_0) \tag{5}$$

Then the end of the pursuit is possible.
Where

$$h_x(x^0,t)=\sum_{k=0}^{n_1-1}t^kE_{\frac{1}{\alpha}}(At^{\alpha};k+1)x_k^0,\ h_y(y^0,t)=\sum_{l=0}^{n_2-1}t^lE_{\frac{1}{\beta}}(Bt^{\beta};l+1)y_l^0.$$

Proof of the theorem. By virtue of the condition of theorem (5), we have $-\Pi_1h_x(x^0,\tau_0)+\Pi_2h_y(y^0,\tau_0)\in W(\tau_0)$. Hence (4), measurable which

$$-\Pi_1h_x(x^0,\tau_0)+\Pi_2h_y(y^0,\tau_0)=\int_0^{\tau_0}w(r)dr. \tag{6}$$

Whence, by virtue of (4), we find

$$w(r)+\Pi_2(\tau_0-r)^{\beta-1}E_{\frac{1}{\beta}}(B(\tau_0-r)^{\beta};\beta)\upsilon(r)\in-\frac{1}{\tau_0}M_1+\hat{u}(\tau_0-r),\ 0\le r\le\tau_0.$$

Hence, by virtue of the measurability condition (6), it follows that there are measurable functions

$$d(r)\in-\frac{1}{\tau_0}M_1,\ \Pi_1(\tau_0-r)^{\alpha-1}E_{\frac{1}{\alpha}}(A(\tau_0-r)^{\alpha};\alpha)u(r)\in\hat{u}(\tau_0-r),\ w(r)+$$
$$\Pi_2(\tau_0-r)^{\beta-1}E_{\frac{1}{\beta}}(B(\tau_0-r)^{\beta};\beta)\upsilon(r)=d(r)+\Pi_1(\tau_0-r)^{\alpha-1}E_{\frac{1}{\alpha}}(A(\tau_0-r)^{\alpha};\alpha)u(r),\ 0\le r\le\tau_0. \tag{7}$$

We define a measurable function to (7). $x(t), y(t)$, relevant functions u,υ, we have

$$-\Pi_1x(\tau_0)+\Pi_2y(\tau_0)=-\Pi_1h_x(x^0,\tau_0)+\Pi_2h_y(y^0,\tau_0)-\int_0^{\tau_0}\Pi_1(\tau_0-s)^{\alpha-1}E_{\frac{1}{\alpha}}(A(\tau_0-s)^{\alpha};\alpha)u(s)ds+$$
$$\int_0^{\tau_0}\Pi_2(\tau_0-s)^{\beta-1}E_{\frac{1}{\beta}}(B(\tau_0-s)^{\beta};\beta)\tilde{\upsilon}(s)ds=\int_0^{\tau_0}d(s)ds\in-M_1.$$

From here $\Pi_1x(\tau_0)-\Pi_2y(\tau_0)\in M_1$, that is trajectory out of the position x^0, y^0, at the moment $t=\tau_0$ on M.

4. Discussion

Movement pursuing player

$$D^{\pi} x = u, \ \|u\| \le \rho, \rho > 0, \tag{8}$$

movement fleeing player determined

$$D^{e} y = \upsilon, \ \|\upsilon\| \le \sigma, \sigma > 0 \tag{9}$$

where $e = 2,71828...$ – limit value $\lim_{n\to\infty}\left(1+\frac{1}{n}\right)^n$. $u = u(t)$, $\|u\| \le \rho$, $\upsilon = \upsilon(t)$, $\|\upsilon\| \le \sigma$, $M = \delta S$, $\delta > 0$, S – unit ball of R^m. Π – R^m to R^m is the identity matrix. This means that the game, is considered over if the conditions are met $x - y \in M$, that is $\|\Pi x - \Pi y\| = \|x - y\| \le \delta$, $\delta > 0$.

In our example $E_{\frac{1}{\pi}}(At^{\pi};\pi) = \frac{1}{\Gamma(\pi)} \mathrm{I}$ and $E_{\frac{1}{e}}(\mathrm{B}t^{e};\mathrm{e}) = \frac{1}{\Gamma(\mathrm{e})}\mathrm{I}$. (8), (9) as $x(0) = x_0^0$, $\dot{x}(0) = x_1^0$, $\ddot{x}(0) = x_2^0$, $\dddot{x}(0) = x_3^0$, $y(0) = y_0^0$, $\dot{y}(0) = y_1^0$, $\ddot{y}(0) = y_2^0$ respectively. We denote

$$x^0 = (x_0^0,\ x_1^0,\ x_2^0,\ x_3^0),\ y^0 = (\mathrm{y}_0^0,\ y_1^0,\ y_2^0),$$

$$h_x(x^0,\mathrm{t}) = x_0^0 + t\,x_1^0 + \frac{t^2}{2}x_2^0 + \frac{t^3}{6}x_3^0,\quad h_y(\mathrm{y}^0,\mathrm{t}) = y_0^0 + t\,y_1^0 + \frac{t^2}{2}y_2^0.$$

Let us check the condition of the theorem and calculate the sets $\hat{u}(r) = \Pi r^{\alpha-1} E_{\frac{1}{\alpha}}(Ar^{\alpha};\alpha)P$, $\hat{\upsilon}(r) = \Pi r^{\beta-1} E_{\frac{1}{\beta}}(\mathrm{B}\,r^{\beta};\beta)Q$, $\hat{w}(r) = \hat{u}(r) \overset{*}{-} \hat{v}(r)$. By virtue of the conditions of the problems, we have

$$\hat{u}(r) = \Pi r^{\pi-1} E_{\frac{1}{\pi}}(Ar^{\pi};\pi)P = r^{\pi-1}\frac{1}{\Gamma(\pi)}\mathrm{I}\rho S,\ \hat{\upsilon}(r) = \Pi r^{e-1} E_{\frac{1}{e}}(\mathrm{B}\,r^{e};\mathrm{e})Q = r^{e-1}\frac{1}{\Gamma(\mathrm{e})}\mathrm{I}\sigma S,$$

Thus, the set $\hat{u}(r)$ there is a ball of radius $\frac{r^{\pi}}{r\Gamma(\pi)}\rho$, and a lot $\hat{\upsilon}$ ball of radius $\frac{r^{e}}{r\Gamma(\mathrm{e})}\sigma$. We define the set, $\hat{w} = \left[\hat{u} - \frac{1}{t}M\right] \overset{*}{-} \hat{\upsilon}$. By condition example $M = \delta S$, $\delta > 0$, S – there is a ball of radius δ. On the other hand, we already know that $\hat{u}(r)$ and $\hat{\upsilon}(r)$ balls of the corresponding radius. From this we get that $\hat{w}(r,\mathrm{t})$ - ball radius, $\frac{\delta}{t} + r^{\pi-1}\frac{1}{\Gamma(\pi)}\rho - \frac{r^{e}}{r\Gamma(\mathrm{e})}\sigma$. And the set defined by the formula (4), $W(\mathrm{t})$ there is a ball of radius

$$\gamma = \frac{t^{\pi}\rho}{\Gamma(\pi+1)} - \delta + \frac{t^{e}}{\Gamma(1+\mathrm{e})}\sigma \tag{10}$$

In order for the value $\gamma(\mathrm{t})$ was positive in front of everyone $t > 0$, it is sufficient that the following inequalities hold

$$\frac{t^{\pi}}{\Gamma(\pi+1)}\rho - \frac{t^{e}}{\Gamma(\mathrm{e}+1)}\sigma + \delta > 0 \tag{11}$$

This inequality is equivalent to the following

$$\delta > \frac{t^{e}}{\Gamma(\mathrm{e}+1)}\sigma - \frac{t^{\pi}}{\Gamma(\pi+1)}\rho \tag{12}$$

Formulas (15), (16) and (17) enable us to find the values of the constants δ, ρ, σ . To solve question, we investigate the function

$$f(t)=\frac{t^{e}}{\Gamma(e+1)}\sigma-\frac{t^{\pi}}{\Gamma(\pi+1)}\rho, \quad t>0,$$

Find the critical points of this function

$$f'(t)=\frac{e\,t^{e-1}}{\Gamma(e+1)}\sigma-\frac{\pi\,t^{\pi-1}}{\Gamma(\pi+1)}\rho=0,\ \frac{e\,t^{e-1}}{e\Gamma(e)}\sigma-\frac{\pi\,t^{\pi-1}}{\pi\Gamma(\pi)}\rho=0,\ t_0=\left(\frac{\sigma\Gamma(\pi)}{\rho\Gamma(e)}\right)^{\frac{1}{\pi-e}}.$$

Now let us prove that the found critical point $t=t_0$ function $f(t)$ reaches its maximum. For this we calculate $f''(t)$, and show that it is at the point $t=t_0$ negatively.

$$f''(t)=\frac{(e-1)\,t^{e-2}}{\Gamma(e)}\sigma-\frac{(\pi-1)t^{\pi-2}}{\Gamma(\pi)}\rho=\frac{\sigma}{\Gamma(e-1)}t^{e-2}-\frac{\rho}{\Gamma(\pi-1)}t^{\pi-2}.$$

When $t=t_0$ we get the following

$$f''(t_0)=\frac{\sigma}{\Gamma(e-1)}t_0^{\,e-2}-\frac{\rho}{\Gamma(\pi-1)}t_0^{\,\pi-2}=\left[\frac{\sigma\Gamma(\pi)}{\rho\,\Gamma(e)}\right]^{\frac{\pi-2}{\pi-e}}\frac{\rho}{\Gamma(\pi-1)}\left(\frac{e-1}{\pi-1}-1\right)<0.$$

Thus, we have shown that $f''(t_0)<0$. Means $t=t_0$ for function $f(t)$ maximum point, so this

$$\max_{t>0} f(t)=f(t_0).$$

Now let's calculate the value $f(t_0)$

$$f(t_0)=\frac{t_0^{\,e}}{\Gamma(e+1)}\sigma-\frac{t_0^{\,\pi}}{\Gamma(\pi+1)}\rho=\left[\frac{\sigma\Gamma(\pi)}{\rho\,\Gamma(e)}\right]^{\frac{\pi}{\pi-e}}\frac{\rho(\pi-e)}{\Gamma(\pi)\pi e}>0.$$

Condition $\gamma(t)>0$ with all $t>0$, is executed if for parameters $\delta,\ \sigma,\ \rho$ conditions are met

$$\delta>\left[\frac{\sigma\Gamma(\pi)}{\rho\,\Gamma(e)}\right]^{\frac{\pi}{\pi-e}}\frac{\rho}{\Gamma(\pi)}\left(\frac{1}{e}-\frac{1}{\pi}\right).$$

If this condition is satisfied, then the radius of the ball $W(t)$ always positive for everyone $t>0$, and the aspirant to infinity grows without limit. The quantity $T(x^0, y^0)$ is now defined as the minimum value t, at which the ball $W(t)$ contains point

$$h_x(x^0,t)-h_y(y^0,t)=\left(x_0^0+t\,x_1^0+\frac{t^2}{2}x_2^0+\frac{t^3}{6}x_3^0\right)-\left(y_0^0+t\,y_1^0+\frac{t^2}{2}y_2^0\right).$$

Thus, the quantity $T(x^0, y^0)=t_1>0$ is a number satisfying equality

$$\left\|h_x(x^0,t)-h_y(y^0,t)\right\|=\frac{\rho t^{\pi}}{\Gamma(\pi+1)}-\frac{\sigma t^{e}}{\Gamma(e+1)}+\delta.$$

5. Conclusion

Research shows that the theory of fractional differential games is currently in its infancy. Because many terms, concepts are not fully defined even in fractional differential equations. For example, so far no consensus has been reached on the geometric or physical meaning of the fractional operator. In the theory of whole-order games, on the contrary, there are ways to solve the problem of chasing, escaping. For example, the first and second methods of Pontryagin were developed to solve the problem of persecution. These methods are now sufficiently simplified and easy to apply to examples.

Therefore, it is necessary to create the above methods that can be used to solve fractional differential games. Our article is one of the first studies conducted in this regard. At the end of the article, an example devoted to the application of the method proposed by us is a special form of fractional differential game, i.e. the order of equations is irrational numbers.

References

[1] Kilbas, A.A., Srivastava, H.M. and Trujillo, J.J. 2006. *Theory and Applications of Fractional Differential Equations*. (Amsterdam: North-Holland/American Elsevier) p 500.
[2] Miller K.S., Ross B. 1993 *An introduction to the fractional calculus and fractional differential equations* (New York: Wiley & Sons) 384 p.
[3] Samko S.G., Kilbas A.A., Marichev O.I. 1987. *Integrals and derivatives of fractional order and some of their applications* (Minsk: Science and Technology) p. 688
[4] Letnikov A.V., Chernykh V.A. 2011 Fundamentals of fractional calculus (*Moscow: Neftegaz*) p. 430.
[5] J. He, 1998 Approximate analytical solution for seepage flow with fractional derivatives in porous media (*Comput. Methods. Appl. Mech. Eng*, 167) pp. 57-68.
[6] S. Momani, N. T. Shawagfeh, 2006. Decomposition method for solving fractional Riccati differential equations (*Appl. Math. Comput*, 182), pp. 1083-1092.
[7] Mainardi F 1997 *Fractional calculus: some basic problems in continuum and statistical mechanics, Fractals and fractional calculus in continuum mechanics* (Springer-Verlag, New York) pp 291-348.
[8] Pontreagin L. 1980 Linear differential games of pursuit *(Mat. Sb. Vol 112)* pp 307-330.
[9] Mamatov M, Sobirov K 2020 Journal of Mathematical Sciences (US), **245(3)**, pp 332–340.
[10] Tukhtasinov M, Mamatov M 2009 Differential Equations, **45(3)**, pp 439-444.

Nonparametric Estimation of Multivariate Extreme Value Copulas with Known and Unknown Marginal Distributions

Samia AYARI[1,2,3,*] and Mohamed BOUTAHAR[2]

[1]Institut des Hautes Etudes de Tunis, Université de Tunis, Tunisie
[2]Institut de Mathématiques de Marseille, Université d'Aix-Marseille, France
[3]BESTMOD, Institut Supérieur de Gestion de Tunis, Université de Tunis, Tunisie

[*]Corresponding author email: samia.ihec@hotmail.fr

Abstract. The purpose of this paper is estimating the dependence function of multivariate extreme values copulas. Different nonparametric estimators are developed in the literature assuming that marginal distributions are known. However, this assumption is unrealistic in practice. To overcome the drawbacks of these estimators, we substituted the extreme value marginal distribution by the empirical distribution function. Monte Carlo experiments are carried out to compare the performance of the Pickands, Deheuvels, Hall-Tajvidi, Zhang and Gudendorf-Segers estimators. Empirical results showed that the empirical distribution function improved the estimators' performance for different sample sizes.
Keywords: Multivariate extreme value copulas, dependence function, nonparametric estimation, empirical distribution function, dependence function, Monte-Carlo simulation.

1. Introduction

The last decades have seen an increase in statistical modeling research of extreme values and rare events. The extreme value theory was developed more than nine decades. It's a branch of statistics that consists on analyzing the stochastic behavior of extreme values in order to estimate the probability of occurrence of extreme events. The cornerstone of extreme value theory is the Fisher-Tippet's theorem [7].

The extreme value analysis began with univariate data. Univariate extreme value theory assumes unrealistic hypotheses such as the independence of observations. In addition, most real extreme value applications are inherently multivariate by nature [16, 1]. This explains why recent researches in extreme value theory are more focusing on multivariate extreme value analysis [2, 6, 15]. Particularly, the copula theory has played a relevant role in modeling multivariate extreme value distributions [8, 17]. Copula is a mathematical model which can decompose a multivariate distribution into two components: the marginal distribution functions, on one hand, and a mathematical function that links them by modeling their dependence, on the other hand. Copula is the joint distribution of multivariate data with uniform marginal distribution functions. Extreme-value copulas are considered a relevant type of copulas [11]. Nonparametric estimation of multivariate copulas consists on the dependence function estimation. This line of research began with the works of Pickands [12] who developed an estimator of the dependence function for bivariate extreme value copulas. Several studies have criticized the work of Pickands because his nonparametric estimator doesn't satisfy any property of a dependence function. To circumvent the drawbacks of the Pickands estimator, several estimators were proposed in the literature namely the Deheuvels estimator [4], the CFG estimator [3] and the Hall-Tajvidi estimator [10].

Content from this work may be used under the terms of the Creative Commons Attribution 3.0 licence. Any further distribution of this work must maintain attribution to the author(s) and the title of the work, journal citation and DOI.
Published under licence by IOP Publishing Ltd

As shown above, the majority of researches on multivariate extreme value theory didn't use more than two random variables. In the frequent case, when we attempt to model extreme values behavior, more than two random variables should be considered. That's why Falk and Reiss [5] generalized the bivariate extreme value copulas results to n random variables.
Multivariate extensions of the nonparametric dependence function estimators were carried out by Zhang et al. [18]. An ordinary least squares estimator of the dependence function of multivariate extreme value distributions was developed by Gudendorf and Segers [9].
The previous dependence function estimators, usually, assume that marginal distribution functions are generalized extreme value distributions. This assumption is unrealistic because marginal distributions are rarely known in practice. In this paper, we aim at estimating the dependence structure when the marginal distributions are unknown. We will propose the Pickands, Hall-Tajvidi, Deheuvels, Zhang and Gudendorf-Segers estimators in a new way which consists on replacing the Gumbel extreme value distribution by the empirical distribution function.
The paper is structured as follows. In section 2, we will deal with nonparametric estimators of the dependence function for multivariate extreme value copulas with known margins. In section 3, an overview of the empirical distribution function is described. A Monte-Carlo simulation is conducted, in section 4, to check the estimators' performance of multivariate extreme values copulas with known and unknown marginal distributions.

2. Nonparametric Estimation of Multivariate Extreme Value Distributions

Given a sample (X_{i1},...,X_{ip}), i = 1,...,n, of independent and identically distributed (i.i.d) random vectors with marginal distributions $F_j(x)$ of X_{ij}, j∈{1,...,p} and a multivariate extreme value copula C presented as follows:

$$C(U_1 \dots, U_p) = C\,(F_1(X_1),\dots,F_p(X_p))$$
$$= \exp\left\{\sum_{j=1}^{p} \lg(u_j)\, A\Big(\frac{\lg(u_1)}{\sum_{j=1}^{p} \lg(u_j)}, \dots, \frac{\lg(u_{p-1})}{\sum_{j=1}^{p} \lg(u_j)}\Big)\right\} \quad (1)$$

The dependence function A (s_1, s_2... s_{p-1}) is defined in the (p-1) dimensional unit simplex:
S_{p-1}= {(s_1, s_2,..., s_{p-1}): $s_j \geq 0$; $\sum_{j=1}^{p-1} s_j \leq 1$ for $j = 1, \dots, p-1$}.
It's a convex function such as max(s_1, s_2, ..., s_{p-1})≤ A(s_1, s_2,..., s_{p-1})≤ 1.
Zhang et al. [18] generalized the Pickands, Hall-tajvidi and Deheuvels estimators to the multivariate case; he also developed a new estimator of the dependence function which is the generalization of the CFG estimator. The estimators are presented as follow:
The Pickands estimator:

$$\hat{A}^P(s_1, s_2,\dots, s_{p-1}) = \frac{n}{\sum_{i=1}^{n} min_{j=1}^{p} \frac{Y_{ij}}{s_j}},\ s \in S_p \quad (2)$$

The Hall-Tajvidi estimator:

$$\hat{A}^{HT}(s_1, s_2,\dots, s_{p-1}) = \frac{n}{\sum_{i=1}^{n} min_{j=1}^{p} \frac{\frac{Y_{ij}}{\grave{Y}_j}}{s_j}},\ s \in S_p \quad (3)$$

The Deheuvels estimator:

$$\hat{A}^D(s_1, s_2,\dots, s_{p-1}) = \frac{n}{\sum_{i=1}^{n} min_{j=1}^{p} \frac{Y_{ij}}{s_j} - n\sum_{i=1}^{n} s_j \grave{Y}_j + n},\ s \in S_p \quad (4)$$

Where $\grave{Y}_j = \frac{\sum_{i=1}^{n} Y_{ij}}{n}$, j=1,..., p.
The Zhang estimator:

$$\hat{A}^Z(s_1, s_2,\dots, s_{p-1}) = \frac{\prod_{j=1}^{p} (\sum_{i=1}^{n} Y_{ij})^{s_j}}{\sum_{i=1}^{n} min_{j=1}^{p} \frac{Y_{ij}}{s_j}},\ s \in S_p \quad (5)$$

where, $Y_{ij} = -\log(F_j(X_{ij}))$, $Y_j = \frac{1}{n}\sum_{i=1}^{n} Y_{ij}$, $1 \leq j \leq p$, $\sum_{j=1}^{p} s_j = 1$.
Gudendorf and Segers [9] modeled the relationship between the dependent variable $(-lg\,\xi_i(s) - \gamma;\ i = 1, \ldots, n;\ s \in S_p)$ and the p regressors $(-lg\,\xi_i(e_j) - \gamma;\ i = 1, \ldots, n; j = 1, \ldots, p;\ s \in S_p)$ in the form of linear regression:

$$(-lg\,\xi_i(s) - \gamma) = U_{n,0}(s) + \sum_{j=1}^{p} U_{n,j}(s)\,(-lg\,\xi_i(e_j) - \gamma) + \varepsilon_i(s), i=1,\ldots,n,\ s \in S_p, \quad (6)$$

where

- $\xi_i(s) = \bigwedge_{j=1}^{p} \frac{Y_{ij}}{s_j}$, whatever i=1,…,n and s$\in S_p$,
- γ is the Euler–Mascheroni constant,
- $e_j = \{0, \ldots, 1, 0, \ldots, 0\} \in \mathbb{R}^p$ are standard unit vectors,
- $\varepsilon_i(s)$ is the error term.

The ordinary least squares method consists on estimating the (p + 1) regression coefficients $\widehat{U} = (U_{n,0}(s), U_{n,1}(s), \ldots, U_{n,p}(s))$ that minimize the sum of squared residuals:

$$\widehat{U} = \arg min \sum_{i=1}^{n} \varepsilon_i^2(s)$$

$$= arg\ min \sum_{i=1}^{n} ((-lg\,\xi_i(s) - \gamma) - U_{n,0(s)} + \sum_{j=1}^{p} U_{n,j}(s)(-\ lg\,\xi_i(e_j) - \gamma))^2, s \in S_p \quad (7)$$

So, the ordinary least square estimator of the dependence function $\log \hat{A}_G^{OLS}(s)$ is the estimated intercept $\widehat{U}_{n,0}(s)$:

$$\widehat{U}_{n,0}(s) = (-lg\,\xi_i(s) - \gamma) - \sum_{j=1}^{p} \widehat{U}_{n,j}(s)(-\ lg\,\xi_i(e_j) - \gamma)$$

$$= lg\hat{A}(s) - \sum_{J=1}^{p} \widehat{U}_{n,j}(s) lg\hat{A}\,(e_j)$$

$$= lg\hat{A}_G^{OLS}(s) \quad (8)$$

3. Overview of the Empirical Cumulative Distribution Function

Statistics aim to understand and to model the stochastic behavior of random variables in order to predict accurately their potential occurrences in the future. There is two branches of statistics: the parametric statistics and the nonparametric statistics. Parametric statistics make some assumptions about the data and all the future inferences are based on these assumptions. If the assumptions are true, then, the resulting inferences are correct. Whereas, if they are false, the statistical analysis will be wrong. In practice, it is often difficult to find a suitable parametric model for real data. That's why researches are more oriented towards nonparametric statistics. Nonparametric approach makes few or no assumptions about the data. Using the empirical distribution function represents an alternative to overcome the limitations of parametric statistics.

The cumulative distribution function F(x) defines the probability that a random variable X takes values less than or equal to x. While, the empirical distribution is the proportion of observations of a random sample which are less than or equal to a value x; it's is a step function that jumps by $\frac{1}{n}$ at each data point x_i.

Suppose $X_1, X_2, \ldots, X_n$, are i.i.d random variables, the empirical distribution function is defined by :

$$F_n(x) = \frac{1}{n}\sum_{i=1}^{n} \mathbb{I}(X_i \leq x) \quad (9)$$

Where $\mathbb{I}$ is the indicator function.

The empirical distribution function has the following properties:

- It is an increasing and right-continuous function.
- $\lim_{n \to -\infty} F_n(x) = 0$ and $\lim_{n \to +\infty} F_n(x) = 1$.

- $E[F_n(x)] = F(x)$ and $var[F_n(x)] = \frac{1}{n}F(x)(1 - F(x))$.

According to the Chebyshev's inequality, for any real number $\varepsilon > 0$, we have:

$$\lim_{n\to\infty} P[|F_n(x)-F(x)| \geq \varepsilon] = 0 \quad (10)$$

This fact implies that $F_n(x) \xrightarrow{prob}$ to $F(x)$ as $n \to \infty$ for any fixed value of x.
A strong proof of convergence is given by the strong law of large numbers, for any $x \in \mathbb{R}$:

$$F_n(x) \xrightarrow{a.s} F(x) \quad (11)$$

The strongest convergence result is given by the Glivenko--Cantelli theorem, it proves the uniform convergence of $F_n(x)$ to (x), , for any $x \in \mathbb{R}$:

$$\sup |F_n(x) - F(x)| \xrightarrow{a.s} 0 \quad (12)$$

Finally, the Dvoretzky-Kiefer-Wolfowitz inequality gives another uniform property of $F_n(x)$ such that:

$$P(sup_{x\in\mathbb{R}}|F_n(x)-F(x)| > \varepsilon) \leq 2e^{-2n\varepsilon^2}, \forall n \in \mathbb{N}, \varepsilon > 0 \quad (13)$$

All the properties presented above show that the empirical distribution function is a reliable tool in nonparametric statistics.
In the multivariate case, the empirical distribution function is presented as follows:
Suppose $(X_{i1},\ldots,X_{ip})$, $i = 1,\ldots,n$, are i.i.d random vectors , the multivariate empirical distribution function is defined by:

$$F_{i,Xi1,\ldots,Xip}(x_1,\ldots,x_p) = \frac{1}{n}\sum_{i=1}^{n} \mathbb{I}(X_{i1} \leq x_1, \ldots, X_{ip} \leq x_p). \quad (14)$$

4. Monte Carlo Simulation Study

In this study, we focus on comparing some nonparametric dependence function estimators for multivariate extreme value copulas in the case of known and unknown marginal distributions. Firstly, we assume that margins are Gumbel extreme value distributions. Secondly, we substitute the extreme value marginal distributions by the empirical distribution functions.
In the first step, we simulate trivariate extreme value distributions having logistic dependence functions which is described as follows [14]:

$$A(s_1,s_2) = \{\alpha^r s_1^r + \beta^r s_2^r\}^{\frac{1}{r}} + \{\alpha^r s_2^r + \beta^r s_3^r\}^{\frac{1}{r}} + \{\alpha^r s_1^r + \beta^r s_2^r\}^{\frac{1}{r}}$$
$$+\{\alpha^r s_1^r + \beta^r s_3^r\}^{\frac{1}{r}} + \mu\{s_1^r + s_2^r + s_3^r\}^{\frac{1}{r}} + 1 - \alpha - \beta - \mu \quad (15)$$

where $s_3 = 1 - s_1 - s_2$.
The symmetric logistic dependence function is considered in our study with $r = 3$, $\alpha = \beta = 0$ and $\mu = 1$. The simulation is repeated using different sample sizes and the Pickands, Deheuvels, Hall-tajvidi, Zhang and Gudendorf-Segers estimators are compared. The mean integrated square errors MISE is used to assess the estimators' performance.

$$\text{MISE} = \iint_{(s_1,s_2)\in S_2} E\,[\hat{A}(s_1,s_2) - A(s_1,s_2)]^2 \partial_{s_1}\partial_{s_2} \quad (16)$$

where $\hat{A}(s_1, s_2)$ is the estimated dependence function and $A(s_1, s_2)$ is the observed dependence function given by the R evd package. In this study, the MISE is approximated by the Monte Carlo approach:

$$\text{MISE}\ (i) \cong \frac{1}{N^2}\sum_{j=1}^{N_1}\sum_{k=1}^{N_j}[\hat{A}(s_{1,j}, s_{2,k}) - A(s_{1,j}, s_{2,})]^2 \quad (17)$$

with $N_1 = N+1$, $N_j = N_1 - j + 1$ and N is the simplex parameter.
The MISE results are presented in table 1.

Table 1. MISE for different sample sizes in the case of Gumbel marginal distributions.

MISE× 10^5	**n=25**	**n=50**	**n=100**	**n=200**
Pickands	1164.02	555.92	262.07	130.36
Deheuvels	152.6	74.46	37.83	18.87
Hall-Tajvidi	26.37	13.38	6.76	3.41
Zhang	43.88	22.04	10.9	5.46
Gudendorf-Segers	18	8.26	3.93	1.91

As claimed in the bivariate case by Hall and Tajvidi [10], the Hall-Tajvidi estimator has better performance than the Pickands estimator in terms of MISE for different sample sizes. However, unlike the bivariate studies, the generalization of the Hall-Tajvidi estimator to the multivariate case allowed him to be a competitor estimator that has better performance than the CFG estimator generalized to the multivariate case.

Moreover, empirical results show that the Pickands estimator has lower performance than the Deheuvels and the Zhang estimators for different sample sizes. It's clear that Gudendorf-Segers estimator has the best performance over the other estimators for different sample sizes.

The MISE results of the estimators when substituting the Gumbel extreme value distribution by the empirical distribution function are presented in table 2.

Table 2. MISE for different sample sizes in the case of empirical marginal distributions.

MISE× 10^5	**n=25**	**n=50**	**n=100**	**n=200**
Pickands	141.16	53.11	20.69	8.06
Deheuvels	40.25	18.84	9.29	4.5
Hall-Tajvidi	32.73	16.87	8.63	4.42
Zhang	32.76	16.7	8.5	4.41
Gudendorf-Segers	28.18	13.81	7.02	3.39

As in the bivariate case [13], the Deheuvels and the Hall-Tajvidi estimators have higher performance than the Pickands estimator in modeling multivariate dependence structure. Furthermore, the Pickands estimator is more efficient in the multivariate case than the bivariate case. The robustness of the Hall-Tajvidi estimator is confirmed for estimating both multivariate and bivariate extreme value copulas. In addition, empirical results show that the Hall-Tajvidi and the Zhang estimators have similar results in terms of MISE for different sample sizes. Concerning the Gudendorf-Segers estimator, it has better performance than the other estimators for different sample sizes.

Our second interest is comparing the MISE results of table 2 with those obtained by the parametric approach which assumes that marginal distributions are known (table 1). Replacing the Gumbel extreme value distributions by the empirical distribution functions, the Pickands estimator performance is improved significantly for different sample sizes. Moreover, the Deheuvels estimator is more efficient with empirical margins than Gumbel marginal distributions. Regarding the Zhang estimator, it's ameliorated slightly when Gumbel distributions are replaced by the empirical distribution functions. Lastly, unlike the other estimators, we note that Gudendorf- Segers and Hall-Tajvidi estimators have better performance with known distributions than unknown ones for different sample sizes.

5. Conclusion

In this study, we developed some nonparametric estimators of the dependence function for trivariate extreme value copulas in two cases. Firstly, we assume Gumbel extreme value margins. Secondly, we substitute the marginal distributions by the empirical distribution functions. Simulation study shows that the nonparametric approach improved the estimators' performance for different sample sizes. This result was predicted because classical estimators assume unrealistic assumption consisting on Gumbel extreme value margins, so the estimator performance will depend on the estimation accuracy of the Gumbel extreme value distribution parameters (scale, shape and location). We can conclude that the rank-based estimators can be used as a reliable technique for modeling the dependence function of multivariate extreme value copulas in different fields such as stock markets, air pollution and hydrology.

References

[1] Ayari S and Boutahar M, 2017, multivariate extreme value theory and application to environment, International Journal of Management and Applied Science, vol 3 pp 11-13.

[2] Beck N., Genest, C Jalbert J and Mailhot M, 2020, Predicting extreme surges from sparse data using a copula - based hierarchical Bayesian spatial model *Environmetrics*, vol *31*, p e2616.

[3] Capéraa P, Fougères A L, Genest C, 1997, A nonparametric estimation procedure for bivariate extreme value copulas, *Biometrica*, vol 84, pp 567-577.

[4] Deheuvels P, 1991, On the limiting behavior of the Pickands estimator for bivariate extreme-value distributions, *Statistics & Probability Letters*, vol 12, pp 429-439.

[5] Falk M and Reiss R D, 2005, On Pickands coordinates in arbitrary dimensions, *Journal of Multivariate Analysis*, vol 92, pp 426-453.

[6] Ferreira H, 2011, Dependence between two multivariate extremes, *Statistics and Probability Letters*, vol 81, pp 586-591

[7] Fisher R A and Tippett L H C, 1928, Limiting forms of the frequency distribution of the largest and smallest member of a sample, *Proc. Cambridge Philosophical Society*, vol 24, pp 180–190.

[8] Frees E W and Valdez E A, 1998, Understanding relationships using copulas, *North American Actuarial Journal*, vol 2, pp 1-25.

[9] Gudendorf G and Segers J, 2012, Nonparametric estimation of multivariate extreme-value copulas, *Journal of Statistical Planning and Inference*, vol 142, pp 3073-3085.

[10] Hall P and Tajvidi N 2000, Distribution and dependence-function estimation for bivariate extreme value distributions, Bernouilli, Vol 6, pp 835-844.

[11] Nelsen R B, 2007, *An introduction to copulas* Springer Science & Business Media.

[12] Pickands J, 1981, Multivariate extreme value distributions, *Proc. of the 43rd Session of the International Statistical Institute* Buenos Aires, vol 49, pp 859-878.

[13] Segers J, 2007, Nonparametric inference for bivariate extreme value copulas in *Topics in extreme values*, Nova Science Publishers New York, pp 181-203.

[14] Stephenson A, 2003, Simulating multivariate extreme value distributions of logistic Type, *Extremes* vol 6, pp 49-59.

[15] Stuart C G J and Tawn A, 1991, Modeling extreme multivariate events, *Journal of the Royal Statistical Society Series B (Methodological)*, vol 53, pp 377-39

[16] Tawn J A, 1994, Applications of Multivariate Extremes in *Extreme Value Theory and its Applications*, Kluwer, pp 249-268.

[17] Tawn J A, 1990, Modelling multivariate extreme value distributions, *Biometrica*, vol 77, pp 245–253.

[18] Zhang D, Wells M T and Peng L, 2008, Nonparametric estimation of the dependence function for a multivariate extreme value distribution, *Journal of Multivariate Analysis*, vol 99, pp 577-588.

The Mathematical Analysis and Classification Research of an Iris Data Set Using Binary Tree and Grey Relation Grade

Chiang Ling Feng*

Department of Electric Engineering, ChienKuo Technology University, Taiwan

*Corresponding author email: acclaim0629v@gmail.com

Abstract. The data from an Iris flower database is studied. The Iris database is the most commonly used database for machine learning algorithms. The Iris database was developed by Ronald Aylmer Fisher in 1936. The Iris database has 150 records in three categories: Iris Sentosa, Iris Versicolor and Iris Virginic. The database has four attributes: sepal length, sepal width, petal length and petal width. For the machine learning algorithm, 150 Iris flower databases are used. Of the 150 Iris in the Iris database, 80% are used as the training set and the remaining 20% Iris as the test set. In machine learning, to perform classification and discrimination is a complicated and difficult thing. In this study, a grey relation grade is used to extract the main features of the Iris flower and a Binary Tree [1] is used to classify the Irises. The results show that for the same specific attributes, grey relation grade extracts the main attributes and can be used in combination with a binary for classification.
Keywords: Grey theory, grey relation grade, binary tree.

1. Introduction

A back propagation neural network BPN uses forward propagation and backward propagation. Forward propagation feeds training data into the network to calculate the error between the result and the corresponding target. Backward propagation adjusts the network weight according to the error value that is calculated using forward propagation, in order to correct the network to give a result that is within the smallest error range after many training cycles. In one study [2], the parameters are first initialized and the weights and biases are randomly established. The hidden layer node is 6 and the maximum number of iterations is 10,000. This study determined that there are four data classification errors. After further adjustment of the model, the training accuracy was improved. The disadvantage of this method is that there is overfitting and the calculation time is very long. Another study [3] achieves an accuracy of only 96.67% using a hidden layer node of 3, 5000 repeated operations and a learning rate of 0.01. Using a neural network, the classification of the Iris database requires a very low learning rate and a large number of repeated operations. This is closely related to the number of hidden nodes and the training data set and the test data set. The control factors are also too complicated. T One study [4] that uses the data from the Iris flower database shows that the sepal width and sepal length of these 150 Iris are not correlated, but the petal width and petal length are highly correlated. This study uses grey theory to verify the results of this study [4].

The nearest neighbor method uses the vector space model to classify. The concept is that the cases of the same category have a high degree of similarity to each other. The possible classification of cases of unknown categories can be evaluated by calculating the similarity with cases of known categories. The

Content from this work may be used under the terms of the Creative Commons Attribution 3.0 licence. Any further distribution of this work must maintain attribution to the author(s) and the title of the work, journal citation and DOI.
Published under licence by IOP Publishing Ltd

disadvantage of the k-nearest neighbor algorithm is that it is very sensitive to the local structure of the data.

Grey relation grade is a quantitative analysis method that is used to compare the relationship between data series. It reflects the degree of relationship between the essential curves of the data series. If the relationship between a data series and a reference data series is larger, the development trend for the data series has a closer relationship with the reference data sequence.

This study uses 150 data points from the Iris flower database and the fuzzy characteristics and classification of uncertain attributes is used. The paper is organized as follows. In the second section, we introduce the grey relation grade method. In the third section, we analyse the characteristics of Iris flowers among these 150 Iris data. We found the import influence factors among these characteristics using grey relation grade method. In fourth section, we also use the grey relation grade method combined with the classification method of binary tree to perform the flower classification. Conclusions are drawn in section five.

2. Grey Relational Grade

Grey relational analysis [5, 6] is a quantitative measurement method for a dynamic process in a system. It measures the degree of convergence between the entire data set using the development trend for multiple attributes. If the original data sequence has m attributes and each attribute original data sequence has n element values, then:

$$\begin{aligned} X_1^{(0)} &= \left\{x_1^{(0)}(1), x_1^{(0)}(2), \ldots\ldots, x_1^{(0)}(n)\right\} \\ X_2^{(0)} &= \left\{x_2^{(0)}(1), x_2^{(0)}(2), \ldots\ldots, x_2^{(0)}(n)\right\} \\ &\vdots \\ X_m^{(0)} &= \left\{x_m^{(0)}(1), x_m^{(0)}(2), \ldots\ldots, x_m^{(0)}(n)\right\} \end{aligned} \tag{1}$$

Grey correlation uses the proximity of the curve for each data set to determine whether the connection between each is close and which attributes are main influencing factors or secondary influencing factors. There is usually insufficient information so it is difficult to determine the relationship between attribute variables using only the data. The largest correlation between each factor is determined by calculating the relationship between each factor. To determine the degree of correlation for the data set, the correlation coefficient is calculated. The data sequence unit for each attribute in the data set may be different so the measurement unit must be determined before calculating relevance. Let

$$X_i^{(1)} = \left\{\frac{x_i^{(0)}(k) - \min\limits_k x_i^{(0)}(k)}{\max\limits_k x_i^{(0)}(k) - \min\limits_k x_i^{(0)}(k)}, k = 1,2, \ldots\ldots, m\right\} \tag{2}$$

$X_1^{(1)}$ tis the reference sequence and other data sequences are comparison data sequences. Define

$$\Delta_i(k) = \left|x_1^{(1)}(k) - x_i^{(1)}(k)\right|, i = 1,2, \ldots\ldots, m \tag{3}$$

and

$$V_{max} = \max_i \max_k \Delta_i(k) \tag{4}$$

$$V_{min} = \min \min_k \Delta_i(k) \tag{5}$$

Then, the correlation coefficient can be defined as

$$\gamma_{1,i} = \frac{V_{min} + \zeta V_{max}}{\Delta_i(k) + \zeta V_{max}} \tag{6}$$

Where ζ is the identification coefficient with a value between 0 and 1. The value of coefficient ζ is usually 0.5 so the correlation coefficient between attribute $i = 2,3, \ldots\ldots, m$ and attribute 1 can be calculated. If $X_j^{(1)}$ is used as the attribute reference data sequence, the correlation coefficient between it and other attributes can also be calculated.

3. The Classification of Iris Data by Binary Tree

Using the classification algorithm of binary tree, we must first find out the targets that are easier to classify. After finding out, we then analyze the important influencing attributes of these goals. After finding these important attributes, you can find the attributes and critical values that can easily divide the remaining targets into different binary trees. This step continues until all important attributes have been traversed. After that, look for all possible attributes, and give these attributes a weighting factor to add as the critical value for its segmentation target. This algorithm is give in Figure 1.

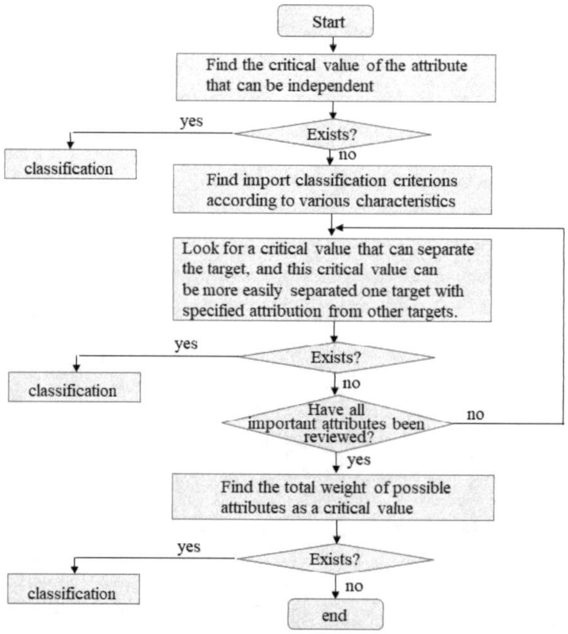

Figure 1. The generalized algorithm for binary tree.

The Iris Flower dataset [7] contains three species: Iris Sentosa, Iris Versicolor and Iris Virginica. Each type of Iris flower has four distinguishing features: sepal width, sepal length, petal width and petal length. Each species of iris flower features a specific range of sizes but not all parameters are used as features to create a decision tree for the available data using these groupings. This is because there are two reasons for making the decision that not all are used as features. First, from the petal features, both the width and length indicate the range that can be used to differentiate between types of Iris Sentosa, Iris versicolor and Iris Virginica. Second, the sepal feature shows a number that does not provide a significant differentiating number, this is because between species have almost the same range, namely the sepal length for Iris Sentosa species is in the size range between 4.3 - 5.8, versicolor 4.9 - 7.0, and Iris Virginica species between 4.9. - 7.9. Based on the size of the sepal width, the type of Iris Sentosa has a size range between 2.3 - 4.4, versicolor 2.0 and 3.4 and the type of Iris Virginica between 2.2 - 3.8. This shows that with this range it could be a certain range between the two types of Iris flowers. For example, the size of the sepal length is 5.0, it could be between the three types of Iris flowers, so it requires other parameters to determine the difference. See Table 1.

Table 1. Data range of various characteristics of each type of Iris.

	sepal length	sepal width	petal length	petal width
Iris Setosa	4.3~5.8	2.3~4.4	1.0~1.9	0.1~0.6
Iris Versicolor	4.9~7.0	2.0~3.4	3.0~5.1	1.0~1.8
Iris Virginica	4.9~7.9	2.2~3.8	4.5~6.9	1.4~2.5

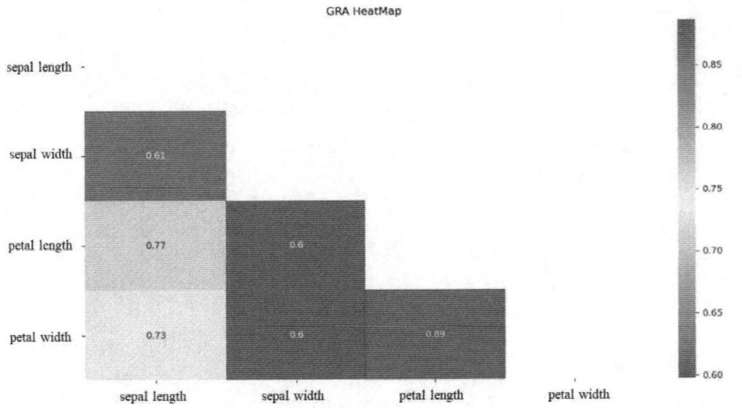

Figure 2. Iris database correlation coefficient.

From Figure 2, we can see that the characteristics of iris flowers can be determined using the petal length and the petal width. If the petal width is not greater than 0.6, Iris Sentosa can be distinguished from the other two Iris species. However, the sepal features have values in a similar range so cannot be used to differentiate between the three species. The sepal length for Iris Sentosa is 4.3 - 5.8, for Iris Versicolor is 4.9 - 7.0 and for Iris Virginica is 4.9. - 7.9.

Iris Sentosa has a sepal width of 2.3 - 4.4, Iris Versicolor has as sepal width of 2.0 - 3.4 and Iris Virginica has as sepal width of 2.2 - 3.8. These values can be used to differentiate between two types of iris flowers. An iris with a sepal length is 5.0 it could belong to any of the three species, so other parameters are required for differentiation. Figure 2 shows that petal width is one of the most informative attributes, so this is the first classification criterion because there are significant differences between each iris species in terms of this measure. From Table 1, if the petal width is less than 0.6, it is classified as Iris Sentosa type. If the petal width is greater than 1.8, it belongs is classified as Iris Virginica. For a petal width of 1.8, Table 10 shows that the species can be differentiated using the sum of the sepal length and the petal length. The decision binary tree to determine the species of iris is shown on the right side of Figure 3.

If the petal width is not greater than 1.8, there are 49 types of Iris Versicolor and 5 types of Iris Virginica. Using the attributes of these data, Iris Versicolor can be clearly distinguished if the petal length is less than 4.9. For other Iris species with a petal width of not less than 4.9, the related characteristics is classified as follows. If the petal length is not less than 5.6, Iris Virginica can be distinguished. Irises with a petal length of less than 5.6 are classified as Iris Virginica if the petal width is not greater than 4.5. If the petal width is not equal to 1.5, the species is classified as Iris Versicolor; otherwise, the species is determined using the grey theory correlation test. The final binary tree is shown in Figure 3.

4. Numeric Result

In the Iris data sheet, Iris is divided into three species of Iris Sentosa, Iris versicolor and Iris Virginica. Each species of Iris has four parameters as distinguishing characteristics, which are based on sepal length, sepal width, petal length and petal width. After calculating the petal width petal width, it is found that there are 10 data for the second category with petal width = 1.5 and 2 data for the third category with petal width = 1.5. Therefore, when the petal width of Iris is 1.5, it can be classified into Iris Versicolour and Iris Virginica. The classification of petal width of petal width = 1.5 is ambiguous. Therefore, we study the classification of Iris flowers by grey relation grade using the attribute petal width of petal width = 1.5 in this section.

According to the grey relation grade method, the data in Table 2 is formed under the condition of petal width = 1.5. It can be seen from Figure 4 that the degree of correlation between the petal length and the category attributes of Iris is 0.78.

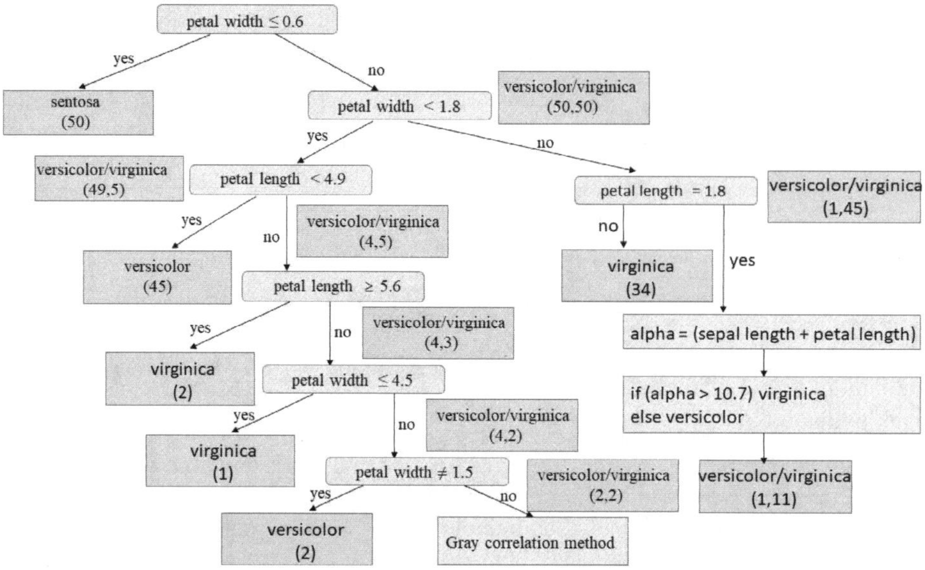

Figure 3. The complete Binary tree classification map of Iris.

Table 2. Attribute relationship table of Iris petal width=1.5.

item	sepal length	sepal width	petal length	petal width	species
1	6.4	3.2	4.5	1.5	Iris Versicolor
2	6.9	3.1	4.9	1.5	Iris Versicolor
3	6.5	2.8	4.6	1.5	Iris Versicolor
4	5.9	3	4.2	1.5	Iris Versicolor
5	5.6	3	4.5	1.5	Iris Versicolor
6	6.2	2.2	4.5	1.5	Iris Versicolor
7	6.3	2.5	4.9	1.5	Iris Versicolor
8	6	2.9	4.5	1.5	Iris Versicolor
9	5.4	3	4.5	1.5	Iris Versicolor
10	6.7	3.1	4.7	1.5	Iris Versicolor
11	6	2.2	5	1.5	Iris Virginica
12	6.3	2.8	5.1	1.5	Iris Virginica

Based on Table 2, set the petal length = 5 Iris flower category attribute to Iris versicolor. It can be seen that the correlation degree between the petal length and the category attributes of Iris is 0.75. Based on Table 2 the petal length = 4.9 Iris flower category attribute is classified as Iris Verginica. It can be seen that the relation grade between the petal length and the Iris category attribute is 0.71.
According to the above research, based on the condition of petal width = 1.5 and the Iris with petal length = 5 is classified into Iris Virginica, the correlation degree is 0.78. Based on the condition of petal width = 1.5 and the Iris with petal length = 5 is classified into Iris Versicolor, the correlation degree is 0.75. Based on the condition of petal width = 1.5 and the Iris with petal length = 4.9 is classified into Iris Virginica, the relation grade is 0.71. Therefore, the most suitable classification is that the Irises with petal length = 5 and petal length = 5.1 belong to Iris Virginica, and the Irises with petal length = 4.9 belong to Iris Versicolor (grey relation grade changed from 0.71 to 0.78).

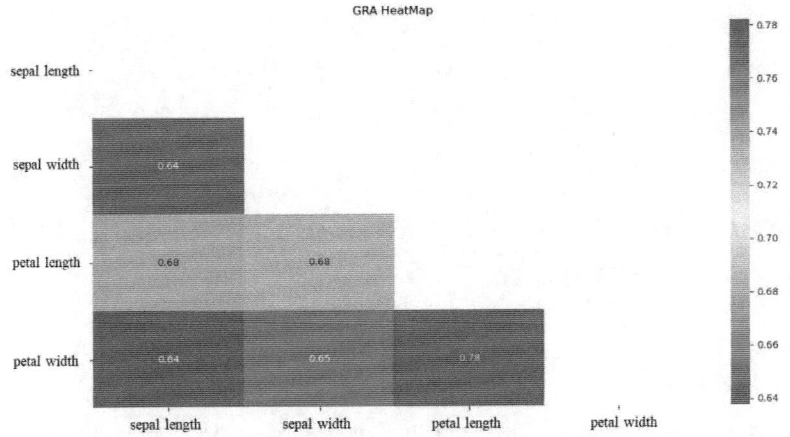

Figure 4. The relation grade table of the Iris database in Table 2.

5. Conclusion

From the discussion of the above examples, it is difficult to assign a main attribute for the Iris data. In the Iris Setosa variety, it can be clearly separated from the other two varieties by the petal width. As for Iris Versicolor and Iris Virginica, the attribute values of the two varieties overlap each other. For iris varieties with petal length = 1.5, we found that the two attributes of petal length and petal width can be used to separate iris varieties through the grey correlation method. For iris varieties with petal length = 1.8, the critical value can be set to the sum between sepal length and petal length through the algorithm shown in Figure 1 to easily separated iris varieties.

References

[1] Adrijan Božinovski and Nevena Ackovska, "The Binary Tree Roll Operation: Definition, Explanation and Algorithm", International Journal of Computer Applications (0975 – 8887) Vol. 46, No.8, pp. 40-47, May 2012.

[2] https://blog.csdn.net/weixin_33724659/article/details/89700732?utm_medium=distribute pc_relevant_download.none-task-blog-BlogCommendFromBaidu-22.nonecase&depth_1-utm_source=distribute.pc_relevant_download.none-task-blog-BlogCommendFromBaidu-22.nonecas.

[3] https://www.jianshu.com/p/52b86c774b0b.

[4] https://github.com/codingXiang/BPN.

[5] Ke H.F, and Chen B.L., "Novel gross error detection approaches of small samples based on *GM*(1,1) model, Vol. 25(1), pp. 44-53, 2013.

[6] Liu S.F., and Lin Y., Grey Systems: Theory and Applications, Berlin: Springer-Verlag, 2011.

[7] https://www.kaggle.com/saurabh00007/Iriscsv.

Explicit Exact Solutions of Space-time Fractional Drinfel'd-Sokolov-Wilson Equations

Hongkua Lin*

Department of Mathematics and Information Science, Guangzhou University, Guangzhou

*Corresponding author email: linhongkua@126.com

Abstract. The space-time fractional Drinfel'd-Sokolov-Wilson equations (DSWEs) has attracted many researchers' attention in recent years. In this study, combining the (G'/G,1/G)-expansion method and a fractional wave transformation, we derive abundant explicit exact solutions of the DSWEs with the conformable fractional derivative. All of the resulting solutions include triangle, hyperbolic and rational function type. It shows this technique is effective and reliable to find exact solutions of other similar nonlinear fractional partial differential equations (NFPDEs).
Keywords: Conformable fractional derivative, exact solutions, (G'/G,1/G)-expansion method.

1. Introduction

As a basic model equation, nonlinear partial differential equations (NPDEs) appear in many science fields, such as acoustics, fluid mechanics, chemical engineering and biological science. It is of great physical significance and difficult to seek exact solutions of NPDEs.

In the past decades, the rapid development of computer science provides more possibilities for solving NPDEs. Many powerful methods and techniques, such as Hirota's bilinear method [1], F-expansion method [2], the extended hyperbolic function expansion method [3], the (G'/G,1/G)-expansion method [4] and so on, have been developed.

Due to its important physical significance in various fields, such as visco-elasticity, anomalous diffusion and electrical engineering, NFPDEs have attracted more attention. The exact solutions of NFPDEs can reflect the global properties of physical processes and test the effectiveness of numerical calculation. Therefore, it is an important work to seek exact solutions of NFPDEs. In recent years, some methods for solving NPDEs have been improved to find exact solutions of NFPDEs, such as Lie-symmetry method [5], polynomial method [6], generalized bifurcation method [7]. For more recent developments in solving fractional partial differential equations, please refer to [8] and references therein.

The classical Drinfel'd-Sokolov-Wilson equations was first derived by Drinfel'd and Sokolov, which has been studied extensively. In recent years, due to its crucial and wide application in engineering and physics, the DSWEs has attracted many researchers' attention. Such as S.Chen [6] obtained four solutions via the polynomial method. Z.Wen [7] found the new so-called M/W-shaped solitary wave solution by the generalized bifurcation method. A lot of study for the DSWEs have been reported, which applied the Lie symmetry analysis, the exp-function method and so on [5,9].

The DSWEs is written as

Content from this work may be used under the terms of the Creative Commons Attribution 3.0 licence. Any further distribution of this work must maintain attribution to the author(s) and the title of the work, journal citation and DOI.
Published under licence by IOP Publishing Ltd

$$\begin{cases} D_t^{\alpha} u + pvD_x^{\beta} v = 0, \\ D_t^{\alpha} v + ruD_x^{\beta} v + svD_x^{\beta} u + qD_x^{3\beta} v = 0, \end{cases} \tag{1}$$

where p,q,r,s are non-zero parameters and $0<\alpha,\beta<1$. The conformable derivative appeared in Eq (1) is first provided by Khalil in references [10]. Supposing $f:(0,+\infty)\to R$, the conformable derivative of f is defined as

$$D_t^{\alpha} f(t) = \lim_{\varepsilon\to 0}\frac{f\left(t+\varepsilon t^{1-\alpha}\right)-f(t)}{\varepsilon},$$

where $\alpha\in(0,1)$. If $f(t)$ and $g(t)$ are differentiable, some significant characters for conformable derivative are listed as follows

$$D_t^{\alpha}\left(t^c\right) = ct^{c-\alpha},\quad c\in R,$$

$$D_t^{\alpha}(f\circ g)(t) = t^{1-\alpha} g'(t) f'(g(t)).$$

This paper aim to construct its exact solutions of the DSWEs by the extended (G'/G,1/G)-expansion method. In Section 2, we introduce a framework of the method. In Section 3, we obtain a series of exact analytical solutions of the DSWEs. The last section is conclusion.

2. The Outline of the Extended (G'/G,1/G)-expansion Method

Considering the following linear ODE

$$G'' + \lambda G = \mu, \tag{2}$$

we let

$$\phi = \frac{G'}{G}, \varphi = \frac{1}{G}. \tag{3}$$

Then, uniting Eq (2) and Eq (3), we yield

$$\phi' = -\phi^2 + \mu\varphi - \lambda, \varphi' = -\phi\varphi. \tag{4}$$

There are some solutions for Eq (2) to be discussed and we set A_1, A_2 are two arbitrary constants.

Case1. When $\lambda<0$

$$G(\xi) = A_1\sinh(\sqrt{-\lambda}\xi) + A_2\cosh(\sqrt{-\lambda}\xi) + \frac{\mu}{\lambda}, \tag{5}$$

and we have

$$\varphi^2 = \frac{-\lambda}{\lambda^2\sigma+\mu^2}\left(\phi^2 - 2\mu\varphi + \lambda\right), \tag{6}$$

where $\sigma = A_1^2 - A_2^2$.

Case2. When $\lambda>0$

$$G(\xi) = A_1\sin(\sqrt{\lambda}\xi) + A_2\cos(\sqrt{\lambda}\xi) + \frac{\mu}{\lambda}, \tag{7}$$

and we have

$$\varphi^2 = \frac{\lambda}{\lambda^2\rho-\mu^2}\left(\phi^2 - 2\mu\varphi + \lambda\right), \tag{8}$$

where $\rho = A_1^2 + A_2^2$.

Case3. When $\lambda=0$

$$G(\xi) = \frac{\mu}{2}\xi^2 + A_1\xi + A_2, \tag{9}$$

and we have

$$\varphi^2 = \frac{1}{A_1^2 - 2\mu A_2}\left(\phi^2 - 2\mu\varphi\right), \tag{10}$$

Give a NFPDEs in the form

$$P\left(u, D_t^{\alpha}u, D_x^{\beta}u, D_t^{2\alpha}u, D_x^{2\beta}u, \cdots\right) = 0, \tag{11}$$

where $u = u(x,t)$, P is a polynomial of u its conformable derivatives.

Step1. Firstly, we introduce a fractional wave transformation

$$u(x,t) = u(\xi), \xi = k(\frac{x^{\beta}}{\beta} + \frac{ct^{\alpha}}{\alpha}), \tag{12}$$

where c, k are non-zero constants. After above transformation, we reduced Eq (11) to the following nonlinear ODE

$$Q\left(u, u', u'', u''', \cdots\right) = 0, \tag{13}$$

where Q is a polynomial in u and its derivatives, especially the prime indicate as the derivation with respect to ξ.

Step2. We suppose the solution for Eq (13) can be expressed as a finite polynomial in ϕ and φ

$$u(\xi) = \sum_{i=0}^{N} a_i \phi^i + \sum_{j=1}^{N} b_j \phi^{j-1} \varphi, \tag{14}$$

where $a_i (i = 0, \cdots, N), b_j (j = 1, \cdots, N)$ are constants to be determined later, and $a_N^2 + b_N^2 \neq 0$. We can determine N in Eq (14) by using the homogeneous balance between the highest order derivatives and the nonlinear terms.

Step3. We substitute Eq (14) into Eq (13), along with Eq (4) and Eq (6) (Case 1 is for an example). Then the left-hand side of Eq (13) becomes a polynomial in ϕ and φ, in which the degree of φ is not larger than 1. When we made each coefficient of the polynomial into zero, a system of algebraic equations in $a_i (i = 0, \cdots, N), b_j (j = 1, \cdots, N), k, c, \lambda, \mu, \sigma$ were yield.

Step4. After solving above algebraic system with the aid of Maple program, we substitute the results and the general solution of Eq (5) into Eq (14), then the exact solution of Eq (11) can be obtained immediately.

Step5. Similar to Step 3 and Step 4, substituting Eq (14) into Eq (13), and uniting Eq (4) and Eq (8) (or Eq (10)), we obtain the exact trigonometric function solutions (or rational function).

3. Application of the Method

On applying the following transformation

$$u(\xi) = u(x,t), v(\xi) = v(x,t), \xi = k(\frac{x^{\beta}}{\beta} + \frac{ct^{\alpha}}{\alpha}), \tag{15}$$

where c, k are non-zero constants, then Eq (1) becomes the following ODEs

$$cku' + kpvv' = 0, \tag{16}$$

$$ckv' + kruv' + ksvu' + k^3 qv''' = 0. \tag{17}$$

Integrating Eq (16) once with regard to ξ and taking integration constant to zero yields

$$cu + \frac{1}{2} pv^2 = 0. \tag{18}$$

Substituting Eq (16) and Eq (18) to Eq (17), then integrating Eq (17) once with regard to ξ and let integration constant to g, we have

$$6c^2 v - p(r + 2s)v^3 + k^2 cqv'' + g = 0. \tag{19}$$

By balancing v'' with v^3 in Eq (19), the value of N is equal to 1. And we suppose the form of exact solution for Eq (19) is

$$v(\xi) = a_0 + a_1 \phi(\xi) + b_1 \varphi(\xi), \tag{20}$$

where the constant coefficients a_0, a_1, b_1 are determined later and satisfy $a_1^2 + b_1^2 \neq 0$.

Case1. When $\lambda < 0$, an algebraic system in $a_0, a_1, b_1, k, c, \lambda, \mu, \sigma, g$ is provided as an example.

$$\phi^0:\frac{2p(r+2s)b_1^3\lambda^3\mu}{(\lambda^2\sigma+\mu^2)^2}+\frac{6ck^2qb_1\lambda^2\mu}{\lambda^2\sigma+\mu^2}+\frac{3p(r+2s)\lambda^2a_0b_1^2}{\lambda^2\sigma+\mu^2}-p(r+2s)a_0^3+6c^2a_0+6ck^2qa_0+g=0,$$

$$\varphi:\frac{p(r+2s)b_1^3\lambda^4\sigma}{(\lambda^2\sigma+\mu^2)^2}-\frac{3p(r+2s)b_1^3\lambda^2\mu^2}{(\lambda^2\sigma+\mu^2)^2}-\frac{6p(r+2s)\lambda a_0b_1^2\mu}{\lambda^2\sigma+\mu^2}$$
$$-\frac{12ck^2qb_1\lambda\mu^2}{\lambda^2\sigma+\mu^2}+6c^2b_1+6ck^2qb_1\lambda-3p(r+2s)a_0^2b_1=0,$$

$$\phi:\frac{3p(r+2s)a_1b_1^2\lambda^2}{\lambda^2\sigma+\mu^2}+6c^2a_1+12ck^2qa_1\lambda-3p(r+2s)a_0^2a_1=0,$$

$$\phi^2:\frac{2p(r+2s)b_1^3\lambda^2\mu}{(\lambda^2\sigma+\mu^2)^2}+\frac{6ck^2b_1\lambda\mu}{\lambda^2\sigma+\mu^2}+\frac{3p(r+2s)a_0b_1^2\lambda}{\lambda^2\sigma+\mu^2}-3p(r+2s)a_0a_1^2=0,$$

$$\phi^3:\frac{3p(r+2s)a_1b_1^2\lambda}{\lambda^2\sigma+\mu^2}+12ck^2qa_1-p(r+2s)a_1^3=0,$$

$$\phi\varphi:\frac{6p(r+2s)a_1b_1^2\lambda\mu}{\lambda^2\sigma+\mu^2}+18ck^2qa_1\mu+6p(r+2s)a_0a_1b_1=0.$$

$$\phi^2\varphi:\frac{p(r+2s)b_1^3\lambda}{\lambda^2\sigma+\mu^2}+12ck^2qb_1-3p(r+2s)a_1^2b_1=0,$$

Then some results are gained by solving this system.

Result 1.1. When $p(r+2s)>0$, we have

$$a_0=0,a_1=\pm 2k^2q\left[\frac{-6\lambda}{p(r+2s)}\right]^{1/2},b_1=0,$$
$$\lambda=\lambda,\mu=0,\sigma=\sigma,c=-2k^2q\lambda,k=k.$$

Result 1.2. Suppose c,k,λ satisfies that $cp(c-k^2q\lambda)(r+2s)>0$, we have

$$a_0=\pm\left[\frac{2c(c-k^2q\lambda)}{p(r+2s)}\right]^{1/2},a_1=0,b_1=\pm\frac{6ck^2q\mu}{[2cp(r+2s)(c-k^2q\lambda)]^{1/2}},$$
$$\lambda=\lambda,\mu=\mu,\sigma=\frac{\mu^2(k^2q\lambda+2c)}{2\lambda^2(k^2q\lambda-c)},c=c,k=k.$$

Result 1.3. When $p(r+2s)>0$, we have

$$a_0=0,a_1=\pm k^2q\left[\frac{-3\lambda}{2p(r+2s)}\right]^{1/2},b_1=\pm k^2q\left[\frac{3(\lambda^2\sigma+\mu^2)}{2p(r+2s)}\right]^{1/2},$$
$$\lambda=\lambda,\mu=\mu,\sigma>-\frac{\mu^2}{\lambda^2},c=-\frac{1}{2}k^2q\lambda,k=k.$$

From the Results 1.1-1.3, we obtain hyperbolic function solutions for Eq (1) .

Theorem 1. (1) Suppose that $p(r+2s)>0$, we got

$$v_1(\xi)=\pm 2k^2q\left[\frac{-6\lambda}{p(r+2s)}\right]^{1/2}\cdot\frac{A_2\sinh(\sqrt{-\lambda}\xi)+A_1\cos h(\sqrt{-\lambda}\xi)}{A_1\sinh(\sqrt{-\lambda}\xi)+A_2\cos h(\sqrt{-\lambda}\xi)},$$

where $\xi=\frac{kx^\beta}{\beta}-\frac{2k^3q\lambda t^\alpha}{\alpha}$, and k,A_1,A_2 are arbitrary constants.

(2) Suppose $cp(c-k^2q\lambda)(r+2s)>0$, we have

$$v_2(\xi)=\pm\left[\frac{2c(c-k^2q\lambda)}{p(r+2s)}\right]^{1/2}\pm\frac{6\mu ck^2q\lambda}{[2cp(r+2s)(c-k^2q)]^{1/2}(\lambda A_1\sinh\sqrt{-\lambda}\xi+\lambda A_2\cosh\sqrt{-\lambda}\xi+\mu)},$$

Where $\xi=\frac{kx^\beta}{\beta}+\frac{ckt^\alpha}{\alpha}$, c,k are arbitrary constants, and A_1,A_2 satisfy $A_1^2-A_2^2=\frac{\mu^2(k^2q\lambda+2c)}{2\lambda^2(k^2q\lambda-c)}$.

(3) Suppose that $p(r+2s)>0$, we got

$$v_3(\xi)=\pm k^2q\lambda\left(\left[\frac{-3\lambda}{2p(r+2s)}\right]^{1/2}\frac{A_2\sinh\sqrt{-\lambda}\xi+A_1\cosh\sqrt{-\lambda}\xi}{\lambda A_1\sinh\sqrt{-\lambda}\xi+\lambda A_2\cosh\sqrt{-\lambda}\xi+\mu}+\left[\frac{3(\lambda^2\sigma+\mu^2)}{2p(r+2s)}\right]^{1/2}\frac{1}{\lambda A_1\sinh\sqrt{-\lambda}\xi+\lambda A_2\cos h\sqrt{-\lambda}\xi+\mu}\right)$$

where $\xi=\frac{kx^\beta}{\beta}-\frac{k^3q\lambda t^\alpha}{2\alpha}$, k is arbitrary constants, and A_1,A_2 satisfy $A_1^2-A_2^2>-\frac{\mu^2}{\lambda^2}$.

Case2. When $\lambda>0$, solving an algebraic system in $a_0,a_1,b_1,k,c,\lambda,\mu,\rho,g$ some results are yield.

Result 2.1. When $p(r+2s)<0$, then

$$a_0=0,a_1=\pm 2k^2q\left[\frac{-6\lambda}{p(r+2s)}\right]^{1/2},b_1=0,$$

$$\lambda=\lambda,\mu=0,\rho=\rho,c=-2k^2q\lambda,k=k.$$

Result 2.2. Suppose c,k,λ satisfies that $cp(c-k^2q\lambda)(r+2s)>0$, we have

$$a_0=\pm\left[\frac{2c(c-k^2q\lambda)}{p(r+2s)}\right]^{1/2},a_1=0,b_1=\pm\frac{6ck^2q\mu}{[2cp(r+2s)(c-k^2q\lambda)]^{1/2}},$$

$$\lambda=\lambda,\mu=\mu,\rho=-\frac{\mu^2(k^2q\lambda+2c)}{2\lambda^2(k^2q\lambda-c)},c=c,k=k.$$

Result 2.3. When $p(r+2s)<0$, then

$$a_0=0,a_1=\pm k^2q\left[\frac{-3\lambda}{2p(r+2s)}\right]^{1/2},b_1=\pm k^2q\left[\frac{3(\mu^2-\lambda^2\rho)}{2p(r+2s)}\right]^{1/2},$$

$$\lambda=\lambda,\mu=\mu,\rho<\frac{\mu^2}{\lambda^2},c=-\frac{1}{2}k^2q\lambda,k=k.$$

Combining the Result 2.1-2.3, we gain the triangle function solution for Eq (1).

Theorem 2. (1) Suppose that $p(r+2s)<0$, we got

$$v_4(\xi)=\pm 2k^2q\left[\frac{-6\lambda}{p(r+2s)}\right]^{1/2}\cdot\frac{A_1\cos(\sqrt{\lambda}\xi)-A_2\sin(\sqrt{\lambda}\xi)}{A_1\sin(\sqrt{\lambda}\xi)+A_2\cos(\sqrt{\lambda}\xi)},$$

Where $\xi=\frac{kx^\beta}{\beta}-\frac{2k^3q\lambda t^\alpha}{\alpha}$, and k,A_1,A_2 are arbitrary constants.

(2) Suppose $cp(c-k^2q\lambda)(r+2s)>0$, we have

$$v_5(\xi)=\pm\left[\frac{2c(c-k^2q\lambda)}{p(r+2s)}\right]^{1/2}\pm\frac{6\mu ck^2q\lambda}{[2cp(r+2s)(c-k^2q)]^{1/2}(\lambda A_1\sin\sqrt{\lambda}\xi+\lambda A_2\cos\sqrt{\lambda}\xi+\mu)},$$

Where $\xi=\frac{kx^\beta}{\beta}+\frac{ckt^\alpha}{\alpha}$, c,k are arbitrary constants, and A_1,A_2 satisfy $A_1^2+A_2^2=-\frac{\mu^2(k^2q\lambda+2c)}{2\lambda^2(k^2q\lambda-c)}$.

(3) Suppose that $p(r+2s)>0$, we got

$$v_6(\xi)=\pm k^2q\lambda\left(\left[\frac{-3\lambda}{2p(r+2s)}\right]^{1/2}\frac{A_1\cos\sqrt{\lambda}\xi-A_2\sin\sqrt{\lambda}\xi}{\lambda A_1\sin\sqrt{\lambda}\xi+\lambda A_2\cos\sqrt{\lambda}\xi+\mu}+\left[\frac{3(\lambda^2\sigma+\mu^2)}{2p(r+2s)}\right]^{1/2}\frac{1}{\lambda A_1\sin\sqrt{\lambda}\xi+\lambda A_2\cos\sqrt{\lambda}\xi+\mu}\right)$$

Where $\xi = \frac{kx^{\beta}}{\beta} - \frac{k^3 q\lambda t^{\alpha}}{2\alpha}$, k is arbitrary constants, and A_1, A_2 satisfy $A_1^2 + A_2^2 < \frac{\mu^2}{\lambda^2}$.

Case 3. When $\lambda = 0$, solving an algebraic system in $a_0, a_1, b_1, k, c, \lambda, \mu, A_1, A_2, g$, we obtain the following results

Result 3.1. When $p(r+2s) > 0$, then

$$a_0 = \pm c\left[\frac{2}{p(r+2s)}\right]^{1/2}, a_1 = 0, b_1 = \pm 3k^2 q\mu\left[\frac{2}{p(r+2s)}\right]^{1/2},$$

$$\mu = \mu, c = c, k = k, A_1 = A_1, A_2 = \frac{2cA_1^2 - 3k^2 q\mu}{4c\mu}.$$

Theorem 3. Suppose that $p(r+2s) > 0$, we got

$$v_7(\xi) = \pm\left[\frac{2}{p(r+2s)}\right]^{1/2}\left(c + \frac{12ck^2 q\mu^2}{2c\mu^2\xi^2 + 4c\mu A_1\xi + 2cA_1^2 - 3k^2 q\mu}\right),$$

Where $\xi = \frac{kx^{\beta}}{\beta} - \frac{ckt^{\alpha}}{\alpha}$, and c, k, A_1 are arbitrary constants.

Remark 1. In this section, we only give the exact solution of $v(\xi)$, whereas the exact solution of $u(\xi)$ can be obtained by Eq (18), i.e

$$u(\xi) = -\frac{1}{2c}pv^2(\xi).$$

4. Conclusion

As we know, the (G'/G,1/G)-expansion method is a common method for solving NFPDE and variable exact solutions can be obtained. In this study, it has been successfully extended to find exact solutions for the DSWEs with conformable derivative. We gain abundant explicit exact solutions including hyperbolic, trigonometric and rational function type. It indicates this method is effective and reliable mathematical tool for seeking the exact solutions of other NFPDEs.

5. Acknowledgments

This research was financially supported by Innovation Research for Postgraduate of Guangzhou University(2019GDJC-M30).

References

[1] R. Hirota. The Direct Method in Soliton Theory, Cambridge: Cambridge University Press, 2004.

[2] Y. B. Zhou, M. L. Wang, Y. M. Wang. Periodic wave solutions to a coupled KdV equations with variable coefficients, Phys. Lett. A., 308(2003), 31-36.

[3] Y. D. Shang. The extended hyperbolic function method and exact solutions of the long short wave resonance equations, Chaos Solitons Fractals., 36(2008), 762-771.

[4] L. X. Li, E. Q. Li, M. L. Wang. The (G'/G,1/G)-expansion method and its application to travelling wave solutions of the Zakharov equations, Appl. Math. J. Chinese Univ. Ser. B, 25(2010),454-462.

[5] R. Sahadevan, P. Prakash. On Lie symmetry analysis and invariant subspace methods of coupled time fractional partial differential equations, Chaos Solitons Fractals, 104(2017),107-120.

[6] S. Q. Chen, Y. Liu, L. X. Wei, et al. Exact solutions to fractional Drinfel'd-Sokolov-Wilson equations, Chinese J. Phys., 56(2018),708-720.

[7] Z. S. Wen. The generalized bifurcation method for deriving exact solutions of nonlinear space-time fractional partial differential equations, Appl. Math. Comput., 366(2020), 124735, 10 pp.

[8] Y. M. Zhao, Y. H. He, H. Z. Yang. The two variable $(\phi' / \phi, 1/\phi)$-expansion method for solving the time-fractional partial differential equations, AIMS Mathematematics, 5(2020), 4121-4135.
[9] O. Guner, A. Bekir, H. Bilgil. A note on exp-function method combined with complex transform method applied to fractional differential equations, Adv. Nonlinear Anal., 4(2015),201-208.
[10] R. Khalil, M. A. Horani, A. Yousef, M. Sababheh. A new definition of fractional derivative, Comput. Appl. Math., 264(2014), 65-70.

Discrete Schemes for a Class of Semilinear Hyperbolic Eqution

Jing Xu and Xiaoli Jiang*

College of Mathematical Sciences, Bohai University, Jinzhou, Liaoning, China

*Corresponding author email: jxls309@163.com

Abstract. This paper focuses on the numerical algorithm of a class of semilinear hyperbolic equation. The dissipation term and external force source term existing in the equation enhance the nonlinearity of the model and make the nonlinear effect complicated. However, this nonlinear effect has a great influence on the discrete scheme, which cannot be neglected. Discretizing the nonlinear terms to ensure the validity of the scheme is the core issue in this paper. An effective scheme based on linearization techniques and iteration theory is proposed. It is based on the finite difference method. The efficiency of the proposed schemes was verified via some numerical examples showing that they compare well with existing methods.

Keywords: Finite difference method, semilinear hyperbolic equation, nonlinear source term, dissipation term.

1. Introduction

Semilinear hyperbolic equation is the most basic and important nonlinear differential equation, which is constantly applied in modern science and technology, engineering model, and daily life. For example, the weather forecast, seawater intrusion, groundwater pollution prevention, waste pollution, crystal growth, and even nuclear weapons and a series of questions about the national economy and people's livelihood, all rely on solving the corresponding nonlinear hyperbolic equation[1]. In the numerical solution of differential equations, the difference method is a very mature method compared with other methods. The method, which has been widely used in the regular long-wave equation, convection-diffusion equation, KG equation, KdV equation, Schrödinger equation and so on. In recent years, mathematical workers represented by Zhou Yulin put forward the discrete functional analysis theory, and some nonlinear problems were extensively discussed in literature [2]-[6] by using the nonlinear fixed-point theory. The relevant results can also be referred to the research and discussion in literature [7]-[12], which will not be explained in detail here. The finite volume method (FVM) is mainly applied to the problems of wind load, hydraulic transmission, gas flow and heat transfer in power machinery in civil engineering[13].

In the finite volume method, volume integrals in a partial differential equation that contain a divergence term are converted to surface integrals, using the divergence theorem. These terms are then evaluated as fluxes at the surfaces of each finite volume. Because the flux entering a given volume is identical to that leaving the adjacent volume, these methods are conservative. Another advantage of the finite volume method is that it is easily formulated to allow for unstructured meshes. It has been widely developed and applied in the field of computational fluid dynamics, which is mainly attributed to its local conservation and flexibility of mesh generation[14]. Since the FVM maintains the conservation of mass on each subdivision element, it can construct a high-precision format, and its conver-

Content from this work may be used under the terms of the Creative Commons Attribution 3.0 licence. Any further distribution of this work must maintain attribution to the author(s) and the title of the work, journal citation and DOI.
Published under licence by IOP Publishing Ltd

gence analysis can also be obtained in the framework of the finite element method. Therefore, it has attracted more and more attention in many practical calculations and theoretical studies.

There are few algorithms and numerical analyses for nonlinear development equations, especially for the nonlinear dissipative hyperbolic equation in this paper. Due to the nonlinear effect caused by the nonlinear factors contained in it, there is no mature technology to deal with the dispersion of the nonlinear term. Specifically, it is reflected in two difficulties: one is how to discretize the second-order term of time to ensure the stability and convergence of the formal; Second, the external force term is a nonlinear function of time and space. Therefore, this paper will make different improvements under the framework of finite difference, and propose a new numerical discrete scheme of differential equations to resolve the two difficulties mentioned above.

The following semilinear hyperbolic equation with dissipative and source terms is discussed

$$u_{tt} - u_{xx} + \gamma u_t = u|u|,\ (x,t) \in [a,b] \times [0,\mathrm{T}), \tag{1}$$

where the boundary condition is

$$u(a,t) = u(b,t) = 0, \tag{2}$$

the initial condition is the numerical solution of

$$u(x,0) = u_0(x), \tag{3}$$

$$u_t(x,0) = u_1(x), \tag{4}$$

where γ is a positive constant, $u_0(x)$ and $u_1(x)$ are continuous in interval $[a,b]$.

2. Finite Difference Scheme

2.1. The Proposal of the difference scheme

In this section, we will use finite difference method to construct a new discrete format, and then prove the existence and uniqueness of solution for difference scheme, and the truncation error is about time and space variables of the second-order accuracy, and give some examples to verify the correctness of the corresponding conclusion. The following symbols are first given

$$x_i = ih + a,\ t_n = n\tau,\ 0 \le i \le M = (b-a)/h,\ 0 \le n \le N = T/\tau,$$

where u_i^n is an approximation of u on the interval $[a,b]$, define

$$\delta_t^2 u_i^n = \frac{u_i^{n-1} - 2u_i^n + u_i^{n+1}}{\tau^2},\ \delta_x^2 u_i^n = \frac{u_{i+1}^n - 2u_i^n + u_{i-1}^n}{h^2},\ \delta_t u_i^n = \frac{u_i^{n+1} - u_i^{n-1}}{2\tau},\ \overline{u}_j^n = \frac{u_j^{n+1} - u_j^{n-1}}{2},$$

where C represents a positive constant but may take different values in different places.

A three-level and two-order linear implicit difference scheme for problem (1) - (4) is constructed as follows

$$\delta_t^2 u_i^n - \delta_x^2 \overline{u}_i^n + \gamma \delta_t u_i^n = u_i^n |u_i^n| + O(h^2 + \tau^2),\ i = 1,2,\dots,M-1, n = 1,2,\dots,N. \tag{5}$$

Ignoring the infinitesimal in the above formula, the following difference schemes are obtained

$$\begin{aligned} & -\frac{1}{2h^2} u_{i-1}^{n+1} + \left(\frac{1}{h^2} + \frac{1}{\tau^2} + \frac{\gamma}{2\tau}\right) u_i^{n+1} - \frac{1}{2h^2} u_{i+1}^{n+1} \\ & = \left(\frac{2}{\tau^2} + |u_i^n|\right) u_i^n + \frac{1}{2h^2} u_{i-1}^{n-1} - \left(\frac{1}{h^2} + \frac{1}{\tau^2} - \frac{\gamma}{2\tau}\right) u_i^{n-1} + \frac{1}{2h^2} u_{i+1}^{n-1}. \end{aligned} \tag{6}$$

From boundary conditions (2) and initial conditions (3), (4), we derive

$$u_0^n = u_M^n = 0, \tag{7}$$

as well as

$$u_i^0 = u_0(ih), \tag{8}$$

$$\delta_t u_i^0 = u_1(ih). \tag{9}$$

Suppose that $u^k = \left[u_1^k, u_2^k, \ldots, u_{M-1}^k\right]^T$, applying difference schemes to each spatial node (6), and all the equations obtained are written in matrix form, we have $Au^{n+1} = d$, this is an algebraic system about u^{n+1}, where

$$A = tri\left[-\frac{1}{2h^2}, \frac{1}{h^2} + \frac{1}{\tau^2} + \frac{\gamma}{2\tau}, -\frac{1}{2h^2}\right]_{(M-1)\times(M-1)},$$

where $tri\left[a_1, a_2, a_3\right]_{(M-1)\times(M-1)}$ represent a tridiagonal matrix of $(M-1)\times(M-1)$, each row of this matrix has only three values of a_1, a_2 and a_3, located on the rows of the main diagonal and the two secondary diagonals. Next, the reversibility of matrix A will be defined and proved in **Lemma 1**.

2.2. *Existence and Uniqueness of Solutions of Difference Scheme*

Lemma 1 The coefficient matrix A is reversible.

Proof: The eigenvalues of coefficient A can be obtained from the following formula (see [12])

$$\lambda_j = \left(\frac{1}{h^2} + \frac{1}{\tau^2} + \frac{\gamma}{2\tau}\right) + 2\sqrt{\left(-\frac{1}{2h^2}\right)\left(-\frac{1}{2h^2}\right)}\cos\frac{j\pi}{M}, j = 1 \ldots M-1,$$

it has been simplified to become the following equality

$$\lambda_j = \left(\frac{1}{h^2} + \frac{1}{\tau^2} + \frac{\gamma}{2\tau}\right) + \frac{1}{h^2}\cos\frac{j\pi}{M}, j = 1 \ldots M-1,$$

then

$$\lambda_j = \left(\frac{1}{h^2} + \frac{1}{\tau^2} + \frac{\gamma}{2\tau}\right) + \frac{1}{h^2}\cos\frac{j\pi}{M} \geq \frac{1}{\tau^2} + \frac{\gamma}{2\tau},$$

because γ is a positive constant, whatever value j takes, the eigenvalue of a coefficient A is positive (non-zero), so coefficient matrix A is reversible.

Theorem 1 The solutions of difference scheme (5) - (9) exist and it is unique.

Proof: According to the difference scheme (9), one has

$$\delta_t u_i^0 = \frac{u_i^1 - u_i^0}{2\tau} = u_1(ih),$$

then we have

$$u_i^1 = 2\tau u_1(ih) + u_i^0,$$

so, we can see from equation (9)

$$u_i^0 = u_0(ih),$$
$$u_i^1 = 2\tau u_1(ih) + u_0(ih),$$

further, u^2 can be deduced from the difference scheme (5) - (9), suppose that u^{n-1}, u^{n-2} is known. By solving the equation group $Au^n = b$, we can get the value of the problem at the time level $t_n = n\tau$, where b can be contained in u^{n-1}, u^{n-2} and is known. **Lemma 1** shows that the solution of the scheme (5) - (9) exists and is unique for any γ and coefficient matrix A.

2.3. *Stability of Solution of Difference Scheme*

The stability of the difference scheme is discussed below. Let $\overline{u}_i^n$ be an approximate solution of the difference scheme (5) - (9), we consider the truncation error of the difference scheme as follows

$$\rho_i^n = u_i^n - \overline{u}_i^n, i = 0, 1, \ldots, M, \text{ n}=0, 1, \ldots, N,$$

according to the above definition, the rounding error equation is derived from equation (5).

$$-\frac{1}{2h^2}\rho_{i-1}^{n+1}+\left(\frac{1}{h^2}+\frac{1}{\tau^2}+\frac{\gamma}{2\tau}\right)\rho_i^{n+1}-\frac{1}{2h^2}\rho_{i+1}^{n+1}$$

$$=\frac{2}{\tau^2}\rho_i^n+\frac{1}{2h^2}\rho_{i-1}^{n-1}-\left(\frac{1}{h^2}+\frac{1}{\tau^2}-\frac{\gamma}{2\tau}\right)\rho_i^{n-1}+\frac{1}{2h^2}\rho_{i+1}^{n-1}+u_i^n\left|u_i^n\right|-\overline{u}_i^n\left|\overline{u}_i^n\right|,$$

combining with Taylor expansion, we can obtain that when $h,\tau\to 0$, $\rho_i^n=O(h^2+\tau^2)$. Refer to the proof of theorem 8 in reference [15], we can get the following convergence and stability.

Theorem 2 The solution of the finite difference scheme (5) - (9) converges to the solution of the initial-boundary value problem (1) - (4) with order $O(h^2+\tau^2)$ by the L_∞ norm.

Theorem 3 The finite difference scheme (5) - (9) is stable by the L_∞ norm.

2.4. Simulation Example

In this section, several numerical examples are given to verify the properties of the formats presented above. Firstly, we present some experiments in two cases to compare the stability of discrete form with the classical form. In those experiments, the horizontal axis shows the spatial variable, the vertical axis denotes the time variable, Z-axis represents the change of numerical solution.

Example 1

Consider the following one-dimensional semilinear wave equation

$$u_{tt}(x,t)-u_{xx}(x,t)+u_t(x,t)=|u(x,t)|u(x,t),(x,t)\in(0,1)\times(0,15),$$

$$u(0,t)=u(1,t)=0,\ u(x,0)=e^{bx},\ u_t(x,0)=0$$

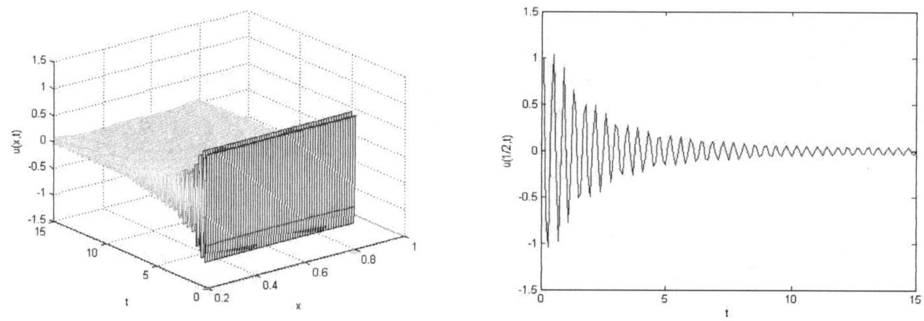

Figure 1. ($b=1$) The figure on the left shows the numerical solution of the equation $u(x,t)$, the right graph shows the change at the midpoint of the space $u(1/2,t)$.

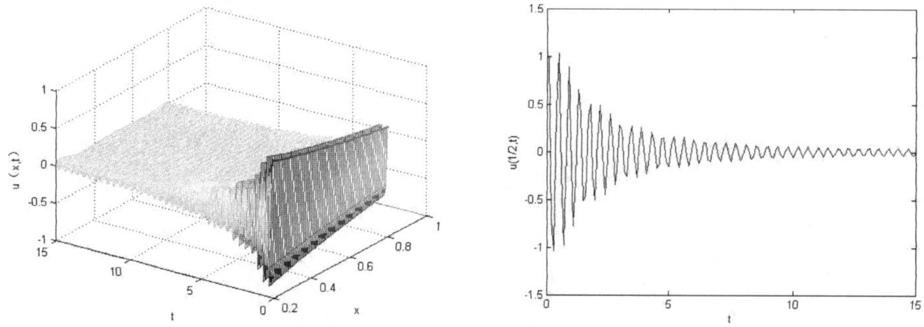

Figure 2. ($b=-1$) The figure on the left shows the numerical solution of the equation $u(x,t)$, the right graph shows the change at the midpoint of the space $u(1/2,t)$.

From figure 1 and figure 2, it is easy to see that the numerical solution of the equation is fast approaching zero, which implies that our discrete form has a stronger convergence ability than the classical discrete form. In addition, the two curves are both stables with increasing time.

Example 2

Consider the numerical solution of the initial-boundary value problem for the following one-dimensional wave equation.

$$u_{tt}(x,t)-u_{xx}(x,t)+u_t(x,t)=|u(x,t)|u(x,t)+t\sin x+\sin x-t^2\sin^2 x,$$
$$u(0,t)=u(2\pi,t)=0,\ u(x,0)=0,\ u_t(x,0)=\sin x,\ (x,t)\in[0,2\pi]\times[0,8).$$

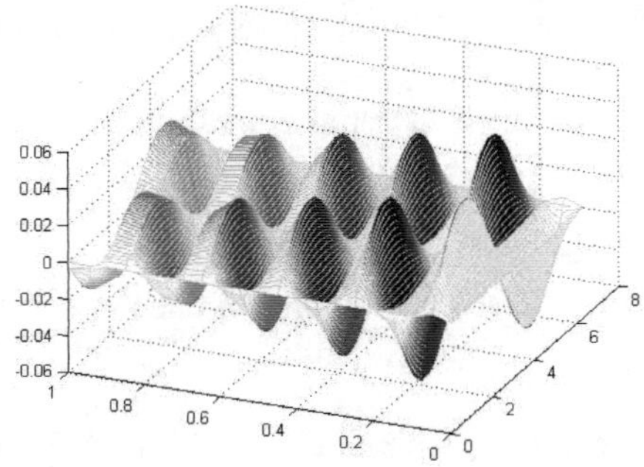

Figure 3. Numerical solution $u(x,t)$ of the equation.

Figure 3 show that if the adjustment function about time and space is added to the discrete scheme, the stability and convergence of the discrete system will become worse, and the numerical solution will blow up at some time.

3. Conclusion

In this paper, finite difference method is used to solve semi-linear wave systems with dissipative terms. Firstly, a new discrete scheme is constructed using the finite difference method, which proves the existence and uniqueness of the solution of the scheme. The truncation error is second-order for both time and space. Some examples are given to verify the correctness of the corresponding conclusions. The scheme also successfully solves the nonlinearity brought about by the source term and ensures the validity and feasibility of the calculation. In addition, the numerical results of the discrete system are given while using the above method to obtain appropriate discrete schemes. The correctness and effectiveness of the method can be seen intuitively by comparing with the results of the existing literature.

Acknowledgments

This work was supported by the National Natural Science Foundation of China (11871117, 61976027), the Natural Science Foundation of Liaoning Province of China (LQ2019008, 20180551262). wishing to acknowledge assistance or encouragement from colleagues, special work by technical staff or financial support from organizations should do so in an unnumbered Acknowledgments section immediately following the last numbered section of the paper.

References

[1] Wang P 2000 New conservative difference schemes for a coupled nonlinear schrodinger system. Applied Mathematics and Computation vol 217 pp 1604-1619

[2] GUO B Y1984 Discussing discrete energy method. Numerical Mathematics A Journal of Chinese Universities vol 6 pp 327-344

[3] Zhu Y 1990 Application of discrete functional analysis to the finite difference method. Inter Acadie Publishers vol8 pp 159-183

[4] Browder F E 1965 Existence and uniqueness theorems for solutions of nonlinear boundary value problems. Application of Nonlinear Partial Differential Equations vol 17 pp 24-49

[5] KRIVIS G D A 1993 Finite difference discretization of the cubic schrodinger equation. Ima Journal of Numerical Analysis vol 13pp 115-124
[6] Peranich L S A 1987 Finite difference scheme for solving a nonlinear schrodinger equation with a linear damping term. Journal of Computational Physics vol 68 pp 501-505
[7] Sun Z 1995 A second-order accurate linearized difference scheme for the two-dimentional cahn-hilliard equation. Mathematics Computation vol64 pp 1463-1471
[8] Sun Z, Shen L 2003 Long time asymptotic behavior of solution of implicit difference scheme for a semi-linear parabolic equation. Journal of Computational Mathematics vol 21 pp 671-680
[9] ZHOU Y 2003 Numerical solutions of differential equations. Hebei Education Press
[10] LI R X 2007 Foundation of finite volume method. National Defense Industry Press vol 3 pp 1-30
[11] RU C M 2002 Simple Analysis of differential equation numerical method. Journal of Jinan Institute of Education vol8 pp 49-50
[12] Smith G D 1993 Numerical solution of partial differential equations, finite difference methods. Oxford University vol5 pp 78-82
[13] TAE G H, DAEWOOK K, IL H J 1998 Global existence and uniform decay rates for the semilinear wave equation with damping and source terms. Computers & Mathematics with Applications vol 4 pp 815-828
[14] Eymard R and Gallouet T 2000 Handbook of numerical analysis. Elsevier vol 7 pp 713-1020
[15] Jiang X L and Wang X F 2017 A conservation difference scheme of generalized boussinesq equation. Discrete Dynamics in Nature and Society pp 1-9.

The Explicit Formulae of the Determinants of Specific Pentadiagonal Toeplitz Hessenberg Matrices

Jishe Feng[1,*] and Hongtao Fan[2,3]

[1]Department of Mathematics, Longdong University, Qingyang, Gansu, China
[2]College of Science, Northwest A&F University, Yangling, Shaanxi, China
[3]School of Mathematics and Statistics, Lanzhou University, Lanzhou, Gansu, China

[*]Corresponding author email: gsfjs6567@126.com

Abstract. The pentadiagonal Toeplitz matrix is a special kind of sparse matrix widely used in linear algebra, combinatorics, computational mathematics, and has been attracted much attention. We use the determinants of two specific Hessenberg matrices to represent the recurrence relations to prove two explicit formulae to evaluate the determinants of specific pentadiagonal Toeplitz matrices proposed in a recent paper [3]. Further, four new results are established.

1. Introduction

Pentadiagonal Toeplitz matrices often appear in statistics, matrix algebra, boundary value problems, etc., and are significant in several research problems. For further details, consult [1-7] and the references therein. In a recent paper evaluating determinants of pentadiagonal Toeplitz matrices, Andelic and Fonseca [3] proposed two conjectures. This paper aims to prove the conjectures above and establish four new formulas for the determinants of other pentadiagonal Toeplitz matrices.

We follow the notations introduced in [3]. All mentioned entries in matrices are assumed as 1s, whereas all the unmentioned entries are to be assumed as zero. The following are the two explicit formulas for determinants of specific pentadiagonal Toeplitz matrices proposed in [3], which are proved herein.

Theorem 1 For all $n \geq 4$ the determinants of the pentadiagonal Toeplitz matrix is given by

$$|A_n| = \begin{vmatrix} 1 & 1 & & & 1 \\ 1 & \ddots & \ddots & & \\ 1 & \ddots & \ddots & \ddots & \\ & \ddots & \ddots & \ddots & 1 \\ & & 1 & 1 & 1 \end{vmatrix}_{n\times n} = \begin{cases} 1, & if\ n \equiv 0 \pmod 4 \\ 2, & if\ n \equiv 1 \pmod 4 \\ -1, & if\ n \equiv 2 \pmod 4 \\ 0, & if\ n \equiv 3 \pmod 4 \end{cases} \quad (1)$$

Theorem 2 For all $n \geq 4$, the determinant of the pentadiagonal Toeplitz matrix is given by

Content from this work may be used under the terms of the Creative Commons Attribution 3.0 licence. Any further distribution of this work must maintain attribution to the author(s) and the title of the work, journal citation and DOI.
Published under licence by IOP Publishing Ltd

$$|B_n|=\begin{vmatrix}1 & 1 & & 1 & \\ 1 & \ddots & \ddots & & 1 \\ 1 & \ddots & \ddots & \ddots & \\ & \ddots & \ddots & \ddots & 1 \\ & & 1 & 1 & 1\end{vmatrix}_{n\times n}=\begin{cases}0, & if\ n\equiv 0\ (\mathrm{mod}\ 4)\\ 2, & if\ n\equiv 1\ (\mathrm{mod}\ 4)\\ 3, & if\ n\equiv 2\ (\mathrm{mod}\ 4)\\ 1, & if\ n\equiv 3\ (\mathrm{mod}\ 4)\end{cases} \tag{2}$$

The remainder of this paper is organized as follows. In Section 2, we recall two lemmas regarding Hessenberg matrices and the linear recurrence sequence and prove a theorem establishing a Toeplitz Hessenberg determinants formula. In Sections 3 and 4, we use the Laplace expansion formula and additional preliminary results to verify Theorems 1 and 2. In Section 5, we prove four new results for pentadiagonal Toeplitz matrices.

2. Preliminaries

Lemma 3 [1,5] Assume that $r_1, r_2, \cdots, r_k$ are the complex roots of the characteristic equation

$$r^k - c_1 r^{k-1} - c_2 r^{k-2} - \cdots c_k = 0$$

of the linear recurrence relation

$$a_n = c_1 a_{n-1} + c_2 a_{n-2} + \cdots + c_k a_{n-k}.$$

If $\alpha_1, \alpha_2, \cdots, \alpha_k$ are arbitrary complex constants, then there is

$$a_n = \alpha_1 r_1^n + \alpha_2 r_2^n + \cdots + \alpha_k r_k^n.$$

Lemma (Laplace Expansion) 4 [5] For a $n\times n$ matrix $A=(a_{ij})_{n\times n}$, the determinant of matrix $|A|$, is equal to

$$|A| = a_{i1}A_{i1} + a_{i2}A_{i2} + \cdots + a_{in}A_{in}$$

Where A_{ij} is the cofactor, which is the products of $(-1)^{i+j}$ and the determinant of the $(n-1)\times(n-1)$ matrix obtained by deleting all the elements in the ith row and the jth column in the matrix A.

Theorem 5 Let

$$|H_n|=\begin{vmatrix}1 & 1 & & 1 & \\ 1 & \ddots & \ddots & & 1 \\ 1 & \ddots & \ddots & \ddots & \\ & \ddots & \ddots & \ddots & 1 \\ & & 1 & 1 & 1\end{vmatrix}_{n\times n}.$$

Then, there exists the recurrence relation

$$|H_n| = |H_{n-1}| - |H_{n-2}| + |H_{n-3}| \tag{3}$$

and the explicit formula

$$|H_n| = \frac{1}{2} + \left(\frac{1}{4} - \frac{i}{4}\right)i^n + \left(\frac{1}{4} + \frac{i}{4}\right)(-i)^n \tag{4}$$

Proof. Applying the Laplace expansion along with the first row of determinant $|H_n|$,

$$|H_n|=\begin{vmatrix}1 & 1 & & & \\ 1 & \ddots & \ddots & & \\ 1 & \ddots & \ddots & \ddots & \\ & \ddots & \ddots & \ddots & 1 \\ & & 1 & 1 & 1\end{vmatrix}_{n\times n}=\begin{vmatrix}1 & 1 & & & \\ 1 & \ddots & \ddots & & \\ 1 & \ddots & \ddots & \ddots & \\ & \ddots & \ddots & \ddots & 1 \\ & & 1 & 1 & 1\end{vmatrix}_{(n-1)\times(n-1)}-\begin{vmatrix}1 & 1 & & & \\ 1 & \ddots & \ddots & & \\ 0 & \ddots & \ddots & \ddots & \\ & \ddots & \ddots & \ddots & 1 \\ & & 1 & 1 & 1\end{vmatrix}_{(n-1)\times(n-1)}$$

$$=\begin{vmatrix} 1 & 1 & & & \\ 1 & \ddots & \ddots & & \\ 1 & \ddots & \ddots & \ddots & \\ & \ddots & \ddots & \ddots & 1 \\ & & 1 & 1 & 1 \end{vmatrix}_{(n-1)\times(n-1)} - \begin{vmatrix} 1 & 1 & & & \\ 1 & \ddots & \ddots & & \\ 1 & \ddots & \ddots & \ddots & \\ & \ddots & \ddots & \ddots & 1 \\ & & 1 & 1 & 1 \end{vmatrix}_{(n-2)\times(n-2)} + \begin{vmatrix} 1 & 1 & & & \\ 1 & \ddots & \ddots & & \\ 1 & \ddots & \ddots & \ddots & \\ & \ddots & \ddots & \ddots & 1 \\ & & 1 & 1 & 1 \end{vmatrix}_{(n-3)\times(n-3)}$$

We obtain a recurrence relation (3).

Applying Lemma 3 to (3), we obtain an equation $x^3 - x^2 + x - 1 = 0$. Its roots are $1, i, -i$, where $i^2 = -1$.

By evaluating the determinants, $|H_4|, |H_5|$ and $|H_6|$ via elementary row operations, we obtain $|H_4| = 1$, $|H_5| = 1$ and $|H_6| = 0$.

Assuming that $|H_n| = \alpha_1\lambda_1^n + \alpha_2\lambda_2^n + \alpha_3\lambda_3^n$, for $n = 4,5,6$. We obtain the system of equations,

$$\begin{cases} \alpha_1 + \alpha_2 i^4 + \alpha_3(-i)^4 = 1 \\ \alpha_1 + \alpha_2 i^5 + \alpha_3(-i)^5 = 1 \\ \alpha_1 + \alpha_2 i^6 + \alpha_3(-i)^6 = 0 \end{cases}$$

Simply solving the equation, we obtain $\alpha_1 = \frac{1}{2}, \alpha_2 = \frac{1}{4} - \frac{i}{4}, \alpha_3 = \frac{1}{4} + \frac{i}{4}$, thereby completing the proof of formula (4).

Theorem 6 Let

$$|T_n| = \begin{vmatrix} 1 & 1 & & & \\ 1 & \ddots & \ddots & & \\ & \ddots & \ddots & \ddots & \\ & & \ddots & \ddots & 1 \\ & & & 1 & 1 \end{vmatrix}_{n\times n}.$$

Then, there exists a recurrence relation

$$|T_n| = |T_{n-1}| - |H_{n-2}| \tag{5}$$

and an explicit formula

$$|T_n| = \frac{3-\sqrt{3}i}{6}\left(\frac{1+\sqrt{3}i}{2}\right)^n + \frac{3+\sqrt{3}i}{6}\left(\frac{1-\sqrt{3}i}{2}\right)^n \tag{6}$$

$$= \begin{cases} 1, & \text{if } n \equiv 0,1 \ (mod\ 6), \\ 0, & \text{if } n \equiv 2,5 \ (mod\ 6), \\ -1, & \text{if } n \equiv 3,4 \ (mod\ 6). \end{cases}$$

Proof. Applying the Laplace expansion along with the first row of determinant $|T_n|$, we obtain a recurrence relation (5).

Applying Lemma 3 to (5), we obtain the equation, $x^2 - x + 1 = 0$. Its roots are $\frac{1+\sqrt{3}i}{2}$ and $\frac{1-\sqrt{3}i}{2}$, where $i^2 = -1$.

Evaluating the determinants $|T_3|, |T_4|$ via elementary row operations, we obtain $|T_3| = -1$ and $|T_4| = -1$.

Assuming that $|T_n| = \alpha_1\lambda_1^n + \alpha_2\lambda_2^n$, for $n = 3,4$, we obtain the system of equations,

$$\begin{cases} \alpha_1(\frac{1+\sqrt{3}i}{2})^3+\alpha_2(\frac{1-\sqrt{3}i}{2})^3=-1 \\ \alpha_1(\frac{1+\sqrt{3}i}{2})^4+\alpha_2(\frac{1-\sqrt{3}i}{2})^4=-1 \end{cases}$$

Solving the equations, we obtain $\alpha_1=\frac{3-\sqrt{3}i}{6}, \alpha_2=\frac{3+\sqrt{3}i}{6}$, thereby completing the proof of formula (6).

3. Proof of Theorem 1

For $n \geq 4$, along with the elements in the first row of the determinant $|A_n|$, we use the Laplace expansion to obtain

$$|A_n|=\begin{vmatrix} 1 & 1 & & & 1 \\ 1 & \ddots & \ddots & & \\ 1 & \ddots & \ddots & \ddots & \\ & \ddots & \ddots & \ddots & 1 \\ & & 1 & 1 & 1 \end{vmatrix}=\begin{vmatrix} 1 & 1 & & & \\ 1 & \ddots & \ddots & & \\ 1 & \ddots & \ddots & \ddots & \\ & \ddots & \ddots & \ddots & 1 \\ & & 1 & 1 & 1 \end{vmatrix}-\begin{vmatrix} 1 & 1 & & & \\ 1 & \ddots & \ddots & & \\ 0 & \ddots & \ddots & \ddots & \\ & \ddots & \ddots & \ddots & 1 \\ & & 1 & 1 & 1 \end{vmatrix}+(-1)^{n+1}\begin{vmatrix} 1 & 1 & 1 & & \\ 1 & \ddots & \ddots & \ddots & \\ & \ddots & \ddots & \ddots & 1 \\ & & \ddots & \ddots & 1 \\ & & & 1 & 1 \end{vmatrix}$$

Where the first determinant is the determinant of the lower Hessenberg matrix $|H_{n-1}|$, and the last is the determinant of an upper Hessenberg matrix, which is equal to the determinant of the lower Hessenberg matrix, $|H_{n-1}|$. By expanding the second determinant along its first column, we obtain

$$|A_n|=|H_{n-1}|-(|H_{n-2}|-|H_{n-3}|)+(-1)^{n+1}|H_{n-1}|$$

Using formulae (3) and (4), we obtain

$$\begin{aligned}|A_n|&=|H_n|+(-1)^{n+1}|H_{n-1}| \\ &=\frac{1}{2}(1+(-1)^{n+1})+\frac{1-i}{4}(i^n+(-i)^{n-1})+\frac{1+i}{4}((-i)^n+i^{n-1}).\end{aligned}$$

We complete the proof.

4. Proof of Theorem 2

For $n \geq 4$, along with the elements in the first row of the determinant $|B_n|$, we use the Laplace expansion to obtain

$$|B_n|=\begin{vmatrix} 1 & 1 & 1 & & \\ 1 & \ddots & \ddots & & 1 \\ 1 & \ddots & \ddots & \ddots & \\ & \ddots & \ddots & \ddots & 1 \\ & & 1 & 1 & 1 \end{vmatrix}=\begin{vmatrix} 1 & 1 & & & 1 \\ 1 & \ddots & \ddots & & \\ 1 & \ddots & \ddots & \ddots & \\ & \ddots & \ddots & \ddots & 1 \\ & & 1 & 1 & 1 \end{vmatrix}-\begin{vmatrix} 1 & 1 & & & 1 \\ 1 & \ddots & \ddots & & \\ 0 & \ddots & \ddots & \ddots & \\ & \ddots & \ddots & \ddots & 1 \\ & & 1 & 1 & 1 \end{vmatrix}+(-1)^{n+1}\begin{vmatrix} 1 & 1 & 1 & & 1 \\ 1 & \ddots & \ddots & \ddots & \\ & \ddots & \ddots & \ddots & 1 \\ & & \ddots & \ddots & 1 \\ & & & 1 & 1 \end{vmatrix}$$

Using formulas (3), (4), we obtain

$$\begin{aligned}|B_n|&=|A_{n-1}|-(|H_{n-2}|-|H_{n-3}|+(-1)^{n+1}|H_{n-3}|)+(-1)^n(|H_{n-2}|-|H_{n-3}|+(-1)^{n+1}) \\ &=1+(1-2(-1)^n)|H_n|+2(-1)^n|H_{n-1}| \\ &=1+(1-2(-1)^n)(\frac{1}{2}+\frac{1-i}{4}(i^n+(-i)^{n-1})+\frac{1+i}{4}((-i)^n+i^{n-1}).\end{aligned}$$

This completes the proof.

5. Four New Results

Using similar arguments as previously, we obtain the following results:

1. For $n \geq 5$, the determinant of the pentadiagonal Toeplitz matrix is given by

$$|C_n| = \begin{vmatrix} 1 & 1 & & 1 & & & \\ 1 & \ddots & \ddots & & 1 & & \\ 1 & \ddots & \ddots & \ddots & & 1 & \\ & \ddots & \ddots & \ddots & \ddots & & \\ & & \ddots & \ddots & \ddots & 1 & \\ & & & 1 & 1 & 1 & \end{vmatrix} = \begin{cases} 7, & if\ n \equiv 0 \pmod 4 \\ 1, & if\ n \equiv 1 \pmod 4 \\ 2, & if\ n \equiv 2 \pmod 4 \\ 4, & if\ n \equiv 3 \pmod 4 \end{cases}$$

$$= 3 + (-1)^{n-1} |H_{n-2}| + (1 + 3(-1)^n) |H_{n-4}|$$

2. For $n \geq 6$, the determinant of the pentadiagonal Toeplitz matrix is given by

$$|D_n| = \begin{vmatrix} 1 & 1 & & 1 & & & & \\ 1 & \ddots & \ddots & & 1 & & & \\ 1 & \ddots & \ddots & \ddots & & 1 & & \\ & \ddots & \ddots & \ddots & \ddots & & 1 & \\ & & \ddots & \ddots & \ddots & \ddots & & \\ & & & \ddots & \ddots & \ddots & 1 & \\ & & & & 1 & 1 & 1 & \end{vmatrix} = \begin{cases} 2, & if\ n \equiv 0 \pmod 4 \\ 4, & if\ n \equiv 1 \pmod 4 \\ 1, & if\ n \equiv 2 \pmod 4 \\ -1, & if\ n \equiv 3 \pmod 4 \end{cases}$$

$$= -1 + |C_{n-1}| - |B_{n-2}| + |A_{n-3}| + (-1)^{n-2} |H_{n-6}|$$
$$= 1 + 2(-1)^n |H_{n-3}| + (1 + 2(-1)^n) |H_{n-4}|.$$

By similar arguments as the previous, we obtain the following results for pentadiagonal Toeplitz matrices in which two subdiagonals are continuous in the upper triangular matrix.

3. For $n \geq 4$, the determinants $|E_n|$ and $|F_n|$ of the hexagonal Toeplitz matrices are given by

$$|E_n| = \begin{vmatrix} 1 & 1 & & & & 1 & 1 \\ 1 & \ddots & \ddots & & & & 1 \\ 1 & \ddots & \ddots & \ddots & & & \\ & \ddots & \ddots & \ddots & \ddots & & \\ & & \ddots & \ddots & \ddots & \ddots & \\ & & & \ddots & \ddots & \ddots & 1 \\ & & & & 1 & 1 & 1 \end{vmatrix} = |F_n| = \begin{vmatrix} 1 & 1 & & & & 1 & 1 & \\ 1 & \ddots & \ddots & & & & 1 & 1 \\ 1 & \ddots & \ddots & \ddots & & & & 1 \\ & \ddots & \ddots & \ddots & \ddots & & & \\ & & \ddots & \ddots & \ddots & \ddots & & \\ & & & \ddots & \ddots & \ddots & & 1 \\ & & & & & 1 & 1 & 1 \end{vmatrix}$$

$$= -|T_{n-3}| + (-1)^{n-2}$$

$$= \begin{cases} 2, & if\ n \equiv 0 \pmod 6 \\ 0, & if\ n \equiv 1,4 \pmod 6 \\ 1, & if\ n \equiv 2 \pmod 6 \\ -2, & if\ n \equiv 3 \pmod 6 \\ -1, & if\ n \equiv 5 \pmod 6. \end{cases}$$

4. For $n \geq 7$, the determinant of the hexagonal Toeplitz matrix is given by

$$|G_n| = \begin{vmatrix} 1 & 1 & & 1 & 1 & & \\ 1 & \ddots & \ddots & & 1 & 1 & \\ 1 & \ddots & \ddots & \ddots & & 1 & 1 \\ & \ddots & \ddots & \ddots & \ddots & & 1 \\ & & \ddots & \ddots & \ddots & \ddots & \\ & & & \ddots & \ddots & \ddots & 1 \\ & & & & 1 & 1 & 1 \end{vmatrix} = \begin{cases} -2, & if\ n \equiv 0 \pmod 6 \\ 4, & if\ n \equiv 1 \pmod 6 \\ -3, & if\ n \equiv 2 \pmod 6 \\ 2, & if\ n \equiv 3 \pmod 6 \\ -4, & if\ n \equiv 4 \pmod 6 \\ 3, & if\ n \equiv 5 \pmod 6. \end{cases}$$

$$= -|T_{n-3}| - 3(-1)^n.$$

5. For $n \geq 6$, the determinant of the heptadiagonal Toeplitz matrix is given by

$$|K_n| = \begin{vmatrix} 1 & 1 & & & 1 & 1 & 1 \\ 1 & \ddots & \ddots & & & 1 & 1 \\ 1 & \ddots & \ddots & \ddots & & & 1 \\ & \ddots & \ddots & \ddots & \ddots & & \\ & & \ddots & \ddots & \ddots & \ddots & \\ & & & \ddots & \ddots & \ddots & 1 \\ & & & & 1 & 1 & 1 \end{vmatrix} = \begin{cases} 3, & if\ n \equiv 0 \pmod 4 \\ 0, & if\ n \equiv 1 \pmod 4 \\ 1, & if\ n \equiv 2 \pmod 4 \\ 2, & if\ n \equiv 3 \pmod 4 \end{cases}$$

6. A Maple example

In our modern world, we can use a computer to evaluate mathematic values. For example, by the mathematic software Maple, we use the following procedure to evaluate the sequences of the determinant $|K_n|$, for $n = 6, 7, \cdots, 30$.

```
with(LinearAlgebra):
for n from 6 to 30 do
    k:=n-3:
    K1:=n-2:
    K2:=n-1:
    DM:=Matrix(n,n):
    for i from 1 to n do DM[i,i]:=1: end do:
    for i from 1 to n-1 do DM[i,i+1]:=1: DM[i+1,i]:=1: end do:
    for i from 1 to n-2 do DM[i+2,i]:=1: end do:
    for i from 1 to n-k do DM[i,i+k]:=1: end do:
    for i from 1 to n-K1 do DM[i,i+K1]:=1: end do:
    for i from 1 to n-K2 do DM[i,i+K2]:=1: end do:
    DM;
    Determinant(DM);
    print("n:=",n);
end do;
```

References

[1] Everest G van der Poorten A Shparlinski I and Ward T 2003 Recurrence Sequences *American Mathematical Society* pp1-12

[2] Jishe Feng 2016 Hessenberg Matrices On Fibonacci And Tribonacci Numbers *Ars Combinatoria* pp117-124.

[3] Milica Andeli´c Carlos M da Fonseca 2020 Some determinantal considerations for pentadiagonal matrices, *Linear and Multilinear Algebra*

[4] Cahill N D D'Errico J R Narayan D A Narayan J Y 2002 Fibonacci determinants *The college mathematics Journal* pp 221-225

[5] Sun Z 1999 Linear recursive sequences and powers of matrices *Fibonacci Quarterly*, pp 339-351

[6] Jia J Yang B Li S 2016 On a homogeneous recurrence relation for the determinants of general pentadiagonal Toeplitz matrices *Comput Math Appl* pp 1036–1044

[7] Jia J Jiang Y A 2013 structure preserving matrix factorization for solving general periodic pentadiagonal Toeplitz linear systems *Comput Math Appl* pp 965–974

Numerical Simulation of Oil -Water Flow Velocity Field and Flow Pattern in Horizontal Wells and Near Horizontal Wells

Yingying Ma[2], Hongwei Song[1,2,*], Changqi Zhao[2], Ran Wei[2] and Lihuizi Sun[2]

[1]Key Laboratory of Exploration Technologies for Oil and Gas Resources (Yangtze University) Ministry of Education, Hubei Wuhan 430100 China
[2]School of geophysics and oil resources, Yangtze University, Hubei Wuhan 430010, China

[*]Corresponding author email: 94528846@qq.com

Abstract. Oil-water flow widely exists in oilfield development. Due to the gravity differentiation, the oil-water flow in low-flow horizontal wells has a clear characteristics of stratified flow. With the increase of flow rate, the stratified characteristics are not obvious, which leads to the difficulty of multiphase flow phase separation flow interpretation in oilfield. In this paper, the oil-water flow in horizontal wells is taken as the research object. The VOF model of Fluent software is used to study the relationship between velocity field and flow pattern distribution characteristics with water cut, well deviation angle and total flow. The results show that with the increase of water cut, the oil-water separation level gradually moves up, and the velocity of water phase is greater than that of oil phase. When the well deviation angle changes slightly, the flow stratification of oil-water changes sharply, and the flow velocity in the pipeline also changes. When the total flow rate is lower than 200 m^3/d, the oil-water phases have obvious stratified flow characteristics. With the increase of flow rate, the oil-water interface fluctuates. The average velocity of oil and water is not much different.

1. Introduction

In the 1980s, the development and application of horizontal well technology in the world has entered a new stage. The technical level has been significantly improved and the cost has been greatly reduced. It has become one of the main technical means for the development of oil and gas reservoirs. Velocity measurement in the detection of horizontal well flow parameters is an extremely important research content. The accurate measurement of velocity can be used to analyze the output of each perforated zone in the well and provide a basis for the establishment of a multiphase flow interpretation model. Since the gravity direction perpendicular to the well axis, the horizontal section oil-water phase and velocity by the vertical well as a symmetrical field distribution becomes asymmetrical distribution, which makes it difficult to interpret production logging data. Therefore, it is necessary to study the characteristics of oil-water flow velocity field in horizontal wells. On the basis of summarizing the previous research results, Trallero[0] carried out the experimental study of oil-water flow in a small-diameter low-flow horizontal tube. The oil-water flow in the horizontal tube was divided into stratified flow (oil and water have a clear interface) and dispersed flow (one phase of the fluid dispersed into another phase of the continuous fluid as droplets). The stratified flow usually occurs when the total flow is low, while the dispersed phase usually occurs when the total flow is high. When the total flow rate is medium, the mixing mode of these two flow patterns may appear. Due to the complex downhole environment, there are three kinds of fluid flow stages. When the well deviation angle is 90°, the density of oil is smaller at the top, and the density of water is larger at the bottom. Since the viscosity of oil is greater than that of water, the friction that hinders the flow is generated, and the velocity of oil is slightly lower than that of water. When the well deviation angle is greater than 90°, the oil

Content from this work may be used under the terms of the Creative Commons Attribution 3.0 licence. Any further distribution of this work must maintain attribution to the author(s) and the title of the work, journal citation and DOI.
Published under licence by IOP Publishing Ltd

gathers at the top of the pipeline, and the buoyancy of oil drags the water at the bottom to flow upward, and the speed of the oil is greater than that of the water. When the well deviation angle is less than 90°, because the density of water is greater than that of oil, the downward flow rate of water is greater than that of oil.
For low-flow horizontal wells, the flow pattern is mainly stratified flow. As the total flow rate gradually increases, the oil-water flow interface is disturbed. The oil-water flow velocity changes with the change of water cut and well deviation angle, which makes it very difficult to interpret fractional flow of multiphase flow. In this paper, the flow state of oil and water in horizontal wells is the research object. The VOF multiphase flow model of Fluent software is used to numerically simulate the oil-water flow under different water cut, well deviation angle and total flow rate are studied. The variation characteristics of velocity field are analyzed, which provides the basis for the interpretation of production logging data.

2. Establishment of Oil - water Two Phase Numerical Simulation Model for Horizontal Well

According to the tubing or casing string of oil wells produced in the oil field and the multiphase flow simulation experiment device of Yangtze University Production Logging Center, in order to make the numerical simulation can be compared with the oil wells and the experimental situation, this paper selected a section of pipeline with a length of 20 m and a diameter of 124 mm as the research object. The Design Modeler is used to establish the three-dimensional geometric model and mesh the pipeline. The sweep method is used to generate 1000 units on the side. All the grid types of the circular surface are triangular free surface grid types, as shown in Figure 1.

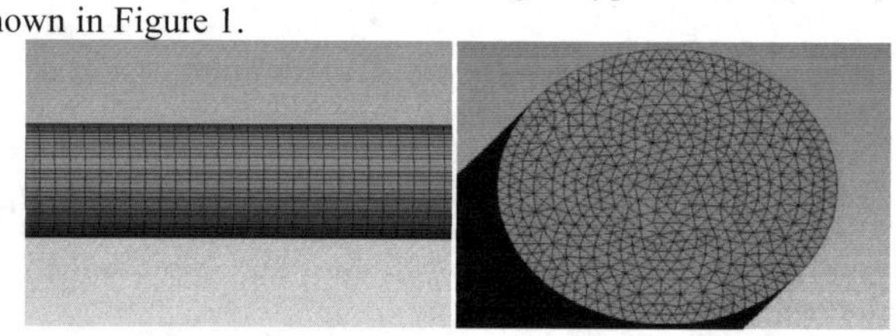

Figure 1. The grid division model of horizontal.

The fluid in the pipeline is tap water and industrial white oil. The viscosity of white oil is 0.00292 kg / m·s and the density is 826.3 kg / m³. The density of water is 998.2 kg / m³ and the viscosity is 0.001003 kg / m·s. Numerical simulation total flow range is 100 ~ 600 m³/d, flow design point is 100,200,300,400,600 m³/d. The range of water cut is 20 ~ 80 %, with 20 % as the step length, and the water cut of 50 % is selected for comparison. The variation range of well deviation angle is 85° – 95°, and the well deviation angles are 85°, 87°, 93°, 95°.Boundary conditions are set. The right end of the pipeline is the inlet of the oil-water mixture, and is set as the velocity inlet boundary condition, which is suitable for incompressible fluid. The pipeline is 20 m long, and it can be considered that the outlet is a fully developed flow, and is set as outflow. In order to ensure the smooth progress of the simulation, the obtained results are accurate and feasible. In the whole simulation process, the control variables are only assumed to be points, ignoring the influence of condensate gas escaped from crude oil in the gathering and transportation process, and the influence of temperature change on the viscosity and density of crude oil and water. At the same time, the pipeline wall is also identified as a non-slip wall.

3. Numerical Simulation

3.1. VOF Multiphase Flow Model

The continuity equation as equation (1) shows :

$$\nabla \cdot \vec{u} = 0 \tag{1}$$

Momentum equation as equation (2) shows :

$$\rho\frac{\partial \vec{u}}{\partial t}+\rho\vec{u} \cdot \nabla\vec{u}=-\nabla\vec{p}+\rho\vec{g}+\nabla \cdot \mu(\nabla\vec{u} + \nabla^{T}\vec{u}) + \vec{f} \tag{2}$$

The VOF model phase equation as equation (3) shows:

$$\frac{\partial\alpha}{\partial t}+\vec{u}\cdot\nabla\alpha=0 \tag{3}$$

In the formula, ρ is the fluid density (kg/m³). t is time (s). v is velocity vector (m /s). μ is dynamic viscosity (kg/m·s). f is surface tension (N/m³).

3.2. Turbulence Model k-ε Model

Turbulent pulsating kinetic energy equation (4):

$$\rho=\frac{\partial}{\partial t}(\rho k)+\frac{\partial}{\partial X_i}(\rho k u_i)=\frac{\partial}{\partial X_j}\left[\left(\mu+\frac{\mu_i}{\sigma_k}\right)\frac{\partial k}{\partial X_j}\right]+G_k+G_b-Y_M+S_k \tag{4}$$

$$\frac{\partial}{\partial t}(\rho\varepsilon)+\frac{\partial}{\partial X_i}(\rho\varepsilon u_i)=\frac{\partial}{\partial X_j}\left[\left(\mu+\frac{\mu_i}{\sigma_\varepsilon}\right)\frac{\partial\varepsilon}{\partial X_j}\right]+C_{1\varepsilon}\frac{\varepsilon}{k}(G_k+C_{3\varepsilon}C_B)-C_{2\varepsilon}\rho\frac{\varepsilon^2}{k}+S_\varepsilon$$

Formula: ρ is fluid density (kg/m³). k is turbulent kinetic energy(J).μ_i and $\mu_\tau$are average speed.σ_k and σ_ε are the turbulent Prandtl number of k equation and ε equation respectively. $G_k$is the generation term of turbulent kinetic energy caused by the average velocity gradient.$G_b$is term for the generation of turbulent kinetic energy caused by buoyancy. Y_M is contributed to pulse expansion in compressible turbulence. ε is turbulent dissipation rate(%). μ is turbulent viscosity(Pa·s). t is time(s). X_i *and* X_j is space coordinates. $C_{1\varepsilon}=1.44, C_{2\varepsilon}=1.92, C_{3\varepsilon}=1$ are all empirical constants.

This paper mainly studies the flow pattern and velocity field of oil-water flow in long circular tube. VOF method can be used to deal with the problem of complex free surface, can only use a function to represent the free surface of various complex changes, can effectively capture the interface between oil and water movement, where the turbulence model selection k-ε model, given the ' turbulence intensity ' and ' hydraulic diameter '. Since the Reynolds number of each working condition is calculated to know that the flow under each simulation condition is in the stratified flow state, the stratified flow model is set in fluent.

4. Numerical Simulation Results and Analysis

4.1 Variation of Oil-water two-phase Flow and Velocity Field in Horizontal Wells under Different Water Cuts

The total flow rate of simulation is 100 m³/d, and the inlet water cut changes from 20% to 80%. The water cut difference of each working condition is 20%. At the same time, the working condition with water cut of 50% is selected to carry out the numerical simulation of flow pattern and velocity field under different water cut conditions. The distribution maps of oil-water two-phase flow and velocity field under different water cut are shown in Figures 2 – 6.

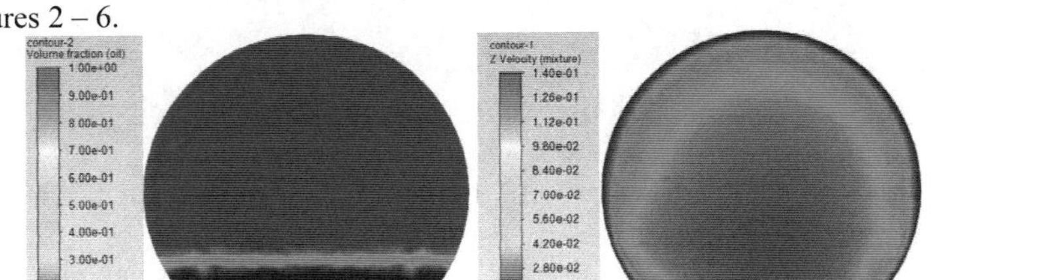

(A) flow pattern map (B) Velocity field imaging figure

Figure 2. When the total flow of oil and water is 100 m³/day, the water cut is 20%, and the well deviation is 90°, the horizontal well flow pattern distribution map and velocity field distribution map.

1) As shown in Figure 2, when the water cut is 20 %, it can be seen from the cross section that the water holdup is greater than 20 %. According to the velocity field, it can be seen that the average velocity of the lower water phase is higher than that of the upper oil phase, the high value area of the velocity field is in the lower part of the pipeline.

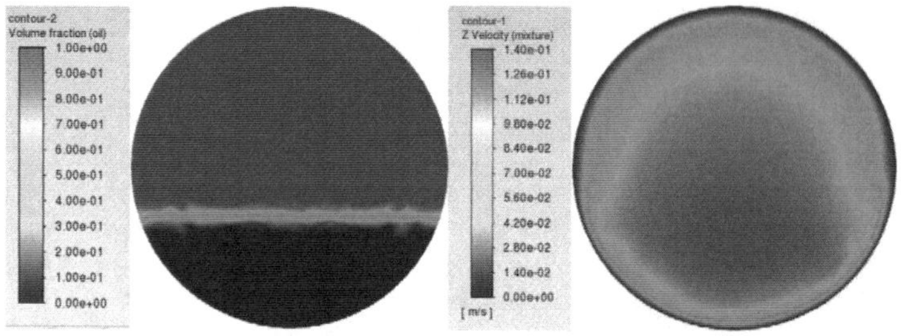

(A) low pattern map (B) Velocity field imaging figure

Figure 3. When the total flow of oil and water is 100 m^3/day, the water cut is 40%, and the well deviation is 90°, the horizontal well flow pattern distribution map and velocity field distribution map.

2) As shown in Figure 3, when the water cut is 40 %, it can be seen from the cross section that the water holdup is less than 40 %, and the water cut is greater than the water holdup. According to the velocity field, it can be seen that the average velocity of the lower water phase is higher than that of the upper oil phase. The high value area of the velocity field is in the lower part of the pipeline.

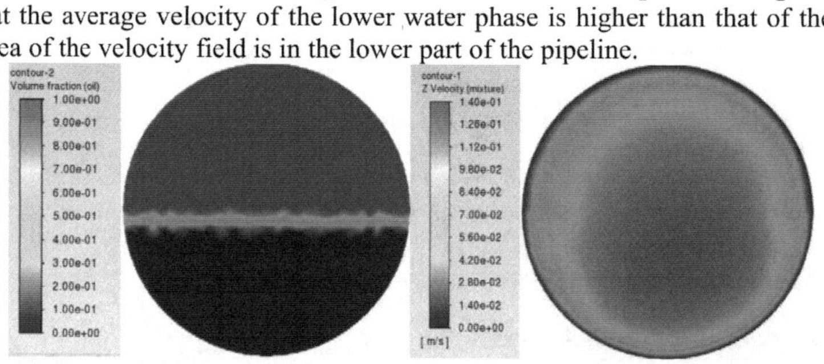

(A) low pattern map (B) Velocity field imaging figure

Figure 4. When the total flow of oil and water is 100 m^3/day, the water cut is 50%, and the well deviation is 90°, the horizontal well flow pattern distribution map and velocity field distribution map.

3) As shown in Figure 4, it can be seen from the cross section that when the water cut is 50 %, the oil and water in the wellbore account for half each. According to the velocity field, it can be seen that the average velocity of the lower water phase is higher than that of the upper oil phase, the high value range of the velocity field becomes larger.

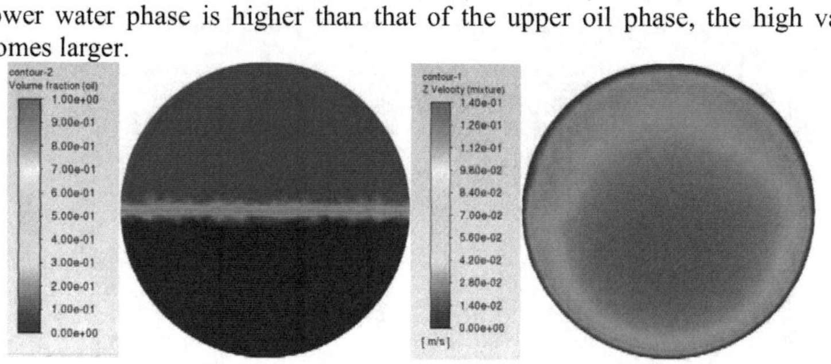

(A) low pattern map (B) Velocity field imaging figure

Figure 5. When the total flow of oil and water is 100 m^3/day, the water cut is 60%, and the well deviation is 90°, the horizontal well flow pattern distribution map and velocity field distribution map.

4) As shown in Figure 5, when the water cut is 60 %, it can be seen from the cross section that the water holdup is less than 60 %. According to the velocity field, it can be seen that the average velocity of the lower water phase is higher than that of the upper oil phase. The high value area of the velocity field begins to move gradually to the top of the pipeline.

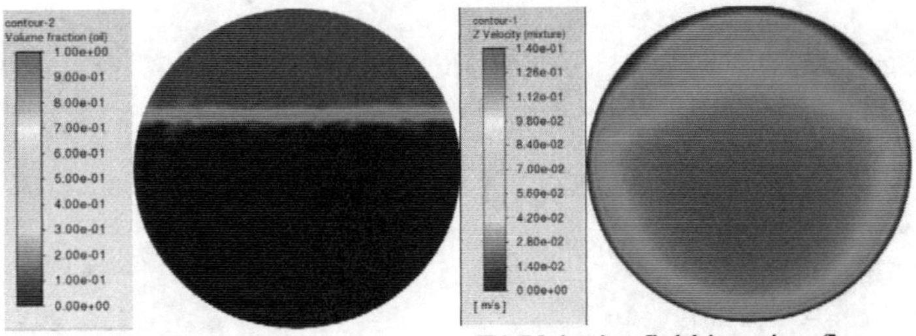

(A) low pattern map (B) Velocity field imaging figure

Figure 6. When the total flow of oil and water is 100 m^3/day, the water cut is 80%, and the well deviation is 90°, the horizontal well flow pattern distribution map and velocity field distribution map.

5) As shown in Figure. 6, when the water cut is 80 %, the water in the wellbore accounts for the majority. It can be seen from the cross section that the water holdup is less than 80 %. According to the velocity field, it can be seen that the average velocity of the lower water phase is higher than that of the upper oil phase. The high value area of the velocity field is in the lower water phase.

With the increase of water cut, the oil-water flow tends to be stable over time after passing through the inlet section. When the water cut is 50 %, it is stratified flow due to the small initial inlet velocity, and the viscous force plays a major role. Under the action of viscous force, the velocity of water is slightly lower than that of oil, and the water holdup is slightly higher than the oil holdup.

4.2 Variation of Oil-water two-phase Flow and Velocity Field in Near Horizontal Wells under Different Well Deviation Angles

Due to the uneven distribution of each phase fluid in horizontal wells and low angle inclined wellbores, the oil-water flow structure is also very sensitive to wellbore inclination, so the influence of different inclination on the flow pattern is explored. The Fluent software is used to simulate the pipeline with the total flow rate of 100 m^3/d, the water cut of 60 % and the well inclination of 85°, 87°, 93°, 95°. The distribution maps of oil-water two-phase flow and velocity field under different well deviation angles are shown in Figure 7-10.

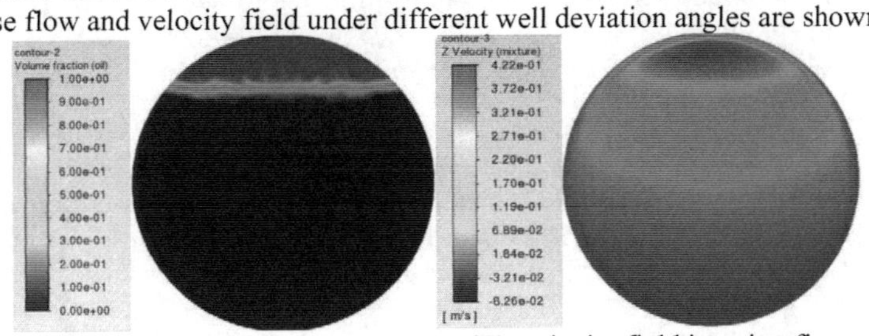

(A) flow pattern map (B) Velocity field imaging figure

Figure 7. When the total flow of oil and water is 100 m^3/day, the water cut is 60%, and the well deviation is 85°, the horizontal well flow pattern distribution map and velocity field distribution map.

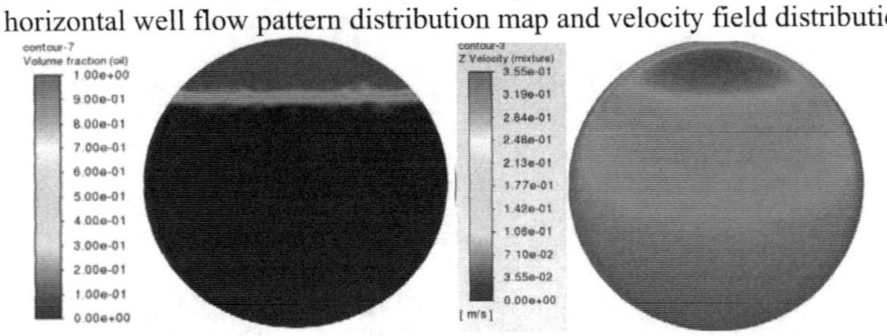

(A) flow pattern map (B) Velocity field imaging figure

Figure 8. When the total flow of oil and water is 100 m^3/day, the water cut is 60%, and the well deviation is 87°, the horizontal well flow pattern distribution map and velocity field distribution map.

1)As shown in Figure 7-8, it can be seen from the flow pattern that the water holdup is greater than 60 %, and the water cut is less than the water holdup. According to the velocity field, the velocity field of oil-water flow is symmetrically distributed. The average velocity of oil phase at the top is greater than that of water phase at the bottom. The high value area of velocity is distributed at the top of the pipeline section.

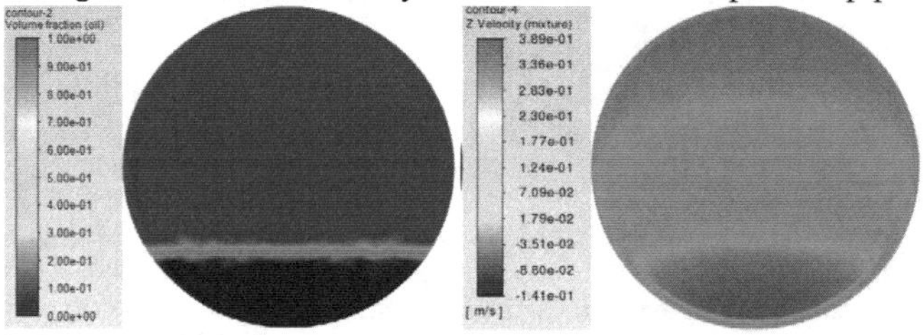

(A) flow pattern map (B)Velocity field imaging figure

Figure 9. When the total flow of oil and water is 100 m^3/day, the water cut is 60%, and the well deviation is 93°, the horizontal well flow pattern distribution map and velocity field distribution map.

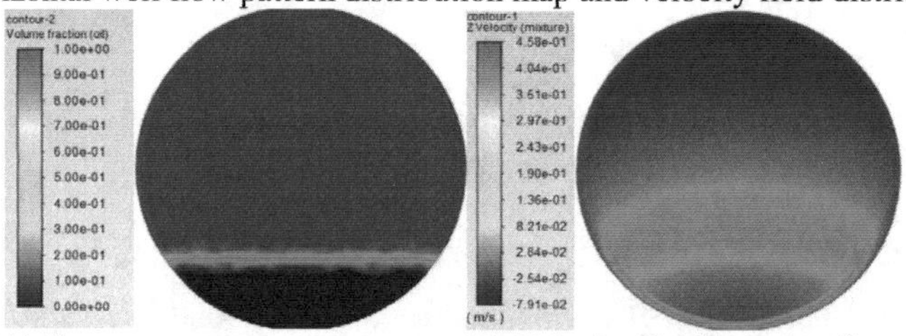

(A) flow pattern map (B) Velocity field imaging figure

Figure 10. When the total flow of oil and water is 100 m^3/day, the water cut is 60%, and the well deviation is 95°, the horizontal well flow pattern distribution map and velocity field distribution map.

5) As shown in Figure 9-10, it can be seen from the cross section of the flow pattern that the water holdup is less than 60 %. According to the velocity field, the oil-water flow velocity field is symmetrical distribution, and the high velocity area is distributed at the bottom of section.

Due to the small initial velocity, the phase interface is a slightly mixed laminar flow and laminar flow. In the uphill stage, the velocity of the upper oil phase is greater than that of the lower water phase. Compared with the horizontal, the water holdup is greater than the oil holdup, and this phenomenon is more obvious with the increase of well deviation angle. However, in the downhill stage, due to the density difference, gravity plays a leading role, resulting in the average velocity of water phase being greater than that of oil phase.

4.3 Variation of Oil-water two-phase Flow and Velocity Field in Horizontal Wells under Different Flow Rates

Since the flow pattern and velocity field distribution in the two-phase pipe changed with the change of flow rate, the total flow rates of 100 ~ 600 m^3/d were selected for the numerical simulation. The distribution maps of oil-water two-phase flow and velocity field with different total flow rates are shown in Figure 11 – 15.

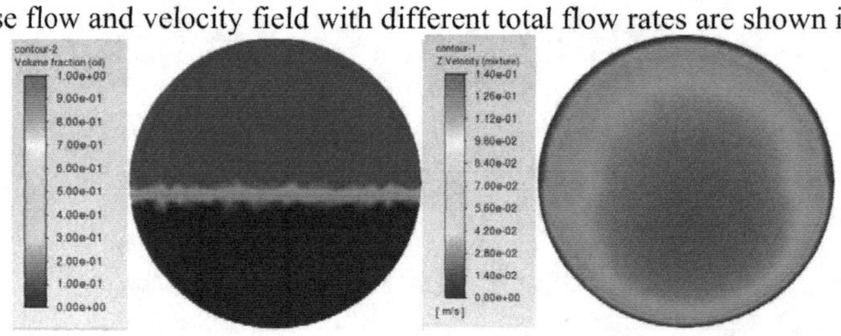

(A) flow pattern map (B) Velocity field imaging figure

Figure 11. When the total flow of oil and water is 100 m^3/day, the water cut is 50%, and the well deviation is 90°, the horizontal well flow pattern distribution map and velocity field distribution map.

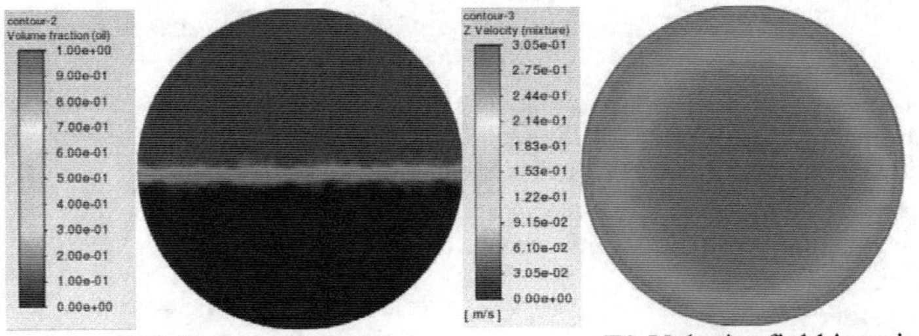

(A) flow pattern map (B) Velocity field imaging figure

Figure 12. When the total flow of oil and water is 200 m^3/day, the water cut is 50%, and the well deviation is 90°, the horizontal well flow pattern distribution map and velocity field distribution map.

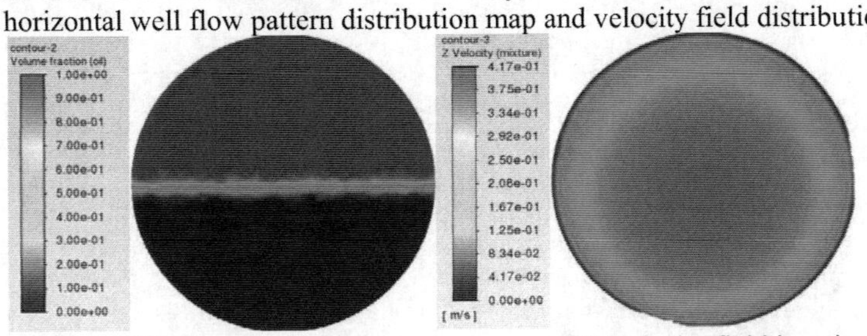

(A) flow pattern map (B) Velocity field imaging figure

Figure 13. When the total flow of oil and water is 300 m^3/day, the water cut is 50%, and the well deviation is 90°, the horizontal well flow pattern distribution map and velocity field distribution map.

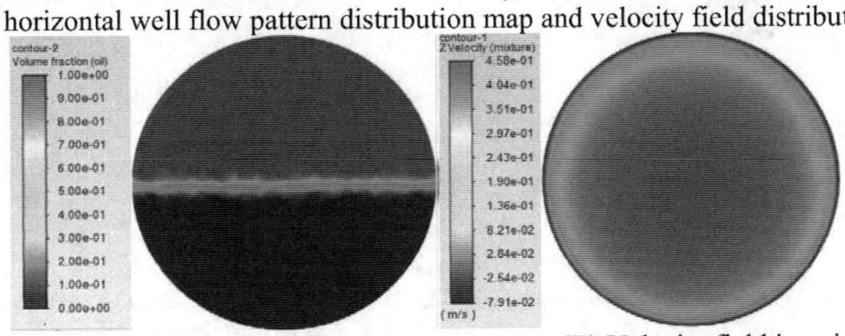

(A) flow pattern map (B) Velocity field imaging figure

Figure 14. When the total flow of oil and water is 400 m^3/day, the water cut is 50%, and the well deviation is 90°, the horizontal well flow pattern distribution map and velocity field distribution map.

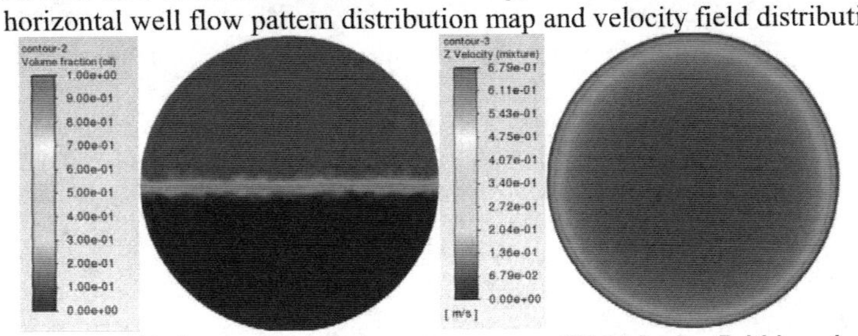

(A) flow pattern map (B) Velocity field imaging figure

Figure 15. Flow pattern and velocity field distribution of horizontal wells with total flow rate of 600 m^3/d and water cut of 50 % at well deviation angle 90°.

From the figure, it can be concluded that the viscous force plays a major role when the flow rate of is 100 m^3/d at well deviation angle 90°. When the flow rate is 200 m^3/d, the viscous effect gradually weakens due to the increase of flow rate and begins to be less than the influence of turbulence on the flow pattern of oil-water flow. At the entrance section, the ' turbulence ' effect is more obvious. When the steady state is reached, the length of the entrance increases, and the interface between oil and water is not stable. When the velocity reaches 300 m^3/d, the effect of turbulence is further enhanced. When the velocity reaches 400 m^3/d, the turbulence is enhanced to disturb a small part of water to the top of the pipeline, and the oil phase is dispersed to the upper part of the pipeline. When the velocity reaches 600 m^3/d, the oil-water flows in a long part of the pipe, and the oil-water flow at the outlet is stratified flow. As the total flow of oil and water increases, the high-speed center area of the velocity field in the tube gradually shifts to the center, and the high-speed center area basically coincides with the center at 600 m^3/d, and the velocity field is axisymmetric.

5. Conclusions and Understanding

In this paper, VOF model in Fluent software is used for numerical simulation of oil-water flow in horizontal wells under different water cut, well deviation angle and total flow rate. The results show that:

(1)When the well deviation angle and water cut are constant, the oil-water two-phase flow is a stratified flow with a clear interface at low flow rate. As the total flow of oil and water increases, the oil-water flow interface is disturbed. The high speed center area of velocity field in pipe gradually shifts to the center of pipe, and the velocity field changes from asymmetric distribution to symmetric distribution.

(2)When well deviation and total flow remain unchanged, the flow pattern is stratified flow at low flow rate, and the velocity of lower water phase is greater than that of upper oil phase. As the water cut increases, the high-speed center area of velocity field gradually shifts to the upper part of the pipeline.

(3)When the water cut and total flow rate are constant, the upper oil phase velocity is greater than the lower water phase in the uphill stage, the high speed area of velocity field is mainly distributed in the oil phase. In the downhill stage, due to the density difference, the circulation flow of oil is easy to occur. The high-speed region of the velocity field is mainly distributed in the water phase, and the velocity field is axisymmetric.

(4)Fluent software can simulate the oil-water flow flow in horizontal wells. The numerical simulation results of this paper have certain guiding significance for the development of field array flow imaging production logging instruments and data interpretation, but it needs to be adjusted to conform to the experimental results.

Acknowledgments

Project supported by the National Natural Science Foundation of China (No. 41474115) and the Open Fund of the Key Laboratory of the Ministry of Petroleum Resources and Exploration Technology (Yangtze University) (K2018-2)

References

[1] Song Hongwei, Guo Haimin, Guo Shuai, Shi Hangyu. Measurement method of oil-water flow stratified flow in horizontal wells [J].Oil exploration and development, 2020, 47(03): 573-582.

[2] Chen Yong. Numerical simulation of oil-water flow pattern in horizontal pipe [J].Industry and Technology Forum 2014, 13 (20): 64-65.

[3] Bian Xiaohang, Liu Junfeng, Ye Tianming, Zhang Zhengchao, Xu Zhang, Cai Hongting. Numerical simulation of oil-water flow pattern in deviated wells with different diameters [J]. Logging technology, 2016, 40 (4): 399-403.

[4] Zhang Yu, Liu Xingbin, Liu Xin, Jiang Zhaoyu, Li Lei. Research and simulation of oil-water flow pattern in horizontal wells [J]. New technology and new process, 2015, 12: 84-86.

[5] Yang Mei, Wu Xiling, Wang Zhilei, Liu Zaibin, etc. oil-water flow pattern measurement experiment of horizontal wells [J]. Logging technology, 2008, 35 (5): 398 – 402.

[6] Zhang Hongbing, Chen Lulu, etc.Numerical simulation of oil-water laminar velocity field in horizontal wells [J].Journal of Hehai University (Natural Science Edition) 2009.37 (4): 415-418.

[7] Xu Xiaoxuan, Huang Satellite, Chen Wenmei, Gong Jing. Research Progress of oil-water flow in Horizontal Pipe [J] Chemical Machinery 2006.33 (1): 59-64.

[8] Trallero J L Sarica C,et a1.A Study of Oil/Water Flow Patterns in Horizontal Pipes[C]//SPE 36609,SPE Annual Technical Conference and Exhibition, Denver, Colorado, 1996,Oct.

[9] Rashmi G Walvekar, Thomas S Y Choong, etc. Numerical Study of Dispersed Oil-water Turbulent Flow in Horizontal Tube [J]. Journal of Petro leum Science and Engineering, 2009, 65: 123—128.

Numerical Simulation of Granular Column Collapse with Fractal Particle Size Distribution Using Discrete Element Method

Zhiqiang Lai[1,2], Li Pan[1,2], Zhongmei Wang[1,2,*], Jiayi Wang[1,2] and Yanfen Ren[1,2]

[1]Yellow River Institute of Hydraulic Research, Yellow River Conservancy Commission, Zhengzhou 450003, China
[2]Key Laboratory of Lower Yellow River Channel and Estuary Regulation, Ministry of Water Resources, Zhengzhou 450003, China

[*]Corresponding author email: yrccwzmei@126.com

Abstract. This study conducts numerical simulations of the granular column collapse with Fractal Particle Size Distributions (FPSDs) via the Discrete Element Method (DEM) and investigated kinetic behaviours of dry granular flows. The aim of this paper is to explore the effects of the fractal dimension of FPSD on the kinetics of dry granular flows. When the fractal dimension of the flows consisting of granular materials increases, the horizontal particle translational velocities become greater and the mobility improves, whereas the particle rotational velocities decrease. Meanwhile, the change in the potential energy increases, and the particle kinetic energy in the rotational form reduces; thus, the particle kinetic energy in the translational form increases. The reducing particle rotational movement may be related to the reducing particle shearing behaviours because only the contact shearing can affect particle rotational motion. In conclusion, a larger fractal dimension of FPSD of a dry granular flow leads to a longer spreading distance and a smaller rotational velocity.
Keywords: Granular column collapse, particle behaviours, flow mobility, flow dynamics.

1. Introduction

1.1. Background
Complicated flow dynamics of granular media, e.g., rock avalanches, debris flows, landslides, and grains in the pharmaceutical and food industries, are typically investigated using the granular column collapse on a transversal wall. This model is effective and has been widely used by many researchers. Figure 1 illustrates the granular column collapse movement model. By moving the wall immediately in the longitudinal orientation or slowly in the transversal orientation, the collapse or quasi-static movement of granular column can be modelled.

1.2. Related Work
Many investigations have been performed via the granular column collapse model on a transversal wall. Geometric configuration parameters own the power-law function relationship with the initial geometric parameters [1-4]. The energy evolution during the collapse was obviously influenced by the aspect ratio [5]. How the factors exert effects on the dynamics of granular columns, e.g. the granular column initial porosity [6], mono-dispersed particle size [7], particle shape [8], particle friction

Content from this work may be used under the terms of the Creative Commons Attribution 3.0 licence. Any further distribution of this work must maintain attribution to the author(s) and the title of the work, journal citation and DOI.
Published under licence by IOP Publishing Ltd

coefficient [9], particle restitution coefficient [10], particle stiffness [11] and the fluid situation [12], have also been examined in detail. Deep insights into the movement mechanisms of the granular column collapse have been obtained.

Although there are many studies about the granular column collapse mechanisms, the investigations on mechanisms of how the FPSD has effects on the granular column collapse model are limited. The existing studies focused on the granular column movement with mono-dispersed [13], bi-dispersed [14], tri-dispersed [15-16] and poly-dispersed mixtures with non-fractal size distributions [17]. However, Crosta et al. [18] found that the rock blocks size distribution in natural rock avalanche deposits is fractal from field measurements. Lai et al. [19] have explored the mechanisms of how FPSD influences the runout distance and the kinetics of different sized grains. However, how FPSD affects particle motion and energy conversion in translational and rotational forms.

We analyzed the collapse processes of the granular columns with varying fractal dimensions of FPSD through DEM. The details about our DEM simulations were then given. We attempted to use the evolutions of particle motion and energy conversion in translational and rotational forms to explore how the flow mobility changes. The motive is to understand the reason why the flows consisting of granular media can own tremendous mobility.

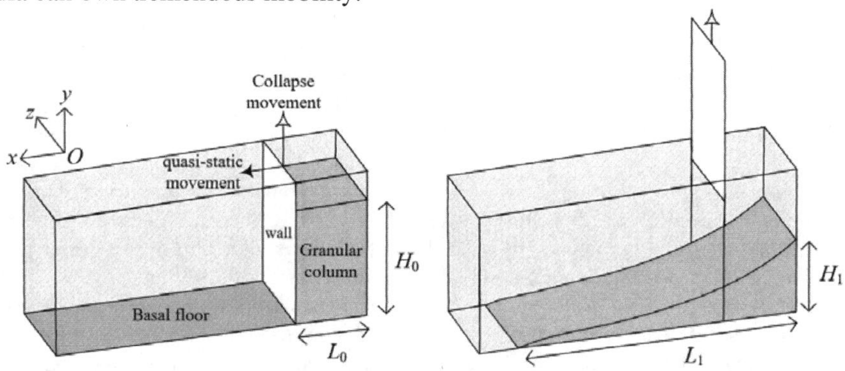

Figure 1. Diagram of granular column collapse movement model.

2. DEM Numerical Model Setup

2.1. The Meaning of FPSD

We employed the method employed by the references [20-21] to generate the numerical samples with FPSDs. The mathematical meaning of FPSD is that the number, $N(d)$, of d-sized particles is 2^D times the $2d$-sized number. The parameter D indicates the fractal dimension. Then, we can use the following equation to represent the flows with FPSD.

$$N(d) = N_0(d/d_0)^{-D} \tag{1}$$

where N_0 is the d_0-sized particle number. On the other hand, the greater value of D means that the flow owns the larger proportion of small sized particles. If the Particle Size Distributions (PSD) of the flow is fractal, it means that the PSD is self-similar from different scales. The other uniform, multi- and poly-dispersed types of distributions don't own this kind of character.

2.2. Granular Column Collapse Processes Simulations with Different FPSDs Using DEM

The DEM has been commonly employed to model the movement process of the flows consisting of granular media [22]. The model in DEM consists of grains and walls. The number of the grains and walls is finite. It is based on a Lagrangian theory where particles are discrete and their movement trajectories are recorded according to the solutions of the Newton's motion equations. The DEM descriptions in detail can be seen in Itasca [23] and Kermani et al. [24] and hence are not presented herein.

The numerical samples with D = 1.0, 1.5, 2.5 and 3.5 are built according to Lai et al. [19] using DEM code. First, we utilized the disks as the particles and their thicknesses are unit. We employed six kinds of particles with varying sizes, which are small-sized particles with 0.2 and 0.4 m radii, medium-sized particles with 0.6 and 0.8 m radii, and large-sized particles with 1.0 and 1.2 m radii, respectively. Small-sized, medium-sized and large-sized particles are in blue, red and yellow, respectively. The reason why these sizes are defined is that they can describe the common rock grains in nature according to Crosta et al. [18] and the acceptable computational efficiency of DEM.

We obtained the particles with specified sizes in DEM via the equation (1). Figure 1 gives the FPSDs of the numerical samples. Then, a rectangular reservoir is built. The length of this reservoir is 40.0 m and its depth is 80.0 m, respectively. The particles with different sizes are randomly generated there. The numbers of particles are specified according to Figure 2. A linear contact model with the viscous damping in the longitudinal and transversal orientations and the slip determination represents the collision and shearing behaviours among the particles. Finally, the particles fall to accumulate stably in the rectangular under the applied gravity. All the simulation parameters are used from the reference of Lai et al. [19], which are shown in Table 1. Figure 3 illustrates the setup of the samples with varying D.

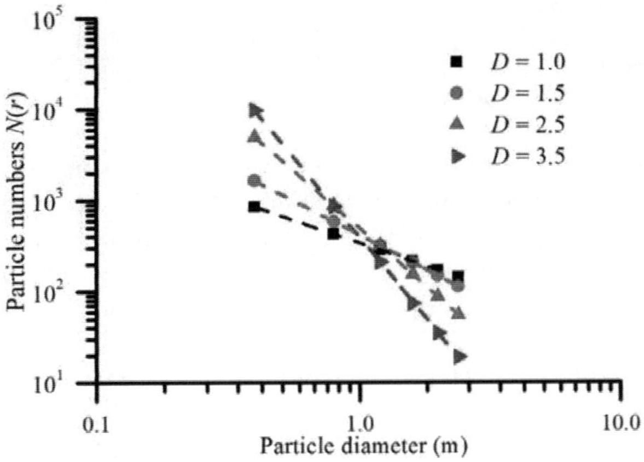

Figure 2. The FPSDs of the samples in DEM.

Table 1. DEM simulation parameters.

Parameter	Value
Sample length L_0	40.0 m
Sample height H_0	67.0 m
Density of particles ρ	2500 kg / m^3
Stiffness of particles and walls	$k_n = k_t = 10^{10}$ N / m
Friction of particles μ_{particle}	0.6
Friction of the left and right wall sides μ_{lateral}	0
Friction of the bottom wall side μ_{bottom}	0.6
Gravity g	9.81 m/s^2
Damping β_{n} in the normal orientation	0.12
Damping β_{t} in the transversal orientation	0.12

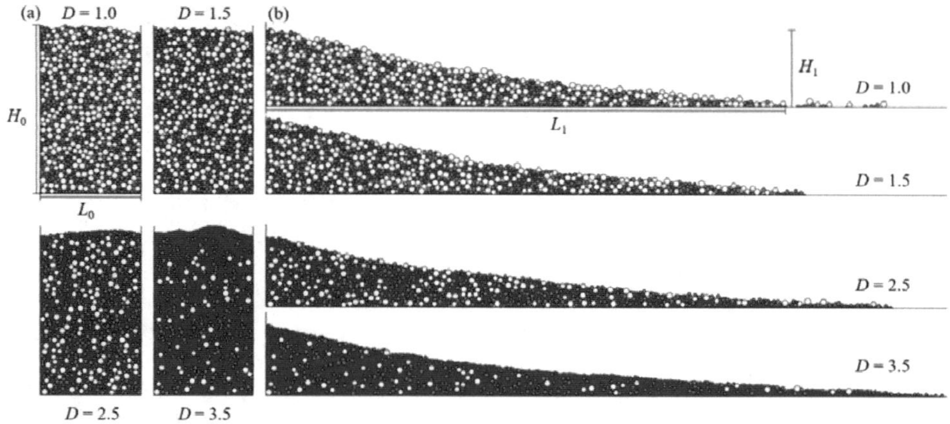

Figure 3. The initial samples and the accumulation shapes for the flows with D = 1.0, 1.5, 2.5 and 3.5 based on DEM.

3. DEM Simulation Results

3.1. Motion Kinematics

The right walls of the initial granular columns with varying D are deleted to trigger the collapse. The friction of the walls is set to the same value of the particles. Due to the friction, the main granular flow becomes still and the velocities in the translational and rotational forms are close to zero gradually. Figure 3 shows the accumulation shapes for the flows with the four varying D. The velocities of the grains in the frontal part of the granular flow are larger than the others, thus making them move away from the main body as shown in the granular flow with D = 1.0. The same trend can be seen for the other three samples. The particles that accumulate away from the main body are deleted because only the accumulation shapes of the main bodies of the flows are focused in our study.

The longitudinal and transversal orientations are built as x and y axes, individually. Then, the translational average values of $\overline{v}_x$ and $\overline{v}_y$ in the longitudinal and transversal orientations, and the average values of $\overline{\omega}$ in the rotational form are defined:

$$\overline{v}_x=\sum_{i=1}^{n} v_x^i \Big/ n \quad \overline{v}_y=\sum_{i=1}^{n} v_y^i \Big/ n \quad \overline{\omega}=\sum_{i=1}^{n} \omega^i \Big/ n \quad t^*=t\Big/\sqrt{H_0/g} \tag{2}$$

The motion velocity distributions of $\overline{v}_x$ and $\overline{v}_y$ for the samples with D = 1.0 and 3.5 during the movement are given in Figure 4. The definition of t^* is chosen according to the literature of Jing et al. [11]. According to Figure 4, $\overline{v}_x$ shows a greater peak than $\overline{v}_y$, and the peaks of $\overline{v}_x$ and $\overline{v}_y$ occur at different values of t^*. Besides, $\overline{v}_x$ increases significantly as D increases from 1.0 to 3.5, whereas $\overline{v}_y$ increases only slightly, and $\overline{\omega}$ decreases obviously.

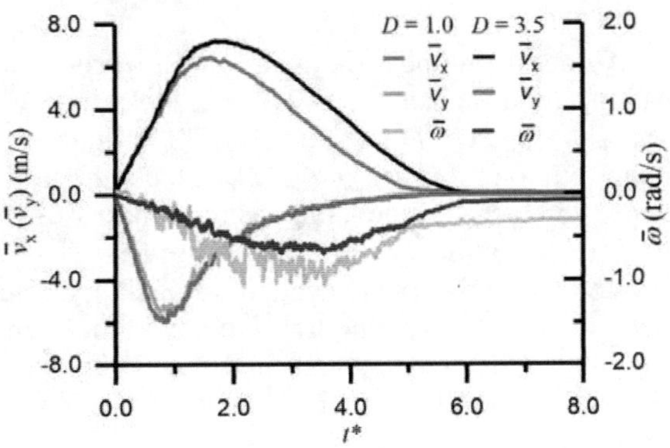

Figure 4. The motion velocity $\overline{v}_{\mathrm{x}}$ and $\overline{v}_{\mathrm{y}}$ distributions in translational forms and the motion velocity $\bar{\omega}$ in rotational forms over t^{*} for the samples with D = 1.0 and 3.5.

3.2. *Energy Conversion*

We use the normalized energy $E^{\mathrm{t}*}$, $E^{\mathrm{r}*}$ and $E^{\mathrm{p}*}$ in the translational, rotational and potential forms to compare the effects of D. The energies in the translational, rotational and potential forms are divided by the initial potential energy $E_0^{\mathrm{p}}=\sum_{i=1}^{n} m^i g H_0^i$, where m^i is the particle i mass, and h_0^i is the initial vertical distance from the bottom wall.

Figure 5 gives the evolutions of $E^{\mathrm{t}*}$, $E^{\mathrm{r}*}$ and $E^{\mathrm{p}*}$ and the average values are given in parentheses. Compared with the values of $E^{\mathrm{r}*}$, $E^{\mathrm{t}*}$ of the flows with D = 1.0 and 3.5 are much greater. The evolutions of $E^{\mathrm{t}*}$ for the flows D = 1.0 and 3.5 are almost the same until t^* = 1.0 . This is in accordance with the time when the differences appear in $\overline{v}_{\mathrm{x}}$ and $\overline{v}_{\mathrm{y}}$. It is shown that $E^{\mathrm{r}*}$ reduces and $E^{\mathrm{t}*}$ enlarges with the increasing D.

Figure 5 also gives the evolution of $E^{\mathrm{p}*}$ during the movement. The values of $E^{\mathrm{p}*}$ at t^* = 6.0 are shown in brackets. It can be seen that the evolutions of $E^{\mathrm{p}*}$ for the flows D = 1.0 and 3.5 are almost the same until t^* = 1.0 and then the difference occurs with the continuation of the movement. According to the values in brackets, $E^{\mathrm{p}*}$ enlarges as D increases. This means that more energy in potential forms is transformed into the energy in translational forms and the energy dissipated by particle friction and collision behaviours.

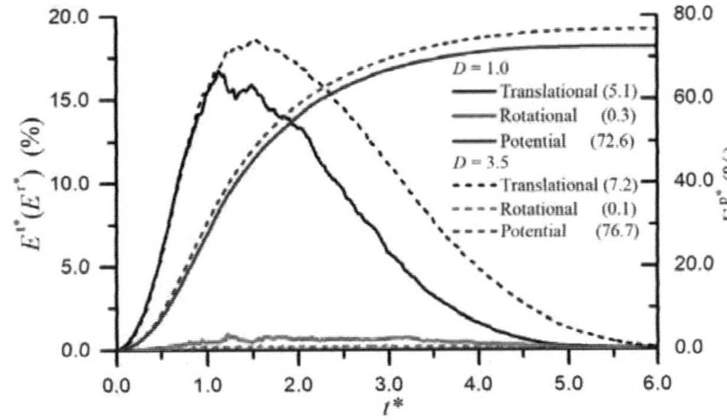

Figure 5. The normalized energy distributions of $E^{\mathrm{t}*}$, $E^{\mathrm{r}*}$ and $E^{\mathrm{p}*}$ in the translational, rotational and potential forms of the samples with D = 1.0 and 3.5.

4. Conclusions

The granular flow is a type of flow including rock avalanches, debris flows, landslides in nature and grains in the pharmaceutical and food industries [25-27]. Based on DEM, we performed the granular column collapse simulations with different *D* and analysed the influences of *D* on the velocities and energy in the translational and rotational forms. The FPSD affects the flow mobility significantly. As the fractal dimension *D* becomes greater, the spreading distance increases.

When *D* becomes greater, the runout distance becomes greater which represents the increasing mobility. During the process of the entire movement, the translational velocity in the longitudinal orientation increases at the beginning and then the translational velocity in the transversal orientation becomes much greater at the following spreading and final stages. The velocity in the rotational form reduces obviously. This is directly responsible for the increasing runout distance of the flows with the inreasing *D*.

More energy in potential forms is transformed into the energy in translational forms and the energy dissipated by particle friction and collision behaviours with the increasing *D*. Therefore, the particle velocities in the translational form increase. However, the particle velocities in the rotational form reduce. Only the particle shearing behaviours can exert the effects on the particle movement in the rotational motion. Therefore, we reckon that the decreasing particle shearing behaviours may be responsible for the decreasing particle rotational motion.

Acknowledgments

This work is supported by the National Key R&D Program of China (Grant No. 2018YFC1508403), the National Natural Science Foundation of China (Grant No. 51909102), the Central Public-Interest Scientific Institution Basic Research Fund (Grant No. HKY-JBYW-2021-01, -2020-05 and -2019-02), the Key Commonwealth Project of Henan Province (Grant No. 201300311600), the Science and Technology Development Project of Henan Province (Grant No. 212300410200).

References

[1] Lajeunesse E, Mangeney-Castelnau A, Vilotte J P. 2004. Spreading of a granular mass on a horizontal plane. Physics of Fluids, 16(7), 2371–2381.
[2] Lajeunesse E, Monnier J B, Homsy G M. 2005. Granular slumping on a horizontal surface. Physics of Fluids, 17(10), 103302-103316.
[3] Lube G, Huppert H E, Sparks R S J, Freundt A. 2005. Collapses of two-dimensional granular columns. Physical Review E, 72(4), 041301.
[4] Lube G, Huppert H E, Sparks R S J, Hallworth M A. 2004. Axisymmetric collapses of granular columns. Journal of Fluid Mechanics, 508, 175–199.
[5] Utili S, Zhao T, Houlsby G T. 2015. 3D DEM investigation of granular column collapse: Evaluation of debris motion and its destructive power. Engineering Geology, 186, 3-16.
[6] Fern E J, Soga K. 2016. The role of constitutive models in MPM simulations of granular column collapses. Acta Geotechnica, 11(3), 659-678.
[7] Cabrera M, Estrada N. 2019. Granular column collapse: Analysis of grain-size effects. Physical Review E, 99(1), 012905. https://doi.org/10.1103/PhysRevE.99.012905.
[8] Tapia-McClung H, Zenit R. 2012. Computer simulations of the collapse of columns formed by elongated grains. Physical Review E, 85(6), 061304. https://doi.org/10.1103/PhysRevE.85.061304.
[9] Mast C M, Arduino P, Mackenzie-Helnwein P, Miller G R. 2015. Simulating granular column collapse using the Material Point Method. Acta Geotechnica, 10(1), 101-116.
[10] Staron L, Hinch E J. 2007. The spreading of a granular mass: role of grain properties and initial conditions. Granular Matter, 9(3-4), 205-217.
[11] Jing L, Yang G C, Kwok C Y, Sobral Y D. 2018. Dynamics and scaling laws of underwater granular collapse with varying aspect ratios. Physical Review E, 98(4), 042901.
[12] Yang G C, Jing L, Kwok C Y, Sobral Y D. 2020. Pore-Scale Simulation of Immersed Granular

Collapse: Implications to Submarine Landslides. Journal of Geophysical Research: Earth Surface, 125, e2019JF005044.

[13] Ng C W W, Choi C E, Liu L H D, Wang Y, Song D, Yang N. 2017. Influence of particle size on the mechanism of dry granular run-up on a rigid barrier. Géotechnique Letters, 7(1), 1-11.

[14] Degaetano M, Lacaze L, Phillips J C. 2013. The influence of localised size reorganisation on short-duration bidispersed granular flows. European Physical Journal E, 36(4), 36.

[15] Zhou W, Lai Z, Ma G, Yang L, Chen Y. 2016. Effect of base roughness on size segregation in dry granular flows. Granular Matter, 18(4), 83.

[16] Zhou W, Lai Z, Yang L, Ma G, Chen Y, Qi T. 2017. Influence of base roughness on kinematic and mechanical characteristics of debris flows. In Proceedings of the 7th International Conference on Discrete Element Methods, 188, 1047–1054. https://doi.org/10.1007/978-981-10-1926-5_109.

[17] Cagnoli B, Romano G P. 2012. Effects of flow volume and grain size on mobility of dry granular flows of angular rock fragments: A functional relationship of scaling parameters. Journal of Geophysical Research, 117, B02207-B02219.

[18] Crosta G B, Frattini P, Fusi N. 2007. Fragmentation in the Val Pola rock avalanche, Italian alps. Journal of Geophysical Research, 112, F01006.

[19] Lai Z, Vallejo L E, Zhou W, Ma G, Espitia J M, Caicedo B, Chang X. 2017. Collapse of Granular Columns with Fractal Particle Size Distribution: Implications for Understanding the Role of Small Particles in Granular Flows, Geophysical Research Letters, 44(24), 12181-12189.

[20] Palmer A C, Sanderson T J O. 1991. Fractal crushing of ice and brittle solids. In Proceedings of the Royal Society of London A: Mathematical, Physical and Engineering Sciences., 433(1889), 469–477.

[21] Hooke R L, Iverson N R 1995. Grain-size distribution in deforming subglacial tills: Role of grain fracture. Geology, 23(1), 57.

[22] Cundall P. A., and Strack O. D. L. (1979). "A discrete numerical model for granular assemblies." Geotechnique, 29(1), 47–65.

[23] Itasca. (2008). Particle flow code, PFC3D 4.0, Itasca Consulting Group, Inc., Minneapolis.

[24] Kermani E, Qiu T, Li T. 2015. Simulation of collapse of granular columns using the discrete element method. International Journal of Geomechanics, 15(6), 04015004.

[25] Aaron J, & Hungr O. 2016. Dynamic analysis of an extraordinarily mobile rock avalanche in the Northwest Territories, Canada. Canadian Geotechnical Journal, 53(6), 899–908.

[26] Robinson T R., Davies, T. R., Reznichenko, N. V., & De Pascale, G. P. (2015). The extremely long-runout Komansu rock avalanche in the Trans Alai range, Pamir Mountains, southern Kyrgyzstan. Landslides, 12(3), 523–535.

[27] Lai Z, Chen D, Jiang E, et al. Effect of fractal particle size distribution on the mobility of dry granular flows[J]. AIP Advances, 2021, 11(9): 095113.

Key-Point Interpolation: A Sparse Data Interpolation Algorithm based on B-splines

Bolun Wang[1,2], Xin Jiang[1,2,3 *], Guanying Huo[1,2,3], Cheng Su[1,2,3], Dongming Yan[4], and Zhiming Zheng[1,2]

[1]Key Laboratory of Mathematics, Informatics and Behavioral Semantics (LMIB), School of Mathematical Science, Beihang University, Beijing, China
[2]Peng Cheng Laboratory, Shenzhen, Guangdong, China
[3]Zhengzhou Aerotropolis Institute of Artificial Intelligence, Zhengzhou, Henan, China
[4]National Laboratory of Pattern Recognition, Institute of Automation, Chinese Academy of Sciences, Beijing, China

[*]Corresponding author email: 09436@buaa.edu.cn

Abstract. B-splines are widely used in the fields of reverse engineering and computer-aided design, due to their superior properties. Traditional B-spline surface interpolation algorithms usually assume regularity of the data distribution. In this paper, we introduce a novel B-spline surface interpolation algorithm: KPI, which can interpolate sparsely and non-uniformly distributed data points. As a two-stage algorithm, our method generates the dataset out of the sparse data using Kriging, and uses the proposed KPI (Key-Point Interpolation) method to generate the control points. Our algorithm can be extended to higher dimensional data interpolation, such as reconstructing dynamic surfaces. We apply the method to interpolating the temperature of Shanxi Province. The generated dynamic surface accurately interpolates the temperature data provided by the weather stations, and the preserved dynamic characteristics can be useful for meteorology studies.

1. Introduction

B-splines are widely used in the fields of reverse engineering and computer-aided design, due to its superior properties such as global smoothness, local support, etc. [1][2][3][17]. Traditional B-spline surface interpolation algorithms usually assume that the data points are distributed in grids or in rows. However, the majority of actual measured data are naturally sparse and non-uniform, such as the data obtained from LIDAR scanners and weather stations [6]. To interpolate sparse data points, traditional algorithms usually introduce degrees of freedom, and require a large number of control points, which may lead to longer runtime and larger memory usage.

In this paper, we introduce a novel B-spline surface interpolation algorithm: KPI, which interpolates sparsely and non-uniformly distributed data points. Our algorithm is a two-stage method [11]. Firstly, we generate a gridded dataset to meet the requirements of B-spline interpolation. Then, we use our Key-Point Interpolation method to generate the control points. Our algorithm can be extended to higher dimensions by properly assigning the key points, thus can be used to construct dynamic surfaces [3][5][7]. We use our method to interpolate the temperature of Shanxi Province. The result shows that our method generates a dynamic surface which interpolates the sparse temperature data

Content from this work may be used under the terms of the Creative Commons Attribution 3.0 licence. Any further distribution of this work must maintain attribution to the author(s) and the title of the work, journal citation and DOI.
Published under licence by IOP Publishing Ltd

accurately, and the dynamic characteristics preserved in the 3 dimensional B-splines can be useful for meteorology studies.

2. Related Work

Previous works usually assume that the data points are distributed in grids or in rows. [14] and [2] use averaging method to calculate the knot vectors from the data points. If the data points are sparsely distributed, these methods will create a large number of control points. [7] designs an algorithm to interpolate sparse data points. In each iteration, [7] doubles the density of the knots, until the data points are interpolated accurately, but the number of required control points highly depends on the distribution of the data points. [13] and [15] use lofting method to obtain the interpolation surfaces. In [13], a sufficient condition for the existence of the interpolation surface is introduced. By employing a flexible knot insertion method, [13] can reduce the control points effectively. [15] further improved the surface smoothness by dividing the curves into groups. Lofting methods can be efficient when interpolating large number of data points, but they usually require that the data points are sampled in rows.

Kriging [10][16] is a standard method to predicate space structure and randomicity of parameters from sparsely sampled data points in geostatistics, by performing optimal linear unbiased interpolation estimates. We use Kriging to construct a gridded dataset out of the input sparse data, thus it can meet the requirements of our KPI control points solving procedure.

3. KPI Method

Let $Q=\left\{q_0,q_1,\cdots,q_{m_1}\right\}$ be a point set in $\mathbb{R}^d$, and $\tilde{Q}=\left\{\tilde{q}_0,\tilde{q}_1,\cdots,\tilde{q}_{m_2}\right\}$ be a subset of Q. Let f_1 be a function on the set Q, and f_2 be a function on $\tilde{Q}$. Suppose the valve of f_2 on $\tilde{Q}$ is given. $f_1(\tilde{q}_i)=f_2(\tilde{q}_i),\forall\tilde{q}_i\in\tilde{Q}$. The purpose of the interpolation is to find the function f_1 on Q with the given value on $\tilde{Q}$. In this paper, a 3 dimensional B-spline function M which reconstructs dynamic surfaces is the function f_1 over Q, while the data points $\tilde{Q}$ is given as the form of sparsely sampled data. Our KPI method is introduced as the following 2 steps:

3.1. Dataset Generation

In this step, we firstly parametrize the dataset, then generate a gridded dataset using Kriging. Constrained Delaunay Triangulation (CDT) method [12] is employed to generate a triangle mesh for each surface. However, if the border of each surface is over complicated (for example, when interpolating the temperature or the topographic information bounded by the border of a certain region), the number of border points could be a few orders of magnitude higher than the sparse data, which may lead to degenerated triangles. To avoid that, we simplify the border until the number of border points is comparable with the number of sparse data points. We use Dmitry and Zsolt's method [8] to detect feature points of the border, and use the feature points as the vertices of the simplified border. Then triangulate the sparse data Q and feature points $\tilde{B}$ of the border B to obtain the triangle mesh $M=(v,E,S)$ of each surface, where $v=\tilde{Q}\cup\tilde{B}$ represents the vertex sets and E presents the edges of the triangulation mesh. We apply the method into a temperature data interpolation problem. Figure 1 is an example of border simplification and triangulation of the weather stations in Shanxi Province, China. In figure 1 (a), the red dots on the border are the selected feature points, (b) and (c) show the simplification result of the border. Figure 1 (d) shows the triangulation result.

As the triangulation M is obtained, we use mean value coordinates method [9] to parametrize the data. Since the data points are randomly distributed, the parameters are extremely non-uniform and dense.

Here we provide a perturbation method to reduce the number of parameters, while maintaining the topology of the triangle mesh.

We use two parameters C_1 and C_2 to bound the perturbation in u and v direction, separately. The parameters for each vertex v_i are denoted as $r_i = (u_i, v_i)$. In the parameter domain $M' = (r, E', S')$, for the u parameters $u_i \neq 0, u_{i+1} \neq 1$, if $|u_{i+1} - u_i| < C_1$, we merge the two u parameters by setting $u_{i+1} \leftarrow u_i$; For each $v_j \neq 0$, $v_{j+1} \neq 1$, if $|v_{j+1} - v_j| < C_2$, set $v_{j+1} \leftarrow v_j$. In this way, any two u (or v) parameters closer than C_1 (or C_2) will be merged. Iterate this procedure with increased C_1 and C_2 values, the number of parameters can be progressively reduced.

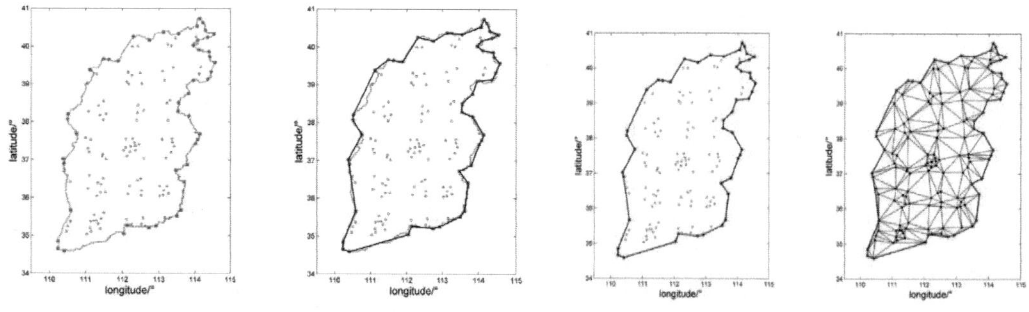

(a) (b) (c) (d)

Figure 1. Border simplification and triangulation of the data points.

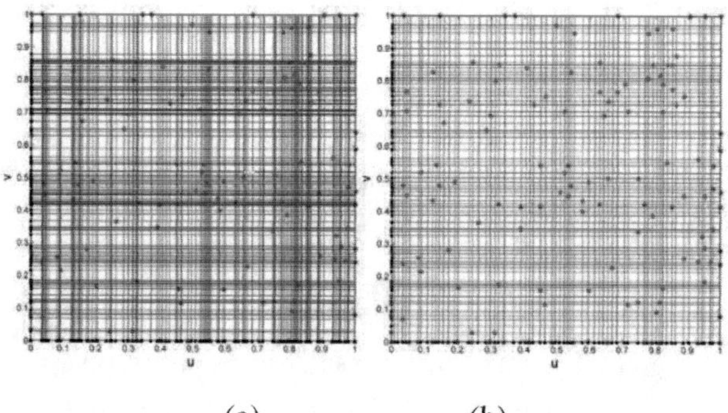

(a) (b)

Figure 2. Original u and v parameters and parameter perturbation result.

To avoid parametrization failures (i.e., there is no one-one mapping between the old and new parametrization), we add an extra check when iteratively reducing the parameters. We assume that for a triangle $T_{i,j,k}$ defined by its three vertices v_i, v_j and v_k whose parameters are r_i, r_j and r_k. After perturbing at least one of the 3 parameters of the vertices, the new parameters become $\tilde{r}_i, \tilde{r}_j$ and $\tilde{r}_k$. We check if the orientation of the 2D triangle in the parametric domain is the same as before, by checking if $sign\left(\det\begin{pmatrix} e_{ij} \\ e_{ik} \end{pmatrix}\right) = sign\left(\det\begin{pmatrix} \tilde{e}_{ij} \\ \tilde{e}_{ik} \end{pmatrix}\right)$, where $e_{pq} = r_q - r_p$, $(p, q \in \{i, j, k\})$, the function $sign(x)$ returns the sign (+1, -1, or 0) of the scalar value x, which is the determinate of a 2×2 matrix. After perturbation, if the orientation of any triangle doesn't hold, we withdraw the perturbation and continue to perturb another pair of u or v parameters. Here we choose $C_1 = C_2 = 0.5L$ as the initialized perturbation bound, where $L = \arg\min_{i,j=1\cdots m} |r_i, r_j|$.

Iterating under the orientation check, the number of parameters can be reduced progressively. Generally, within a few iterations, the number of the parameters can be reduced to a suitable quantity. The generated parameters are denoted as $\bar{U} = \{\bar{u}_0, \bar{u}_1, \cdots, \bar{u}_{m_1}\}$ and $\bar{V} = \{\bar{v}_0, \bar{v}_1, \cdots, \bar{v}_{m_2}\}$. In figure 2 (a), the original parameters are very dense and non-uniform. After perturbation (figure 2 (b)), the parameters are reduced, and become more uniform.

To ensure the smoothness of the interpolation surface, our method generates a gridded dataset using Kriging interpolation method [10]. The generated dataset $Q_{i,j}, i = 0, \cdots, m_1, j = 0, \cdots, m_2$ of each time step is composed of two parts: the original sparse data and the Kriging generated data.

3.2. *Control Points Solving based on KPI*

A 3 dimensional B-spline function [4] can be expressed as follows:

$$M(u,v,t) = \sum_{i=0}^{n} \sum_{j=0}^{m} \sum_{k=0}^{l} N_{i,p}(u) N_{j,q}(v) N_{k,r}(t) P_{i,j,k}, \tag{1}$$

Where $P_{i,j,k}$ are the control points, $N_{i,p}(u)$, $N_{j,q}(v)$, and $N_{k,r}(t)$ are B-spline basis functions of degree p, q, and r. The knot vectors are U, V and T. Solving the control points of the dynamic surface can be achieved by 3 levels of iterations of curve interpolations. In the first level, the data points to be fitted are the sparse data points and the Kriging generated points. In the second and the third level, the data points are the control points of the curves of the previous levels. We use the following equation to generate a KPI curve:

$$\min f = \sum_{i=0}^{m_1} (Q_{i_1} - C(\bar{u}_{i_1}))^2$$
$$s.t. Q_{i_2} = C(\bar{u}_{i_2}), i = 0, 1, \cdots, m_2, \tag{2}$$

where $Q_{i_1}, i = 0, 1, \cdots, m_1$ are the non-key points, $Q_{i_2}, i = 0, 1, \cdots, m_2$ are the key points. $\bar{u}_{i_1}$ and $\bar{u}_{i_2}$ are the parameters of Q_{i_1} and Q_{i_2}. The curve $C(u)$ interpolates the key points, and approximates the points generated by Kriging. It is crucial to determine the key points when iteratively constructing interpolation curves. The key points are determined as follows:

Step 1. For the first level, where the data points are the sparse points and the Kriging generated points, we directly set the sparse data points as the key points.

Step 2. For the second or the third level, the data points are the control points of the previous level. Without loss, we assume one key point in the previous level is $Q_{r,k}$, fitted by the curve $C_k(u) = \sum_{j=0}^{m} b_j N_{i,p}(u)$ and the parameter of $Q_{r,k}$ is $\bar{u}_r$. Then, according to the local support property of B-spline surface, there are at most $p+1$ non-zero B-spline bases $N_{i,p}(\bar{u}_r)$. Then, we set the $p+1$ corresponding control points b_j ($N_{j,q}(\bar{u}_r) \neq 0$) as the key points of the next level.

There are many methods [2] to ensure the equality constraints in equation (2) can be satisfied. We iteratively inserting knots, until at most one key point whose parameter falls into each interval (e.g., interval $[u_i, u_{i+1})$) of the knot vector of the curve. The generated knot vectors guarantee that the equality constraints can be satisfied.

4. Application

We interpolate the temperature of Shanxi Province, China, using our KPI algorithm. The data contains temperature data for each hour provided by 98 weather stations, from 1:00, October 15, 2017 to 24:00, October 17, 2017. Figure 1 shows the result of border simplification. The original border which is presented by 6102 points (shown in Figure 1(a)) is simplified into a polygon with 49 vertices, while maintaining the basic shape of the original border. The data was provided by China Meteorological Data Sharing Service System (http://data.cma.cn/en). To parameterize the data points, the 98 station points and 49 vertices of the simplified border are triangulated with Constrained Delaunay Triangulation method in a two-dimensional plane. The triangulation result is shown in Figure 1(b, c, d). Using mean value coordinates, the triangle mesh is parametrized into a quad parametrization domain shown in Figure 2(a). The number of u parameters and v parameters are 124 and 120, respectively. From Figure 2(a), we can see that the parameters are extremely dense and non-uniform. Process the parameters with our perturbation method. With 3 iterations, the numbers of u and v parameters are reduced to 66 and 68, respectively (shown in Figure 2(b)).

By Applying Kriging, the gridded data points $Q_{i,j,k}\left(i=0,\cdots,65, j=0,\cdots,67, k=0,\cdots,71\right)$ are generated. A dynamic surface which contains $41\times45\times74$ control points is constructed (figure 3). The dynamic surface accurately interpolates the temperature data (with the maximum interpolation error 9.95×10^{-14}). The surface is smooth since the Kriging generated data follows the trend of the sparse data. Figure 3 (a), Figure 3 (b) and Figure 3 (c) are the temperature interpolation results at 4:35, October 15, 2017, 14:00, October 15, 2017 and 10:15, October 16, 2017.

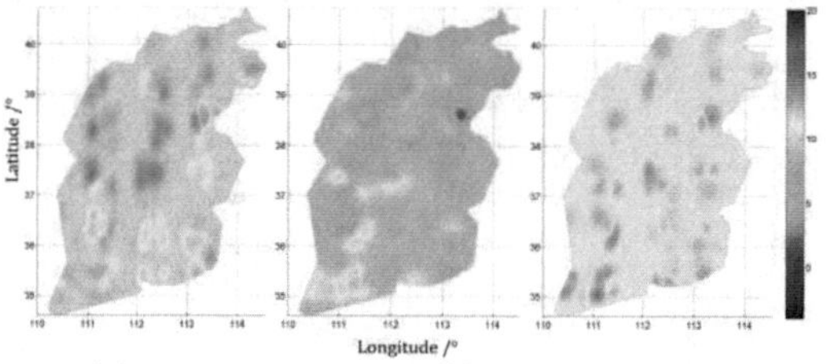

(a) (b) (c)

Figure 3. Temperature interpolation results of Shanxi Province with KPI method

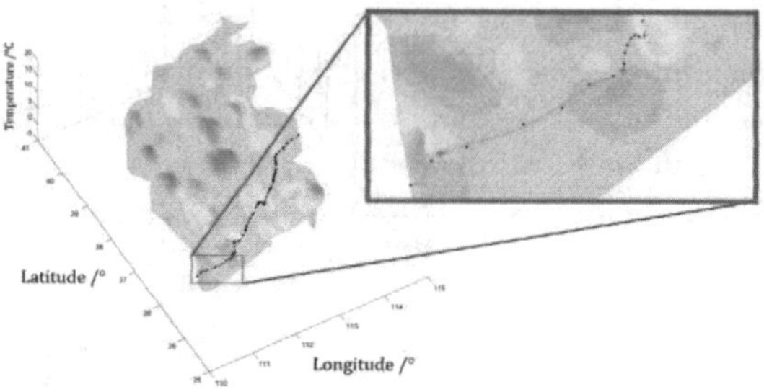

Figure 4. An iso-u curve which interpolates the sparse points (red crosses) and approximates the Kriging generated points (black dots).

Figure 4 shows an iso-u curve, where the red crosses are the original data provided by the weather stations and the black dots denote the data points generated by Kriging. It can be seen that our KPI method interpolates the original data points precisely, while approximates the Kriging generated data points.

5. Conclusion

In this paper, a KPI method for dynamic B-spline surface reconstruction is introduced. As a two-stage B-spline interpolation method, our method firstly constructs a gridded dataset, and then generates control points to interpolate the sparse data points. To construct the dataset, the sparsely distributed data points of each time step are triangulated. Then the meshes are parameterized in quad-domains using mean value coordinates method. We also design a parameter perturbation method which reduces the parameters to reduce the computation of control points solving. The control points are solved by iteratively assigning key points and solving curve KPI. Our method can generate a smooth dynamic surface which interpolates sparsely distributed data points given in a time sequence.

Acknowledgments
This work has been supported by the National Key Research and Development Program of China (Grant No.2020YFA0713700)) and the Natural Science Foundation of China (Grants No. 12001028, No. 61772523).

References

[1] Park H, Jung H B, Kim K (2004) A new approach for lofted B-spline surface interpolation to serial contours. The International Journal of Advanced Manufacturing Technology 23(11-12):889-895
[2] Piegl L, Tiller W (2012) The NURBS book. Springer Science & Business Media
[3] Koch K R (2010) Three-dimensional NURBS surface estimated by lofting method. The International Journal of Advanced Manufacturing Technology 49(9-12):1059-1068
[4] Koch K, Schmidt M (2011) N-dimensional B-spline surface estimated by lofting for locally improving IRI. Journal of Geodetic Science 1(1):41-51
[5] Huang Y, Qian X (2007) Dynamic B-spline surface reconstruction: Closing the sensing-and-modeling loop in 3D digitization. Computer-Aided Design 39(11):987-1002
[6] Zhang R, Huang Z, Li J, et al (2013) Interpolation Technique for Sparse Data Based on Information Diffusion Principle-Ellipse Model. Journal of Tropical Meteorology 19(1):59
[7] Lee S, Wolberg G, Shin S Y (1997) Scattered data interpolation with multilevel B-splines. IEEE transactions on visualization and computer graphics 3(3):228-244
[8] Chetverikov D (2003) A simple and efficient algorithm for detection of high curvature points in planar curves. International Conference on Computer Analysis of Images and Patterns:746-753
[9] Floater M S (2003) Mean value coordinates. Computer aided geometric design 20(1):19-27
[10] Matheron G (1963) Principles of geostatistics. Economic geology 58(8):1246-1266
[11] Schumaker, L. L. . (1982). Fitting surfaces to scattered data. approximation theory ii.
[12] Chew, L. P. . (1989). Constrained delaunay triangulations. Algorithmica, 4(1), 97-108.
[13] Wang W, Zhang H, Park H, et al (2008) Reducing control points in lofted bspline surface interpolation using common knot vector determination. Computer-Aided Design 40(10-11): 999-1008.
[14] Kineri Y, Wang M, Lin H, et al (2012) B-spline surface fitting by iterative geometric interpolation/approximation algorithms. Computer-Aided Design 44(7):697-708.
[15] Wang W, Tian T, Li S (2011) B-spline surface skinning to non-parallel sectional curves, in: Proceedings of the 10th International Conference on Virtual Reality Continuum and Its Applications in Industry, pp. 569-572.
[16] Shi G. (2013) Data mining and knowledge discovery for geoscientists. studies in big data.
[17] Liang F, Zhao J, Ji S, et al (2017) A novel knot selection method for the error-bounded B-spline curve fitting of sampling points in the measuring process. Measurement Science and Technology 28(6):065015.

An Echo Sequence Profile Image Based Ship Target Classification Algorithm for Low-Resolution Radar

Duoduo Hang[1,*], Ji Zhang[1], Chuanwen Chang[2], Wei Zhu[2], Beibei Wu[2], Hua Zhang[1] and Junbo Wang[1]

[1]National Mobile Communications Research Laboratory, Southeast University, Nanjing, China
[2]Nanjing Les Electronic Equipment Co., ltd, Nanjing 210014, China

[*]Corresponding author email: hangduoduo@seu.edu.cn

Abstract. Ship target classification is of great significance in both military and civilian fields. We propose a ship target classification algorithm for low-resolution radars with echo sequence profile images. This algorithm can be realized in the following steps. First, we collect radar profile image data. We use five perspectives of a radar target, including target shape, Radar Cross Section (RCS), echo amplitude, motion attribute, and features of two-dimensional grayscale maps, to extract eight-dimensional feature vectors. The proposed algorithm uses the Support Vector Machine (SVM) as the classifier, and the parameters of the classifier are optimized by either grid search or the Particle Swarm Optimization (PSO) algorithm. The proposed algorithm is verified through real data classification tests.
Keywords: Low-resolution radar, target classification, feature extraction, support vector machine.

1. Introduction

Radar Automatic Target Recognition (RATR) is a technique to achieve automatic determination of target attributes, types, or categories by extracting target features based on radar echo sequences. It is found that most of the existing radars are narrow-band and low-resolution radars, which cannot provide enough information to make accurate recognition on target types [1]. Low-resolution radar is characterized by wider pulse width, narrower bandwidth, and poor radial range resolution [2]. Compared with the one-dimensional echo sequence, the two-dimensional profile image consists of both distance and azimuth, which not only provides the echo amplitude information but also reflects its correlation information with azimuth angle. Therefore, two-dimensional radar profile images are commonly used for low-resolution radar target recognition.

The selection of features with high separability and stability plays a crucial role in the classification capability. [3] classified ship targets with di erent working conditions based on Radar Cross Section (RCS). [4] combined three di erent methods to classify island targets and ship targets using waving values of target profile images. [5] and [6] used RCS and spectral entropy to tri-chotomize ground targets for tri-classification. [7] used five-dimensional feature vectors such as central moments and skewness of two-dimensional profile images for tri-classification of ship targets. Numerous studies have shown that extracting more features to form feature vectors usually leads to higher classification accuracy. Moreover, the commonly used methods include the k-Nearest Neighbor (kNN) algorithm [8], the Support Vector Machine (SVM) algorithm [8,9], the neural network [2], or a combination of them [10]. Among them, SVM has a simple structure with better generalization ability in small data set training.

Content from this work may be used under the terms of the Creative Commons Attribution 3.0 licence. Any further distribution of this work must maintain attribution to the author(s) and the title of the work, journal citation and DOI.
Published under licence by IOP Publishing Ltd

In order to overcome the problems of the poor radial resolution, feature aliasing of low-resolution radar, and improve the recognition accuracy, we propose a complete ship target classification algorithm via echo sequence profile images for low-resolution radar. First, we collect radar profile image data and construct data set. Five perspectives of radar, including target shape, Radar Cross Section (RCS), echo amplitude, motion attribute, and features of two-dimensional grayscale maps, are used to extract eight-dimensional feature vectors. Finally, we use SVM to recognize and classify the targets. The experiments on real data prove the e ectiveness of the algorithm.

2. Principle of Radar Echo Sequence Profile Image

The radar beam scanning process is shown in figure 1. The radar target echo sequence profile image is a two-dimensional image formed by arranging the echo sequences during the scanning process in azimuthal order and then projecting them to azimuth and distance. The projection width of the echo sequence in azimuth and distance is called azimuthal broadening and distance broadening, respectively.

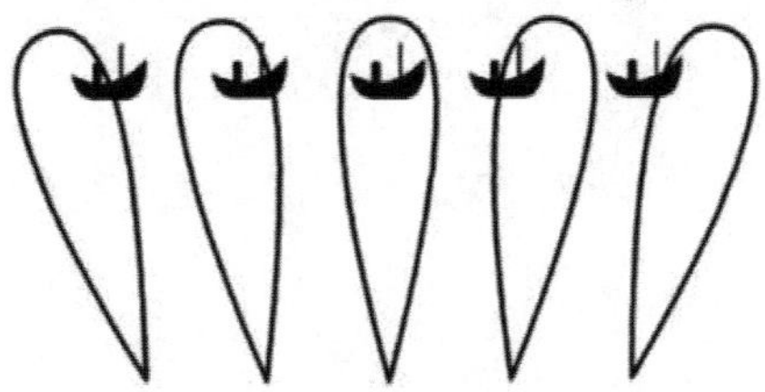

Figure 1. Radar beam scanning.

Compared with the target recognition method that uses one-dimensional echo sequences [2, 9], the target recognition method based on two-dimensional profile images can not only provide echo amplitude information but also reflect the correlation information between echo sequences and azimuths. Thus, profile images provides favourable auxiliary information for target classification.

3. Principle of Radar Echo Sequence Profile Image

The selection of features or feature vectors with high separability and stability plays a crucial role in the classification ability. In this paper, eight features are extracted from five perspectives for target recognition. The eight features are illustrated as follows.

3.1. Shape Features

The target shape is the most basic feature of the profile image. In this paper, we use area (number of non-zero echo points) and lateral length as the first feature and the second feature to characterize the target shape, which can be given respectively as

$$f_1 = \sum_i \{y_i \neq 0\} \tag{1}$$

$$f_2 = \theta_T R \tag{2}$$

where, y_i denotes the size of the i-th echo value, θ_T is the broadening caused by the lateral length of the target, and R denotes the distance between the target and the radar.

3.2. RCS Feature

Let the radar cross-sectional area of the target be σ, the echo power P_r received by the radar satisfies

$$P_r = \frac{P_t G A_e \sigma}{(4\pi)^2 R^4} = \frac{1}{k}\frac{\sigma}{R^4} \tag{3}$$

where, P_t is the radar transmit power, A_e is the effective receiving area of the antenna, G is the antenna gain, R is the distance between the target and the radar, and k is a constant. From (3), it can be derived that the target intercept area, σ, is

$$\sigma = kP_r R^4 \tag{4}$$

Therefore, the third feature of this paper is chosen as

$$f_3 = 10\log_{10}\left(P_r R^4\right) \tag{5}$$

3.3. Echo Amplitude

Under normal conditions, the echo amplitude peak of a large ship is larger than that of a small ship when the distance is the same. The fourth feature used in this paper is

$$f_4 = \max\left\{y_i\right\} \tag{6}$$

3.4. Motion Attribute

The speed is the main motion attribute of a ship target and the measurement of the target velocity is directly achievable by in-service radars. The fifth feature used in this paper is defined as

$$f_5 = v \tag{7}$$

where v is the target velocity.

3.5. Features of Two-dimensional Grayscale Maps

The profile images can be processed as general two-dimensional grayscale maps after normalization and image size complementation. Denote by $H(i,j)$ the pixel value of the i-th row and j-th column of the grayscale image, the sixth feature used in this paper is the image energy, which is given as

$$f_6 = \sum_{i,j} H^2(i,j) \tag{8}$$

Let (x,y) be the coordinate of an object point represented by the image and (x_c, y_c) be the centre of the object, then the $p+q$ order central moment is given as

$$\mu_{p,q} = \sum_{(x,y)\in R} (x-x_c)^p (y-y_c)^q \tag{9}$$

By normalizing these central moments, seven invariant moments can be obtained [11]. The seventh and the eighth features are the first two invariant moments, which can be given respectively as

$$f_7 = \mu_{2,0} + \mu_{0,2} \tag{10}$$

$$f_8 = \left(\mu_{2,0} - \mu_{0,2}\right)^2 + 4\mu_{1,1}^2 \tag{11}$$

4. Ship Target Recognition Process Based on Eight Features and SVM

The steps of ship target recognition in this paper are shown in figure 2.

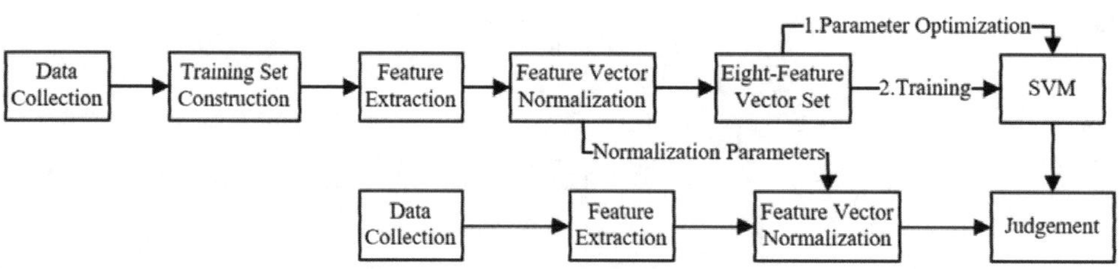

Figure 2. Ship target identification steps.

For classifier training, we firstly collected the data of radar echo sequences from an available radar station. The data set is constructed based on the echo data and Automatic Identification System (AIS)

Journal of Physics: Conference Series **2068** (2021) 012011 doi:10.1088/1742-6596/2068/1/012011

messages. Then, eight features are extracted to form a feature vector set. After that, the feature vectors are normalized, and the normalized parameters are saved. Finally, the SVM classifier is optimized and trained based on the normalized feature vector set and the trained model is saved.

For ship target recognition, the target data of the ship to be measured are collected firstly. Then, eight features are extracted to form a feature vector and normalized by the saved normalization parameters. After that, the trained SVM classifier is used to classify the ship.

In this paper, we collect radar data from 3-12 kilometres and classify ships into seven categories by length. The training and test sets are divided by 8:2. The data set construction and optimization methods of SVM parameters are introduced in Sections 4.1 and 4.2, respectively.

4.1. Data Set Construction

The process of radar target data set construction is: using the C-band radar that has been set up, the data collector is activated to record the radar echo data. At the same time, the AIS message acquires the target position, motion status, target type, and other information. The original data is stored in a DAT file. The data format is shown in figure 3, where the first 42 bytes are information data, including Maritime Mobile Service Identify (MMSI), start/end position, start/end distance, echo data quantity, and so on. Echo data consists of echo sequences of m azimuths, each azimuth has n data by distance, arranged in the format of "bearing (2 bytes) + echo sequence (n bytes)".

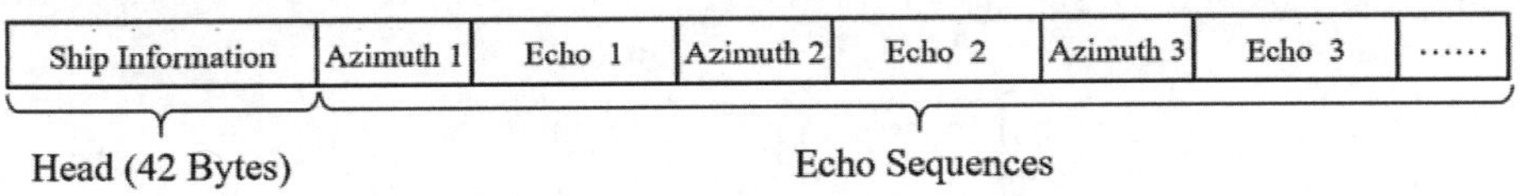

Figure 3. Data acquisition format.

Based on the unique MMSI number of each ship, information such as the length and the type of the ship can be obtained. The echo sequences are arranged by azimuth to obtain two-dimensional echo sequence profile image data. Figure 4 gives an example of a two-dimensional echo sequence profile image. In this paper, the raw data are grouped by MMSI number as training and testing data sets.

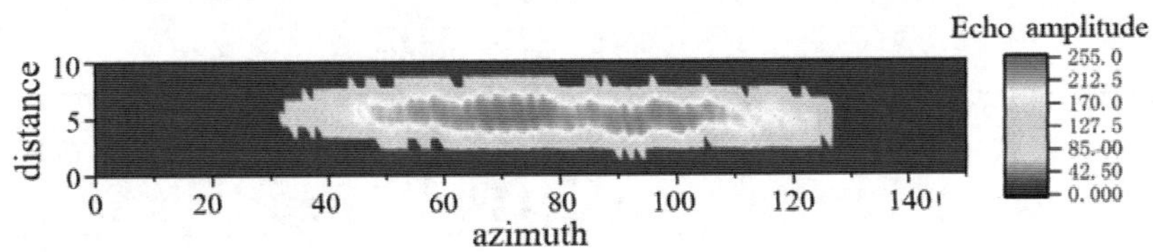

Figure 4. An echo sequence profile image (ship length=165m, distance=8km).

4.2. Classifier Parameter Optimization

Due to the advantages in generalizability and complexity, a Radial Basis Function (RBF) based SVM classifier is used. The parameters to be optimized are the penalty factor C and the RBF kernel parameter γ. We use grid search and PSO to optimize the parameters and make performance comparison.

4.2.1. Grid search. The process of grid search has the following steps. First, we set up the parameter grid, where each point represents a parameter combination. Then, the score for each combination is obtained in turn by enumeration. Finally, the parameter combination with the highest score is outputted. The advantage of the grid search is that the structure is simple and the search is comprehensive while the disadvantage is that the accuracy is limited by the grid interval and each optimization must search all the grid points, which makes it lack flexibility.

4.2.2. PSO algorithm. The PSO algorithm was proposed by J. Kennedy and R. C. Eberhart in 1995. The basic principle is to iterate from a number of random positions to find the individual optimum and the global optimum. The advantages of the PSO algorithm are high accuracy and fast convergence while the disadvantage is that it is easy to fall into the local optimum.

The velocity vector and position vector of the i-th particle are represented by V_i and X_i, respectively, and the (t+1)-th iteration is formulated by (12) and (13)

$$V_i(t+1)=\omega V_i(t)+c_1 r_1\left(P_{bi}(t)-X_i(t)\right)+c_2 r_2\left(G_b(t)-X_i(t)\right) \tag{12}$$

$$X_i(t+1)=X_i(t)+V_i(t+1) \tag{13}$$

where ω is the inertia weight, c_1 and c_2 are the learning factors, r_1 and r_2 are uniformly distributed random numbers in the range of 0 to 1, P_{bi} is the individual optimal position of the i-th particle, and G_b is the global optimal position.
We use the accuracy of the cross-validation as the fitness of particles.

5. Experiment with Real Data

5.1. Scene Parameters

In this paper, narrowband low-resolution radar operating on C-band is used to identify ship targets longer than 15m in 3-12 kilometres of sea area. The radar parameters are shown in table 1. The classification rules and corresponding data size are shown in table 2.

Table 1. Radar parameters.

Technical specifications	Parameters
Operating frequency band	C-band
Operating frequency range	4-8GHz
Measurement range	0.05km-105km
Scanning bandwidth	4MHz
Distance Resolution	37.5m
Sampling frequency	10MHz
Distance unit	15m

Table 2. Classification parameters.

	Category	Category	Category	Category	Category	Category	Category
Ship	≤30	30-60	60-100	100-140	140-170	170-200	>200
Data Size/pc	394	324	1059	3832	1512	2675	549

5.2. Effect of Different Feature Combinations

By combining different features to compose multidimensional feature vectors, a multi-angle reflection of features of the target can be achieved. The combination of features used in existing studies is usually single in type or small in number, which makes it difficult to achieve good results when the number of classification categories is much larger. In this paper, eight-dimensional feature vectors of area, lateral length, RCS, echo amplitude peak, velocity, image energy, Hu1, and Hu2 are used for target recognition. The feature combinations proposed in this paper are compared with some of the feature combinations used in other existing works of literature (see table 3). Combination 1 is the three-dimensional feature vector of lateral length, RCS, and echo amplitude peak proposed in [12]. Combination 2 is the five-dimensional feature vector of the central moment, skewness, energy, entropy, and Hu2 used in [7].

Table 3. Classification accuracy with different feature combinations.

Features	Combination 1	Combination 2	Combination 3 (proposed)
Accuracy	51.04%	52.73%	90.48%

Combination 1 uses a small number of features and combination 2 uses the grayscale image characteristics without considering typical radar features such as RCS. Thus, in the classification of seven categories by length, both of them have low accuracy. The eight-dimensional feature combination proposed in this paper has the highest accuracy of 90.48%.

5.3. Effect of Different Parameter Optimization Algorithms

By default, SVM classifier parameters are usually set to $\gamma = 1/N$, $C = 1$, where N is the number of classification categories. In this paper, parameter optimization is performed in two ways: grid search and PSO. For grid search, we set the grid to be $\log(\gamma) \in \{-3,-2.5,-2,-1.5,-1,-0.5,0,0.5,1,1.5,2\}$, $\log(C) \in \{0,1,2,3,4,5\}$. For the PSO algorithm, the maximum number of iterations is set to be 40.

From table 4, it can be proved that when using the SVM classifier, the classification accuracy is largely affected by the classifier parameters. The recognition accuracy after optimizing the parameters with either PSO or grid search is improved by more than 20% compared with before. With similar optimization time, using PSO may obtain slightly higher accuracy than grid search.

Table 4. Accuracy with different optimization methods.

Method	No optimization	Grid Search	PSO algorithm
Accuracy	69.00%	89.27%	90.48%

6. Conclusion

In this paper, a ship target classification algorithm based on echo sequence profile image and SVM classifier was proposed for low-resolution radar. The effectiveness of the algorithm was verified by real data. Meanwhile, the importance of the optimization of SVM classifier parameters on the recognition accuracy was verified. The PSO algorithm led to the highest accuracy, slightly higher than the method of the grid search, and much larger than using default parameters.

References

[1] S. Wang and X. Tian, “A decision method for narrowband radar ship target classification,” Radar Science and Technology, vol. 14, no. 002, pp. 159-162, 2016.

[2] Y. Shao, H. Wang, H. Zhang, and H. Chen, “Low-resolution radar target recognition based on waveform features,” Naval electronic countermeasures, no. 04, pp. 62-65, 2015.

[3] Y. Ji, T. Liu, G. Xu, Y. Shi, Y. Zhang, and J. Yang, “A method for ship target recognition based on rcs statistical features,” Ship science and technology, vol. 040, no. 007, pp. 129-132, 2018.

[4] S. Wu, C. Xie, J. Li, Y. Sang, and Z. Liu, “Target detection near island shores based on contour image processing,” Tactical missile technology, vol. No.201, no. 03, pp. 83-90, 2020.

[5] Z. Chen, H. Gu, W. Su, and Z. Wang, “Improved support vector machine low resolution radar target classification algorithm,” systems engineering and electronics, no. 11, pp. 2456-2462, 2017.

[6] Z. Chen and H. Gu and W. Su and Z. Wang, “Low-resolution radar ground target classification based on feature probability distribution,” systems engineering and electronics, vol. 000, no. 002, pp. 274-280, 2016.

[7] C. Wang, M. Kang, and M. Hao, “Research on ship target recognition method based on silhouette image,” modern radar, vol. 34, no. 4, pp. 45-47, 2012.

[8] T. Luo, “Study on feature extraction and classification of low-resolution radar motion targets,” Ph.D. dissertation, Nanjing University of Technology, 2019.

[9] Q. Li, B. Li, W. Hu, C. Zhao, and L. Liu, “Low-resolution radar-based target classification and recognition technology for sea surface ships,” Modern Radar, vol. 34, no. 12, pp. 45-49, 2012.

[10] S. A. Wagner, “Sar atr by a combination of convolutional neural network and support vector machines,” IEEE Transactions on Aerospace and Electronic Systems, vol. 52, no. 6, pp. 2861-2872, 2016.

[11] M. K. Hu, “Visual pattern recognition by moment invariants,” Information Theory, IRE Transactions on, vol. 8, no. 2, pp. 179-187, 1962.

[12] Q. Lin, W. Hu, H. Yu, and L. Zhang, “Research on low resolution radar echo sequence profile image target classification method,” modern radar, vol. 27, no. 3, pp. 24-24, 2005.

A Classification Algorithm: Data Mining and Mathematical Model

R Cheng*, X Kong, M Yu and N Wang

Beijing Polytechnic, Beijing, 100028

*Corresponding author email: tianxiaocheng0925@126.com

Abstract. In this paper, we propose a classification algorithm based on Recency-Frequency-Monetary (RFM) model and K-means data mining method. In addition, the designed algorithm is verified by the experiments on the member data in a large shopping mall. The experiments results show that the proposed algorithm can provide an accurate classification of the members. Finally, some marketing strategies for different classes of members are given according to the classification results.
Keywords: Classification algorithm, RFM model, K-means data mining method.

1. Introduction

Data classification refers to the process of distinguishing and classifying data using certain principles and methods according to its attributes or characteristics, and establishing a certain classification system and order. As a common data mining method, data classification is generally used to improve management ([1],[2]). Therefore, the classification problem has been a popular research topic in recent years, and many studies have proposed various methodologies for the problem ([3]-[5]). Because of the speed and accuracy in classification, k-means is considered as a popular classification method ([6],[7]). For example, based on a real-world customer data of an enterprise, Wu et al. [8], designed a classification strategy according to K-means method; Zhang et al. [9], cluster highly-correlated neighboring samples into multi-dimensional vectors using K-means method; The weighted k-means method is applied to forecast rainfall by identifying related meteorological data in Zhang et al. [10].

In this paper, we proposed a classification algorithm based on k-means method and RFM model. Next, the algorithm is used to make a scientific classification of the mall members as a real-data based case study. In the case study, we first find the proper number of classes of the members. And then, the classification results are compared between the algorithms using different normalization methods. Finally, we find the proper classification algorithm for the mall members. It must be mentioned that the classification method proposed in this paper is not limited to the classification of members, but can also be used for graduate employment analysis and student mental health analysis.

The remainder of this paper is organized as follows. In Section 2, we provide an introduction of k-means algorithm and the relevant normalization methods. In Section 3, a real-data based case study is conducted by using the propose algorithm. In Section 4, the experiment results are provided.

2. Classification Algorithm

K-means is one of the classical clustering algorithms in partition methods [11]. Because of its high efficiency, this algorithm is widely used in clustering large-scale data. This paper designs a classification algorithm for mall members by using k-means clustering method.

Content from this work may be used under the terms of the Creative Commons Attribution 3.0 licence. Any further distribution of this work must maintain attribution to the author(s) and the title of the work, journal citation and DOI.
Published under licence by IOP Publishing Ltd

The basic idea of K-means clustering method is to find K clusters by iteration with the goal of minimizing the loss function corresponding to the clustering result. Here, the loss function can be defined as the sum of distance between each sample and the centre node of the cluster, which can be formulated as

$$F(c,\mu)=\sum_{j=1}^{N}\left\|x_j-\mu_{c_j}\right\|_2 \tag{1}$$

where x_j is the jth sample of the member consumption data, c_j is the cluster to which x_j belongs, μ_{c_j} refers to the centre node of c_j, and N is the number of samples.

The detailed steps of K-means classification algorithm are listed in Table 2.

Table 1. K-means classification algorithm.

1: Selection K samples from the N samples as the centre nodes $\mu_1,\mu_2,\cdots,\mu_K$, set M, the maximum number of iterations, ε, the accuracy requirements, and $D'=\mathrm{Inf}$;
2: While $m<M$ & $D'<\varepsilon$ //Iterate until the maximum number of iterations is reached or the accuracy requirements are met;
3: set $C_i=\varnothing(1\le i\le K)$ //Initially, all clusters are empty;
4: for $j=1:N$ // N is the number of samples;
5: $d_{ji}=\left\|x_j-\mu_i\right\|_2$ //Calculate the distance between x_j and μ_i, i.e., d_{ji};
6: $\pi_j=\arg\min_{i\in\{1,2,\cdots,K\}} d_{ji}$ //Find the cluster closest to x_j, i.e., C_{π_j};
7: $C_{\pi_j}=C_{\pi_j}\bigcup\{x_j\}$ // Classify x_j into the closest cluster C_{π_j};
8: end for
9: for $i=1:M$ //Update centre node of the cluster;
10: $\mu_i'=\left|C_i\right|^{-1}\sum_{x_j\in C_i}x_j$ //generate new centre node of the cluster;
11: if $\mu_i'\ne\mu_i$, then
12: $D=D+\left\|\mu_i'-\mu_i\right\|_2$, $\mu_i=\mu_i'$ // Accumulation the distance between μ_i and μ_i', and update centre node
13: end if
14: end for
15: $D'=D$

K-means is essentially a data classification method based on Euclidean distance measurement. The dimensions with big difference on mean and variance will have a decisive impact on the clustering results of data. Therefore, it is very important to normalization the data (specifically the characteristics of each dimension) before clustering. There are two types of common normalization methods, which are for data with big difference on mean and variance respectively [12].

- **Min-Max normalization**

This method generally maps the data to a specified range to remove the influence of dimension and unit, which is usually for data with big difference on mean, specifically is formulated as,

$$x_i'=(x_i-x_{\min})/(x_{\max}-x_{\min}) \tag{2}$$

According to this method, the data of each dimension is mapped to interval [0,1], where $x_i'=0$ when $x_i=x_{\min}$ and $x_i'=1$ when $x_i=x_{\max}$.

- **Z-score normalization**

The normalization method prefers to the case with data having big difference on variance, which is modelled as

$$x_i' = (x_i - \mu) / \sigma \tag{3}$$

where μ is the mean value and σ is the standard deviation value of the data.

3. Case study: Classification of Members in A Shopping Mall

Here we use the proposed classification algorithm to figure out the members classes in a shopping mall based on classical RFM model. RFM model is commonly used for member clustering, the principle of which is to classify elements with last consumption time (Recency), consumption frequency (Frequency), and consumption amount (Monetary) as indicators. Therefore, we take *Recency*, *Frequency* and *Monetary* as the classification index by applying RFM model in this paper.
Firstly, we pre-process the data to remove the default data. Through statistical analysis, there are great differences in the mean and variance of the data corresponding to the three indicators, so the data need to be normalized before classification. After that, the number of classifications needs to be determined, therefore, the value of loss function $F(c,u)$ with different classification numbers (elbow figure) is generated (see Figure 2).

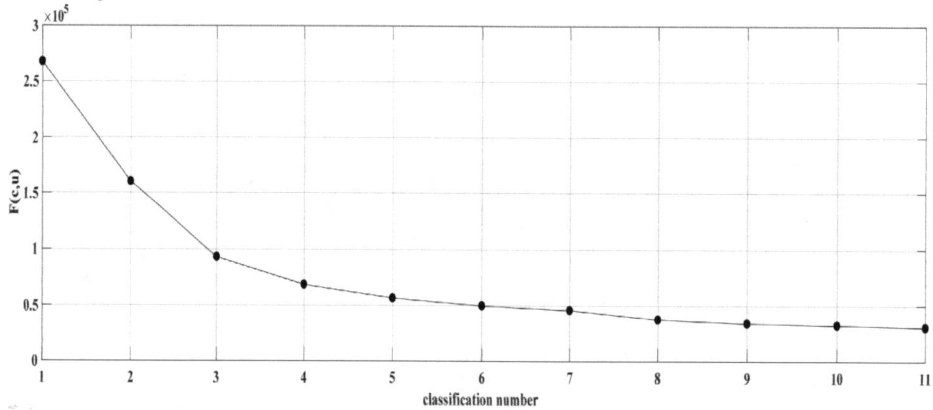

Figure 1. Value of loss function $F(c,u)$ with different classification numbers.

When the cluster number K is less than the real number of classifications, the increase of K will greatly increase the aggregation degree of each cluster, so the decrease of $F(c,u)$ will be large. When K reaches the real number of clusters, the return of aggregation degree obtained by increasing K will decrease rapidly, so the decrease of $F(c,u)$ will decrease sharply, and then tend to be flat with the continuous increase of K , In other words, the figure of $F(c,u)$ and K is the shape of an elbow, and the K value corresponding to this elbow is the real cluster number of the data. As shown in Figure 2, K shall be taken as 5. Therefore, next, the shopping mall members will be divided into five classes.

- Classification results with Min-Max normalization

Firstly, we processing the data by using the Min-Max normalization method, and then obtain the classification results as shown in Figure 3 and the tables (Table 3 and Table 4). Figure 3 shows the class to which each member belongs. And Table 3 lists the center nodes of each class (the values of the nodes are transformed by Min-Max normalization). From the classification results shown in Figure 3 and table 3, it can be seen that variable *Recency* plays a major role in the classification process, while the other two variables *Monetary* and *Frequency* have little impact on the classification results, which can be verified by the analysis of variance (see Table 4). In Table 4, F is the ratio of the sum of squares of deviations between and within groups to the degree of freedom. The greater its value, the greater the proportion of variance between groups, that is, the more significant the corresponding variables have on the classification results. As is shown in the table, variable *Frequency* has a much larger *F* compared with the other two. Consequently, the classification with Min-Max normalization is not appropriate for the member data of the shopping mall.

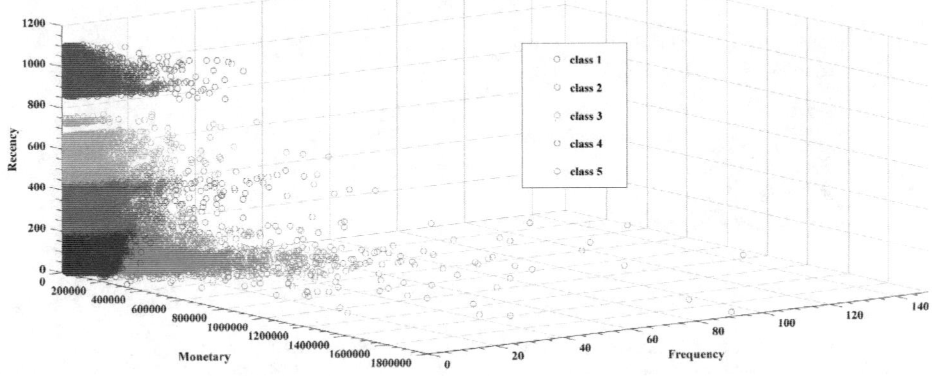

Figure 3. Classification results with Min-Max normalization.

Table 3. Center nodes of Classification results with Min-Max normalization.

Variable	Cluster				
	1	2	3	4	5
Recency	973.5	309.7	573.6	80.8	35.6
Frequency	2.6	3.2	2.7	4.0	23.7
Monetary	6650	9360	6871	11685	104856

Table 4. Analysis of variance for the classification results with Min-Max normalization.

Variable	*F*
Recency	10286
Frequency	605386
Monetary	23279

• Classification results with Z-score normalization

Then, a classification with Z-score normalization method is carried out. And the results of the classification are shown in Figure 4, Table 5 and 6. Though the *F* value of *Frequency* is still the largest, the gap with the other two is no longer obvious, which means that the classification with Z-score normalization method is effective (see Table 6). From Figure 4, we can clearly see to which class each member belongs. Furthermore, Table 5 gives the center nodes of each class. As shown in Figure 4 and Table 5, there are five classes of the members: 1) Class 1 are the members with the highest *Monetary* and *Frequency* but lowest value of *Recency*, which is named premium member. This kind of members do not need special marketing means to stimulate consumption, but only need to recommend some products according to their needs. 2) Members in class 4 with the second higher *Monetary*, *Frequency* and the second lower *Recency* are called key member. 3)Class 5 are formed by members with a medium level of the three variables, which here we call normal member. In order to stimulate the consumption of these two types of members, shopping malls can take some small preferential activities and publicity such as point exchange to improve their consumption level. 4) because of the lower *Monetary*, *Frequency* and the higher *Recency*, class 2 is identified as inactive member. 5) class 3 is regarded as dead member due to the lowest *Monetary*, *Frequency* and the highest *Recency*. As shown in the table, such kind of members have not consumed in the mall for more than two years. Because the cost of activating such members is high and the return is small, activating such members is of little significance.

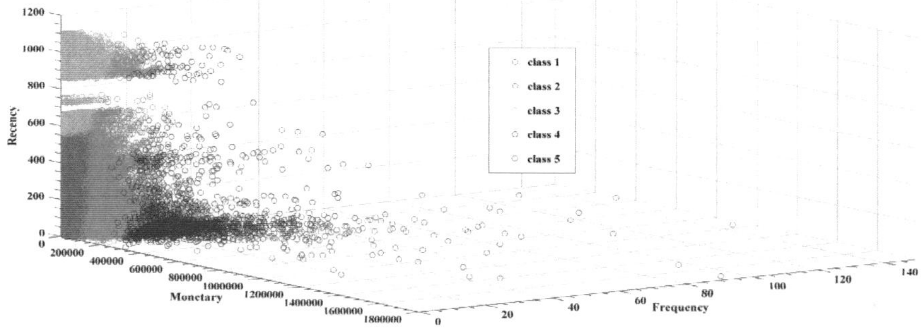

Figure 4. Classification results with Z-score normalization.

Table 5. Center nodes of Classification results with Z-score normalization.

	Cluster				
	1	2	3	4	5
Recency	41	226	885	112	153
Frequency	54.4	2.5	2.2	28.3	12.4
Monetary	523351	6157	5276	142499	39437

Table 6. Analysis of variance for the classification results with Z-score normalization.

Variable	*F*
Recency	64487
Frequency	81641
Monetary	68277

4. Conclusion

With the purpose of improving the membership management of the shopping malls, this paper designed a classification algorithm based on the RFM model and k-means method. By applying the classification algorithm to the case of a shopping mall, the validity of the proposed method is verified. The results show that the members of the mall are divided into 5 classes according to three variable indicators, which are called premium member, key member, normal member, inactive member and dead member respectively. Membership classification can provide theoretical support for managers to formulate targeted membership management strategies. For example, the members most in need of promotion are the *normal members* who have the moderate consumption level in the mall. In contrast, *premium members* with the highest level of consumption do not need to spend too much effort on management. For *inactive members*, a series of promotional offers are required to activate them. In addition, removing "*dead members*" while developing new members.

Furthermore, the application of the algorithm proposed in this paper is not limited to the classification of shopping mall members, and can also be used in other cases. For example, this classification method can be used for mental health assessment of college students. Moreover, analysis of graduate employment can also be carried out by using this classification algorithm.

Acknowledgments
This research is supported by Foundation Sciences of Beijing Polytechnic under grant 2020Z109-KXZ.

References

[1] Guan J, Tang C, Ou J 2019 The portrait depiction of the market members based on data mining *International Journal of Pattern Recognition and Artificial Intelligence* 34 p 2059024

[2] Deng M 2019 Analysis of "member portraying for a large department store" *International Mathematical Modeling and Its Application* 8 pp 66-70

[3] Adeniyi DA, Wei Z, Yang Y 2016 Automated web usage data mining and recommendation system using k-nearest neighbor (KNN) classification method *Applied Computing and Informatics* 12 pp 90-108

[4] Dong X, Qian L, Guan Y, Huang L, Yu Q, Yang J 2016 A multiclass classification method based on deep learning for named entity recognition in electronic medical records *New York Scientific Data Summit (NYSDS)* pp 1-10

[5] Wang A, Xue D, Wu H, Iwahori Y 2020 LiDAR data classification based on improved conditional generative adversarial networks *IEEE Access* 8 pp 209674-209686

[6] Shahri A, Hosseini M, Phalp K, Taylor J, Ali R 2019 How to engineer gamification *Journal of Organizational and End User Computing* 31 pp 39 - 60

[7] Namasudra S, Roy P 2018 PpBAC: popularity based access control model for cloud computing *Journal of Organizational and End User Computing* 30 pp 14 - 31

[8] Wu J, Shi L, Lin W, Tsai SB, Xu G 2020 An empirical study on customer segmentation by purchase behaviors using a RFM model and k-means algorithm *Mathematical Problems in Engineering* 6 pp 1-7

[9] Lu Z, Pang X, Ozolins O, Udalcovs A, Chen J 2018 K-means clustering based multi-dimensional quantization scheme for digital mobile fronthaul *Optical Fiber Communication Conference* pp 1 - 3

[10] Zhang P, Cao W, Li W 2020 Surface and high-altitude combined rainfall forecasting using convolutional neural network *Peer-to-Peer Networking and Applications* pp 1-13

[11] Li H and Wang S 2019 Modeling and verification of mall member portraits based on RFM model *Journal of Zhengzhou Railway Vocational and technical College* 3 pp 14-16

[12] Henderi H 2021 Comparison of min-max normalization and z-score normalization in the K-nearest neighbor (KNN) algorithm to test the accuracy of types of breast cancer *International Journal of Informatics and Information Systems* 4 pp 13-20

Numerical Modeling of A New Virtual Trajectory Password Architecture

Yuanyuan Chen[1,*], Jing Chen[2] and Mingzhu Li[3]

[1,3]Haikou University of Economics, Haikou, China
[2]Qingdao City University, Qingdao, China

[*]Corresponding author email: 252009674@qq.com

Abstract. With the development of digital technology, Computer technology, communication technology and multimedia technology gradually infiltrate into each other and become the main core of information technology. In information technology, digital, text, graphics, images, sound, video, animation and other information carriers are spread through computers and the Internet. Due to the openness, sharing, dynamic and other characteristics of the Internet, information security is threatened and interfered. Information security has become a strategic issue that people must pay attention to, which is related to social stability, economic development and national security. This paper studies a new kind of virtual track cryptography, which breaks through the traditional character memory method and transforms it into memory graphics by using mathematical modeling, and changes the surface memory method from memorizing numbers and characters to memorizing trajectory graphics, so as to make the traditional process of people memorizing passwords easier.
Keywords: Virtual trajectory, password, security, algorithm, numerical modelling.

1. Introduction
In the era of rapid development of the Internet, it is quite common to use mobile phones for online shopping, information inquiry, bank transfer, account recharge and other services. When users register and log in, they need to register account password for information security, that is, use password for security verification. Password is a tool to protect one's own privacy, property, information and important data. However, some criminals break the password, peep and steal other people's important information and data through technical means, causing data and property losses [1]. Due to the openness, sharing and dynamic characteristics of the Internet, information security is threatened and interfered[2] .Information security has become a strategic issue that people must pay attention to, which is related to social stability, economic development and national security. At the same time, in the verification operation through the mobile phone, it is often required to remember the passwords of various operations. There are many kinds of passwords, which are easy to cause memory confusion [3-4]. In view of this, people urgently need a password authentication method which can help memorize passwords, prevent peeping, and is difficult to decipher and protect password information security.

2. The Research Significance
There are several factors to develop this kind of virtual trajectory password.

2.1 The First Factor
At present, all kinds of widely used software passwords often use short 6-digit password (such as bank password) or the so-called security password with letters and numbers as the verification and login

Content from this work may be used under the terms of the Creative Commons Attribution 3.0 licence. Any further distribution of this work must maintain attribution to the author(s) and the title of the work, journal citation and DOI.
Published under licence by IOP Publishing Ltd

password[5]. This kind of short password is also vulnerable to the computing power of modern ordinary computers, and the super long password is not easy to remember.

2.2 The Second Factor
At present, in order to make the password not easy to be cracked illegally, people use many modern technology methods, such as displacement keyboard, "two-dimensional code" scanning verification, third-party platform verification, fingerprint verification, mobile phone verification code and so on[6]. But in the process of data storage and network transmission, it is inevitable to be illegally monitored, intercepted and stolen.

2.3 The Third Factor
With the rapid development of China's economy, network information security is an inescapable topic, and password security is the first problem. To develop a new password verification system and to simplify the long password memory method is in line with the current needs of China's national conditions.
In this paper, a new architecture of " virtual trajectory password" is studied, which breaks through the traditional method of memorizing characters, changes it into memorizing graphics, and automatically converts it into actual character cipher by our self-designed system, and changes the familiar surface memory method from memorizing numbers and characters to memorizing graphic trajectory, So that the process of people memorizing passwords becomes easier.

3. The Architecture Design
"New virtual trajectory password", hereinafter referred to as virtual trajectory password, provides a new method for people in password verification login means. It can help people remember and use complex super long password, and has the characteristics of preventing peeping and being difficult to decipher. The specific design is as follows.

3.1 Design of Encoder
In order to facilitate the user to remember the password, it is convenient for the user to remember the simple "input seed code" by generating a constantly changing random reference graph - "code disk". This seed code is a constant graphic code during the user's input and verification, and this simple seed code, is the lead that the system uses to generate the actual, extra-long, encrypted, "code base" covering the entire ASCII code table, which is the actual transmission and storage password.
It has designed a 36-digit alphanumeric figure with 6 rows and 6 columns. We call it the "code disk ", and the user's virtual trajectory password is to use the code salver as a reference graph to facilitate memory password. The graph of code salver is shown in Figure. 1.

Figure 1. A random code salver diagram.

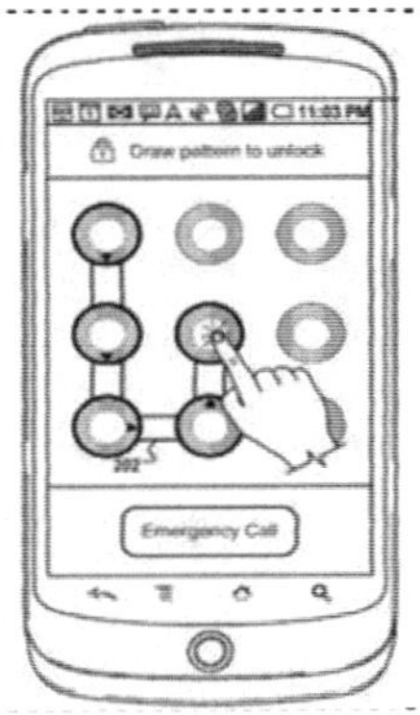

Figure 2. The diagram of mobile phone dial unlocking.

3.2 The Design Thought of Code Disk

In real life, the passwords that most people use are basically relatively simple, short passwords with definite meaning, this is because people overestimate the difficulty of hacking and decoding their passwords and underestimate the risk that their own passwords will be compromised. Therefore, the following aspects are not considered when using the password: ① the composition of the elements of the password (code base), and the use of a simple number type of password in the majority. ② Password length (number of digits) 6-8 passwords in the majority. Random scrambled passwords and passwords with more than eight digits are rare because they are hard to remember. However, these are also relatively secure passwords.

Inspired by the mobile phone dial unlocking method (as shown in Figure. 2), it is noted that the password is graphics. As we all know, it is easier for people to memorize specific graphics than abstract numbers and characters. However, this method is not very private when it is used, and it is easy to be peeped, and the number of its code base is only 9, and the generated password is not difficult to crack. Can we design a trajectory code that is easy to remember and can't be seen by others? The answer is yes. The design of the virtual trajectory password method, that is to solve the password "easy to remember" and "super long" contradiction.

Instead of memorizing a character password, people can simply remember a "virtual trajectory" on a code disk. When the system is running, click the password input box, and the system will automatically pop up the "code disk".

3.3 The Generation of "Code disk"

A 6 by 6 code disk is a two-dimensional matrix, and we define a two-dimensional array to describe it. Each array element can appear randomly in (0-36) code bases. The following is the instantaneous data of a random code disk, as shown in Figure.3.

Scd(1,1)=4,Scd(1,2)=t,Scd(1,3)=j,Scd(1,4)=0,Scd(1,5)=q,Scd(1,6)=6,Scd(2,1)=y,Scd(2,2)=b,Scd(2,3)=x,Scd(2,4)=m,Scd(2,5)=3,Scd(2,6)=1,Scd(3,1)=i,Scd(3,2)=n,Scd(3,3)=7,Scd(3,4)=h,Scd(3,5)=l,Scd(3,6)=a,Scd(4,1)=v,Scd(4,2)=c,Scd(4,3)=f,Scd(4,4)=u,Scd(4,5)=k,Scd(4,6)=g,Scd(5,1)=8,Scd(5,2)=9,Scd(52,3)=p,Scd(5,4)=z,Scd(5,5)=o,Scd(5,6)=5,Scd(6,1)=s,Scd(6,2)=w,Scd(6,3)=d,Scd(6,4)=2,Scd(6,5)=e,Scd(6,6)=r.

$$\mathrm{Scd} = \begin{pmatrix} 4 & t & j & 0 & q & 6 \\ y & b & x & m & 3 & 1 \\ i & n & 7 & h & l & a \\ v & c & f & u & k & g \\ 8 & 9 & p & z & o & 5 \\ s & w & d & 2 & e & r \end{pmatrix}$$

Figure 3. Two dimensional matrix of code disk elements.

Figure 4. The yellow is the virtual track of user memory.

The code disk is generated according to the continuous elements (code base) according to the randomly generated position, so as to ensure that all code base elements are placed on all code disks, and there are no duplicate elements, so as to ensure that when the selected password trajectory overlaps on the code disk, the system can still correctly distinguish without ambiguity path.

3.4 Two Dimensional Matrix Encoder and One Dimensional Matrix Encoder

Two dimensional matrix: the elements of the matrix are two-dimensional vectors, that is, two-dimensional matrix. We usually see code disks, such as 3 * 3 two-dimensional code disks on mobile phones. As shown in the figure below, the code disk contains 9 code bases.

$$A=\begin{pmatrix} A(1,1) & A(1,2) & A(1,3) \\ A(2,1) & A(2,2) & A(2,3) \\ A(3,1) & A(3,2) & A(3,3) \end{pmatrix}$$

One dimensional matrix: the elements of the matrix are one-dimensional vectors, that is, one-dimensional matrix. We can convert the two-dimensional matrix into one-dimensional matrix. For example, the above two-dimensional matrix into one-dimensional matrix can be changed into:

$$A=\left[A(1,1),A(1,2),A(1,3),A(2,1),A(2,2),A(2,3),A(3,1),A(3,2),A(3,3) \right]$$

It can be transformed into:

$$A=a(1),a(2),a(3),a(4),a(5),a(6),a(7),a(8),a(9)$$

3.5 The Formation of Virtual Trajectory

As we can see from a random code disk in Figure. 4, each element in the array is the character displayed on the code disk. If the user selects an element and inputs it, the system will get its position on the code disk. When continuous or discontinuous points are selected, there will be a certain position trajectory. These selected positions are combined into a continuous (or discontinuous) position sequence to obtain its motion trajectory on the code disk, which is called "virtual trajectory".

4. The Generation of Actual Password and the Principle of Encryption

The password finally generated in the system is not stored in the form of the location of the trajectory pattern matrix, but only as a "seed", which is stored in the form of the password sequence generated by a series of mathematical transformations. When the track password is obtained by the system, it becomes a super-long password which covers the storage and transmission of the entire ASCII code table after encryption and base expansion.

We know that when the password is long enough and the code base is large enough, it is difficult to decipher (High timeliness cost) even if it is intercepted by others .This greatly enhances the security of the password. At the beginning of the design of "virtual track password", there is no limit on the number of digits of the password, which is also the case in the user's practical operation to realize the design idea. The walking track of the virtual track password is allowed to walk at will on the code disk continuously or discontinuously, cross or not. In order to enhance the security and reliability of virtual trajectory cryptography, we adopt the basic theory of chaotic cryptography; The excellent characteristics such as randomness, ergodicity, certainty and sensitivity to initial conditions are reflected in the chaotic characteristics in the range of its value distribution, and this chaotic characteristic just provides a guarantee for us to generate a super secure cipher sequence.

Logistic chaotic model (also known as insect-population model) was originally a mathematical model to describe the reproduction of biological population in ecology, but later evolved into a mathematical model to study random phenomena. See formula 1.

$$x_{n+1} = ux_n(1 - x_n) \tag{1}$$

The letter n is a natural number (1,2,3...) , when X1 is fixed, the initial value is selected, the value of u is used to control the change of the whole function value, and an infinite repeatable random number sequence can be obtained through infinite iterations.

By observing Figure. 5, in the chaotic interval, we can see that a large number of initial values of real numbers can be selected differentially when u is between real numbers [1,4]. When the initial value is selected, a large number of random numbers can also be obtained when u is between 3.8-4.0 (except for some regions).

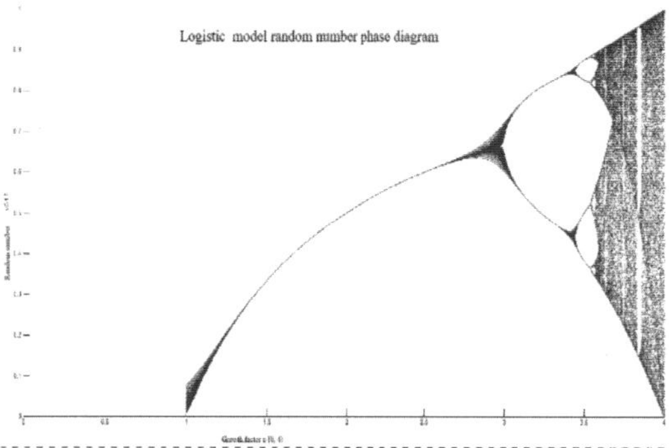

Figure 5. Logistic chaotic model.

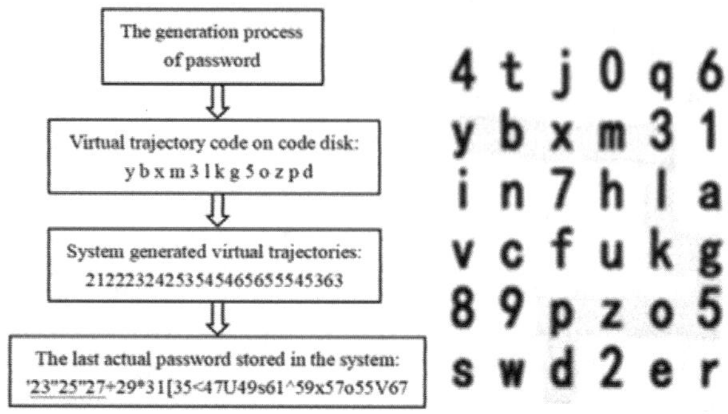

Figure 6. The diagram of an unlimited password generation process.

The generation of virtual trajectory password is the password sequence generated by using the chaotic sequence generated by the chaotic system and the reproducible random number generated by the user's seed trajectory code (initial value) after the selected formula transformation in the client, as shown in Figure. 6.

In password verification, the same seed can be used to regenerate the previous password sequence through the chaotic system, and the same password stream can be generated through the same reverse process to verify the user; the seed cipher sequence input by the user is not transmitted, but the super long generated cipher is transmitted. The encryption end and decryption end of the chaotic cipher system are two independent and identical chaotic systems.

The chaotic stream encryption algorithm uses the perturbation algorithm based on the chaotic sequence itself to overcome the problem caused by its limited precision effect. At the same time, the coding algorithm and nonlinear transformation are used to further improve the randomness of the chaotic system and the output sequence. Theoretically, the period of the random sequence generated by the algorithm can reach at least 128^{600}.

5. Conclusion

This paper designs a new architecture of virtual trajectory password, which breaks through the traditional method of memorizing characters and changes it into memorizing graphics. By generating a constantly changing random reference graph "code disk" as a means of auxiliary memory, it helps human beings remember some special extended passwords that cannot be memorized routinely. In the aspect of password verification and login, it provides a new method for people, which makes it easier for people to memorize the traditional process of password and complete it by memorizing the trajectory graph.

Acknowledgments

This research was supported by Hainan Provincial Natural Science Foundation of China (620QN285), Haikou University of Economics Scientific Research Project (HJKY (ZD) 20-10).

References

[1] Tsai, Jia-Lun; Wu, Tzong-Chen; Tsai, Kuo-Yu. New dynamic ID authentication scheme using smart cardsInternational Journal of Communication Systems, 2010

[2] Xie, Qi. A new authenticated key agreement for session initiation protocol International Journal of Communication Systems, 2012.

[3] S.Kumari; M.K.Gupta; M.Kumar. "Cryptanalysis and security enhancement of Chen et al. "s remote user authentication scheme using smart card," Central European Journal of Computer Science, 2012.

[4] Chen, Chun;;He, Daojing; Chan, Sammy;;Bu, Jiajun;;Gao, Yi;;Fan, Rong. Lightweight and provably secure user authentication with anonymity for the global mobility network International Journal of Communication Systems, 2011.

[5] Cuiping Ma, Jiahui Liu.” Design and Implementation of a Hash Algorithm Based on Chaotic Logistic Equation”. Harbin University of Science and Technology.
[6] Li Zhengtao. User Identity Authentication Technology for Mobile Social Network Applications, Nanjing University of Posts and Telecommunications, 2017.

The Uniqueness of Solution to the Inverse Eigenvalue Problem for an Arrow-shaped Generalised Jacobi Matrix

Hongliang Huang, Qike Wang, Zhibin Li and Lidong Wang*

School of Public Basic and Applied Statistics, Zhuhai College of Science and Technology, Zhuhai, Guangdong

*Corresponding author email: wld0707@163.com

Abstract. This paper studies the inverse eigenvalue problem for an arrow-shaped generalised Jacobi matrix, inverting matrices through two eigen-pairs. In the paper, the existence and uniqueness of the solution to the problem are discussed, and mathematical expressions as well as a numerical example are given. Finally, the uniqueness theorem of its matrix is established by mathematical derivation.

Keywords: Arrow-shaped generalised Jacobi matrix, inverse eigenvalue problem, uniqueness theorem.

1. Introduction

The forward problem, in a system with both input and output, is to yield the output with a known input, whereas the inverse problem looks for the input with a given output[1].

The inverse eigenvalue problem of matrices[1, 2] has important applications in a lot of disciplines and fields, such as automatic control, system parameter identification, structural design and so on[3]. In vibration mechanics, for example, the frequency of a discrete system corresponds to the eigenvalues of matrices, and the modal corresponds to eigenvectors. In this way, the question of how we find out the original vibration system with given frequencies and modals is thus transformed into the matrix inverse problem and solved.

This paper investigates the inverse eigenvalue problem of an arrow-shaped generalised Jacobi matrix, clarifies the existence and uniqueness of the solution, and has it verified by a numerical example.

2. The Question

Let arrow shaped generalized Jacobi matrix[4, 5]

$$J=\begin{pmatrix} a_1 & b_1 & & & b_1 \\ c_1 & a_2 & b_2 & & b_2 \\ & \ddots & \ddots & \ddots & \vdots \\ & & c_{n-2} & a_{n-1} & b_{n-1} \\ c_1 & c_2 & \cdots & c_{n-1} & a_n \end{pmatrix}$$

k and l are known constants, $c_i = kb_i + l(i = 1,2,\cdots,n-2)$.

Problem A: Given two distinct real numbers λ,μ and two dimensions of non-zero vectors $x=(x_1,x_2,\cdots,x_n)^{\mathrm{T}}$, $y=(y_1,y_2,\cdots,y_n)^{\mathrm{T}}$ to find N-order arrow shaped generalized Jacobi matrix J. Let $Jx=\lambda x$, $Jy=\mu y$.

Content from this work may be used under the terms of the Creative Commons Attribution 3.0 licence. Any further distribution of this work must maintain attribution to the author(s) and the title of the work, journal citation and DOI.
Published under licence by IOP Publishing Ltd

3. Existence and Uniqueness of Solution

3.1 Set up Equations
Let

$$x_0 = x_{n+1} = y_0 = y_{n+1} = c_0 = c_n = b_0 = b_n = 0 \tag{1}$$

$$E_i = \begin{vmatrix} x_i & y_i \\ x_n & y_n \end{vmatrix} (i = 1,2,\cdots,n-1) \tag{2}$$

$$D_i = \begin{vmatrix} x_i & y_i \\ x_{i+1} & y_{i+1} \end{vmatrix} (i = 1,2,\cdots,n-2) \tag{3}$$

$$d_i = \sum_{j=1}^{i} \left(\prod_{s=j}^{i-1} \frac{kD_s}{D_s + E_s} \right) x_j y_j \left(i = 1,2,...,n-2\right) \tag{4}$$

$$e_i = \sum_{j=1}^{i-1} \left(\prod_{s=j+1}^{i-1} \frac{kD_s}{D_s + E_s} \right) D_j \left(i = 2,3...,n-2\right) \tag{5}$$

where $e_1 = 0$.

Let (λ, x) and (μ, y) be two characteristic pairs of J, and the equations are obtained.

$$c_{i-1}x_{i-1} + a_i x_i + b_i x_{i+1} + b_i x_n = \lambda x_i (i = 1,2,\cdots,n-2) \tag{6}$$

$$c_{n-2}x_{n-2} + a_{n-1}x_{n-1} + b_{n-1}x_n = \lambda x_{n-1} (i = n-1) \tag{7}$$

$$\sum_{i=1}^{n-1} c_i x_i + a_n x_n = \lambda x_n (i = n) \tag{8}$$

$$c_{i-1}y_{i-1} + a_i y_i + b_i y_{i+1} + b_i y_n = \mu y_i (i = 1,2,\cdots,n-2) \tag{9}$$

$$c_{n-2}y_{n-2} + a_{n-1}y_{n-1} + b_{n-1}y_n = \mu y_{n-1} (i = n-1) \tag{10}$$

$$\sum_{i=1}^{n-1} c_i y_i + a_n y_n = \mu y_n (i = n) \ . \tag{11}$$

3.2 Solution When $i = 1,2,\cdots,n-2$

When $i = 1,2,\cdots,n-2$, according to equations (6) and (9), it is obvious that

$$b_i (x_i y_{i+1} - x_{i+1} y_i + x_i y_n - x_n y_i) = (\mu - \lambda) x_i y_i + c_{i-1}(x_{i-1} y_i - x_i y_{i-1}) \tag{12}$$

According to equation (2) and equation (3), formula (12) is sorted out

$$b_i (D_i + E_i) = (\mu - \lambda) x_i y_i + kb_{i-1} D_{i-1} + lD_{i-1} \tag{13}$$

According to the recurrence law, it is integrated with equation (4) and equation (5).

$$b_i (D_i + E_i) = (\mu - \lambda) d_i + le_i (i = 1,2,\cdots,n-2) \ . \tag{14}$$

3.3 Solution when $i = n-1$

When $i = n-1$, according to formula (7) and formula (10), we got the following formula.

$$b_{n-1}(x_{n-1} y_n - x_n y_{n-1}) = (\mu - \lambda) x_{n-1} y_{n-1} + c_{n-2}(x_{n-2} y_{n-1} - x_{n-1} y_{n-2}) \tag{15}$$

Namely

$$b_{n-1} E_{n-1} = (\mu - \lambda) x_{n-1} y_{n-1} + c_{n-2} D_{n-2} \ . \tag{16}$$

3.4 Solution when $i = n$

When $i = n$, there was a difference between equation (8) and equation (11).

$$c_{n-1}(x_n y_{n-1} - x_{n-1} y_n) = (\mu - \lambda) x_n y_n + (y_n \sum_{i=1}^{n-2} c_i x_i - x_n \sum_{i=1}^{n-2} c_i y_i) \tag{17}$$

The following formula is obtained by formula (2).

$$-E_{n-1} c_{n-1} = (\mu - \lambda) x_n y_n + (y_n \sum_{i=1}^{n-2} c_i x_i - x_n \sum_{i=1}^{n-2} c_i y_i) \tag{18}$$

Next, the uniqueness theory of arrow shaped generalized Jacobi matrix is proved by iterative method[6].

3.5 Uniqueness Theorem of Arrow-shaped Generalized Jacobi Matrix

Theorem Let λ and μ be two different real numbers, x , y are two dimensional non-zero real vectors, and k and l be two known real constants. If the following conditions are met:

(1) $D_i + E_i \neq 0 (i = 1, 2, \cdots, n-2)$; (2) $D_i \neq 0 (i = 1, 2, ..., n-1)$;

(3) $c_i = kb_i + l (i = 1, 2, \cdots, n-1)$; (4) $E_{n-1} \neq 0$

Then J of problem A is unique and has the following result.

$$b_i = \frac{(\mu - \lambda) d_i + l e_i}{D_i + E_i}, c_i = kb_i + l (i = 1, 2, \cdots, n-2), \tag{19}$$

$$a_i = \begin{cases} \lambda - \dfrac{c_{i-1} x_{i-1} + b_i (x_{i+1} + x_n)}{x_i}, x_i \neq 0 \\ \mu - \dfrac{c_{i-1} y_{i-1} + b_i (y_{i+1} + y_n)}{y_i}, y_i \neq 0 \end{cases}, (i = 1, 2, \cdots, n-2). \tag{20}$$

$$b_{n-1} = \frac{(\mu - \lambda) x_{n-1} y_{n-1} + c_{n-2} D_{n-2}}{E_{n-1}}, \tag{21}$$

$$a_{n-1} = \begin{cases} \lambda - \dfrac{b_{n-1} x_n}{x_{n-1}} - c_{n-2} \dfrac{x_{n-2}}{x_{n-1}}, x_{n-1} \neq 0, \\ \mu - \dfrac{b_{n-1} y_n}{y_{n-1}} - c_{n-2} \dfrac{y_{n-2}}{y_{n-1}}, y_{n-1} \neq 0. \end{cases} \tag{22}$$

$$c_{n-1} = \frac{(\mu - \lambda) x_n y_n + (y_n \sum_{i=1}^{n-2} c_i x_i - x_n \sum_{i=1}^{n-2} c_i y_i)}{-E_{n-1}}, \tag{23}$$

$$a_n = \begin{cases} \lambda - \dfrac{\sum_{i=1}^{n-1} c_i x_i}{x_n}, x_n \neq 0, \\ \mu - \dfrac{\sum_{i=1}^{n-1} c_i y_i}{y_n}, y_n \neq 0. \end{cases} \tag{24}$$

Proof:

When $i = 1, 2, \cdots, n-2$, The result is obtained from a $D_i + E_i \neq 0 (i = 1, 2, \cdots, n-2)$ and equation (14), that is $b_i = \dfrac{(\mu - \lambda) x_i y_i + kb_{i-1} D_{i-1} + l D_{i-1}}{D_i + E_i} (i = 1, 2, \cdots, n-2)$.

Then $c_i = kb_i + l (i = 1, 2, \cdots, n-2)$ can be obtained by condition (3). According to $D_i \neq 0$, We know that x_i and y_i cannot be 0 at the same time. According to equation (6), when $x_i \neq 0$, it can be concluded that $a_i = \lambda - \frac{c_{i-1}x_{i-1} + b_i(x_{i+1} + x_n)}{x_i} (i = 1, 2, \cdots, n-2)$. When $y_i \neq 0$, it can be concluded that $a_i = \mu - \frac{c_{i-1}y_{i-1} + b_i(y_{i+1} + y_n)}{y_i} (i = 1, 2, \cdots, n-2)$ by equation (9).

When $i = n-1$, From equation $E_{n-1} \neq 0$, we can get $b_{n-1} = \frac{(\mu - \lambda)x_{n-1}y_{n-1} + c_{n-2}D_{n-2}}{E_{n-1}}$ by equation(6). x_{n-1} and y_{n-1} can't be zero at the same time. If $x_{n-1} \neq 0$, we can get $a_{n-1} = \lambda - \frac{b_{n-1}x_n}{x_{n-1}} - c_{n-2}\frac{x_{n-2}}{x_{n-1}}$ by equation(7). If $y_{n-1} \neq 0$, we can get $a_{n-1} = \mu - \frac{b_{n-1}y_n}{y_{n-1}} - c_{n-2}\frac{y_{n-2}}{y_{n-1}}$ by equation(10).

When $i = n$, From condition $E_{n-1} \neq 0$ and equation (18), equation (23) can be obtained. Because of $D_{n-1} \neq 0$, x_n, y_n can't be zero at the same time. If $x_n \neq 0$, substituting c_{n-1} into equation (8), we can get $a_n = \lambda - \frac{\sum_{i=1}^{n-1} c_i x_i}{x_n}$. If $y_n \neq 0$, substituting c_{n-1} into equation (11), we can get $a_n = \mu - \frac{\sum_{i=1}^{n-1} c_i y_i}{y_n}$.

Prove up.

4. Numerical Example

Example Given $\lambda = 1, \mu = 2, x = (1, -1, 0, 1)^T, y = (1, 1, 1, 1)^T, \quad k = -1, l = 3$.

According to the title, it can be easily obtained.

$$E_1 = 0,\ D_1 = 2 \neq 0,\ D_1 + E_1 = 2 \neq 0,\ E_2 = -2,\ D_2 = -1 \neq 0,\ D_2 + E_2 = -3 \neq 0,\ D_3 = -1 \neq 0,$$
$$e_1 = 0,\ e_2 = D_1 = 2,\ d_1 = x_1y_1 = 1,\ d_2 = -2,\ E_{n-1} = E_3 = -1 \neq 0.$$

Then, according to the theorem, the matrix has a unique solution.

When $i = 1, 2$

$$b_1 = \frac{(\mu - \lambda)d_1 + le_1}{D_1 + E_1} = \frac{1}{2},\ c_1 = kb_1 + l = (-1) \times \frac{1}{2} + 3 = \frac{5}{2},\ a_1 = 1.$$
$$b_2 = \frac{(\mu - \lambda)d_2 + le_2}{D_2 + E_2} = -\frac{4}{3},\ c_2 = kb_2 + l = (-1) \times -\frac{4}{3} + 3 = \frac{13}{3},\ a_2 = \frac{13}{6}.$$

When $i = 3$

$$b_3 = \frac{(\mu - \lambda)x_3y_3 + c_2D_2}{E_3} = \frac{13}{3},\ a_3 = -\frac{20}{3}.$$

When $i = 4$

$$c_3 = -\frac{23}{3},\ a_4 = \frac{17}{6}.$$

Therefore

$$J=\begin{pmatrix} 1 & \frac{1}{2} & 0 & \frac{1}{2} \\ \frac{5}{2} & \frac{13}{6} & -\frac{4}{3} & -\frac{4}{3} \\ 0 & \frac{13}{3} & -\frac{20}{3} & \frac{13}{3} \\ \frac{5}{2} & \frac{13}{3} & -\frac{23}{3} & \frac{17}{6} \end{pmatrix}$$

And the following formula holds.

$$Jx=\lambda x, Jy=\mu y.$$

5. Conclusion

By constructing an arrow shaped generalized Jacobi matrix, we solved the inverse problem of its eigenvalues[7], and at the same time studied the uniqueness theorem of matrix. Finally, we proved the theorem through a numerical case.

Acknowledgements
This work was supported by the <Key Natural Science Foundation of Universities in Guangdong Province> under Grant<number2019KZDXM027>.

References

[1] Xinxin Zhao. Inverse Eigenvalue Problem of Generalized Periodic Jacobi Matrix [D]. Dalian Jiaotong University, 2010.
[2] Zhibin Li, Xinxin Zhao, Wei Li. Inverse Eigenvalue Problem for Generalized Jacobi Matrix [J]. Journal of Dalian Jiaotong University, 2008(04):6-10.
[3] Harriman S. Application background of positive and negative eigenvalue problems for matrices [J]. Journal of Xinjiang University (Natural Science Edition), 1997(04):26-35.
[4] Cong Bu. Inverse Eigenvalue Problem of Arrow Matrix [D]. Dalian Jiaotong University, 2012.
[5] Yinfeng Xu. Inverse eigenvalue problems for a special class of matrices [J]. Applied Mathematics, 1993(01):68-75.
[6] Minghui ang. Interval consensus problem of multi-agent systems in accordance with switching protocol [J]. International Journal of Systems Science. 2014(47). 1-11.
[7] Xing Tao Wang, Mei Ling Jin. Inverse Eigenvalue Problem for Quasi-tridiagonal Matrices [J].Bulletin of the Iranian Mathematical Society, 2019(45):1697–1712.

Location Planning of Field Ammunition Depot for Multi-stage Supply Based on Dijstra Algorithm

Xianming Shi, Haobang Liu*, Yuan Li, Bo Zhu and Junpeng Liang

Shijiazhuang Campus, Army Engineering University, Shijiazhuang, China

*Corresponding author email: liuwuliu_2020@tju.edu.cn

Abstract. The location of field ammunition depots is related to the efficiency and effectiveness of ammunition supply and support, and is affected by the changing needs of different stages of operations. In response to complex situations such as damage to transportation roads, repairs, and dynamic changes in demand in the course of combat, the fastest supply speed is the primary goal, while the satisfaction rate, balance, and cost factors of the ammunition supply at each demand point are taken into account to build a multi-stage supply Site selection planning model for field ammunition depot. Use Dijstra algorithm's hierarchical sequence method to determine the shortest supply time, gradually increase the supply time, find the relationship between supply time, satisfaction rate, balance and cost, and find a satisfactory location plan. Finally, the simulation case calculations show that the model can cover some key dynamic changes during wartime, with strong adaptability and more scientific site selection.

1. Introduction

The field ammunition depot is a strong support to ensure the continuous supply of ammunition in wartime, and its setting must consider many factors. In the course of combat, there are often complicated factors such as the enemy's damage to transportation and the dynamic changes in our ammunition demand, which have a greater impact on the punctual supply of ammunition. As an important part of the rapid ammunition supply system, the field ammunition warehouse is responsible for ammunition storage under wartime conditions, and can provide fast and accurate ammunition support for combat troops [1]. It plays a key role in the process of combat. The location and construction of field ammunition depots greatly affect the level of rapid ammunition support and quality assurance. The method of site selection for field ammunition depots and the method of forecasting ammunition demand are both important parts of ammunition guarantee theory, and both greatly improve the on-time supply capability of ammunition [2-3]. In this regard, it is urgent to find a rapid location method for field ammunition depots suitable for multiple stages of the combat process.

The role of the field ammunition depot is similar to that of the distribution center in local logistics, but compared to the distribution center, the field ammunition depot is more flexible and will adjust and change with the battlefield situation. The scientific nature of the location of the distribution center in local logistics can effectively improve the on-time supply of materials. For example, many automobile manufacturers pay more attention to the research on the location of auto parts distribution centers in order to improve the level of parts supply [4]. In addition, many methods have emerged in the research on the location of distribution centers. The more classic location problem is solved by lingo, genetic algorithm, and particle swarm algorithm [5-7]. At present, many more advanced location selection methods have been developed, such as improved immune algorithm, cuckoo search algorithm, etc. [8-9]. However, these methods are mainly aimed at the site selection research under the single-stage

Content from this work may be used under the terms of the Creative Commons Attribution 3.0 licence. Any further distribution of this work must maintain attribution to the author(s) and the title of the work, journal citation and DOI.
Published under licence by IOP Publishing Ltd

situation, and are not applicable to the site selection problem of field ammunition depots under multi-stage and changing conditions. Regarding the location of the emergency material distribution center, although changes in demand have been considered, the relationship between supply time, cost, and satisfaction rate has not been explored [10-12].

Aiming at the problem of field ammunition selection under multi-stage changes, this paper uses the layered sequence method coupled with dijstra algorithm to solve the shortest supply time. The labeling method is used to determine the fastest path from each ammunition demand site to the field ammunition depot and the location of the field ammunition depot. Then adjust the supply time in turn, select the location of the field ammunition depot, explore the relationship between supply time, satisfaction rate, balance and cost, and find a satisfactory location plan.

2. Multi-stage Supply Model of Field Ammunition Location Selection

2.1. Problem Background Description

During the course of the battle, due to changes in the battlefield conditions, the ammunition transportation road in the rear was damaged to varying degrees and the road was rushed to repair by our troops. At the same time, the demand for our ammunition in the course of combat has also shown phase changes. In this case, quickly determine the location of the field ammunition depot from the candidate points of each ammunition depot, and determine the level of the field ammunition depot to maximize the demand for combat ammunition and reduce the waste of resources. The background process is shown in Figure 1.

Among them, the thicker the transportation line indicates that the road transportation capacity is stronger under normal circumstances, and the dashed transportation line indicates that the line is attacked by the enemy, which will cause the traffic capacity of the route to be weakened. The dashed ellipse indicates that there is no ammunition warehouse at this location. At the same time, in order to simplify the model, suppose:

(1) The dynamic demand of ammunition for each combat unit at each stage of the war is known.

(2) The damage and repair of the road are known.

(3) There are sufficient ammunition transportation vehicles, and the transportation cost is linearly related to the load and time.

(4) The opening time of the ammunition warehouse and the time of ammunition loading and unloading can be ignored.

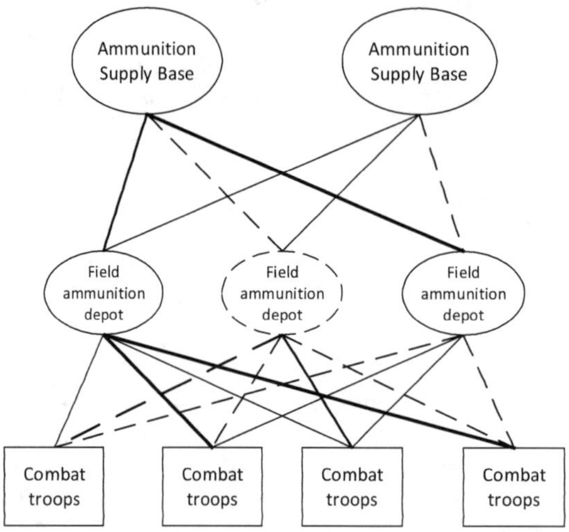

Figure 1. The background map of the site selection of the field ammunition depot in wartime.

2.2. Site Selection Model Construction

Based on the above analysis of the characteristics of wartime ammunition supply, the coupled dijkstra algorithm is used to model the location of the field ammunition depot. The coupled Dijkstra algorithm can use the label to gradually find the location and path of the field ammunition depot that is the fastest from the combat unit, which can be adapted to the solution of this model. After the battle begins, quickly collect road network information in the combat area. include:

(1) $d_{\widehat{ilj}}$ The actual distance from the location node i to the path l to the node j.

(2) v_l The traffic speed when the road is undamaged.

(3) ρ_l^t The traffic efficiency of road l in the t-stage of the operation. $\rho_l^t \in [0\ 1]$, The lower the ρ_l^t-value, the more severe the road damage. ρ_l^t takes 0, indicating that the road is completely destroyed. ρ_l^t is 1, indicating that the road is not attacked by the enemy. And ρ_l^t will increase by repairing the road.

Due to the high requirements for the on-time supply of ammunition under wartime conditions, the shortest supply time of ammunition should be the primary goal for the site selection of field ammunition depots. In this regard, establish a model with the shortest supply time as the primary goal:

$$\min W_1 = \sum_{i\in Q^t}\sum_{j\in M}\frac{d_{\widehat{ilj}}\,y_{\widehat{ilj}}^t}{v_l\rho_l^t} \tag{1}$$

$$y_{\widehat{ilj}}^t = \begin{cases} 1, & \text{At stage } t, \text{ select route } l \text{ from field ammunition library } i \text{ and deliver it to combat unit } j \\ 0, & \text{At stage } t, \text{ do not select route } l \text{ from field ammunition library } i \text{ and deliver it to combat unit } j \end{cases} \tag{2}$$

Among them, W_1 is the total transportation time of ammunition. Q^t is the assembly of the open ammu nition depot in the first stage, and M is each combat unit. With the advancement of the combat process , the positions of the combat forces, the demand for ammunition, and the transportation road conditions are in a process of constant change. Therefore, the established target model can reflect this situation in wartime.

Within a short period of time after the war, the supply of ammunition could not meet the needs of all combat troops. Therefore, the built model needs to make full use of the ammunition in the ammunition supply base to maximize the ammunition demand of the combat troops. At the same time, in order to avoid a situation where some combat troops are in sufficient supply of ammunition while others are in short supply. In this regard, the supply of ammunition should consider the balance of demand satisfaction. A model that targets the rate of ammunition supply satisfaction and the rate of ammunition supply imbalance should be established:

$$\max W_2 = \frac{\sum_{i\in Q^t}\sum_{j\in M}\sum_{t\in T} S_{ij}^t}{\sum_{j\in M}\sum_{t\in T} D_j^t} \tag{3}$$

$$\min W_3 = \sum_{i\in Q^t}\sum_{j\in M}\sum_{t\in T}\left[\left(\frac{S_{ij}^t}{D_j^t}\right) - \min\left(\frac{S_{ij}^t}{D_j^t}\right)\right] \tag{4}$$

Among them, W_2 is the ammunition demand satisfaction rate, which reflects the satisfaction of the ammunition demand. W_3 is the supply imbalance value, which reflects the imbalance in the supply of ammunition; T is the n stages in the battle process; D_j^t is the demand for ammunition at the combat location j in the T stage. S_{ij}^t is the supply of field ammunition depot i to the combat site j in the T stage.

The location of field ammunition depots should also consider the cost of ammunition supply, which mainly includes: ammunition transportation costs, ammunition depot setup costs, ammunition depot reconstruction costs, and ammunition depot maintenance costs. In this regard, a cost-oriented model can be established.

$$\min W_4 = \sum_{i\in Q^t}\sum_{t\in T}(E_{ef}x_{ib}^t + M_b z_{ib}^t) + \sum_{i\in Q^t}\sum_{j\in M}\sum_{t\in T} wS_{ij}^t d_{\widehat{ilj}}\,y_{\widehat{ilj}}^t \tag{5}$$

Among them, W_4 is the total supply cost; E_{ef} is the construction cost required to expand the e-class ammunition depot to f-class, $\{e,f\}=\{0,1,2,3\}$ and $f \geqslant e$, if there is no field ammunition depot originally set up, then e=0; M_f is the maintenance cost of the f-class ammunition depot, It is related to the size of the field ammunition depot and the amount of ammunition dispatch; w is the cost of transporting a unit amount of ammunition per unit distance.

$$x_{ib}^{t}=\begin{cases}1, & \text{At stage } t\text{, an } f-\text{level ammunition depot needs to be built at } i\\ 0, & \text{Do not use the ammunition depot at } i \text{ at stage } t\text{, or an ammunition depot that satisfies the scale has been built}\end{cases} \tag{6}$$

$$z_{ib}^{t}=\begin{cases}1, & \text{In phase } t\text{, the ammunition depot with the scale of } f \text{ at } i \text{ is activated}\\ 0, & \text{In phase } t\text{, the ammunition depot with the scale of } f \text{ at } i \text{ is not activated}\end{cases} \tag{7}$$

$$s.t\begin{cases} z_{ib}^{t} \geqslant y_{ilj}^{t} \quad \forall i \in U^{t},\ l \in L,\ j \in D, t \in T & (8)\\ 0 < \rho_{l}^{t} \leqslant 100\% \quad \forall\ l \in L,\ t \in T & (9)\\ \sum_{i \in U^{t}} z_{ib}^{t} \geqslant \sum_{i \in U^{t}} x_{ib}^{t} > 0,\ t \in T & (10)\\ \sum_{i \in U^{t}} S_{ijc}^{t} \leqslant Q_{j}^{t} \quad \forall j \in D,\ c \in C & (11)\end{cases}$$

Among them: Formula (8) indicates that the field ammunition warehouse i is activated, and there must be a road from the ammunition warehouse to other locations. If the ammunition depot is not activated, the road from the ammunition depot to other places must not be selected. Equation (9) indicates that the traffic efficiency is between 0 and 100%. Equation (10) indicates that in any period, the number of ammunition depots to be built is greater than or equal to 0, and the number of ammunition depots constructed in this phase must be less than the number of ammunition depots activated in this phase. Equation (11) indicates that the supply of ammunition cannot be higher than the demand for ammunition.

3. Location Model Solution

The punctual supply of ammunition is the key to winning the war, so the ammunition supply time should be as short as possible. At the same time, for the overall stability of the war, it is not only necessary to consider the satisfaction of the ammunition demand of combat troops, but also to consider the balance of ammunition supply. Moreover, due to the limitations of the human and material resources of the ammunition support department, these limitations are converted into cost constraints. The solution process of this paper is as follows:

(1) Collect information at each stage of the combat process, use Dijkstra's method to solve the fastest transportation path from the field ammunition depot to each combat unit, mark each combat unit as the initial point, gradually mark to find the fastest transportation path, and at the same time determine the field that needs to be developed Ammunition warehouse, get the initial opening plan of the ammunition warehouse.

(2) Based on the initial starting plan, the supply time will be adjusted upwards. At the same time, considering the balance of ammunition supply and the cost of supply, and considering the use of the established field arsenal, the initial model is adjusted.

(3) Based on the above analysis, a diagram of the "time-balance-cost" relationship of ammunition supply under wartime conditions is established. According to the existing conditions, the site selection plan of the field ammunition depot should be carried out reasonably.

Journal of Physics: Conference Series **2068** (2021) 012015 doi:10.1088/1742-6596/2068/1/012015

4. Case Simulation Site Selection

4.1. Simulation Case Description

With a simulation combat exercise as the background, the dynamic information of the battlefield is collected to provide design basis for ammunition support. No. 1-2 is the ammunition supply base, No. 3-8 is the waiting point for the field ammunition depot, and No. 9-18 is the combat unit. In order to simplify the calculation, this case has the following assumptions:

(1) Taking into account the convenience of supply, the No. 1 ammunition supply base is only suitable for supplying ammunition to the field ammunition depots at the candidate points 3, 4, and 8, and the No. 2 ammunition supply base is only suitable for the candidate points 5, 7, and 6. The field ammunition storehouse supplies ammunition.

(2) Regardless of the level of ammunition storage, the cost of field ammunition storage is only divided into fixed costs and variable costs.

(3) The transportation cost and management cost per unit of ammunition are only directly proportional to the supply time at this stage.

The ammunition storage volume of each stage of the ammunition supply base 1 and 2 and the ammunition demand of each stage of the combat troops 9-18 are shown in Table 1, and the road traffic conditions of each stage are shown in Table 2:

Table 1. Storage/demand scale.

	Stage 1	Stage 2	Stage 3
Supply base 1	4800	5300	1400
Supply base 2	5300	4500	2800
Combat troops 9	2300	0	0
Combat troops 10	2800	1800	0
Combat troops 11	0	1100	500
Combat troops 12	0	1200	400
Combat troops 13	3200	1000	0
Combat troops 14	0	600	0
Combat troops 15	0	800	200
Combat troops 16	0	0	800
Combat troops 17	0	0	500
Combat troops 18	0	0	800

Table 2. Road traffic time conditions at each stage.

Road	Stage 1	Stage 2	Stage 3	Road	Stage 1	Stage 2	Stage 3
3-9	4	5	3	8-15	4	4	3
4-10	2	6	2	8-11	-	5	4
5-9	6	6	3	9-10	8	6	4
5-13	7	-	5	11-15	9	6	4
6-14	4	3	5	11-12	-	7	3
6-16	6	4	6	11-17	-	3	3
6-18	4	4	4	12-15	-	4	4
7-9	5	5	6	12-18	7	-	2
7-10	3	-	6	14-15	3	8	2
7-13	3	5	6	15-16	3	10	6
7-15	4	6	2	17-18	4	3	2

4.2. Solving the Simulation Case

The coupling Dijkstra algorithm is used to find the fastest path from the field ammunition depot to the combat unit, and the supply time is adjusted upwards to determine the location of the field ammunition depot and its supply route. Consider the situation of demand satisfaction and supply imbalance, remove the site selection plan with long time, low demand satisfaction rate, and high supply imbalance, and further adjust the site selection plan. The site selection plan shown in Table 3 can be obtained:

Table 3. Site selection plan.

Program	Phase 1 Open the ammunition depot number	Phase 2 Open the ammunition depot number	Phase 3 Open the ammunition depot number	Total supply time	Consume guarantee resources	Demand satisfaction	Supply imbalance
1	3, 4, 7	4, 7, 8, 6	8, 7, 6	68	32262	0.9834	0.117
2	4, 7	4, 7, 8, 6	8, 7, 6	69	27442	0.989	0.0727
3	3, 7	4, 7, 8	8, 7, 6	69	32342	0.9613	0.233
4	3, 4, 7	4, 7, 8, 6	8, 6	70	32390	0.9834	0.117
5	4, 7	4, 7, 8, 6	8, 6	71	27570	0.989	0.0727
6	3, 7	4, 7, 8, 6	8, 6	71	27471	0.9613	0.233
7	3, 4, 7	7, 8, 6	8, 7, 6	73	33072	0.9834	0.117
8	4, 7	7, 8, 6	8, 7, 6	74	28092	0.989	0.0727
9	3, 7	7, 8, 6	8, 7, 6	74	27993	0.9613	0.233
10	3, 4, 7	4, 7, 8, 6	7, 8	75	32710	0.9834	0.117
11	3, 4, 7	7, 8, 6	8, 6	75	33040	0.9834	0.117
12	4, 7	4, 7, 8, 6	7, 8	76	27890	0.989	0.0727
13	4, 7	7, 8, 6	8, 6	76	28220	0.989	0.0727
14	4, 7	4, 7, 8	7, 8	85	23904	0.9781	0.2537

From this, the relationship change diagram shown in the following figure can be obtained:

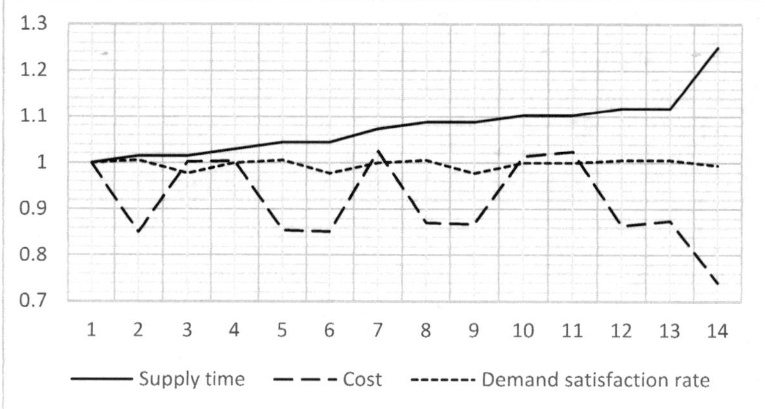

Figure 2. Diagram of relationship changes.

From the above analysis, it can be found that Scheme 2 has a slight increase in time compared with Scheme 1, but the cost is drastically reduced. The reason is that Scheme 2 does not open a field ammunition depot. At the same time, Option 2 is also better than Option 1 in terms of demand satisfaction rate and balance, so Option 2 can be given priority. In addition, from the above research results, it is found that position 7 appears more frequently in various stages, indicating that position 7 is the focus of the construction of field ammunition depots and plays an important role. Key consideration should be given to building a field ammunition depot at position 7.

5. Conclusion

The field ammunition depot is an important part of ammunition protection, and its construction must change with the changes of the battlefield stage to ensure the rapid supply of ammunition. In this regard, from the perspective of the fastest supply time, this article considers the balance of supply and the limitation of guarantee resources, establishes a 0-1 planning model, and uses the layered sequence method coupled with the Dijstra algorithm to solve the model, and uses the Dijstra labeling method to determine operations The fastest distance from the troops to the field ammunition depot, and considering the supply capacity of the field ammunition depot, determine the relationship between the supply time and the consumption of resources, and at the same time find the key construction location, to provide help for the site selection of the field ammunition depot under wartime conditions

Acknowledgments

We are very grateful to our colleagues for their support and help in the process of writing the paper. Whether it is the acquisition of topic ideas or the collection of simulation data, they have contributed a lot to the paper. We would like to express our sincere gratitude to them again.

References

[1] Wang Jinjun, Yi Jianzheng, Duan Zhiqiang, Zhou Zhiqiang, & Wang Bo. (2010). *Research on battlefield survivability countermeasures of field ammunition depot. Equipment Environmental Engineering*, 07(003), 96-98

[2] Shi Xianming, Zhao Rudong, Luo Xingyuan, Zhang Jianjun, & Song Xing. (2020). *Research on conversion coefficient of ammunition in cold plateau area based on fuzzy bp neural network. Firepower and Command Control*, v.45; No. 299(02), 61 -65.

[3] ZHAO, Rudong, SHI, Xianming, WANG, & Qian, et al. (2020). *Bayesian inference for ammunition demand based on Gompertz distribution. Journal of Systems Engineering and Electronics*, v. 31(03), 125-135.

[4] Fang Lijuan. *Research on the location of auto parts supply logistics distribution center based on JIT*. (Doctoral dissertation, Huazhong University of Science and Technology)

[5] Huo Xuemi, & Fu Hang. (2020). *Research on the location of logistics distribution center based on lingo. Logistics Engineering and Management*, v.42; No. 308(02), 117-118.

[6] Wen Xiaozheng, & Zhang Jing. (2018). *Location selection of cold chain distribution center based on genetic algorithm. Journal of Xi'an University of Posts and Telecommunications*, 23(06), 98-102+110.

[7] Zhao Wei, Li Wenfeng, & Liang Xiaolei. (2012). *Research on location selection of multiple distribution centers based on particle swarm optimization. Journal of Wuhan University of Technology (Transportation Science and Engineering Edition)*, 36(3), 501-505

[8] Jia Jing. (2007). Research on the location of emergency medical facilities for public health emergencies. (Doctoral dissertation, Beijing Jiaotong University)

[9] Shang Zhiyong. Research on location selection of distribution center based on improved cuckoo search algorithm. (Doctoral dissertation, Henan University)

[10] Song Yinghua, Sub Beibei, Huo Feizhou, Ning Jingjing, & Fang Danhui. (2019). *Research on rapid location selection of emergency material distribution center considering dynamic demand. Chinese Safety Science Journal*, v. 29(08), 176-181

[11] Liu, J., & Xie, K..(2017). *Emergency materials transportation model in disasters based on dynamic programming and ant colony optimization.* Kybernetes, 46(4), 656-671

[12] Shi Xuan. Dynamic location selection under soft time window constraints-path optimization research [D]. Fuzhou University, 2014.

Design of Fuel Supply Curves for Aircraft Center of Gravity Balance Based on DCT Compression

Yawen Wei[1], Bo Su[1], Xuru Wang[2], Zhen Li[1], Pengfei Wang[1] and Xiangchao Feng[1,*]

[1]Qian Xuesen Laboratory of Space Technology, China Academy of Space Technology, Beijing, China
[2]Beijing Institute of Control Engineering, China Academy of Space Technology, Beijing, China

[*]Corresponding author email: fengxiangchao@qxslab.cn

Abstract. The center of gravity (CG) has a significant influence on the controllability, stability, and fuel efficiency of the aircraft. For the aircraft with multiple fuel tanks, the CG position can be controlled by the fuel quality of different fuel tanks during flight. In this paper, an optimization method based on discrete cosine transform (DCT) compression is proposed to solve the optimization strategy of aircraft multi-tank fuel supply according to the current aircraft mission and engine working requirements. The transform coefficients of the low frequency components are used as design variables to represent the fuel supply curves. The improved KS function is proposed to deal with constraints. The numerical example is demonstrated to verify the effectiveness of the proposed method.

1. Introduction

The CG position is one of the most important parameters of the aircraft. Especially in the flight, monitoring and maintaining the CG are the main challenges of controllability and stability of the aircraft [1–3]. In addition, the CG also plays an important role in determining the fuel efficiency of aircraft [1].

The fuel consumption mode has a great influence on the CG position of an aircraft [4]. For an aircraft with multiple fuel tanks, the CG can be controlled by changing the fuel mass in a specific fuel tank or a group of fuel tanks during flight, so as to suppress the moment caused by the asymmetrical distribution of fuel. Yan et al. [5] proposed a mathematical model of aircraft's CG varying with fuel transfer and a general scheme of the active CG control system. To obtain a better overall solution, Zheng et al. [6] extracted the aircraft centroid balance fuel supply strategy problem into several sub-problems and combined sequential quadratic programming, improved simulated annealing algorithm to solve.

Several difficulties arise in the design of fuel supply curves for the CG balance of an aircraft with multiple fuel tanks. Especially a large number of local constraints need to be considered. The Kreisselmeier-Steinhauser (KS) function [7,8] is one of the most effective approaches to transform a set of local constraints into a single global constraint, which greatly reduces the computational cost of sensitivity analysis and improves convergence. However, the original KS function is a relaxation approach that allows an individual element may slightly superior to the limit value. To address the issue, the improved KS function is proposed in section 3.

Content from this work may be used under the terms of the Creative Commons Attribution 3.0 licence. Any further distribution of this work must maintain attribution to the author(s) and the title of the work, journal citation and DOI.
Published under licence by IOP Publishing Ltd

In this work, the transform coefficients of the low frequency components, as the optimization variables, are used to represent the fuel supply curve. The original KS function is improved to aggregate the constraints of mission and engine working requirements. In fact, the proposed method is a general computational framework for design of one-dimensional curves with a set of complex and strict local constraints. It will be extended in the future to treat more complex situations.

2. Problem Statement

Consider an aircraft with m fuel tanks, several fuel tanks are combined to supply fuel to meet the mission requirements and engine work requirements during the flight. The fuel tanks are all cuboid and fixed inside the aircraft. The inner length, width, and height of the kth fuel tank are $L^{(k)}, W^{(k)}, H^{(k)}$ respectively. In the aircraft coordinate system, the CG position (without fuel) is (0, 0, 0), and the CG position of the kth empty fuel tank is $(x_0^{(k)}, y_0^{(k)}, z_0^{(k)})$. The total mass of the aircraft (without fuel) is M. The fuel supply of the aircraft's fuel tank needs to meet the following constraints:

(1) The fuel supply speed of the kth fuel tank is limited to no more than $v_{\max}^{(k)}$ $(v_{\max}^{(k)} > 0)$. The duration of one fuel supply for each fuel tank is not less than 60 seconds.

(2) The main fuel tank can directly supply fuel to the engine. The auxiliary fuel tank can only supply fuel to the corresponding main fuel tank, not directly to the engine.

(3) Limited by the structure of the aircraft, the number of main fuel tanks and total fuel tanks for supplying simultaneously fuel cannot exceed the limit value.

(4) The total amount of fuel supplied by fuel tanks must at least meet the fuel consumption of the engine.

As an algorithm verification, only the case of the aircraft in level flight is considered in this paper. It is convenient to extend to other situations such as pitch, roll, and yaw by replacing equation (6) with the corresponding situation.

3. Method

3.1. One-dimensional DCT

The discrete cosine transform (DCT) [9] maps a sequence $g_x = \{g_1, g_2, ..., g_{N-1}\}$ in spatial domain into the sequence $G_u = \{G_1, G_2, ..., G_{N-1}\}$ in frequency domain.

$$G_u = \sum_x \left\{ \alpha_u g_x \cos \frac{\pi(2x+1)u}{2N} \right\} \tag{1}$$

While the inverse discrete cosine transform (IDCT) maps G_u back into spatial domain by:

$$g_x = \sum_u \left\{ \alpha_u G_u \cos \frac{\pi(2x+1)u}{2N} \right\} \tag{2}$$

where the term α_u is obtained as follows:

$$\alpha_u = \sqrt{\frac{2-\delta_u}{N}} \tag{3}$$

where $\delta_u = 1$ if $u = 0$, otherwise $\delta_u = 0$.

DCT is one of the most frequently used transformations for image compression. By only a small proportion of G_u corresponding to the low frequency components, the g_x ($where\ u = 1, 2, ..., n_X$; $x = 1, 2, ..., N_X; n_X \ll N_X$) can be obtained from equation (2). And the sharp regions in spatial domain will be filtered. This property is used to filter the regions in the fuel supply curve where the duration of one fuel supply is less than 60 seconds without adding any constraints.

3.2. Formulation of fuel Supply Curve Design Method for Aircraft Fuel Tank Based on DCT

To begin, the n_X transform coefficients of the low frequency components $x_i^{(k)}(i=1,2,...,n_X,k=1,2,...,m)$ are used as the design variables of fuel supply curve of the kth fuel tank. Fuel supply speed curve of each fuel tank $v_j^{(k)}(j=1,2,...,N;\ N\gg n_X)$ is presented using N discrete time points (the time interval is 1 second). By using the IDCT equation (2), the speed in spatial domain $g_j^{(k)}$ can be reconstructed by the design variables $x_i^{(k)}$. Then equation (4) plotted in Figure 1(a) is used to drive $g_j^{(k)}$ approach $\{0,\ v_{\max}^{(k)}\}$ in order to obtain the fuel supply speed curve $v_j^{(k)}$.

$$v_j^{(k)} = v_{\max}^{(k)} \frac{1+tanh(\beta g_j^{(k)})}{2} \tag{4}$$

where β is the scaling parameter. Thus, the most difficult constraints mentioned in section 2 are perfectly embedded in the design variables.

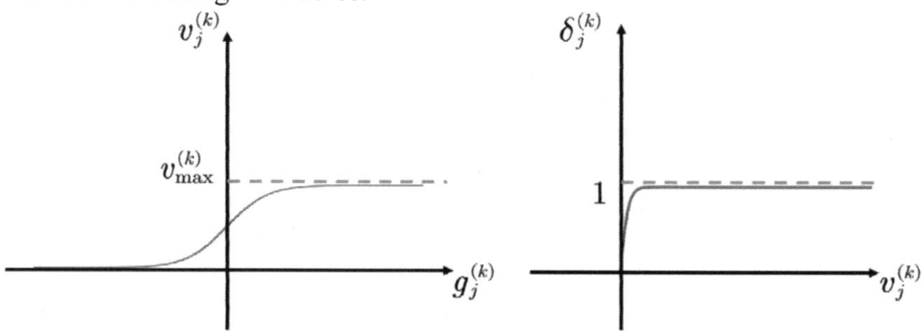

(a) (b)

Figure 1. Illustration of the (a) equation (4) and (b) equation (10).

With the fuel supply speed curve $v_j^{(k)}$, the remaining fuel volume of the pth main tank can be calculated by

$$V_j^{(p)} = V_0^{(p)} - \sum_{t=1}^{j-1} \frac{v_j^{(p)} + v_{j+1}^{(p)}}{2\rho} + \sum_{t=1}^{j-1} \frac{v_j^{(p^0)} + v_{j+1}^{(p^0)}}{2\rho} \tag{5}$$

where p^0 is the number index of the auxiliary fuel tank corresponding to the pth main fuel tank. $V_0^{(p)}$ is the initial fuel volume of the pth fuel tank. ρ is the density of the fuel. As for the calculation of the remaining fuel volume of auxiliary fuel tanks, only the first two terms in equation (5) need to be retained.

Considering the situation of the aircraft in level flight, the CG position of the remaining fuel in the jth second $\vec{r}_{cj}$ can be derived as follows:

$$\vec{r}_{cj} = \left[x_0^{(k)},\ y_0^{(k)},\ z_0^{(k)} - 0.5H^{(k)} + \frac{V_j^{(k)}}{2L^{(k)}W^{(k)}} \right]^T \tag{6}$$

The actual position vector of the CG of the aircraft at the jth second is expressed as:

$$\vec{R}_{cj} = \frac{\sum_{k=1}^{m} \rho V_j^{(k)} \vec{r}_{cj}^{\,(k)}}{M + \sum_{k=1}^{m} \rho V_j^{(k)}} \tag{7}$$

Hence, the functional relationship between the design variables and the aircraft's CG is established. Then the improved KS function is used to aggregate the constraints so that only one constraint is used to deal with the other constraints mentioned in section 2.

For the constraint: the total amount of fuel supplied by fuel tanks v_j^e must at least meet the fuel consumption of the engine v_j^{e*}. By introducing a penalty parameter η, the term $\bar{g}_j$ of KS function [7,8] in equation (8) is improved to equation (9), which is used to suppress that the original KS function allows an individual element to be slightly higher than the limit value.

$$v_{ks} = \frac{1}{p}\ln\left(\frac{1}{N}\sum_{j=1}^{N} e^{p\bar{g}_j}\right) \tag{8}$$

with

$$\bar{g}_j = 1 - \frac{v_j^e - v_j^{e*}}{\eta} \tag{9}$$

By using the improved KS function, the constraints ($v_j^e \geqslant v_j^{e*}, j=1,2,...,N$) are aggregated to only one constraint ($v_{ks} \leqslant 0$).

For the constraint: Limited by the structure of the aircraft, the number of main fuel tanks and all the fuel tanks for supplying simultaneously fuel cannot exceed the limit value. Whether the fuel tank stops supplying fuel or not is identified by equation (10) as shown in Figure 1(b).

$$\delta_j^{(k)} = tanh(\alpha v_j^{(k)}) \tag{10}$$

where α is set to 10.

In this way, the integer constraint can be converted into a continuous constraint function with a sensitivity. Similarly, these integer constraints can be aggregated into $n_{ks}^{main} \leqslant 0$ and $n_{ks}^{all} \leqslant 0$ by the KS equation. And the last constraint, the remaining fuel volume of the fuel tank cannot be less than zero, is also aggregated into $V_{ks} \leqslant 0$ by the KS function.

The objective is to minimize the difference between the actual CG curve and the ideal CG curve. Thus, the optimization problem can be formulated as follows:

$$\begin{aligned} &find\ x_i^{(k)}\ (k=1,2,...,m, i=1,2,...,n_X) \\ &\min\ f = (\| \boldsymbol{R}_c - \boldsymbol{R}_c^* \|_F)^2 \\ &st: \begin{cases} v_{ks} \leqslant 0 \\ n_{ks}^{main} \leqslant 0 \\ n_{ks}^{all} \leqslant 0 \\ V_{ks} \leqslant 0 \quad (j=1,2,...,N) \end{cases} \end{aligned} \tag{11}$$

where $\boldsymbol{R}_c$ is a matrix and the jth row represents $\vec{R}_{cj}$. Correspondingly, $\boldsymbol{R}_c^*$ contains the ideal CG curve of the aircraft.

In this paper, the method of interior-point method of the MATLAB optimization toolbox is employed to update design variables. The sensitivity information of the objective function and constraint functions can be calculated via the chain rule.

4. Numerical Example and Result Analysis

In this section, an aircraft with six fuel tanks is considered as shown in Figure 2. The fuel tanks 1 and 6 are the auxiliary fuel tanks corresponding to the main fuel tanks 2 and 3, respectively. The speed limit $v_{\max}^{(k)}$ of each fuel tank is $\{1.1, 1.8, 1.7, 1.5, 1.6, 1.1\}\,kg/s$, respectively. Up to two tanks can supply fuel to the engine and up to three tanks can supply fuel at the same time.

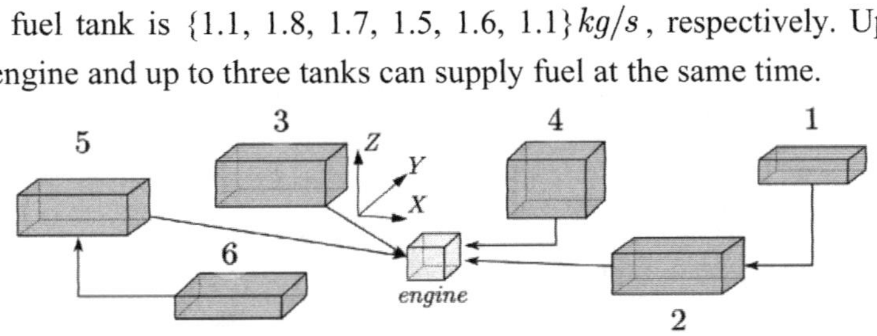

Figure 2. Schematic of fuel tanks of the aircraft.

A two-hour flight mission of the aircraft stipulates the fuel consumption of the engine (the red dotted line plotted in Figure 4(a)) and the ideal CG position (the red dotted line plotted in Figure 3(b)) at each moment during the flight. Since the time interval is 1 second, N is equal to 7200. The number of transform coefficients of the low frequency components is set to $n_X = 20$. The initial guess of each fuel supply curve is chosen as the design variables corresponding to $v_j^{(k)} = 0.5 v_{\max}^{(k)}$.

The entire optimization process is completed in 99 iterations and the objective function converges at 168.71. The final remaining fuel volume of each fuel tank is {0.0016, 0.0079, 0.0440, 0.0137, 0.6097, 0.3001} m^3, which satisfies the constraint that the remaining fuel volume of the fuel tank cannot be less than zero.

As depicted in Figure 3(a), the optimized fuel supply speed curves of fuel tanks are very smooth and the duration of a single fuel supply for each fuel tank is higher than 60 seconds. According to the designed fuel supply curves, the matching of the actual and ideal CG is presented in Figure 3(b). It can be found that actual CG is very close to the ideal one.

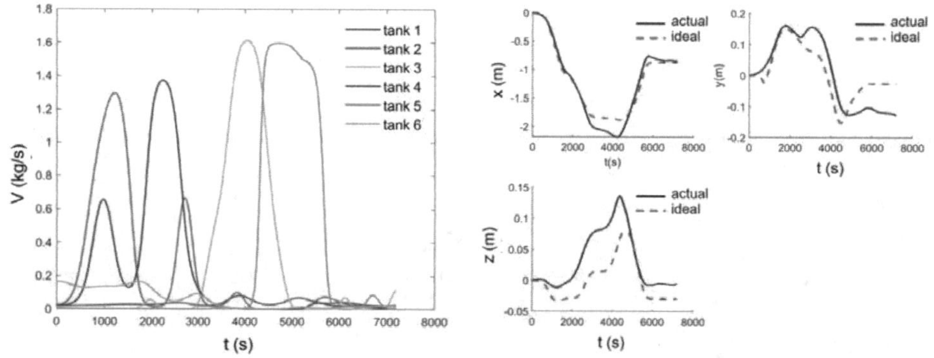

(a) (b)

Figure 3. (a) The optimized fuel supply speed curves of aircraft fuel tanks (b) Comparison of components between the actual and the ideal CG position of the aircraft.

The excess fuel which is higher than the planned fuel consumption needs to be discharged from the aircraft through other devices. As illustrated in Figure 4(a), the total amount of fuel supplied by fuel tanks fully meets the fuel consumption of the engine, and only a small amount of fuel is discharged from the aircraft. It can be concluded from Figure 4 (b) that the number of main fuel tanks as well as the number of all fuel tanks simultaneously supplying fuel are not higher than the limit values of 2 and 3, respectively. The results demonstrate the effectiveness of improved KS function.

Compared with the existing method for the design of the fuel supply speed curve, there are several advantages of the proposed method. First, only a small number of design variables are required. Second, the most difficult constraints (The upper limit of the fuel supply speed and the constraint that the duration of one fuel supply for each fuel tank is not less than 60 seconds) are automatically satisfied by design variables avoiding applying more complex models in constraint functions. This makes the method proposed in this paper superior to the method of dividing the total time into time slices with a length of 60 seconds reported in [6] in terms of algorithmic complexity and computational cost. The third merit of the proposed method is the smoothness of the design results. Frequent switching phenomenon (sharp change of the fuel supply speed curve) [6] is filtered.

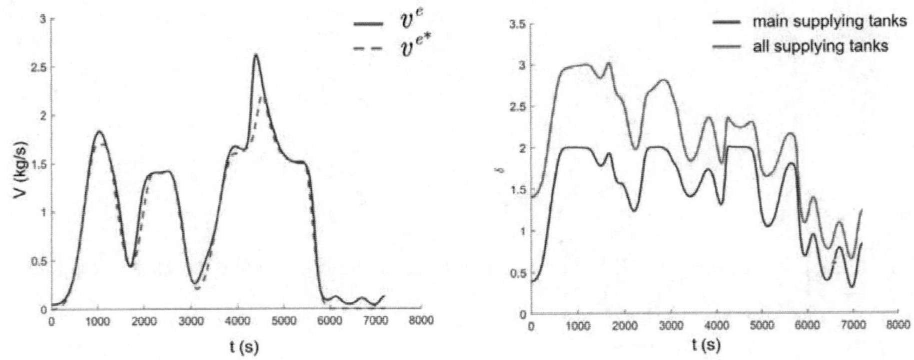

(a) (b)

Figure 4. (a) Comparison of the fuel supplied by the fuel tanks and the fuel consumption of the engine. (b) The number of main tanks and all tanks supplying fuel simultaneously.

5. Conclusion

In this paper, a curve design method based on DCT is proposed to solve the optimization strategy of aircraft multi-tank fuel supply according to the actual aircraft mission and engine working requirements. The KS function is improved to suppress that the original KS function allows an individual element to be slightly higher than the limit value. From the numerical results, the optimized fuel supply curve is very smooth and satisfies the corresponding constraints. Other constraints are also satisfied by the improved KS equation. The numerical results highlighted the advantages of the optimization method.

References

[1] Liu Y, Yang Z, Deng J and Zhu J 2018 Investigation of fuel savings for an aircraft due to optimization of the center of gravity *IOP Conference Series: Materials Science and Engineering* vol 322 p 072018

[2] Miao Y, Wang S and Zhao Y 2012 Study on change of aircraft center of gravity during fuel consumption *IEEE 10th International Conference on Industrial Informatics* pp 86–90

[3] Xiaohong Y 2018 Study on the Center Gravity Control Strategy of Tandem Aircraft Fuel Tank *IET Conference Proceedings* (Institution of Engineering and Technology) pp 169 (5 pp.)-169 (5 pp.)(1)

[4] Tudosie A, NEGREA P and VĂDUVESCU V 2019 METHOD OF CONTROL OF AIRCRAFT CENTER OF GRAVITY BASED ON THE FUEL CONSUMPTION ORDER *Sci. Res. Educ. AIR FORCE* **21** 209–14

[5] Yan J, Zhang J and Li H 2016 Design of aircraft center of gravity control law based on sliding mode control *2016 IEEE International Conference on Aircraft Utility Systems (AUS)* pp 438–42

[6] Zheng Z, Pei D, Zhou J, Chen Y and Pu P 2021 An Optimization Method of Aircraft Centroid Balance Fuel Supply Strategy Based on Sequential Quadratic Programming *2021 4th International Conference on Advanced Electronic Materials, Computers and Software Engineering (AEMCSE)* pp 1332–6

[7] Kreisselmeier G and Steinhauser R 1979 Systematic Control Design by Optimizing a Vector Performance Index *IFAC Proc. Vol.* **12** 113–7

[8] Verbart A, Langelaar M and Keulen F van 2017 A unified aggregation and relaxation approach for stress-constrained topology optimization *Struct. Multidiscip. Optim.* **55** 663–79

[9] Zhou P, Du J and Lü Z 2018 A generalized DCT compression based density method for topology optimization of 2D and 3D continua *Comput. Methods Appl. Mech. Eng.* **334** 1–21

Weapon-Target Assignment Problem Using Modified Water Wave Optimization Algorithm

Yuanfei Wei[1], Qifang Luo[2] and Yongquan Zhou[2,*]

[1]Xiangsihu College of Guangxi University for Nationalities, Nanning, 532100, China
[2]College of Artificial Intelligenc, Guangxi University for Nationalities, Nanning 530006, China

[*]Corresponding author email: yongquanzhou@126.com

Abstract. The weapon-target assignment (WTA) is a classic problem. The WTA mathematical model is that warship formations are reasonably equipped with weapons resources for each weapon system to attack the air threaten targets. The purpose of targets optimization is to maximize combat effectiveness, that is to say, the mathematical expectation is maximum. We adopt the greedy strategy and improved propagation operation is to strengthen the water wave optimization (WWO) search performance. This article elaborates a modified water wave optimization (MWWO) to solve the WTA problem, which can detect optimized allocation decision matrix and search for the maximum mathematical expectation. Based on parameter optimization, the overall performance of the MWWO is more stable, the search speed is accelerated and the accuracy is improved. The experiment results indicate that the MWWO are verified and avoids local optimum, and can be more convenient for solving the WTA and obtain better performance.

1. Introduction

In recent years, heuristic optimization technology develops very rapidly; especially meta-heuristic optimization algorithm has achieved success in solving complex problems. In general, the meta-heuristics algorithms are mainly divided into evolutionary algorithm (EA), swarm-based algorithm, physics-based algorithm. EAs mimic the concept of natural evolution. For example, Genetic Algorithm (GA) [1]. Differential Evolution (DE) [2], Evolutionary Strategy (ES) [3], Evolutionary Programming (EP) [4], and Biogeography-based Optimization (BBO) [5]. Swarm intelligent-based have been widely used in science and engineering. For example, Artificial Bee Colony (ABC) [6], Particle Swarm Optimization (PSO)[7], Cuckoo Search Algorithm (CS) [8], Firefly Algorithm (FA) [9], Bat algorithm (BA) [10], Grey Wolf Optimizer (GWO) [11], Dolphin Echolocation (DE) [12], Whale Optimization Algorithm (WOA) [13], Fruit-fly Optimization Algorithm (FOA) [14], and Chimp Optimization Algorithm (ChOA) [15], and so on. Physics-based algorithm, such as, Gravity Search Algorithm (GSA) [16], Charged System Search (CSS) [17], Magnetic Optimization Algorithms (MOA) [18], Harmony Search (HS) [19], and water wave optimization (WWO) [20] are all physics-based optimization algorithms.

In the military, air strikes and anti-air raids have become the main mode of combat in modern warfare. The weapon-target assignment (WTA) in a modern combat command and decision support system, command automation system is an indispensable key component. The main task of the WTA is to destroy the incoming aircraft, unmanned aerial vehicles, to intercept the enemy cruise missiles, anti-ship missiles and radiation missiles. This requires the commander to consider how to rationally allocate firepower, quickly and efficiently to complete the air defense mission to ensure that the air defense efficiency of warship formation, improve the viability of the warship formation.

Content from this work may be used under the terms of the Creative Commons Attribution 3.0 licence. Any further distribution of this work must maintain attribution to the author(s) and the title of the work, journal citation and DOI.
Published under licence by IOP Publishing Ltd

The WTA [21] is essentially an integer type nonlinear combination optimization decision problem. The WTA based on the purpose of combat, battlefield situation and weapons performance and other factors will be a certain type and quantity of weapon units to a reasonable allocation of certain criteria to attack a certain number of air threaten targets. The result of the optimization is that the combat effectiveness is maximized. The WTA is the key technology in military operations. There are many intelligent technology are applied in many fields of military affairs, such as grey wolf optimizer (GWO) [22], bat search algorithm (BA) [23], flower pollination algorithm (FPA) [24], sine cosine algorithm (SCA) [25], PSO [7], Social spider optimization algorithm(SSO)[26], Quantum-inspired satin bowerbird algorithm(QSBA)[27], water wave optimization (WWO) [20]. To improve WWO performance, the WTA model was established by Ruan [28]. In this paper, the MWWO algorithm is used to solve WTA problems, which not only can balance exploration and exploitation abilities, and the MWWO algorithm finds the optimal decision matrix, so that it has the maximum combat effectiveness in WTA. This indicates the MWWO algorithm has a stable performance and a strong search capability.

2. Related Works

2.1. Problem Formulation

The WTA, using most effective allocation of weapon units to intercept multiple batches of air threaten targets, is a very important issue in modern warfare and an important auxiliary decision in the command automation system. The WTA is a dynamic allocation process in warship formation antiaircraft that can be depicted as Fig.1.

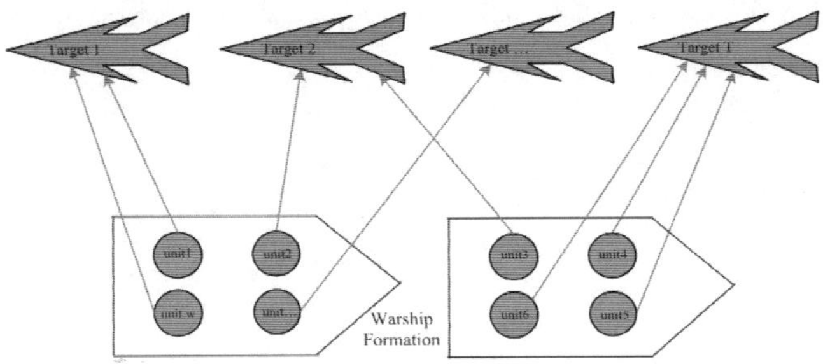

Figure 1. Schematic diagram of WTA in warship formation antiaircraft.

2.1.1. Assumption Description. The WTA is a complex and vital issue, which makes full use of weapon units of different weapon systems to combat air threaten targets and mathematical expectation is maximum. In order to better describe the WTA problem, some assumptions are defined:

***Assumption* 1** There are K different types of weapons systems in warship formation antiaircraft. Each weapon system has a certain number of weapon units and air threaten targets T. A weapon system is distributed effectively for air threaten target in the effective area and the time of action, otherwise no distribution.

***Assumption* 2** The total number of weapon units W and targets T. A weapon system can application multiple weapon units to attack a target.

***Assumption* 3** A weapon unit can only assail an air threaten target once. The number of weapon units distributed for each weapon cannot more than the number of allocated weapon unit resources in the combat time.

***Assumption* 4** The probability of damage P_{ij} that each type weapon system strikes the air threaten targets and the threat coefficient μ_j that menace of target to warship formation are known. Each weapon system damage probability is P_{ij}, which indicates that the ith weapon system engage jth air threaten target.

***Assumption* 5** To achieve maximum operational effectiveness, when faced with multiple batches of air threaten targets, the available resources for each weapon system are fully allocated within effective areas of action and time of action.

2.1.2. Mathematical Model of WTA. There are T air threaten targets and K weapon systems. In the effective warfare time and combat area, the number of resources by a weapon system is W_i. The objective of WTA optimization is that the mathematical expectation of the damage target is maximized by reasonably assigning weapon units to the air threaten targets. Therefore, the mathematical model of the WTA can be established.
Definition 1 The WTA mathematical model is defined:

$$f = \max \sum_{j=1}^{T} \mu_j \left[1 - \prod_{i=1}^{K} (1 - P_{ij})^{x_{ij}} \right] \tag{1}$$

Definition 2 The $P_{ij} \in [0,1]$ is damage probability of the weapon systems, which declares the ith ($i = 1,2,..., K$) weapon system strikes the jth ($j = 1,2,...,T$) target.
Definition 3 The $\mu_j \in [0,1]$ represents threat coefficient matrix for each air threaten target $j = 1,2,...,T$.
Definition 4 The decision matrix $X = [x_{ij}]_{K \times T}$ can be given:

$$X = [x_{ij}]_{K \times T} = \begin{bmatrix} x_{11} & x_{12} & \cdots & x_{1T} \\ x_{21} & x_{22} & \cdots & x_{2T} \\ \vdots & \vdots & \ddots & \vdots \\ x_{K1} & x_{K2} & \cdots & x_{KT} \end{bmatrix} \tag{2}$$

where $x_{ij} \in Z$ $(i = 1,2,..., K; j = 1,2,...,T)$ indicates weapon units number are assigned to the jth air threaten target by ith weapon system.
Definition 5 Constraints conditions stated as:

$$\sum_{i=1}^{K} x_{ij} \ge 1, \ \forall j = 1,2,...,T \tag{3}$$

$$\sum_{j=1}^{T} x_{ij} \ge 1, \ \forall i = 1,2,...,K \tag{4}$$

$$\sum_{i=1}^{K} \sum_{j=1}^{T} x_{ij} = K \tag{5}$$

$$\sum_{j=1}^{T} \mu_j = 1 \tag{6}$$

The Eq.(3) express at least one weapon unit is arranged to each incoming threaten target. Eq.(4) express each weapon unit concentrated against only one incoming threat target. Eq.s(5-6) express parameters x_{ij} and μ_j range, respectively.

2.2. WWO Algorithm
The WWO is obtained the objective function optimal value by wave propagation, refraction and breaking operations. Population initialization, a water wave population is constructed that the wave height h and wavelength λ of each water wave are set to $h_{\max}$ and 0.5, respectively. For the uneven of seabed, the fitness value is variable with wave propagation, refraction and breaking. The fitness value of the long wave with low energy is much smaller than that of the short wave with high energy from Fig.2.

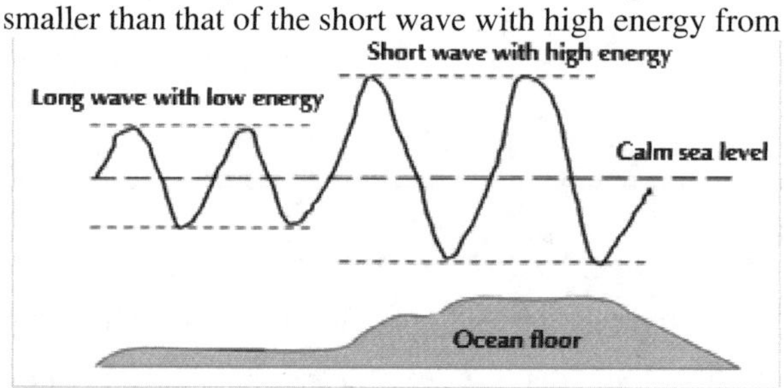

Figure 2. Diagram of the difference between deep wave and shallow water shapes.

2.2.1 Propagation. The each wave is subjected to a propagation in each iteration process. The new wave x' location update:

$$x'(d) = x(d) + rand(-1,1) \cdot \lambda L(d) \tag{7}$$

Here $d \in [1, D]$ is the problem space dimension, $x(d)$ and $x'(d)$ represents the original water wave and the new water wave position, respectively, $rand(-1,1)$ is a random number, $L(d)$ expresses the d-dimension space. If the new wave position is beyond bound, it's randomly generated new position.
In wave propagation, the fitness value of the original wave and that of the new wave are calculated. If $f(x') > f(x)$, x is replaced by x' and wave height of the new wave is $h_{\max}$. Otherwise, x is retained and wave height is reduced by one that represents the loss of water wave energy. The update formula for the wavelength λ is:

$$\lambda = \lambda \cdot \alpha^{-(f(x)-f_{\min}+\varepsilon)/(f_{\max}-f_{\min}+\varepsilon)} \tag{8}$$

where α indicates the attenuation coefficient for the water wave wavelength, $f_{\max}$ and $f_{\min}$ are the maximum and minimum fitness values, respectively, ε is a very small positive integer.

2.2.2 Breaking. The seabed is rugged, the water wave may become much higher and higher, that is to say, the water wave energy increase. Finally, the water wave breaks into a series of solitary waves. Once a new optimal wave is found in the global search space, the new water wave will execute the breaking operation, which improves the solution accuracy. The operation is randomly selected $k \in [1, D]$ dimension from D and each dimension is updated:

$$x'(d) = x(d) + N(0,1) \cdot \beta L(d) \tag{9}$$

Here the parameter $\beta \in [0.001, 0.1]$ is broken wave coefficient. If the fitness value of x^* is much better than all the solitary waves, x^* is remained; otherwise, x^* be replaced.

2.2.3 Refraction. The refraction operation is to remove the energy depleted water waves, which effectively avoids the search stagnation. The wave height gradually decays to zero, meaning that a wave has not been improved after many propagation operations. In that case, the water wave refraction is:

$$x'(d) = N\left(\frac{(x^*(d) + x(d))}{2}, \frac{|x^*(d) - x(d)|}{2}\right) \tag{10}$$

The x^* represents the best wave in the population, in other words, the current position is the optimal solution, $N(\mu, \sigma)$ represents Gaussian random number, μ and σ are mean and standard deviation respectively. After refraction, wave height x' is also reset to $h_{\max}$ and the fitness value is inversely proportional to the wavelength.

$$\lambda' = \lambda \frac{f(x)}{f(x')} \tag{11}$$

In short, we solving the object problem is to conduct by propagation, refraction and breaking operations.

3. Proposed Modified WWO Algorithm

The water wave superior process regarded as the process of the deep water moving to shallow area. The propagation operation makes that the larger fitness walue exploits in a small range and the lower fitness walue explores in large water area. Similar to the PSO [21], the formula (7) is added to the inertia weight to strengthen population search ability. The improved propagation operator can be provided as follows.

$$w = w_{\max} - (w_{\max} - w_{\min})\frac{iter}{iter_{\max}} \tag{12}$$

$$\rho = \frac{1}{1 + e^{a \cdot iter - b}} rand(-1,1) \tag{13}$$

$$x'(d) = w \cdot x(d) + \rho \cdot \lambda L(d) \tag{14}$$

where the $w_{\min} = 0.4$ and $w_{\max} = 1.5$ expression the minimum and maximum inertia weights, respectively, the $iter$ is current iterations number, $a = 0.02$ and $b = 20$ are given constants.

3.1 Greedy Strategy

To get a better optimal solution in the water wave movement process, the greedy strategy is applied to improving the exploitation ablity of the WWO. The $x''_{i,j}$ represents greed solution, the formula (15) and formula (16) are as follows

$$x''_{i,j} = x_{i,j} + \varphi_{i,j} C_{i,j} x_{i,j} \tag{15}$$

$$C_{i,j} = c_{\min} + \exp\left[-c\frac{iter}{iter_{\max}}\right]^d (c_{\max} - c_{\min}) \tag{16}$$

where the $x''_{i,j}$ is the ith greedy solution of the jth decision variables. $\varphi_{i,j} \in [-1,1]$ is a random number, $iter$ is the current iterations, $iter_{\max}$ is the maximum iterations $c_{\min} = 0.1$, $c_{\max} = 0.5$, $c = 60$ and $d = 5$ are related control parameters.

3.2 The MWWO to solve WTA problem

Algorithm 1. The pseudo-code of MWWO to solve the WTA problem

Step 1: Initialize population P of n waves (solutions), the parameters $iter_{\max}$, α, β, λ;

Step 2: Establish the WTA optimization model, define distribution principle, weapon systems K, air threaten targets T, weapon units W, set up constraint conditions according to Eqs. (1), (3), (4), (5) and (6).

Step 3: Compute the fitness function f value of the population by the decision matrix X.

Step 4: When the termination condition is not met do,

Step 5: For each wave $x \in P$ do

Step 5.1: The greedy strategy add to waves based on Eqs. (15) and (16).

Step 5.2: The wave x is subjected to propagation operation to obtain a new wave each individual x' use Eq.s (12), (13) and (14).

Step 5.3: If the fitness function $f(x') > f(x)$ then

Step 5.3.1: If the fitness function $f(x') > f(x^*)$, the wave x' performs breaking operation according to Eq. (9), the current optimal wave x^* will be replaced by wave x'.

Step 5.3.2: The original wave x is replaced by the new wave x' in the population.

Step 5.4: Otherwise, wave height h of wave x is reduced one by one. If $x.h = 0$, the wave x will perform refraction operation using Eqs. (10) and (11).

Step 5.5: Updating the wavelength of all waves using Eq.(8).

Step 6: According to the MWWO update process, a new decision matrix X is obtained and the fitness value of the new population is calculated.

Step 7: Let $iter = iter + 1$.

Step 8: If $iter < iter_{\max}$, go Step 5.

Step 9: Output optimal value f and relevant WTA decision matrix X.

4. Experiment Results and Discussion

4.1. The Solve the WTA Problem Using MWWO

Assume that there are 10 air threaten targets from different directions and 7 different types of weapons systems in warship formation antiaircraft, this are $K = 7$ and $T = 10$. The number of available resources for each weapon system is defined as $W = [4,5,4,5,4,5,4]$ in the specified combat time. For the sake of comparison, targets' threat coefficient μ_j ($j = 1,2,...,T$) and weapons' damage probability P_{ij} ($i = 1,2,...,K$, $j = 1,2,...,T$) are list in Table 1 and 2, respectively.

Table 1. The targets' threat coefficient.

Target	1	2	3	4	5	6	7	8	9	10
Threat coefficient	0.09	0.12	0.14	0.06	0.05	0.10	0.08	0.09	0.15	0.12

Table 2. The weapons' damage probability.

Weapon systems	Target batch									
	1	2	3	4	5	6	7	8	9	10
1	0.82	0.85	0.78	0.75	0.52	0.88	0.44	0.39	0.82	0.56
2	0.56	0.72	0.88	0.46	0.72	0.56	0.68	0.45	0.48	0.75
3	0.45	0.61	0.54	0.73	0.84	0.84	0.78	0.42	0.53	0.65
4	0.56	0.42	0.76	0.84	0.73	0.83	0.86	0.62	0.78	0.82
5	0.45	0.58	0.38	0.44	0.36	0.59	0.78	0.77	0.65	0.81
6	0.46	0.61	0.55	0.68	0.75	0.83	0.73	0.66	0.82	0.48
7	0.66	0.71	0.65	0.44	0.86	0.79	0.44	0.85	0.53	0.39

The proposed MWWO performances are compared with that of the other intelligent algorithms, such as PSO [7], BA [10], GWO [11], FPA [24], SCA [25], and WWO [20]. The primary parameters are given (see Table 3).

Table 3. The seven algorithms parameters

Algorithms	Parameter values
GWO	The population size is 30.
BA	$A=0.25$, $f\in[0,2]$, $r=0.5$, the population size is 30.
FPA	$\rho=0.8$, the population size is 30.
SCA	$r_2\in[0,2\pi]$, $a=2$, $r_4\in[0,1]$, the population size is 30.
PSO	$c_1=1.4962$, $c_2=1.4962$, $\omega=0.7298$, the population size is 30.
WWO	$\lambda=0.5$, $h_{\max}=12$, $\alpha=1.0026$, $\beta\in[0.01,0.25]$, $k_{\max}=\min(12,D/2)$, the population size is 30.
MWWO	$\lambda=0.5$, $h_{\max}=12$, $\alpha=1.0026$, $\beta\in[0.01,0.25]$, $k_{\max}=\min(12,D/2)$, $w_{\min}=0.4$, $w_{\max}=1.5$, $a=0.02$, $b=20$, $c_{\min}=0.1$, $c_{\max}=0.5$, $c=60$, $d=5$;The population size is 30.

The population size is 30, the maximum iterations number is 100 and all algorithms run 30 times independently. The "Best," "Worst," "Mean," and "Std." are four most important evaluation criteria, respectively.

$$X=\begin{bmatrix}3&1&0&0&0&0&0&0&0&0\\0&0&2&0&0&0&0&1&0&2\\0&0&1&1&2&0&0&0&0&0\\0&0&0&2&0&0&2&0&0&1\\0&0&0&0&0&0&1&2&0&1\\0&0&0&0&0&2&0&0&3&0\\0&2&0&0&1&1&0&0&0&0\end{bmatrix} \quad (17)$$

The objective of WTA is to find the best decision matrix X and achieve the optimal combat effectiveness, that is to say, the mathematical expectations of the damage targets is maximized. The weapon systems effectively strikes the air threaten targets, which is reasonably assigned to the weapon units. There are three weapon units be arranged to first target and one weapon unit is arranged to second target in first weapon system, some targets have been stricken by several weapon units. For the third weapon system, there are for weapon units be arranged to third, fourth, fifth and fifth target. For the seventh weapon system, there are second, second, fifth and sixth target can be attacked by three weapon units, separately. To better demonstrate the performance of the MWWO, the comparison result between MWWO and other algorithms are given in Table 4.

Table 4. The comparison results between MWWO and the other six algorithms.

Objective function	Algorithm	Results				
		Best	Worst	Mean	Std	Rank
	GWO	0.773019	0.39	0.628297	0.097394	6
	BA	0.773019	0.39	0.679762	0.106242	7
f	FPA	0.861542	0.6279	0.742109	0.066069	5
	SCA	0.773019	0.6279	0.656924	0.05904	4
	PSO	0.861542	0.861542	0.861542	4.52E-16	1
	WWO	0.861542	0.773019	0.846788	0.033554	3
	MWWO	0.995849	0.955084	0.981475	0.011098	2

From the Table 4, we can obtain that mean fitness function value of the MWWO is the best, which shows the MWWO has strong stable performance and balances exploration and exploitation. The optimal fitness value of the MWWO is better than that of other algorithms, which represents the MWWO has fast convergence speed and accurate accuracy.

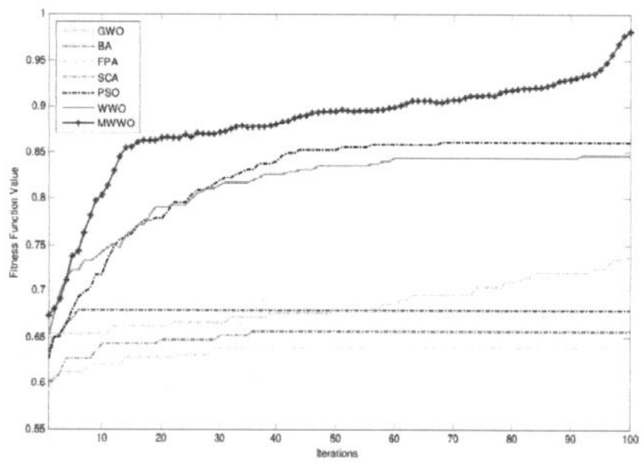

Figure 3. Fitness functions evolution curves

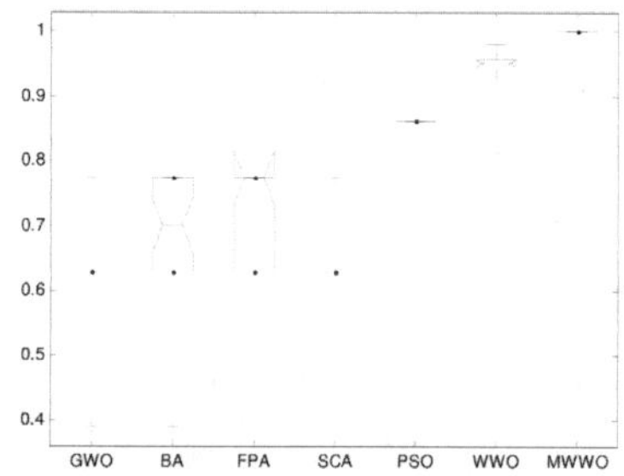

Figure 4. ANOVA test

As is revealed in the Fig.3, the MWWO compared with GWO, BA, FPA, SCA, PSO and WWO algorithms, the MWWO convergence speed is the fastest, that is to say, it achieves the optimal combat effectiveness or obtains maximum mathematical expectation. The MWWO search ability is better than that of the original WWO, because it is adopted improved propagation operator and greedy strategy. So the MWWO calculation accuracy is much better than that of the GWO, BA, FPA, SCA, PSO and WWO algorithms. In Fig.4, we can seen the standard deviation of the MWWO is much worse than that of the PSO, but the MWWO gains maximum combat effectiveness. The standard deviation of the MWWO is better than other algorithms.

5. Conclusion

This paper, an improve MWWO is proposed, which can better balance exploration and exploitation ablity. The optimal fitness value of the MWWO is much better than that of other intelligent optimization algorithms. The MWWO is used to solve the WTA problem, which can find the optimal decision matrix X and maximum mathematic expectation. The experimental results show that proposed MWWO outstanding performance than other the swarm intelligent algorithms in terms of the performance measures.

References

[1] Holland J H. Genetic Algorithms. Scientific American, 1992, 267(1):66-72.
[2] Rainer Storn. Differential Evolution Research-Trends and Open Questions. Studies in Computational Intelligence, 2008, 143:1-31.
[3] Rechenberg, I . Evolutionary Strategy. Computational Intelligence Imitating Life, 1994.
[4] Fogel, Gary B. .Evolutionary Programming. In: Rozenberg G., Bäck T., Kok J.N. (eds) Handbook of Natural Computing. Springer, Berlin, Heidelberg. 2011.
[5] D. Simon. Biogeography-Based Optimization. IEEE Transactions on Evolutionary Computation, 2008, 12(6): 702-713.
[6] Karaboga, Dervis. Artificial Bee Colony Algorithm. Scholarpedia, 2010, 5(3):6915.

[7] Kennedy, J. , R. Eberhart. Particle Swarm Optimization. ICNN95-international Conference on Neural Networks, IEEE, 2002.
[8] Valian, Ehsan, E. Valian. A cuckoo search algorithm by Lévy flights for solving reliability redundancy allocation problems. Engineering Optimization, 2013 45:1273-1286.
[9] Yang, Xin She, X. He. Firefly Algorithm: Recent Advances and Applications. International Journal of Swarm Intelligence , 2013,1:36-50.
[10] Yang X S, Gandomi A H. Bat Algorithm: A Novel Approach for Global Engineering Optimization. Engineering Computations, 2012, 29(5):464-483.
[11] Mirjalili, Seyedali , S. M. Mirjalili, A. Lewis. Grey Wolf Optimizer. Advances in Engineering Software, 2014, 69:46-61.
[12] Kaveh, A., N. Farhoudi. A new optimization method: Dolphin echolocation. Advances in Engineering Software, 2013, 59:53-70.
[13] Mirjalili S, Lewis A. The Whale Optimization Algorithm. Advances in Engineering Software, 2016, 95:51-67.
[14] Xing B., Gao WJ. Fruit Fly Optimization Algorithm. In: Innovative Computational Intelligence: A Rough Guide to 134 Clever Algorithms. Intelligent Systems Reference Library, vol 62. Springer, Cham. 2014.
[15] Khishe, M., M. R. Mosavi. Chimp optimization algorithm. Expert Systems with Applications , 2020,149:113338
[16] Rashedi, Esmat, H. Nezamabadi-Pour, S. Saryazdi . GSA : a Gravitational Search Algorithm. Information Sciences, 2009, 179:2232-2248.
[17] Xing B., Gao W J. Charged System Search Algorithm. In: Innovative Computational Intelligence: A Rough Guide to 134 Clever Algorithms. Intelligent Systems Reference Library, vol 62. Springer, Cham. 2014.
[18] Tayarani-N, M. H., M. R. Akbarzadeh-T. Magnetic Optimization Algorithms a new synthesis. 2008 IEEE Congress on Evolutionary Computation, IEEE, 2008.
[19] Geem, Zong Woo , J. H. Kim , G. V. Loganathan. A New Heuristic Optimization Algorithm: Harmony Search. Simulation, 2001, 2:60-68.
[20] Y. J. Zheng. Water wave optimization: A new nature-inspired metaheuristic. Computers & Operations Research, 2015, 55:1–11.
[21] A. S. Manne. A target-assignment problem. Operations Research, 1958, 6: 346-351
[22] Mirjalili S, Mirjalili S M, Lewis A. Grey Wolf Optimizer. Advances in Engineering Software, 2014, 69(3):46-61.
[23] Yang X S, He X. Bat algorithm: literature review and applications. International Journal of Bio-Inspired Computation, 2013, 5(3):141-149.
[24] Yang X S. Flower pollination algorithm for global optimization, In: Proceedings of the Unconventional Computation and Natural Computation, Lecture Notes Computer Science, 2012, 7445: 240-249.
[25] Mirjalili S. SCA: A Sine Cosine Algorithm for solving optimization problems. Knowledge-Based Systems, 2016, 96:120-133.
[26] Guo Zhou, Yongquan Zhou, Ruxin Zhao. Hybrid social spider optimization algorithm with differential mutation operator for the job-shop scheduling problem. Journal of Industrial & Management Optimization. 2021, 17(2): 533-548.
[27] Sen Zhang, Guo Zhou, Yongquan Zhou, Qifang Luo. Quantum-inspired satin bowerbird algorithm with Bloch spherical search for constrained structural optimization. Journal of Industrial & Management Optimization. doi: 10.3934/jimo.2020130
[28] M. Z. Ruan, Q.M. Li, T.H. Liu, Modeling and optimization on fleet antiaircraft firepower allocation, Acta Armamentarii, 2010, 31 (11):1525–1529.

Analysis of Opening Angle for Radial Magnetic Field Compensation Windings of Double-layer Ferromagnetic Cylindrical Shells

Chao Zuo*, Meng Chen, Hanchen Xiao and Pan Geng

China Wuhan second ship design & research Institute, Wuhan, 430205, China

*Corresponding author email: zuochao1986@qq.com

Abstract. A double-layer ferromagnetic cylindrical shell causes a magnetic anomaly in its surrounding space under the action of the geomagnetic field. Aiming at the problems of poor compensation uniformity and obvious local magnetization caused by the single-branch winding compensation method, a double-branch winding solution is proposed. By establishing a double-layer ferromagnetic cylindrical shell model, the radial magnetization characteristics of the cylindrical shell under the action of the vertical geomagnetic field was analyzed, and the opening angle of double-branch winding was studied. By optimizing the opening angle of compensation windings in the three proposed configuration schemes with respect to the minimum abnormal magnetic field, the optimal ranges of opening angle for the double-branch compensation windings were determined.

1. Introduction

Under the Earth's magnetic field, ferromagnets induce a magnetic field around themselves that exhibits abnormal characteristics. With the development of magnetic detection technology and torpedo and other weapons, the magnetic protection of cylindrical shell structures such as submarines, and underwater unmanned vehicles needs to be considerably improved. The radial induced magnetic field under the vertical magnetization of geomagnetism is crucial to measure the characteristics of magnetic anomalies. And the magnetic field induced by the ferromagnets is also an important direction for degaussing system design[1-3].

A double-branch compensation winding configuration scheme is proposed to achieve the best compensation effect, which can improve the compensation uniformity of the radial magnetic field outside the casing, reduce the peak value of magnetic anomalies observed. In order to achieve the best compensation effect, comparative results of optimal opening angles of double-branch windings under the three proposed configurations were analyzed. The proposed double-branch winding compensation scheme and optimization method can provide guidance for related engineering applications.

2. Radial Magnetic Field Compensation Model of Double-layer Cylindrical Shells

The degaussing windings used for magnetic field compensation are usually laid along the horizontal plane in case of a double-layer cylindrical shells in the vertical geomagnetic field. The single-branch windings are spread along the mid-plane, as shown in figure1 (a), while double-branch windings are parallel to the mid-plane, as shown in figure1 (b).

Content from this work may be used under the terms of the Creative Commons Attribution 3.0 licence. Any further distribution of this work must maintain attribution to the author(s) and the title of the work, journal citation and DOI.
Published under licence by IOP Publishing Ltd

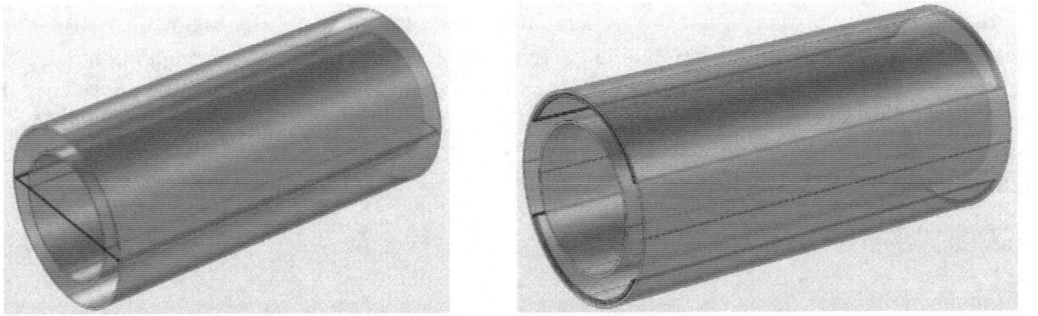

(a) (b)
(a). Single-branch windings (b). Double-branch windings

Figure 1. The schematic of single- and double- branch windings.

Considering that its axial size is considerably larger than its radial size, a double-layer cylindrical shells in a uniform magnetic field can be regarded as an infinitely long structure[4-7]. Therefore, its vertical two-dimensional section can be used for magnetization characteristic analyses. The three proposed configurations of double-branch windings are shown in figure2. figure2(a) shows that the double-branch windings are laid inside the inner shell; figure2(b) shows that the double-branch windings are laid between the inner and outer shells (close to the inner shell); figure2(c) shows that the double-branch windings are laid between the inner and outer shells (close to the outer shell).

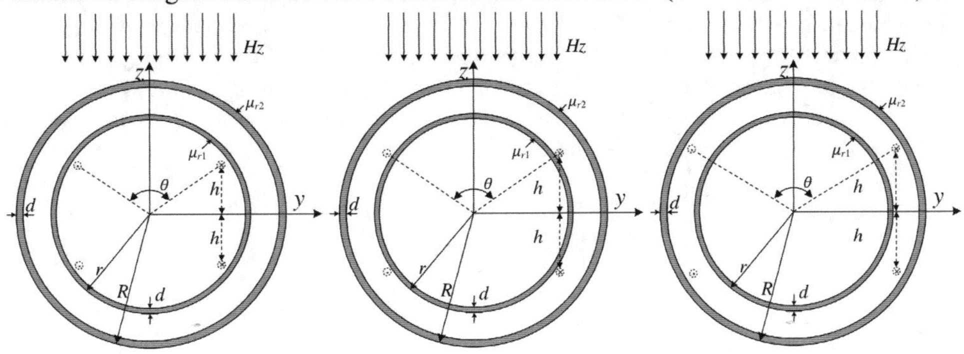

(a) (b) (c)
(a). Inside the inner shell (b). Between the inner and outer shells (close to the inner shell)
(c). Between the inner and outer shells (close to the outer shell)

Figure 2. The three configuration schemes of double-branch windings.

There is no current in the shell domain when no compensation measure is used. The scalar magnetic potential φ exists and the magnetic field strength **H** can be expressed by:

$$\mathrm{H} = -\nabla\varphi \tag{1}$$

When the compensation measures are added as shown in figure2, due to the existence of current, the vector magnetic potential A=[Ax, Ay, Az] should be introduced such that $\mathbf{B} = \nabla \times \mathbf{A}$ and $\nabla \cdot \mathbf{A} = 0$. In the two-dimensional plane field, the vector magnetic potential ***A*** and current density ***J*** are parallel to each other and only exist in the direction perpendicular to the plane, that is, Ay = Az = 0, Ax = A, Jy = Jz = 0, and Jx = J. The relationship between the magnetic potential and current density is as follows:

$$\frac{\partial^2 A}{\partial y^2} + \frac{\partial^2 A}{\partial z^2} = -\mu J \tag{2}$$

Because a ferromagnetic medium exists in the domain, the spatial magnetic field can be expressed as a superposition of three parts:

$$\mathbf{H} = \mathbf{H}_J + \mathbf{H}_M + \mathbf{H}_B \tag{3}$$

where $\mathbf{H_J}$ is the magnetic field intensity generated by the current; $\mathbf{H_M}$ is the magnetic field generated by the ferromagnetic material; and $\mathbf{H_B}$ is the geomagnetic background magnetic field intensity.
For a multi-medium model, the boundary satisfies the continuous tangential component of the magnetic field intensity and the continuous normal component of the magnetic flux density on both sides of the interface. The magnetic field intensity distribution throughout the domain can then be obtained via numerical solutions such as finite elements method.

3. Example Analysis

3.1. Example Model and Optimal Opening Angle

To analyze the relationship between the compensation effect, compensation winding layout, and current configuration, the following example analysis models are established through the finite element simulation software COMSOL Multiphysics.
The section radius of the inner shell is 5 m; relative magnetic permeability, 100; The section radius of the outer shell is 6 m; relative magnetic permeability, 200; inner and outer shell thickness, 100 mm; number of winding layers, 2; the compensation effect is measured for the upper 20 m of the measurement line. The shortest distance between the winding arrangement and shell was fixed at Gap = 100 mm. The overall structure of the model is shown in figure2. Considering the opening angle of the winding with respect to the center of the circle as α, the coordinates of the center of the incoming line on the upper left of the winding are [r*cos(90°-α/2), r*sin(90°-α/2)]; the coordinates of the remaining points can be obtained based on symmetry; and the vertical component of the geomagnetic field is considered as 30000 nT. Analyze the magnetic flux density B of the double-branch windings in the space with different number of ampere-turns IA and open angle α through simulation.
The objective function of the best compensation is the minimum value of the maximum magnetic anomaly on the observation line, as shown below:

$$\begin{cases} Objective = \min\left\{\max\left[abs\left(B_z^i(\alpha, I_A) - B_B^i\right)\right]\right\},\ i \in l \\ Constra\text{int} = \alpha \in \left[10^\circ, 170^\circ\right], step = 0.25^\circ \\ Constra\text{int} = I_A \in [100, 300], step = 0.25^\circ \end{cases} \tag{4}$$

where B_Z^i is the vertical component of the magnetic flux density B at each point on the observation line, which can be solved by finite element simulation; B_B^i is the vertical component of the geomagnetic flux density of the corresponding point, that is, 30000nT in this example; i is the number of each point on the observation line, and l represents the measurement line.
The variables are the relative opening angle α and the number of ampere-turns of the winding. To determine the best solution, a genetic algorithm is selected to realize parameter optimization[8-10]. For the parameter optimization, the number of iterations was set to 50; number of populations per generation, 80; range of the number of ampere-turns for each winder layer, 100 ~ 300; and range of opening angle α, 10~170°. Considering the practical feasibility, both variables are selected in 0.25 steps.
Case 1: The double-branch windings are laid inside the inner shell. The vertical magnetic anomaly increases and then decreases with the increase in the opening angle of the windings. The final optimization result is that the number of ampere-turns for each branch of the double-branch windings is 907, and the opening angle α is 119°; further, the maximum magnetic anomaly value of the measuring line at 20 m was 0.21 nT, as shown in figure3.
Case 2: The double-branch windings are laid between the inner and outer shells (close to the inner shell). The final optimization result is that the number of ampere-turns for each branch of the double-branch windings is 293, and the opening angle α is 118.5°; further, the maximum magnetic anomaly value of the measuring line at 20 m was 0.76 nT, as shown in figure4.
Case 3: The double-branch windings are laid between the inner and outer shells (close to the outer shell). The final optimization result is that the number of ampere-turns for each branch of the double-

branch windings is 278.8, and the opening angle α is 118°; further, the maximum magnetic anomaly value of the measuring line at 20 m was 1.21 nT, as shown in figure5.

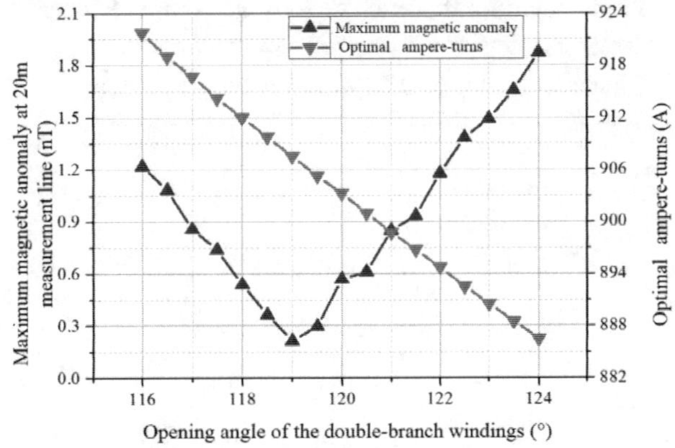

Figure 3. Maximum magnetic anomaly and ampere-turns for different opening angle α in case 1.

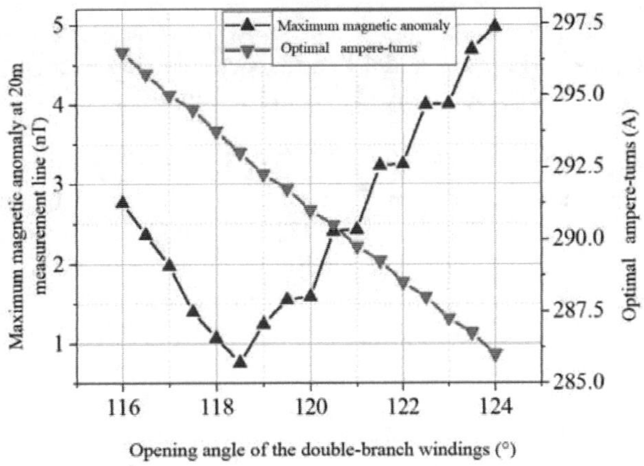

Figure 4. Maximum magnetic anomaly and ampere-turns for different opening angle α in case 2.

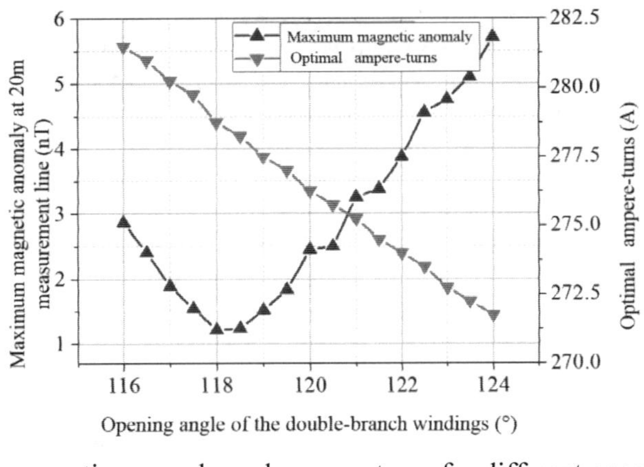

Figure 5. Maximum magnetic anomaly and ampere-turns for different opening angle α in case 3.

3.2. Performance Improvement Analysis

To verify the improvement in the performance of the double-branch windings configuration scheme over that of the traditional method, the current configurations of single- and double-branch windings were optimized for the same double-layer cylindrical shell model. Table 1 compares the compensation effects of the single- and double-branch winding schemes.

Table 1. Comparison of the compensation effects.

	Case	Total ampere-turns(A)	Maximum magnetic field(A/m)	Magnetic anomaly(nT)
Case 1	Single-branch	1559	231	12.5
	Double-branch	1815	86	0.2
	Comparison	16%	-63%	-98%
Case 2	Single-branch	512	74	29.7
	Double-branch	586	39	0.8
	Comparison	14%	-47%	-97%
Case 3	Single-branch	487	48	32.6
	Double-branch	557	24	1.2
	Comparison	14%	-50%	-96%

As summarized in table I, when the double-branch windings scheme is adopted for the three cases, the total ampere-turns increase by 16%, 14% and 14%, respectively, while the maximum magnetic field decreases by 63%, 47% and 50%, respectively. Furthermore, the magnetic anomaly drops by 98%, 97% and 96%, respectively.

Comparing to single-branch winding scheme, the maximum compensation magnetic field for the double-branch winding scheme decreases significantly, due to more distributed current for the double-branch winding scheme. The magnetic anomaly value of double-branch winding scheme drops significantly in comparison with single-branch winding scheme, because the double-branch winding scheme exhibits better fitting freedom of the spatial magnetic field, the optimal compensation effects can be achieved by adjusting the opening angle α.

4. Conclusion

Aiming at the magnetic anomaly induced by the geomagnetic field magnetization for the double-layer ferromagnetic cylindrical shells, the optimal opening angles of the double-branch degaussing winding scheme in three cases were compared and analyzed. The optimal opening angle of the double-branch winding scheme ranges in 118°~119°, in which cases the compensation effects have been significantly improved. For the double-branch winding scheme, the magnetization effect of the ferromagnetic cylindrical shells was weakened due to more distributed current. In addition, the magnetic anomaly after compensation drops over 96% due to better fitting freedom of the spatial magnetic field.

References

[1] W Guanghui Z Hai G Zhengdong et al. 2015 Magnetic anomaly signal characteristics of ship induced magnetism [J]. *Ship Science and Technology* 37 (009): 137-139

[2] Z. Anming et al. 2014 Equivalent calculation of induced magnetic field of ship hull module *Chin. Ship Res.* vol. 9 no. 3 pp. 109-122.

[3] G. Chengbao et al. 2006 numerical analysis of the longitudinal magnetization state of the hollow ferromagnetic cylinder in the geomagnetic field J. *Harbin Eng. Univ.* vol. 27 no. 6 pp. 886-890.

[4] Z. Duanduan et al. 2015 Study on the distribution law of magnetic field around vertical casing in geomagnetic field *Petrol. Mach.* vol. 43 no. 3 pp. 12-16.

[5] O. Chadebec et al. 2000 Modeling of static magnetic anomaly created by iron plates *IEEE Trans. Magn.* vol. 36 no. 4 pp. 667-671.

[6] G. Chengbao et al. 2006 The numerical simulation of the longitudinal magnetic state of a hollow ferromagnetic cylinder in the geomagnetic field J. *Harbin Eng. Univ.* vol. 27 no. 6 pp. 886-890.

[7] N. Guangzheng et al. 2004 numerical calculation of engineering electromagnetic field. *Beijing: Machinery Industry Press* pp. 209-211.
[8] MODAGEKAR Vishal et al. 2016 Optimization in Tri-axial degaussing system design and estimation of degaussing coil currents. *IEEE Transactions on Magnetics* 53(4): 1-12.
[9] Lian Liting et al. 2012 Closed loop control method of winding current based on neural network. *Journal of Naval University of Engineering* 2(21):24-72.
[10] Geng Pan Wang Xiangjun Wang Jianxun Wang Zuoshuai Zuo Chao. 2020 Research on multi-objective optimization design method of naval induced magnetic field compensation[J]. *Journal of Naval University of Engineering* 32(01): 76-81.

Finite Analysis of Support Time for Tunnels in Strain-softening Rock Mass in Consideration of Fictitious Support Pressure

Lan Cui[1,2], Qian Sheng[1,2] and Zhenzhen Niu[3,*]

[1]State Key Laboratory of Geomechanics and Geotechnical Engineering, Institute of Rock and Soil Mechanics, Chinese Academy of Sciences, Wuhan 430071, China
[2]University of Chinese Academy of Sciences, Beijing 100049, China
[3]College of Marine Science and Technology, China University of Geosciences, 388 Lumo Road, Wuhan 430074, China

[*]Corresponding author email: 1202021753@cug.edu.cn

Abstract. This study considers that the support time is quantitatively determined by the production limit of the displacement reduction factor and the support force under the extrusion conditions of the strain-softening rock mass. Therefore, the two indicators of the downlines under the support time are the displacement reduction factor of the support force and the yield limit. Based on the solution of the fictitious pressure proposed in an existing paper, the finite difference method is adopted to investigate the variations of the support force and displacement reduction factor versus the delayed distance considering different support types, initial stresses, and post-peak behaviours. The results show that on the one hand, the delay distance is suggested within 1 R_0 in most tunnel cases; on the other hand, the factors have greater impact on rock-support interactions are rock mass and in-situ stress. Relatively contrast, softening and expansion behavior was not significant enough. Furthermore, it is also very important in composite support systems to assess the proportion of loads shared with the weakest part.
Keywords: Fictitious pressure, strain-softening rock mass, support time, delayed distance, allowable deformation.

1. Introduction

Rock-support interaction is critical to the assessment of the load imposed on the support during tunneling, which is often analysed with convergence-confinement method [1-5]. Improvement of the convergence constraint method yields a two-stage analysis, a method that divides the tunneling process into two stages [6]. It is first assumed that the fictitious support pressure caused by the tunnel surface effect supports the rock mass during the first stage before supporting the installation. As the tunnel surface gradually moves forward, the fictitious support pressure gradually shifts to the support pressure. Through a two-stage analysis method, Einstein and Schwatz [7,8] presented an analytical solution to compute the stresses and displacements on the support interacting with an elastic rock mass under a non-uniform stress field, and discussed the influence of stress reduction factor on the loads and displacements on the support. Cai et al. [9,10] delivered an analytical model to investigate the interaction between the ground and the rock bolt and considered the released stresses at the installation of the rock bolt. Assuming a stress reduction factor, Carranza-Torres and Diederichs [11] computed the force on the liner and the steel set of a composite support by using the commercial finite difference software. Hoek [12] obtained the fictitious support pressure by substituting the released displacement

Content from this work may be used under the terms of the Creative Commons Attribution 3.0 licence. Any further distribution of this work must maintain attribution to the author(s) and the title of the work, journal citation and DOI.
Published under licence by IOP Publishing Ltd

into the strain prediction equation, and evaluated the force in different support types by relieving the fictitious pressure.
However, the fictitious support pressure and the stress release factor in most studies [7-12] are predicted by simplifying the rock mass as elastic, which is not appropriate if more complicated mechanical behaviours of the rock mass are considered. Cui et al. [6] accounted for the strain-softening and dilatant behaviours for rock mass in first stage, but without discussing the rock-support interaction in the second stage. Meanwhile, the support in many studies [7-12] is postulated to be added at the time the tunnel is excavation without any delay. However, effects of the support time on the forces in the support and rock deformation are significant for support safety. Therefore, how to reduce the support load under the extrusion conditions is the top priority. It is recommended to install delay or yield support and permitted rock displacement [12-16]. However, the impact of rock mass mechanical properties is not clear enough, while the delay distance and reserved displacement are difficult to obtain the relevant quantitative analysis.
In this paper, the rock-support interaction is investigated with the two-stage analysis method by incorporating a Finite Difference Method (FDM) solver. It mainly includes the following parts:
(1) The fictitious support pressure presented in Ref. [6] is adopted as the initial pressure on the support at the beginning of the second stage.
(2) Based on the above theory, a planar model with a circular water purification stress field allows the rock-support interaction by gradually loosening the virtual support pressure outside the tunnel from the initial value to zero.
(3) The displacement reduction factor is proposed to evaluate the impact of the support on controlling the rock deformation, yielding the ratio of the supported displacement reduction to the maximum rock deformation.
(4) Combined with considerations of different support types, initial stress, and post-peak behavior, the variation of support force and displacement reduction factors with delay distance is investigated. The two indicators of the support installation delay distance and the allowable deformation are also determined by quantification.

2. Problem Statement

2.1. Basic Assumptions

The following assumptions are made in this study:
(1) The tunnel opening is circular and excavated under a hydrostatic stress field σ_0.
(2) The tunnel problem is reduced to a plane problem by applying virtual support pressure on the periphery of the tunnel. On the plane, radial stress σ_r in the rock body corresponds to smaller main stress σ_3, while tangential stress σ_θ corresponds to large main stress σ_1.
(3) Post-peak stages will involve elastic-perfectly-plasticity (EPP), strain-softening (SS), and elastic – brittle-plasticity (EBP) behavioral [17]. Meanwhile, the rock bulk strength mainly follows the Mohr-Coulomb (M-C) and Hoek-Brown (H-B) failure criteria. Thus, the fault standard is set as follows:

$$f(\sigma_\theta,\sigma_r,\eta)=\begin{cases}\sigma_\theta-\sigma_r-\sigma_{ci}\left[m_b(\eta)\sigma_r/\sigma_{ci}+s(\eta)\right]^{a(\eta)}\\ \sigma_\theta-K_\varphi(\eta)\sigma_r-2C(\eta)\sqrt{K_\varphi(\eta)}\end{cases} \tag{1}$$

In which, for M-C criterion, C is the cohesion and K_φ is the friction coefficient; for the H-B failure criterion, m_b, s, a are the strength constants; the ω is assumed to be a function of the plastic softening parameter η. The relation between η and ω can be expressed as:
for EPP rock mass:

$$\omega(\eta)=\omega^p \tag{2}$$

for SS rock mass:

$$\omega(\eta)=\begin{cases}\omega^{p}-(\omega^{p}-\omega^{r})\dfrac{\eta}{\eta^{*}}, & 0<\eta<\eta^{*}\\ \omega^{r}, & \eta\geq\eta^{*}\end{cases} \tag{3}$$

for the EBP rock mass:

$$\omega(\eta)=\omega^{r} \tag{4}$$

In Equation (2), (3) and (4), η^{*} is the critical plastic softening parameter; ω^{p} is the peak intensity parameter; and ω^{r} is the residual strength parameter. Among them, η^{*} varies by rock, in the EPP rocks, η^{*} is ∞; in the EBP rock mass, η^{*} is 0.

(4) M-C function is selected as the flow potential g, which is described as:

$$g(\sigma_\theta,\sigma_r,\eta)=\sigma_\theta-K_\psi\sigma_r \tag{5a}$$

$$K_\psi=(1+\sin\psi)/(1-\sin\psi) \tag{5b}$$

Where K_ψ is the dilatancy coefficient and ψ is the dilatancy angle.

2.2. Two-stage Analysis Method

As the name suggests, the two-stage analysis method is divided into stage one and two. In the first phase, the rock bunker near the tunnel surface before the scaffold mounting is supported by a fictitious support pressure p_f mobilized by the tunnel surface effect and evenly distributed around the tunnel periphery. The tunneling process is plotted based on this method, as shown in figure 1. In the initial state, the distance of the cross-section A – A' to the tunnel face is x_{ini}, the rock displacement at the tunnel periphery is $u_{0,ini}$, and the fictitious support pressure is $p_{f,ini}$. During the second stage, $p_{f,ini}$ at the cross-section A – A' is unloaded and transferred to the support as the excavation advances. It is seen that the tunnel surface and support together support the rock mass. The pressure p_i can be decomposed into the fictitious support pressure p_f and the support pressure p_s. Therefore, the intermediate and final states of the second phase can be analyzed separately. The case of the factor change in the intermediate state is: x_{ini} and $u_{0,ini}$ increases to x_{int} and $u_{0,int}$ respectively. The p_s was increased from zero to the $p_{s,int}$, the $p_{f,ini}$ drops to the $p_{f,int}$. At the final state, the pressure is fully transferred to the support, $p_{f,int}$ decreases to $p_{f,fin}=0$, and p_s grows to $p_{s,fin}$. Given that the study does not consider the rheological behavior of the rock mass, and the bracket installation at x_{ini}, $p_{s,fin}$ is the final load of each excavation step, and $u_{0,fin}$ is the final rock mass displacement at the periphery of the tunnel.

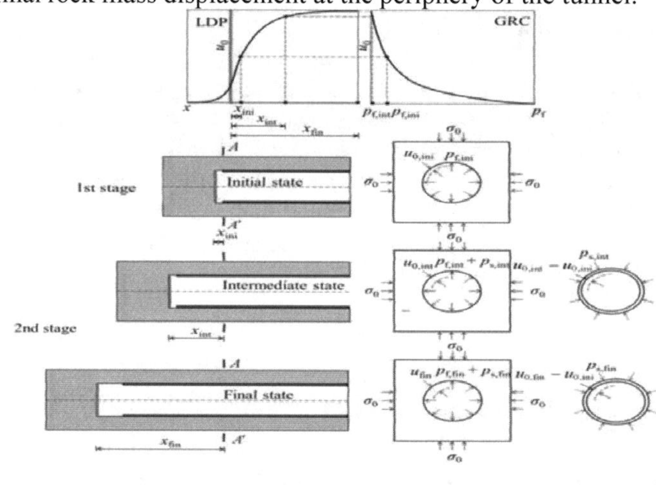

Figure 1. Tunnel excavation process with two-stage analysis method.

3. Finite Difference Analysis

3.1. Model Configuration

A series of FDM analyses simulating the tunnel excavation process are conducted using a commercial finite difference software [18]. 1 / 4 of the correlation regions is modeled based on the symmetry of tunneling through the tunnel to the vertical and horizontal planes (Figure 2 (a) – (b)). A problem slice with a thickness of 0.5m was considered in the out-of-plane direction of the structure. Meanwhile, the displacement in this direction was fixed to follow the planar strain condition. Furthermore, The model size under plane strain conditions was 100m in the vertical and horizontal directions, respectively.

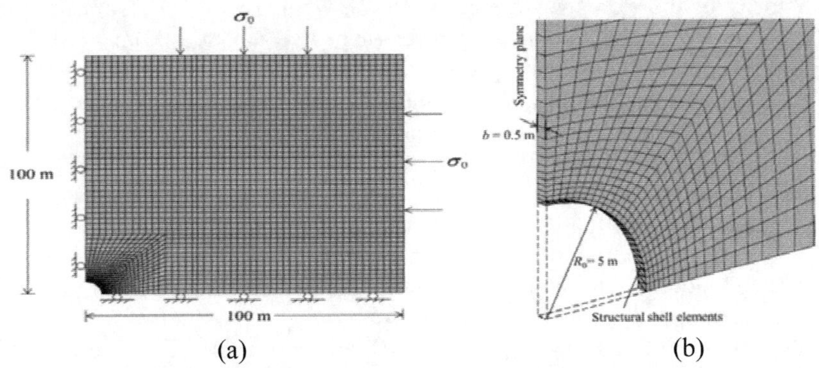

Figure 2. FDM model for the rock-support interaction: (a) Boundary conditions that belong to the model; (b) Meshes around circular openings.

The support is discretised with structural shell elements. This predicts a high interface strength between the support and the rock mass, assuming a rigid contact with the periphery of the tunnel. Three types of supports (I, II, III) are adopted. Support types I and II are comprised of shotcretes with an equivalent thickness t of 0.21 but different elastic moduli E_s, the former is 20 GPa, and the latter is 30 GPa. Support type III can be referred to Ref. [11], Its components include shotcrete and two steel sections along the width of the unit (Type W6 × 25). The elastic modulus E_s of the support type III is derived as 37.69 GPa from the equivalent section approach in Ref. [11].

The data in Table 1 show the properties of the rock mass. From A1 to A3,B1 to B6,C1 to C5,D1 to D4, the study considered different GSI_{max}, η^*, ψ and σ_0 within these intervals. EPP rock mass is assumed in the cases A1 to A3, C1 to C5 and D1 to D4. The peak of the geological strength index is indicated by GSI_{max}, which also means rock mass. Meanwhile, the numerical procedure proposed in Ref. [6] can solve the virtual support pressure $p_{f,ini}$. Values of $p_{f,ini} / \sigma_0$ versus a normalised initial distance to the tunnel face X^*_{ini} with each case are presented in the Appendix A. X^*_{ini} is the ratio of x_{ini} to the radius of the circular opening R_0. The mechanical properties of rock masses from A1 to A3 are displayed in table 1. From cases B1 to B6, and C1 to C5, the basic parameters are: φ is 32.47°, C is 3.08 MPa, φ^r is 24.41°, E_r is 9000 MPa, R_0 is 5 m, σ_0 is 35 MPa. For cases B1 to B6, η^* is 0, 0.005, 0.01, 0.05, 0.5, ∞. For cases C1 to C5, ψ is 0°, 8.12°, 16.24°, 24.36°, 32.47°. The basic parameters for cases D1 to D4 are the same to Case A3. For cases D1 to D4, σ_0 is 5 MPa, 25 MPa, 35 MPa, 50 MPa.

Table 1. Data on rock mass mechanical properties from cases A1 to A3

	GSI_{max}	m_b	s	a	E_r / MPa	p_i / MPa	μ	R_0 / m	σ_{ci} / MPa	σ_0 / MPa	ψ / °
A1	75	6.35	0.0509	0.5	30	0	0.25	5	110	35	0
A2	50	1.71	0.0026	0.5	9000	0	0.25	5	80	35	0
A3	40	0.99	0.0008	0.5	3000	0	0.25	5	50	35	0

3.2. Numerical Procedures

Implementing the FDM analysis can be done by gradually relaxing the virtual support pressure outside the tunnel from the in-situ value to zero. The numerical procedures to implement the FDM analysis are as follows:

(i) Exert $p_{f,ini}$ on the tunnel periphery in the model, and iterate to reach the equilibrium state and obtain $u_{0,ini}$.

(ii) Add shell elements on the tunnel periphery; at excavation step i, x_{ini} increases to $x_{(i)}$, obtain $p_{f,(i)}$ at $x_{(i)}$ and reduce $p_{f,ini}$ to $p_{f(i)}$ to mimic the excavation advance; iterate to reach the equilibrium state.

(iii) Check whether $p_{f,(i)}$ decreases to zero or not; if yes, cease the rock-support interaction and obtain $u_{0,fin}$ and the final thrust force in the shell element N_s, otherwise, set $i = i + 1$ and repeat the iteration (ii).

In the implementation, the interval between $x_{(i)}$ and $x_{(i+1)}$ denotes the excavation step size. As listed in Table A-1, a majority of the step sizes are equal to R_0, whereas several step sizes are smaller than $0.5R_0$. As in Ref. [19], when the step size is lower than or equal to R_0, the LDP remains basically constant. This means various step sizes in Table A-1 exert negligible impact on the FDM analysis, and the analysis is applied to the tunnelling with all the excavation step size smaller than or equal to R_0.

3.3. Delayed Distance and Allowable Deformation

Allowable rock formations are indicated by $u_{0,allow}$. X^*_{del} is reserved for support installation, which is the ratio of the delayed distance x_{del} to the radius of the circular opening R_0. In the LDP, $u_{0,allow}$ and X^*_{del} represent a one-to-one communication. In addition to X^*_{del}, X^*_{ini} also indicates the normalized distance of the bracket mounting to the tunnel surface. Furthermore, $u_{0,ini}$ and $u_{0,allow}$ indicate the initial rock mass displacement at X^*_{ini}. However, it need to note the purpose of estimating X^*_{del} and $u_{0,allow}$, ensuring the safety of support and the effectiveness of controlling rock displacement. X^*_{del} and $u_{0,allow}$ represent relatively narrow ranges of X^*_{ini} and $u_{0,ini}$, whereas X^*_{ini} and $u_{0,ini}$ represent random distance and initial displacement, respectively. The thrust of the concrete in the final state N_s can be used to determine the range of the $u_{0,allow}$ and X^*_{del} used to support the installation, and the displacement reduction factor f_0. f_0 denotes the reduction ratio of rock deformation in terms of the support installation at X^*_{ini}, which is defined as:

$$f_0 = \frac{u_{0,max} - u_{0,fin}}{u_{0,max}} \tag{6}$$

In formula, $u_{0,max}$ represents the maximum rock displacement at the periphery of the tunnel, the parameter can be obtained in the FDM model by applying $p_i = 0$ to the tunnel periphery; figure 3 is the flowchart for the methodology, the changes of f_0 and N_s are negatively correlated with the X^*_{ini}, they increase with decreasing X^*_{ini}, and the upper and lower bounds of X^*_{del} and $u_{0,allow}$ can be obtained by the following steps:

(i) Solve $u_{0,max}$ and the variations of $u_{0,ini}$, $u_{0,fin}$, and N_s versus X^*_{ini} by implementing FDM analysis.

(ii) Obtain the variation of f_0 by implementing $u_{0,fin}$ and $u_{0,max}$ into Eq. (6).

(iii) Solve lower bound of X^*_{del} by the variation of N_s versus X^*_{ini} corresponding to the yield limit of N_s; solve the lower bound of $u_{0,allow}$ by the variation of $u_{0,ini}$ versus X^*_{ini} at the lower bound of X^*_{del}.

(iv) Solve the upper bound of X^*_{del} by the variation of f_0 versus X^*_{ini} corresponding to the required f_0; solve the upper bound of $u_{0,allow}$ by the variation of $u_{0,ini}$ versus X^*_{ini} at the upper bound of X^*_{del}.

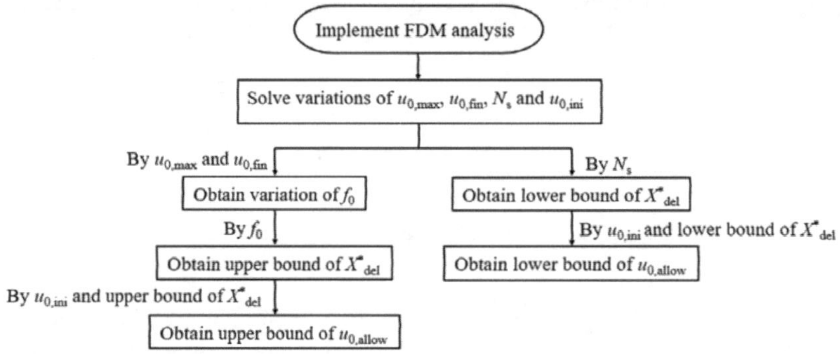

Figure 3. The flowchart for the methodology.

3.4. Verification for FDM Model

$u_{0,max}$ by FDM model and by the numerical approach proposed in Ref. [23] are compared. In Ref. [23], the surrounding rock mass was divided into finite annuli with a constant radial stress increment, and the stress and strain components were solved with the stress equilibrium and displacement compatibility equations by the FDM method. The number of finite annuli is about 5000. It has been compared to a numerical procedure in Ref. [24], the numerical error of the maximum rock deformation is less than 1 %. The comparison of $u_{0,max}$ by the FDM analyses and Ref. [23] for cases A1 to A3, B1 to B6, C1 to C5, D1 to D4 are shown in figure 4. As shown in figure 4, in most cases the difference between the two methods is less than 3 %. This implies a satisfying agreement between the two methods. The proposed method is acceptable.

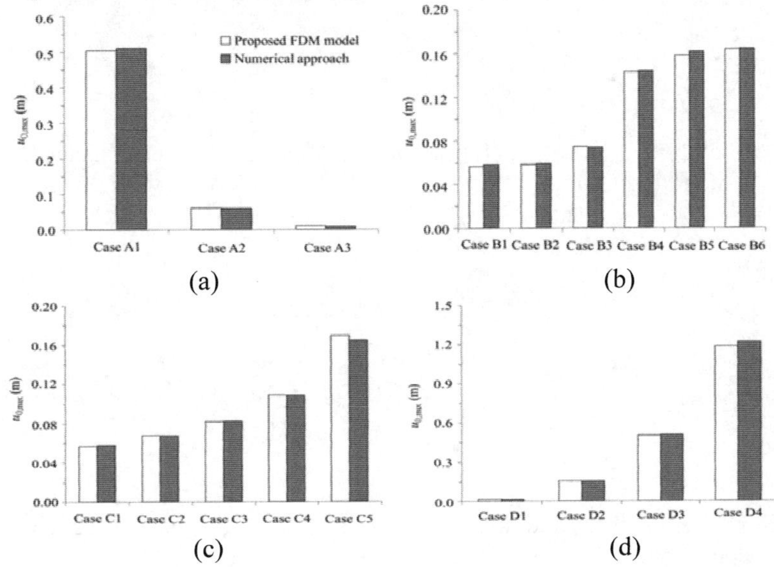

Figure 4. Comparison of $u_{0,max}$ by FDM model and a numerical approach proposed in Ref. [23].
(a) Cases A1 to A3; (b) Cases B1 to B6; (c) Cases C1 to C5; (d) Cases D1 to D4

4. Results and Discussion

4.1. Deformation Reduction Ratio and Thrust force

Figures 5 - 8 plot variation laws of N_s and f_0 with cases A1 to A3, cases D1 to D4, cases B1 to B6 and cases C1 to C5. Various values of GSI_{max} σ_0, η^* and ψ are assigned to the above cases, respectively (see table 2). For cases B1 to B6 with different η^*, and cases C1 to C5 with different ψ, support type I is adopted. For cases A1 to A3 with different GSI_{max}, and cases D1 to D4 with different σ_0, support types II and III are adopted. Compared with the steel sets, the shotcrete is easier to yield while the support is loaded. Hence, for the support type III, N_s shared in the shotcrete alone is discussed. As N_s calculated by the shell element in the FDM model denotes the composite force in the two materials, i.e., concrete and steel sets, N_s shared by the shotcrete should be obtained by the equivalent section approach proposed in Ref. [6].

As observed in figures 5 to 8, with the support type II, N_s and f_0 for the case D4 with σ_0 = 50 MPa are 9 times and 1.5 times as higher as those for case D1 when σ_0 is 5 MPa; N_s and f_0 for the case A3 with GSI_{max} = 40 are 3.67 times and 5 times as higher as those for the case A1 with GSI_{max} = 75. In contrast, with the influences of ψ and η^*, the ranges of N_s and f_0 is relatively small. Therefore, the influences of the rock mass quality and the in-situ stress on the support force and support effect are more remarkable than post-peak behavior such as the softening and dilatancy.

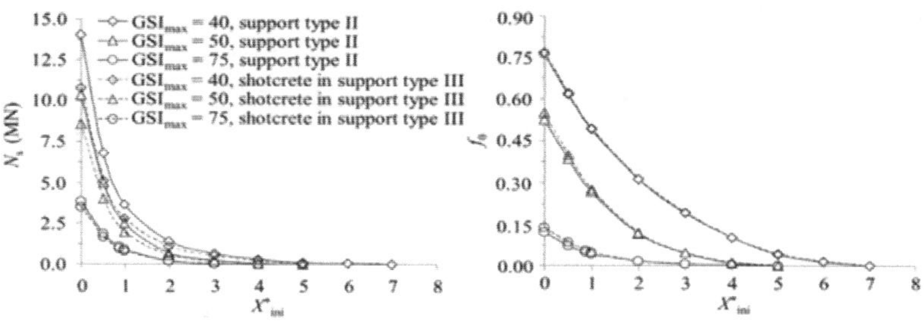

(a) (b)

Figure 5. Variations of thrust force and deformation reduction ratio versus distance to tunnel face with various GSI_{max} for support types II and III. (a) The variation laws of N_s with cases A1 to A3; (b) The variation laws of f_0 with cases A1 to A3.

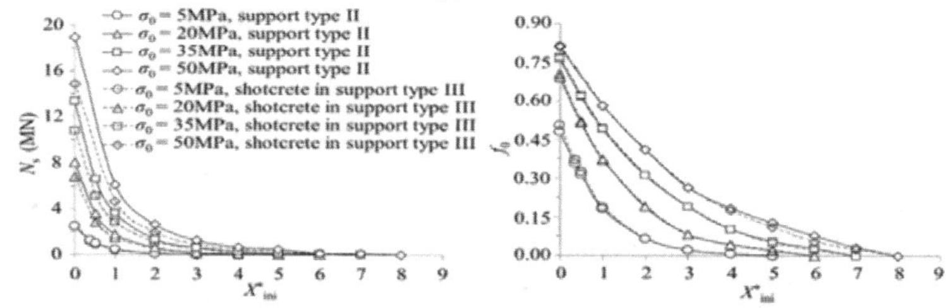

(a) (b)

Figure 6. Variations of thrust force and deformation reduction ratio versus distance to tunnel face with various σ_0 for support types II and III. (a) The variation laws of N_s with cases D1 to D4; (b) The variation laws of f_0 with cases D1 to D4.

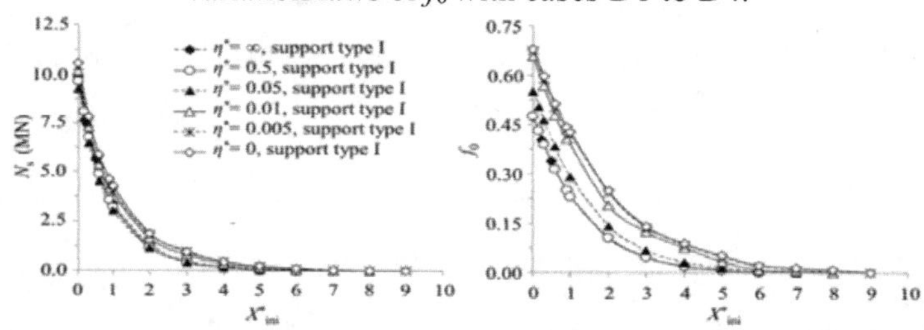

(a) (b)

Figure 7. Variations of thrust force and deformation reduction ratio versus distance to tunnel face with various η^* for support type I. (a) The variation laws of N_s with cases B1 to B6; (b) The variation laws of f_0 with cases B1 to B6.

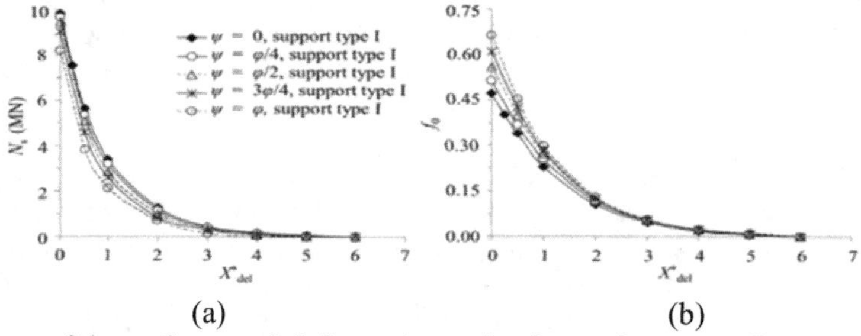

(a) (b)

Figure 8. Variations of thrust force and deformation reduction ratio versus distance to tunnel face with various ψ for support type I. (a) The variation laws of N_s with cases C1 to C5; (b) The variation laws of f_0 with cases C1 to C5.

As shown from figure 5 and 6, N_s is negatively correlated with GSI_{max}, and it declined as GSI_{max} grows. f_0 was positively associated with σ_0, and it decreased with σ_0. The results show that both N_s and f_0 have large values under weak rock mass conditions with high initial stress. The larger f_0 is beneficial to the safety of the tunnel, because the support energy can effectively control the rock mass displacement and make the shotcrete cracks. Therefore, this demonstrates that the yield support is installed when the condition is a weak rock mass with a high initial stress, while the allowed deformation is also retained, the purpose of which is to substantially reduce N_s. Figures 5 - 6 indicates that N_s and f_0 decline with X^*_{ini}. The above shows that the supporting effect limiting the rock displacement will be obvious when the delay distance or the deformation is large. This means that the scaffold mount is needed to be investigated for the optimal allowed deformation or delay distance, thereby guaranteeing a smaller N_s and a larger f_0.

Figures 5 to 8 indicates that f_0 increases with the decreases of GSI_{max} and η^*, and decreases with the decreases of σ_0 and ψ. The reason for presenting such a result is that $u_{0,max}$ will have a major effect on f_0, and it increases as $u_{0,max}$ grows (see Eq. (6)). Apparently, lower GSI_{max} and η^*, greater σ_0 and ψ give rise to a greater $u_{0,max}$. Hence, the trends of f_0 with variations in η^*, σ_0, GSI_{max}, and ψ coincides with those of $u_{0,max}$. In contrast, variations of N_s affected by GSI_{max}, η^*, σ_0 and ψ obviously differentiate from those of f_0. N_s rises with decrease of GSI_{max} and ψ, and increase of σ_0; It will decline, but then rise with η^*. Upon analysis, $u_{0,max}$ and $p_{f,ini}$, (the fictitious support pressure) will have an impact on N_s ,and resulting in such changes.

4.2. Lower and Upper Bounds of Delayed Distance and Allowable Deformation

Assessing the yielding limits of N_s for the concrete in the supports, it can carry out from three aspects of safety factor FS, support area A and concrete compressive strength σ_c [11]:

$$N_s = \frac{\sigma_c A}{\mathrm{FS}} \tag{7}$$

FS of 0.67, 1 and 1.5 are utilised in Ref. [11]; therefore, the yield limits of N_s are 2.8 MPa, 4.2 MPa, and 5.6 MPa by Eq. (7), its use is to evaluate three different lower bounds of X^*_{del} and $u_{0,allow}$. From figures 5 to 8 , the values of X^*_{del} and $u_{0,allow}$ with FS of 0.67, 1.0 and 1.5, and the specific results are shown in Table 2. The displacement reduction factor f_0 determines the upper bound of X^*_{del} and $u_{0,allow}$. f_0 of 30%, 40%, 50% and 60% are presented, different data describe the effects of supporting processes from weak to strong on the control of rock displacement. Furthermore, the blank representative support in Table 2 is very secure. Similarly, as shown from figures 5 to 8, Table 3 shows the value of X^*_{del} and $u_{0,allow}$ when f_0 reaches 30%, 40%, 50% and 60%.

Table 2. Lower bounds of X^*_{del} and $u_{0,allow}$ (a) X^*_{del}; (b) $u_{0,allow}$ (unit: cm)

(a)

	Support type III			Support type II		
	FS = 1.5	FS = 1	FS = 0.67	FS = 1.5	FS = 1	FS = 0.67
σ_0 = 5MPa	—	—	—	0.042	—	—
σ_0 = 20MPa	0.546	0.327	0.060	0.725	0.474	0.240
σ_0 = 35MPa	1.082	0.685	0.407	1.406	0.937	0.590
σ_0 = 50MPa	1.659	1.171	0.854	2.081	1.507	1.067
GSI_{max} = 40	1.079	0.683	0.407	1.406	0.934	0.590
GSI_{max} = 50	0.772	0.511	0.249	0.973	0.661	0.390
GSI_{max} = 75	0.197	—	—	0.270	—	—

(b)

	Support type III			Support type II		
	FS = 1.5	FS = 1	FS = 0.67	FS = 1.5	FS = 1	FS = 0.67
σ_0 = 5MPa	—	—	—	0.472	—	—
σ_0 = 20MPa	7.552	6.173	4.353	8.511	7.110	5.567
σ_0 = 35MPa	27.403	22.234	17.954	30.755	25.708	20.811
σ_0 = 50MPa	67.359	56.773	48.510	76.077	64.236	54.231
GSI_{max} = 40	27.432	22.256	17.952	30.794	25.720	20.825
GSI_{max} = 50	3.849	3.209	2.431	4.217	3.589	2.870
GSI_{max} = 75	0.350	—	—	0.395	—	—

As shown from Table 3, σ_0 = 5 MPa or GSI_{max} = 75, even if the X^*_{del} is reduced to zero, cannot bring the N_s in the support types II and III to the yield limits of FS = 1.0 and 0.67. Among these, when GSI_{max} = 75, in the support type III, the N_s is about 3.5 MPa, At the X^*_{ini} equals 0 for the support type II, the N_s is about 3.75MPa; when σ_0 = 5 MPa, in the support types III, N_s is about 2.3 MPa, and at the X^*_{ini} equals 0 for the support type II, the N_s is also about 2.3 MPa (see figures 5 and 6).
It is shown that the support force is fairly small as GSI_{max} is large but σ_0 is insignificant. Simple tunnel supports such as rockbolts and concrete are expected. Furthermore, since deformation reduction factor is decreased by increasing of the distance to the tunnel face, for smaller initial stress or strong rock mass to effectively control the deformation of rock, the support should be installed immediately. At a particular f_0, the upper bounds of X^*_{del} and $u_{0,allow}$ are almost equivalent for support types II (see table 3). As shown in figures 5 and 6, f_0 versus X^*_{del} change almost consistently in support type III and support type II. The only difference existing between the support types III and II is that the former adds steel groups. Considering the above results, the effect of steel group on limiting rock displacement is not obvious.
Meanwhile, when by the support type II and support type III, The lower bound of the X^*_{del} and $u_{0,allow}$ of the former is significantly larger than the latter. At the tunnel face, N_s of the support type II is 21.79 % higher than that of the support type III for the case A3 when σ_0 is 35 MPa (see figure 6). Thus, the more flexible support suffers larger pressure and can be delayed installed. This is contrary to Convergence Confinement Method. By this method, the stronger support is subjected to heavier load while installing close to the tunnel face. The is due to the fact that: the load on support type III is partially bore by steel set, hence, N_s of concrete in type III is smaller than N_s by the type II with a same concrete type. This follows that it is important to assess the proportion of loads shared to the weakest part in the type III acting as a composite support system.
The upper and lower bounds of X^*_{del} and $u_{0,allow}$ decrease with the increase in GSI_{max} and the decrease in σ_0 (see table 3). This indicates that as the rock mass is more stable, the delayed distance and the allowable deformation should be diminished. In addition, the lower bounds of X^*_{del} and $u_{0,allow}$ grow with the increase of FS. The upper bounds of X^*_{del} and $u_{0,allow}$ grow with the decrease of f_0. Hence, the ranges of X^*_{del} and $u_{0,allow}$ are correlative with the support strength capacity, and the required displacement reduction factor. Furthermore, with a stress level under 50 MPa, the upper bound of X^*_{del} with a larger f_0 (i.e. 50% ~ 60%) is below 1.0 (see table 3). Hence, larger f_0 will be achieved when X^*_{del} is less than 1. On the other hand, while X^*_{del} is 1, p_f/σ_0 for each case is less than 10%. This means the stress relief factor reaches more than 90% when X^*_{del} is 1. Therefore, in order to restrict the rock mass displacement, the delayed distance for support installation is suggested to be within R_0 for most tunnels. It is noted that the blank space indicates that f_0 is smaller than 30%, 40%, 50% or 60% at the tunnel face.

Table 3. Upper bounds of X^*_{del} and $u_{0,allow}$ (a) X^*_{del}; (b) $u_{0,allow}$ (unit: cm)

(a)

	Support type III				Support type II			
	f_0 = 30%	f_0 = 40%	f_0 = 50%	f_0 = 60%	f_0 = 30%	f_0 = 40%	f_0 = 50%	f_0 = 60%
σ_0 = 5MPa	0.630	0.312	0.019	—	0.590	0.256	—	—
σ_0 = 20MPa	1.475	0.961	0.606	0.298	1.463	0.948	0.590	0.276
σ_0 = 35MPa	2.259	1.601	1.038	0.620	2.246	1.595	1.030	0.610
σ_0 = 50MPa	2.946	2.240	1.563	0.990	2.939	2.235	1.563	0.981
GSI_{max} = 40	2.259	1.601	1.038	0.620	2.246	1.595	1.030	0.610
GSI_{max} = 50	0.948	0.538	0.175	—	0.894	0.469	0.090	—

(b)

	Support type III				Support type II			
	f_0 = 30%	f_0 = 40%	f_0 = 50%	f_0 = 60%	f_0 = 30%	f_0 = 40%	f_0 = 50%	f_0 = 60%
σ_0 = 5MPa	1.016	0.760	0.447	—	0.991	0.702	—	—
σ_0 = 20MPa	11.386	9.636	7.902	5.970	11.363	9.589	7.778	5.815
σ_0 = 35MPa	38.244	32.657	27.071	21.304	38.245	32.601	26.829	21.120
σ_0 = 50MPa	92.786	79.194	65.746	52.153	92.678	79.242	65.653	52.065
GSI_{max} = 40	38.241	32.665	26.943	21.221	38.227	32.622	26.841	21.060
GSI_{max} = 50	4.176	3.268	2.182	—	4.087	3.097	1.929	—

5. Conclusion

Due to the delayed effect of support installation, given this, commercial finite difference software is used for two-stage analysis. A displacement reduction factor is mainly introduced, defined as the ratio of displacement reduction to maximum rock mass displacement. Based on the numerical simulations, variations of support force and displacement reduction factor versus the delayed distance considering different support types, the initial stresses, and the post-peak behaviours are investigated. Two indicators of support time refer to the allowable deformation and delay distance of the support installation, and they are both quantitative. The main conclusions are summarised as follows:

(1) The influence of the rock mass quality and the in-situ stresses on the support force and support effect is more remarkable than the post-peak behaviours.

(2) The displacement reduction factor depends primarily on the deformation property of rock mass. As the rock mass quality is weakened, the critical plastic softening parameter decreases, the initial stress and the dilatancy angle grow, the displacement reduction factor becomes greater. Determination of the final force caused by the virtual support pressure and rock deformation characteristics at the main delay distance. The support force changes as follows: it decreases with the expansion angle and then increases with the critical plastic softening parameter.

(3) Steel set has little effect on restraining rock mass displacement, but it plays a key role in bearing support load. When the installation spacing of supports is the same, the load applied by the steel sets inserted to the shotcrete is smaller than that applied by the shotcrete alone, so that the steel sets has a protective effect on the shotcrete.

(4) The upper and lower bounds of the delayed distance and allowable deformation rely heavily on the initial stress, mechanical properties of the rock mass, and rigidity and capacity of the support with a specific case. The delayed distance and allowable deformation diminish as the rock mass quality becomes better and the initial stress decreases. The lower bound grows with the increase in the safety factor of the support; the upper bound grows with a lower displacement reduction factor. In most cases, the delayed distance for the support installation is suggested to be within R_0.

References

[1] Carranza-Torres C., Fairhurst C. Application of the convergence-confinement method of tunnel design to rock masses that satisfy the Hoek-Brown failure criterion. Tunn Undergr Space Technol. 2000; 15(2): 187-213.

[2] Paraskevopoulou C, Diederichs M. Analysis of time-dependent deformation in tunnels using the Convergence-Confinement Method. Tunnelling and Underground Space Technology incorporating Trenchless Technology Research, 2018; 71: 62-80.

[3] Manuel D L F , Taherzadeh R , Sulem J , et al. Applicability of the Convergence-Confinement Method to Full-Face Excavation of Circular Tunnels with Stiff Support System. Rock Mechanics and Rock Engineering, 2019.

[4] Su, Y., Su, Y., Zhao, M. et al. Tunnel Stability Analysis in Weak Rocks Using the Convergence Confinement Method. Rock Mech Rock Eng, 2020; 54: 559–582.

[5] Zhao HB, Li SJ, Chen BR, Yu XB. The Reliability-Based Design Optimization of considering Rock-Support Interaction for Rock Tunnels. Advances in Civil Engineering, 2021.

[6] Cui L, Zheng JJ, Zhang RJ, Lai HJ. A numerical procedure for the fictitious support pressure in the application of the convergence-confinement method for circular tunnel design. Int J Rock Mech Min Sci. 2015; 78: 336-349.
[7] Schwartz CW, Einstein HH. Simplified analysis for tunnel supports. J Geotech Eng Div ASCE. 1979; 104(4): 499 - 518.
[8] Einstein HH, Schwartz CW. Discussion of the article: simplified analysis for tunnel supports. J Geotech Eng Div ASCE. 1980; 106(7): 835 - 838.
[9] Cai Y, Esaki T, Jiang Y J. An analytical model to predict axial load in grouted rock bolt for soft rock tunnelling. Tunn Undergr Space Technol. 2004; 19: 607-618.
[10] Cai Y, Jiang Y J, Djamaluddin I, et al. An analytical model considering interaction behaviour of grouted rock bolts for convergence-confinement method in tunnelling design. Int J Rock Mech Min Sci. 2015; 76: 112-126.
[11] Carranza-Torres C, Diederichs M. Mechanical analysis of circular liners with particular reference to composite supports. For example, liners consisting of shotcrete and steel sets. Tunn Undergr Space Technol. 2009; 24: 506-532.
[12] Hoek E. Tunnel support in weak rock. In: Proceedings of Regional Symposium of Sedimentary Rock Engineering. 1998: 20-22.
[13] Barla G, Bonini M, Semeraro M. Analysis of the behaviour of a yield-control support system in squeezing rock. Tunn Undergr Space Technol. 2011; 26: 146-154.
[14] Dalgic S. Tunnelling in squeezing rock, the Bolu tunnel, Anatolian Motorway, Turkey. Eng Geol. 2002; 67: 73-96.
[15] Tian H M, Chen W Z, Yang D S, et al. Numerical analysis on the interaction of shotcrets liner with rock for yielding supports. Tunn Undergr Space Technol. 2016; 54: 20-28.
[16] Cantieni L, Anagnostou G. The interaction bewteen yeiding supports and squeezing ground. Tunn Undergr Space Technol. 2009; 24: 309-322.
[17] Hoek E, Brown ET. Practical estimates of rock mass strength. Int J Rock Mech Min Sci. 1997; 34(8): 1165-1186. 19
[18] Itasca. FLAC3D. Fast Lagrangian analysis of continua in 3dimensions (computer software). Version4.0 www.itascacg.com. Minneapolis, Minnesota; 2010.
[19] Vlachopoulos N, Diederichs MS. Improved longitudinal displacement profiles for convergence confinementanalysis of deep tunnels. Rock Mech Rock Eng. 2009; 42: 131-146.
[20] Basarir H, Genis M, Ozarslan A. The analysis of radial displacements occurring near the face of a circular opening in weak rock masses. Int J Rock Mech Min Sci. 2015; 76: 112-126.
[21] Basarir H. Analysis of rock-support interaction using numerical and multiple regression modelling. Can Geotech J. 2008; 45: 1-13
[22] Asef M R, Reddish D J, Lloyd P W. Rock-support interaction analysis based on numerical modelling. Geotechnical and Geological Engineering, 2000(1): 23-37.
[23] Cui L, Zheng JJ, Zhang RJ, Dong YK. Elasto-plastic analysis of a circular opening in rock mass with confining stress-dependent strain-softening behaviour. Tunn Undergr Space Technol. 2015; 50: 94-108.
[24] Wang S L, Yin X T, Tang H, Ge X R. A new approach for analyzing circular tunnel in strain-softening rock masses. Int. J. Rock Mech. Min. Sci. 2010; 47: 170-178.

Analysis and Application of Salop Model under Network Effect

Qian Tong*
School of mathematics, Guangdong University of Education, Guangzhou, China

*Corresponding author email: waiteforyou@163.com

Abstract. Based on Salop's ring city model, this paper applies the mathematical model to solve the problem of benefit game about Bertrand competition under the environment of network effect. The equilibrium solutions for both players are given in this paper, and the equilibrium solution of Salop's ring city model depends on the effect of network externalities. The magnitude of network externalities determines the changes in welfare.
Keywords: Salop model, network effect, uneven distribution, theory of games.

1. Introduction

Salop model plays an important role in game theory. Economides once pointed out used duopoly model to solve game problems involving horizontal merger[1]. The conclusions of Mason and Wang are similar to the conclusion that of Economides[2-3]. However, from an intuitive perspective, this conclusion does not match reality. To explain this "paradox", many scholars have modified the research scope and conditions of the Salop circle model.

At present, the research on benefit game theory includes motivation of participants related to their efficiency [4], the impact of product variety changes on the enterprise merger[5], consideration of the effect based on the internal incentive mechanism of the parties to the game[6], application problems with the internal incentive mechanism and effect in M-type enterprises[7], the Salop model-based discussion on the motives and the welfare of the actors under network externalities, and empirical studies on the frequency of the game model[8].

It is worth noting that the above studies do not consider the geographic distribution of the inhabitants or assume that people are evenly distributed with a probability density of 1, which is inconsistent with the actual situation.

In real life, the geographic distribution of residents is uneven, and generally, the pedestrian volume in the downtown areas is much larger than that in the suburbs.

In this context, multiple enterprises that produce homogeneous products are used as study subjects, and the differences in the commodity are the manufacturer location difference and the network effect. Based on the Salop circular city model, this paper studies the effect on the competition behaviour of game players in a network effect environment where people are unevenly distributed. On this basis, equilibrium solutions of the model are discussed, and the corresponding conclusions are obtained.

2. Model Establishment

Suppose an oligarchic competition industry composed of n players is evenly distributed on the ring city with unit perimeter, and Bertrand competition is implemented in this industry (participator

Content from this work may be used under the terms of the Creative Commons Attribution 3.0 licence. Any further distribution of this work must maintain attribution to the author(s) and the title of the work, journal citation and DOI.
Published under licence by IOP Publishing Ltd

i_{n+1} is located at x=0). The probability density distribution function that inhabitants obey is $f(x) = \begin{cases} 1+\cos 2n\pi x, 0 \le x \le 1 \\ 0, \text{other} \end{cases}$, and the total number is normalized to 1.

Each game player i provides homogeneous products $n_i, i = 1,2,...,n$, except for the differences in the location of the actor and network externalities.

To simplify the analysis, this paper assumes that the marginal cost of the product is zero. Because the distribution density of inhabitants is symmetric, only the selection behaviour of users between participants i and $i+1$ needs to be considered.

If the distance between the user and contestant i is x_i, the transportation cost function is quadratic, and the unit transportation cost is τ. In this case, the transportation cost of users at position x_i is τx_i^2 from i and is $\tau(\frac{1}{n} - x_i)^2$ from contender $i+1$.

Due to the network effect, competitors' obtained utility from n_i is divided into two parts (i.e., the return derived from n_i itself $R - p_i - \tau x_{i_i}^2$ and the influence from network externalities $uE(q_i)$). p_i is the price of n_i, q_i is the quantity demanded of n_i, u represents the network effect coefficient of n_i, and $E(q_i)$ represents the value of expectation of n_i.

The author assumes that each participator have rational expectations (i.e., $E(q_i) = q_i$); therefore, for residents who are located between i and $i+1$ before the cooperation between i and $i+1$ at a distance x_i away from i, the utility of n_i and n_{i+1} for the people is

$$U_{x_i}^{i} = R - p_i - \tau x_i^2 + uq_i$$
$$U_{x_i}^{i+1} = R - p_{i+1} - \tau(\frac{1}{n} - x_i)^2 + uq_{i+1} \quad i = 1,2,...,n \tag{1}$$

This paper only considers the cooperation between two actors. If i and $i-1$ work together, the incompatibility between n_i and n_{i-1} is eliminated due to it. Therefore, for the inhabitants located between participators i and $i+1$ after the consociation and at a distance $x_i^{'}$ away from i, the obtained network externality of $n_i(n_{i-1})$ is $u(q_{i-1} + q_i)$, and that of n_{i+1} is uq_{i+1}. Hence, after the cooperation between i and $i+1$, the utility for the people at $x_i^{'}$ for $n_i(n_{i-1})$ and n_{i+1} is

$$\begin{aligned} U_{x_i^{'}}^{i} &= R - p_i - \tau x_i^{'2} + u(q_{i-1} + q_i) \\ U_{x_i^{'}}^{i+1} &= R - p_{i+1} - \tau(\frac{1}{n} - x_i^{'})^2 + uq_{i+1} \end{aligned} \tag{2}$$

3. Equilibrium Solution of the Model

The probability density of the population distribution mentioned in this paper is $f(x) = \begin{cases} 1+\cos 2n\pi x, 0 \le x \le 1 \\ 0, \text{other} \end{cases}$, therefore, each competitor is symmetrical to its population. The fluctuation of the population distribution along the circular city has n cycles, and the circular city has exactly n participants to compete, which is in line with the actual situation.

Since the model adopts a circular structure, $p_{n+1} = p_1$ and $p_n = p_0$ can be set. Before the consociation, the symmetry can indicate that only the demand function of any $n_i, i = 1,2,...,n$ is required.

If inhabitants, who are located between i and $i+1$ at a distance x_i away from i, have no difference in the effectiveness ($U_{x_i}^{i} = U_{x_i}^{i+1}$) between n_i and n_{i+1}, the demand function of n_i is

$$q_i = 2x_i, i = 1,2,...,n \tag{3}$$

After the consociation of i and $i-1$, if the utility for each n_i are indifferent ($U^i_{x_i^{'}} = U^{i+1}_{x_i^{'}}$) between $n_i(n_{i-1})$ and n_{i+1} when residents are located between i and $i+1$ at a distance $x_i^{'}$ away from i, then the demand function of $n_i(n_{i-1})$ is

$$q_i^{'} = 2x_i^{'} + \frac{1}{n} \tag{4}$$

The demand function for each n_i, all the participants who did not participate in the collaboration, is

$$q_{-i}^{'} = \frac{1}{n-2}(1 - \frac{1}{n} - 2x_i^{'}) \tag{5}$$

The following conclusions can be obtained by comparing the demand of i before and after the merger:
Proposition 1: If the distributions of n participators and inhabitants meet the conditions of the model described in this paper before the consociation, then due to symmetry, the demand of each n_i is $q_i = \frac{1}{n}, i = 1,2,...,n$, and the indifference point between n_i and n_{i+1} for the utility is

$$x_i = (p_{i+1} - p_i + \frac{\tau}{n^2} - \frac{u}{n})\frac{n}{2\tau - 2nu} \tag{6}$$

After the cooperation between i and $i-1$, the demand function of n_i and n_{i-1} is

$$q_i^{'} = \frac{1}{n} + (p_{i+1} - p_i + \frac{\tau}{n^2})\frac{n}{\tau - nu},$$

the demand function of each uninvolved is $q_{-i}^{'} = \frac{1}{n-2}\left[1 - \frac{1}{n} - (p_{i+1} - p_i + \frac{\tau}{n^2})\frac{n}{\tau - nu}\right]$, and the indifference point between $n_i(n_{i-1})$ and n_{i+1} for utility is

$$x_i^{'} = (p_{i+1} - p_i + \frac{\tau}{n^2})\frac{n}{2\tau - 2nu} \tag{7}$$

Please see the appendix for the proof.
This conclusion shows that a horizontal collaboration expands the user scale and eliminates the incompatibility between n_i and n_{i-1} of the two involved players. Therefore, after the cooperation, the demand for all n_i is asymmetric.
Before the collaboration, each i simultaneously adopt prices competition strategy to maximize their profits. The profit function of each i can be obtained using symmetry and Equation (3):

$$\pi_i = \max_{p_i}\left\{p_i(p_{i+1} - p_i + \frac{\tau}{n^2} - \frac{u}{n})\frac{n}{\tau - nu}\right\}$$

Based on the first-order condition, the equilibrium of i ($i = 1,2,...,n$) before the merger are

$$p_i^* = \frac{\tau}{n^2} - \frac{u}{n}$$

$$\pi_i^* = \frac{\tau}{n^3} - \frac{u}{n^2} \quad i = 1,2,...,n \tag{8}$$

Therefore, in a market with network externalities, the network effect coefficient is positively correlated with the utility of n_i. Therefore, under competition, each i may adopt a strategy to expand the scale of sales. Equation (5) shows that the strengthening of the network effect might lead to intensified competition, lowered equilibrium solution, and reduced earnings.

After the consociation, the merged and uninvolved sides start competing for incomes. It is worth noting that the remaining $n-2$ uninvolved satisfy symmetry. Based on Equation (4), the equilibrium solutions are

$$p_i^{**} = \frac{(\tau - nu)(n+1)+\tau}{3n^2}$$

$$p_{-i}^{**} = \frac{(\tau - nu)(2n-1)-\tau}{3n^2}$$

$$\pi_i^{**} = \frac{[(\tau - nu)(n+1)+\tau]^2}{9n^3(\tau - nu)}$$

$$\pi_{-i}^{**} = \frac{1}{n-2}\frac{[(\tau - nu)(2n-1)-\tau]^2}{9n^3(\tau - nu)} \tag{9}$$

The comparison of the interests of the involved side before and after the cooperation shows that when $u \in \left(0, \frac{\tau}{n}\right) \cup \left(\frac{\tau}{n} + \frac{\tau}{n(n+1+\sqrt{18})}, \frac{\tau}{n} + \frac{\tau}{n(n+1-\sqrt{18})}\right)$, contestants' profits increase. In the case of network externalities, because a merger between contenders with differentiation of n_i might expand the network size, competitors generally have motives for horizontal mergers.

Therefore, the merger increases the value and demand of n_i , and cause uninvolved side to fall into increasingly unfavourable situations in competition. Combining these two aspects, each participant have a solid motive to implement merger strategies to expand the network externality effects and gain a competitive advantage.

Equations (5) and (6) show that:

when $p_{-i}^{**} < 0$, $u \in \left(\frac{(2n-2)\tau}{(2n-1)n}, \frac{\tau}{n}\right) \cup \left(\frac{\tau}{n} + \frac{\tau}{n(n+1+\sqrt{18})}, \frac{\tau}{n} + \frac{\tau}{n(n+1-\sqrt{18})}\right)$, all uninvolved i could be expelled from the game, and the cooperant i could become a monopolist. In a monopolistic market, for consumers located between i and $i+1$ at a distance $\frac{1}{n}$ away from i, utility of n_i is the lowest due to symmetry. In this case, the equilibrium solution of i are

$$p^M = R - \frac{\tau}{n^2} + u$$

$$\pi^M = R - \frac{\tau}{n^2} + u$$

When $p_{-i}^{**} > 0$, $u \in \left(0, \frac{(2n-2)\tau}{(2n-1)n}\right]$, participants who did not participate in the collaboration can remain in the market. Under this circumstance, the equilibrium solution of them before and after the merger is $\Delta p_{-i} = p_{-i}^{**} - p_i^{*}$:

$$\Delta p_{-i} \geq 0 \ 0 < u \leq \frac{(2n-5)\tau}{(2n-4)n}$$

$$\Delta p_{-i} < 0 \ \frac{(2n-5)\tau}{(2n-4)n} < u \leq \frac{(2n-2)\tau}{(2n-1)n}$$

This calculation also shows that when they remain in the market to participate in the competition, the change in the equilibrium solution of the firm that implements a cooperative strategy is $\Delta p_i = p_i^{**} - p_i^{*} > 0$.

We categorize the participants who practiced the cooperative strategy as category Ⅰ, and the rest as category Ⅱ, the plays choose A as the strategy for raising prices and B as the strategy for raising prices. The above analysis shows that the category Ⅰ might increase their prices to maximize profit.
For the category Ⅱ, when the effect of network externalities is small, i.e., $0 < u \le \frac{(2n-5)\tau}{(2n-4)n}$, they might increase their prices to sell products. When the network externalities are strengthened, i.e., $\frac{(2n-5)\tau}{(2n-4)n} < u \le \frac{(2n-2)\tau}{(2n-1)n}$, they might play strategy B. They might be expelled from the game, when the effect of network externalities is strong, and the network coefficient increases to

$$u \in \left(\frac{(2n-2)\tau}{(2n-1)n}, \frac{\tau}{n}\right) \cup \left(\frac{\tau}{n} + \frac{\tau}{n(n+1+\sqrt{18})}, \frac{\tau}{n} + \frac{\tau}{n(n+1-\sqrt{18})}\right),$$

With the enhancement of network externalities, the utility of n_i increases, the demand for n_i is the key for the contestant.
The above analysis shows that the gains from the expansion of the network effect cannot compensate for the loss, so simply adopting strategy B should be unwise.

4. The Analysis of the Solution

With the enhancement of network externalities, the participant in category Ⅱ must continue to choose strategy B, and when $P_i < c_i$ they all are forced to withdraw from the game. Therefore, changes in W can be discussed based on whether the actor in category Ⅱ are expelled.
W is the sum of profit and surplus CS. From Equation (5),

$$W^* = \sum_{i=1}^{n} \pi_i^* + CS = n\pi_i^* + \int_{-\frac{1}{2n}}^{\frac{1}{2n}} U_{x_i}^i dx = R + \frac{u}{n} - \frac{\tau}{12n^2} \tag{10}$$

When the contestants in category Ⅱ have not been expelled ($u \in \left(0, \frac{(2n-2)\tau}{(2n-1)n}\right]$), the change of W is

$$\Delta W = W^{**} - W^* = \pi_i^{**} + (n-2)\pi_i^{**} + 2\left[\int_{-\frac{1}{2n}}^{x_i'} U_{x_i}^i dx + \int_{x_i'}^{\frac{n-1}{2n}} U_{x_i}^{i+1} dx\right] - W^*$$

$$\cong \frac{7n^2 - 12n + 8}{9n^2} u + \frac{-11n^2 + n + 10}{36n^2}\tau \tag{11}$$

Thus, we have:

$$\Delta W < 0 \ u \in \left(0, \frac{11n^2 - n - 10}{4(7n^2 - 12n + 8)}\tau\right)$$

$$\Delta W \ge 0 \ u \in \left[\frac{11n^2 - n - 10}{4(7n^2 - 12n + 8)}\tau, \frac{2n-2}{(2n-1)n}\tau\right]$$

When the contenders in category Ⅱ are expelled one by one and all the competitors becomes one, the change of W is

$$\Delta W' = \pi^M + 2\int_{-\frac{1}{2n}}^{\frac{n-1}{2n}} U_{x_i}^i dx - W^*$$

$$= \frac{n-1}{n} u - \frac{(n-1)(n-2)}{12n^2}\tau \tag{12}$$

$$\Delta W' < 0 u \in \left(\frac{2n-2}{(2n-1)n}\tau, \frac{n-2}{12n}\tau\right)$$

$$\Delta W' \geq 0 \, n \in \left[\frac{n-2}{12n}\tau, \frac{\tau}{n}\right) \cup \left(\frac{\tau}{n} + \frac{\tau}{n(n+1+\sqrt{18}}, \frac{\tau}{n} + \frac{\tau}{n(n+1-\sqrt{18}}\right)$$

The above analysis shows that the change in W depends on the strength of network externalities. It is worth noting that if $n = 3$ in Equations (10), (11), and (12), the calculation results are exactly consistent with W changes among the three competitors described in Ref.[7].
Therefore, the change in W due to the collaboration of game players under the uneven inhabitants distribution and network effect analysed in this paper is just the generalization of the situation described in Ref.[7]. This paper assumes that residents follow a nonuniform distribution, indicating that W is not affected by the population distribution.

5. Conclusion

This paper studies the impact of a cooperative strategy between i in the context of network externalities and the uneven population distribution. On this basis, the motives of the cooperative strategy and changes in W before and after the tactic are discussed.
The conclusions show that under the environment of network externalities, the participants generally have motives for the cooperative strategy, that can place the competitors who did not adopt the cooperant tactic at a competitive disadvantage.
With the gradual increase in network externalities of n_i, the adverse impact on the participant in category Ⅱ also increases. No matter how network externalities change, the game players in category Ⅰ can obtain the higher return.
When network externalities increase, the contenders in category Ⅱ can be reasonable with choosing strategy B, but when the effect of network externalities become stronger, the advantage is more pronounced.
The players in category Ⅱ keep adopting strategy B, so they might eventually be expelled from the game. The analysis of W reveals that it changes in line with the strength of network externalities.
Based on the Salop circular model, the author discusses the equilibrium solution and the impact of the tactic for collaboration under network externalities.
In contrast, the effects of network externalities and unit transportation costs on the competitive equilibrium of manufacturers under different consumer preferences are not considered, then the model considering these conditions will be studied in the future.

Acknowledgements

This work is supported by Quality engineering project of GDEI (Grant No. 2018jxgg22 and Grant No. 202lzxkc04). The author is very grateful to the referee for valuable comments and suggestions.

References

[1] Economides N 1986 *Minimal and maximal differentiation in Hotelling's duopoly (Operations Research and Management Science vol 6)* chapter 1 pp 67-71.
[2] Mason R 2017 *Network externalities and the coase conjecture (European Economic Review vol 44)* chapter 2 pp 1981-1992.
[3] Wang Haiping and Liu Shulin and Lin Jun 2018 *Competition analysis of duopoly freemium Strategy based on Hotelling Model (Operations Research and Management Science vol 27)* chapter 4 pp 15-21.
[4] Liu Zhiyong and Li Minqiang 2019 *Research on information product versioning strategy under the influence of network externalities (Journal of Management Science vol 18)* chapter 7 pp 13-26.
[5] Xu Bing and Zhu Daoli 2016 *Retailer level competition analysis and supply chain coordination strategy design (Journal of Systems Engineering vol25)* chapter 5 pp 609-615.
[6] Yang Daojian and Bai yin 2019 *Competition and decentralization of core enterprises in supply chain based on Hotelling model (System Engineering -Theory & Practice vol 35)* chapter 12 pp 3025-3037.

[7] Li Jizi and Wang Zhongrui 2018 *Hotelling decision-making model and strategy in supply chain transnational entry mode (Journal of Industrial Engineering and Engineering Management vol 32)* chapter 43 pp 162-170.
[8] Wang Guocai 2018 *Research about horizontal merger of network companies based on Salop model (Journal of Systems Engineering vol 36)* chapter 3 pp 343-349.

Appendix:
Proof of Proposition 1
Proof: Because the density function for the population distribution in this circular city is

$$f(x)=\begin{cases}1+\cos 2n\pi x, 0\le x\le 1\\ 0, \text{other}\end{cases},$$

according to Equations (1) and (3), the indifference point between n_i and n_{i+1} before the merge strategy satisfies:

$$R-p_i-\tau x_i^2+u\int_{\frac{i-1}{n}+x_{i-1}}^{\frac{i}{n}+x_i}(1+\cos 2n\pi x)dx=R-p_{i+1}-\tau\left(\frac{1}{n}-x_i\right)^2+2u\left(\frac{1}{n}-x_i\right)$$

$$\Rightarrow -p_i-\tau x_i^2+u\left[\frac{i}{n}+x_i-\frac{i-1}{n}-x_{i-1}\right]+\frac{u}{2n\pi}\left[\sin 2n\pi\left(\frac{i}{n}+x_i\right)-\sin 2n\pi\left(\frac{i-1}{n}+x_{i-1}\right)\right]$$
$$=-p_{i+1}-\tau\left(x_i^2-\frac{2x_i}{n}+\frac{1}{n^2}\right)+2u\left(\frac{1}{n}-x_i\right)$$

Because each i is symmetric with respect to its population, the indifference distance between any two n_i is equal ($x_i=x_j, i\neq j, i,j=1,2,\ldots,n$).

Using the first - order Taylor series expansion of $\sin 2n\pi\left(\frac{i}{n}+x_i\right)-\sin 2n\pi\left(\frac{i-1}{n}+x_{i-1}\right)$, the equation can be simplified as follows:

$$-p_i-\tau x_i^2+\frac{u}{n}=-p_{i+1}-\tau\left(x_i^2-\frac{2x_i}{n}+\frac{1}{n^2}\right)+2u\left(\frac{1}{n}-x_i\right)$$

$$\Rightarrow x_i=\left(p_{i+1}-p_i+\frac{\tau}{n^2}-\frac{u}{n}\right)\frac{n}{2\tau-2un}$$

From the above equation and Equation (3), we have: $q_i=\frac{1}{n}$, $n=1,2,\ldots,n$.
After the consociation, the indifference point between n_i (n_{i-1}) and n_{i+1} satisfies:

$$R-p_i-\tau x_i'^2+u\int_{\frac{i-2}{n}+x_{i-2}'}^{\frac{i}{n}+x_i'}(1+\cos 2n\pi x)dx=R-p_{i+1}-\tau\left(\frac{1}{n}-x_i'\right)^2+2u\left(\frac{1}{n}-x_i'\right)$$

Similarly, we know:

$$x_i'=\left(p_{i+1}-p_i+\frac{\tau}{n^2}\right)\frac{n}{2\tau-2un}$$

Based on the above equation and Equations (4) and (5), the demand function of n_i for the category Ⅰ choosing the strategy is

$$q_i'=\frac{1}{n}+\left(p_{i+1}-p_i+\frac{\tau}{n^2}\right)\frac{n}{\tau-nu}$$

The function of n_i for the category Ⅱ is

$$q_{-i}'=\frac{1}{n-2}\left[1-\frac{1}{n}-\left(p_{i+1}-p_i+\frac{\tau}{n^2}\right)\frac{n}{\tau-nu}\right]$$

Optimization for Far-distance Cooperative Rendezvous with Multiple Direction-fixed Thrusts

Fei Ren*

School of Mechanical and Automotive Engineering, Qilu University of Technology (Shandong Academy of Sciences), Jinan 250353, Shandong, China

*Corresponding author email: ren87@qlu.edu.cn

Abstract. The continuous-thrust far-distance cooperative rendezvous problem between two spacecraft is investigated. The indirect optimization method, based on Pontryagin's maximum principle (PMP), is applied to optimize fuel consumption. To overcome the difficulty in nonsmooth integration caused by the bang-bang control, the homotopy method is adopted to solve the fuel-optimal problem from a related energy-optimal problem. The quantum-behaved particle swarm optimization (QPSO) algorithm is used to obtain the energy-optimal solutions. The energy-optimal solutions are used as the initial values for the homotopic procedure to obtain the fuel-optimal solutions and optimal bang-bang control law. A hybrid algorithm combined homotopy method and sequential quadratic programming (SQP) algorithm is proposed. This hybrid algorithm can effectively obtain feasible optimal solutions even though the indirect optimization method exhibits a narrow convergence domain. Simulations of high-thrust and low-thrust rendezvous problems are provided and the proposed hybrid algorithm is verified. Moreover, the necessity of radial thrust is investigated.

1. Introduction

A cooperative rendezvous is that two spacecraft with different initial locations and velocities are active and eventually meet at the same location with the same velocity. Many studies on the problems of the impulse or low-continuous-thrust cooperative rendezvous have been made. Prussing and Conway [1] indicated that cooperative rendezvous is desirable if a single active spacecraft has not enough fuel to accomplish the rendezvous. Mirfakhraie and Conway [2] investigated the time-fixed impulsive cooperative rendezvous problem. They indicated that cooperative rendezvous is efficient and time-saving. Crispin and Seo [3] used the direct optimization method to solve the low-continuous-thrust cooperative rendezvous problem.

Presently, three types of methods for solving the trajectory optimization problem are widely used: direct method, hybrid method, and indirect method. The indirect method transforms the optimal control problem into a two-point boundary value problem (TPBVP) [4, 5] based on Pontryagin's maximum principle (PMP) [6], and introduces a co-state variable for each state variable. In theory, the indirect method is superior to the other two methods since the control strategy is automatically determined by the PMP, but not presupposed. However, the bang-bang control law makes the dynamic equation nonsmooth on the right side, the nonsmooth integration problem limits the application of the indirect method. Bertrand and Epenoy [7] proposed a homotopy technique to smooth the dynamic equation. Jiang and Tang [8, 9] studied some low-thrust trajectory optimization problems by using the homotopy technique. The main idea of this technique is to solve the fuel-optimal problem from a related energy-optimal problem, the energy-optimal problem is smooth and easier to be solved, the difficulty of nonsmooth integration is reduced. However, the indirect method exhibits a very small radius of

Content from this work may be used under the terms of the Creative Commons Attribution 3.0 licence. Any further distribution of this work must maintain attribution to the author(s) and the title of the work, journal citation and DOI.
Published under licence by IOP Publishing Ltd

convergence domain, sometimes it is very difficult to find a good initial co-state variable value to solve the shooting function.

Nowadays, artificial intelligence optimization algorithms have been widely used for solving trajectory optimization problems. Feng et al. [10-12] and Yang et al. [13] studied the high thrust rapid rendezvous problems by using the hybrid method. They investigated several intelligence algorithms and proposed a hybrid algorithm combined quantum-behaved particle swarm optimization algorithm (QPSO) [14, 15] and sequential quadratic programming algorithm (SQP) [16]. QPSO is useful for searching for the initial co-state value due to its great global search capability, and SQP can be used to further correct the solution obtained by QPSO.

This paper mainly focuses on the trajectory optimization for continuous-thrust far-distance cooperative rendezvous based on the indirect optimization method. The fuel-optimal control is obtained by using the QPSO algorithm and homotopy-SQP hybrid algorithm. The proposed optimization method can effectively obtain feasible optimal solutions even though the indirect optimization method exhibits a narrow convergence domain. Moreover, the necessity of radial thrust is investigated.

2. Dynamical Equations of Modified Equinoctial Elements

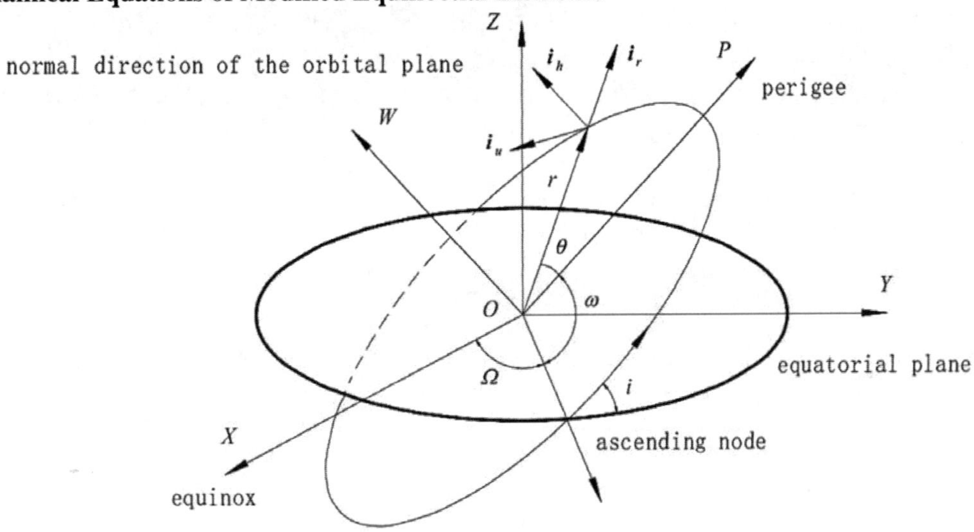

Figure 1. Inertial reference system and plane coordinates system.

The orbital plane coordinate system of the spacecraft is built as shown in figure 1. The origin is located at the spacecraft. i_r, i_u, and i_h are the unit vectors along the coordinate axes. i_r is radial to the center of mass of the Earth. In the orbital plane, i_u is perpendicular to i_r. i_h is along the normal direction. i_r, i_u, and i_h satisfy the right-hand rule.

The dynamical equations for spacecraft are expressed as follows:

$$\begin{cases} \dot{\boldsymbol{x}} = \boldsymbol{M}_r \dfrac{F_r}{m} u_r + \boldsymbol{M}_u \dfrac{F_u}{m} u_u + \boldsymbol{M}_h \dfrac{F_h}{m} u_h + \boldsymbol{D} \\ \dot{m} = -\left(F_r |u_r| + F_u |u_u| + F_h |u_h|\right) / \left(I_{sp} g_0\right) \end{cases} \tag{1}$$

In equation (1), the radial, circumferential, and normal thrusts are denoted as F_r, F_u, and F_h. g_0 is the gravitational acceleration. I_{sp} is the thruster-specific impulse. The engine switch control u_j $(j = r,u,h)$ equals -1 or 0 or 1. $\boldsymbol{x}$ is the matrix of modified equinoctial elements written as $\boldsymbol{x} = [p, f, g, h, k, L]^T$. $\boldsymbol{M}_r$, $\boldsymbol{M}_u$, $\boldsymbol{M}_h$, and $\boldsymbol{D}$ are expressed as follows:

$$\boldsymbol{D}=\begin{bmatrix}0 & 0 & 0 & 0 & 0 & \sqrt{\mu p}\left(w/p\right)^2\end{bmatrix}^T \tag{2}$$

$$\begin{aligned}
\boldsymbol{M}_r &= \begin{bmatrix}0 & \sqrt{\frac{p}{\mu}}\sin L & -\sqrt{\frac{p}{\mu}}\cos L & 0 & 0 & 0\end{bmatrix}^T \\
\boldsymbol{M}_u &= \begin{bmatrix}\frac{2p}{w}\sqrt{\frac{p}{\mu}} & \sqrt{\frac{p}{\mu}}\frac{(w+1)\cos L+f}{w} & \sqrt{\frac{p}{\mu}}\frac{(w+1)\sin L+g}{w} & 0 & 0 & 0\end{bmatrix}^T \\
\boldsymbol{M}_h &= \begin{bmatrix}0 & \sqrt{\frac{p}{\mu}}\frac{g}{w}[k\cos L-h\sin L] & \sqrt{\frac{p}{\mu}}\frac{f}{w}[h\sin L-k\cos L] & \sqrt{\frac{p}{\mu}}\frac{s^2}{2w}\cos L & \sqrt{\frac{p}{\mu}}\frac{s^2}{2w}\sin L & \sqrt{\frac{p}{\mu}}\frac{h\sin L-k\cos L}{w}\end{bmatrix}^T
\end{aligned} \tag{3}$$

where μ is the gravitational constant of the earth. The supplementary equations, s and w, are expressed as $s^2=1+h^2+k^2$, $w=1+f\cos L+g\sin L$. The modified equinoctial elements can be expressed by the classic orbital elements as $p=a(1-e^2)$, $f=e\cos(\omega+\Omega)$, $g=e\sin(\omega+\Omega)$, $h=\tan(i/2)\cos\Omega$, $k=\tan(i/2)\sin\Omega$, and $L=\omega+\Omega+\theta$, where a is the semi-major axis, e is the eccentricity, Ω is the longitude of the ascending node, i is the orbit inclination angle, ω is the perigee amplitude, and θ is the true anomaly. For a near earth orbiting spacecraft, the time unit TU, the distance unit DU, the velocity unit VU, and the mass unit MU are defined as $1\text{TU}=\sqrt{R_e^3/\mu}$, $1\text{DU}=R_e$, 1VU=DU/TU, and 1MU=m_0, where R_e is the equator radius and m_0 is spacecraft's initial mass.

3. Homotopy Method and Optimization Process

The homotopy performance index for one spacecraft is

$$J=\lambda_0\frac{F_r}{I_{\mathrm{sp}}g_0}\int_{t_0}^{t_f}\left[(1-\varepsilon)|u_r|+\varepsilon u_r^2\right]\mathrm{d}t+\lambda_0\frac{F_u}{I_{\mathrm{sp}}g_0}\int_{t_0}^{t_f}\left[(1-\varepsilon)|u_u|+\varepsilon u_u^2\right]\mathrm{d}t+\lambda_0\frac{F_h}{I_{\mathrm{sp}}g_0}\int_{t_0}^{t_f}\left[(1-\varepsilon)|u_h|+\varepsilon u_h^2\right]\mathrm{d}t \tag{4}$$

In equation (4), t_0 denotes the initial time, t_f denotes the terminal time, and λ_0 is a positive numerical multiplier. When ε=1, the problem is an energy-optimal problem, which is easier to be solved because its control is smooth. When ε decreases from 1 to 0, a continuous transformation from the energy-optimal problem (when ε=1) to the fuel-optimal problem (when ε=0) is carried out. Actually, the physical meaning of J is fuel consumption (reduction of mass from t_0 to t_f) if ε=0.
Based on the PMP, Hamilton's function H is given as follows:

$$\begin{cases}H=H_r+H_u+H_h+\boldsymbol{\lambda}^T\boldsymbol{D} \\ H_j=\lambda_0\frac{F_j}{I_{\mathrm{sp}}g_0}\varepsilon u_j^2+\lambda_0(1-\varepsilon)\frac{F_j}{I_{\mathrm{sp}}g_0}|u_j|-\lambda_m\frac{F_j}{I_{\mathrm{sp}}g_0}|u_j|+\boldsymbol{\lambda}^T\boldsymbol{M}_j\frac{F_j}{m}u_j \quad (j=r,\ u,\ h)\end{cases} \tag{5}$$

where $\boldsymbol{\lambda}$ and λ_m are co-states, $\boldsymbol{\lambda}=[\lambda_p,\lambda_f,\lambda_g,\lambda_h,\lambda_k,\lambda_L]^T$. H_r, H_u, and H_h are named radial, circumferential, and normal Hamilton's function. Take H_u for example,

$$H_u=\begin{cases}H_{u-}=\lambda_0\frac{F_u}{I_{\mathrm{sp}}g_0}\varepsilon u_u^2-\lambda_0(1-\varepsilon)\frac{F_u}{I_{\mathrm{sp}}g_0}u_u+\lambda_m\frac{F_u}{I_{\mathrm{sp}}g_0}u_u+\boldsymbol{\lambda}^T\boldsymbol{M}_u\frac{F_u}{m}u_u & (-1\le u_u<0) \\ H_{u+}=\lambda_0\frac{F_u}{I_{\mathrm{sp}}g_0}\varepsilon u_u^2+\lambda_0(1-\varepsilon)\frac{F_u}{I_{\mathrm{sp}}g_0}u_u-\lambda_m\frac{F_u}{I_{\mathrm{sp}}g_0}u_u+\boldsymbol{\lambda}^T\boldsymbol{M}_u\frac{F_u}{m}u_u & (0\le u_u\le 1)\end{cases} \tag{6}$$

Hamilton's function should be minimized by the optimal control. When $u_u\in[-1,\ 0)$,

$$\begin{cases}u_{u-}=-1, & \rho_{u-}<-\varepsilon \\ u_{u-}=0, & \rho_{u-}>\varepsilon \\ u_{u-}=-0.5+\rho_{u-}/2\varepsilon, & |\rho_{u-}|\le\varepsilon\end{cases} \quad \left(\rho_{u-}=1-\frac{\lambda_m}{\lambda_0}-\frac{\boldsymbol{\lambda}^T\boldsymbol{M}_u I_{\mathrm{sp}}g_0}{\lambda_0 m}\right) \tag{7}$$

When $u_u \in [0,\ 1]$,

$$\begin{cases} u_{u+} = 1, & \rho_{u+} < -\varepsilon \\ u_{u+} = 0, & \rho_{u+} > \varepsilon \\ u_{u+} = 0.5 - \rho_{u+}/2\varepsilon, & |\rho_{u+}| \le \varepsilon \end{cases} \quad \left(\rho_{u+} = 1 - \frac{\lambda_m}{\lambda_0} + \frac{\boldsymbol{\lambda}^T \boldsymbol{M}_u I_{\mathrm{sp}} g_0}{\lambda_0 m} \right) \tag{8}$$

Therefore,

$$u_u = \begin{cases} u_{u-}, & H_{u-} < H_{u+} \\ u_{u+}, & H_{u+} < H_{u-} \end{cases} \tag{9}$$

The optimal control in radial and normal directions can be obtained in the same way.
The co-state equations, obtained by computing the partial derivative of equation (5), are

$$\begin{cases} \dot{\boldsymbol{\lambda}} = -\dfrac{\partial H}{\partial \boldsymbol{x}} = -\boldsymbol{\lambda}^T \dfrac{\partial \boldsymbol{M}_r}{\partial \boldsymbol{x}} \dfrac{F_r}{m} u_r - \boldsymbol{\lambda}^T \dfrac{\partial \boldsymbol{M}_u}{\partial \boldsymbol{x}} \dfrac{F_u}{m} u_u - \boldsymbol{\lambda}^T \dfrac{\partial \boldsymbol{M}_h}{\partial \boldsymbol{x}} \dfrac{F_h}{m} u_h - \boldsymbol{\lambda}^T \dfrac{\partial \boldsymbol{D}}{\partial \boldsymbol{x}} \\ \dot{\lambda}_m = -\dfrac{\partial H}{\partial m} = \boldsymbol{\lambda}^T \boldsymbol{M}_r \dfrac{F_r}{m^2} u_r + \boldsymbol{\lambda}^T \boldsymbol{M}_u \dfrac{F_u}{m^2} u_u + \boldsymbol{\lambda}^T \boldsymbol{M}_h \dfrac{F_h}{m^2} u_h \end{cases} \tag{10}$$

The terminal mass co-state $\lambda_m(t_f)$ is zero if the terminal mass is free. Two spacecraft's initial states, $\boldsymbol{x}_1(t_0)$ $=[p_1(t_0), f_1(t_0), g_1(t_0), h_1(t_0), k_1(t_0), L_1(t_0), m_1(t_0)]^T$ and $\boldsymbol{x}_2(t_0) = [p_2(t_0), f_2(t_0), g_2(t_0), h_2(t_0), k_2(t_0), L_2(t_0), m_2(t_0)]^T$, are known. Once the unknown initial co-states, $\boldsymbol{\lambda}_1(t_0) = [\lambda_{p1}(t_0), \lambda_{f1}(t_0), \lambda_{g1}(t_0), \lambda_{h1}(t_0), \lambda_{k1}(t_0), \lambda_{L1}(t_0), \lambda_{m1}(t_0)]^T$ and $\boldsymbol{\lambda}_2(t_0) = [\lambda_{p2}(t_0), \lambda_{f2}(t_0), \lambda_{g2}(t_0), \lambda_{h2}(t_0), \lambda_{k2}(t_0), \lambda_{L2}(t_0), \lambda_{m2}(t_0)]^T$, and the positive numerical multipliers, λ_{01} and λ_{02}, are gotten, integrating equation (1) and equation (10) yields the terminal states and co-states. The control law is obtained by equation (9). The problem is to search the appropriate initial co-states $\boldsymbol{\lambda}_1(t_0)$ and $\boldsymbol{\lambda}_2(t_0)$ and multipliers λ_{01} and λ_{02} to minimize equation (4), meanwhile satisfying the constraint function ψ as follows:

$$\psi(z) = \left[\left| \boldsymbol{x}_1^T(t_f) - \boldsymbol{x}_2^T(t_f) \right|, \lambda_{m1}(t_f), \lambda_{m2}(t_f) \right]^T = \boldsymbol{0} \tag{11}$$

where $z = [\lambda_{01}, \boldsymbol{\lambda}_1(t_0), \lambda_{02}, \boldsymbol{\lambda}_2(t_0)]^T$.
The energy-optimal problem is solved at first, its solution is obtained by using the QPSO algorithm. The fitness function of QPSO, $Fit(z)$, is built as follows:

$$Fit(z) = J_1 + J_2 + P_f \left\| \psi(z) \right\|^2 \tag{12}$$

The punishment coefficient P_f is 10 in this paper. Then the energy-optimal solutions are used as the initial values for the homotopic procedure. The implementation procedure of the optimization is listed as follows:
(1) Use QPSO algorithm to solve z for $\varepsilon=1$.
(2) Use SQP algorithm to successively solve z for a series of decreased ε_n, and $\varepsilon_n=10^{-n/10}$ (n=1, 2, …, N). The solution of step (1) can be used as the initial value to solve z for $\varepsilon=\varepsilon_1$. Then, z for $\varepsilon=\varepsilon_n$ is solved by using z for $\varepsilon=\varepsilon_{n-1}$ as the initial value.
(3) z of the fuel-optimal problem is obtained when ε_n approaches zero. Then the optimal control law can be gotten according to equation (9). N is set as 40 in this paper, the solution of z for $\varepsilon=10^{-4}$ will be obtained eventually.

4. Simulation Results

The high-thrust and low-thrust cooperative rendezvous cases are simulated in this section. The working conditions, including engine-specific impulse (I_{sp}), initial mass (m_0), and thrust magnitude setting, are the same for two spacecraft. m_0 is set as 1000kg and I_{sp} is set as 400s for all cases. The terminal orbit is not specified. Table 1 shows the initial states of two spacecraft. Three nonplanar rendezvous cases are simulated in Case 1, 2, and 3. In Case 1, the radial, circumferential, and normal thrust magnitudes are

20N, 40N, and 40N. In Case 2, the thrust magnitudes are reduced to 2N, 4N, and 4N. In Case 3, the thrust magnitudes are 2N, 4N, and 4N, and the time is limited, it is required that the rendezvous time should be less than 180TU. Two coplanar rendezvous cases are simulated in Case 4 and 5. Only circumferential thrust is considered for the coplanar rendezvous problem. In Case 4, the thrust magnitude is set as 10N. In Case 5, the thrust magnitude is set as 40N. Two spacecraft's terminal states are shown in Table 2. The fuel and time consumptions are shown in Table 3. The flight trajectories of two spacecraft and the time history of the thrusts (control laws) for Case 1, 2, and 3 are shown in figures 2, 3, and 4.

Table 1. Two spacecraft's initial orbital parameters.

	Spacecraft No.	a/km	e	Ω	i	ω	θ	m_0/kg
Nonplanar rendezvous	1	12000	0.2	0.3	0.4	0.6	-0.5	1000
	2	14000	0.3	0.3	0.3	0.8	0.6	1000
Coplanar rendezvous	1	12000	0.3	0.1	0.1	0	-0.5	1000
	2	13000	0.3	0.1	0.1	0	1	1000

Table 2. Orbital parameters of two spacecraft after cooperative rendezvous.

	Spacecraft No.	a/km	e	Ω	i	ω	θ	m/kg
Case 1	1	13332.30	0.264	0.300	0.352	0.708	1.031	871.54
	2	13372.04	0.263	0.302	0.352	0.707	1.029	900.02
Case 2	1	13080.09	0.252	0.302	0.351	0.701	5.212	876.44
	2	13060.11	0.252	0.303	0.347	0.702	5.209	892.47
Case 3	1	13011.48	0.262	0.277	0.358	0.772	4.057	838.18
	2	13009.53	0.262	0.277	0.354	0.772	4.057	828.80
Case 4	1	12664.56	0.326	0.100	0.100	0.004	0.306	969.38
	2	12666.45	0.328	0.100	0.100	0.004	0.304	975.87
Case 5	1	12664.69	0.330	0.100	0.100	-0.043	3.070	969.91
	2	12664.53	0.330	0.100	0.100	-0.043	3.070	975.42

Table 3. Fuel and time consumptions.

	Fuel consumption	Time consumption
Case 1	228.45kg	20.52TU (16531s)
Case 2	231.10kg	347.54TU (279917s)
Case 3	333.02kg	176.97TU (142541s)
Case 4	54.74kg	52.37TU (42183s)
Case 5	54.67kg	25.49TU (20532s)

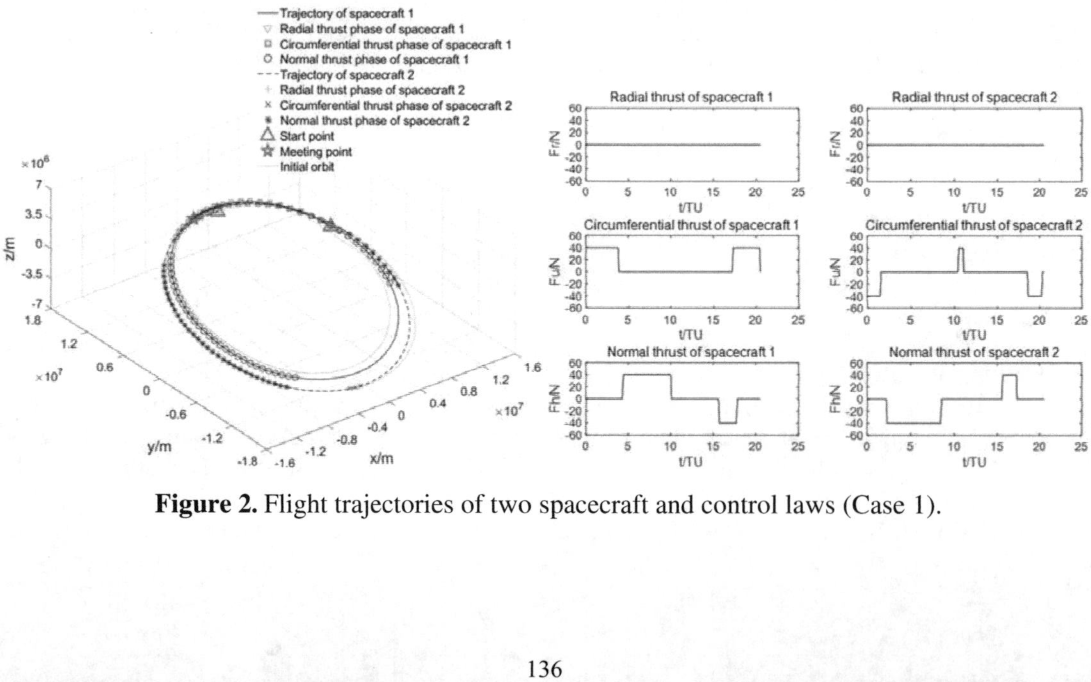

Figure 2. Flight trajectories of two spacecraft and control laws (Case 1).

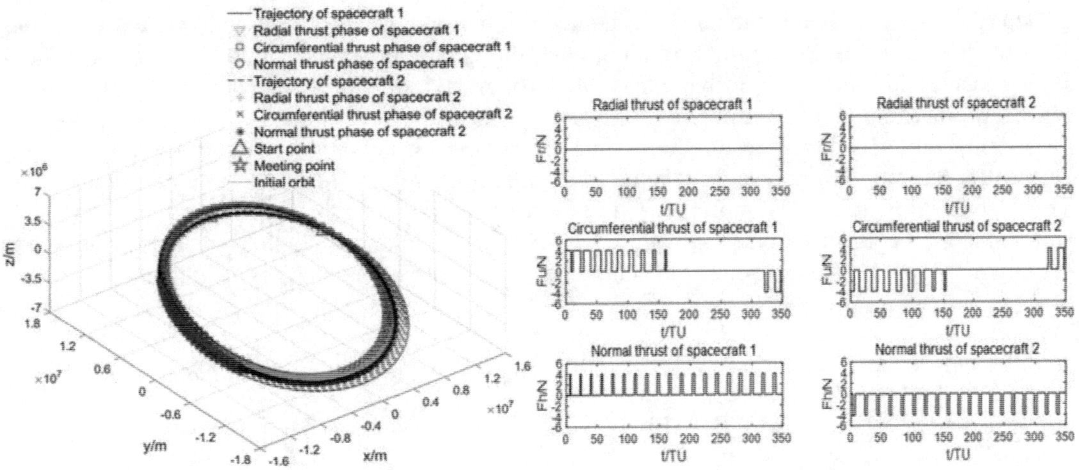

Figure 3. Flight trajectories of two spacecraft and control laws (Case 2).

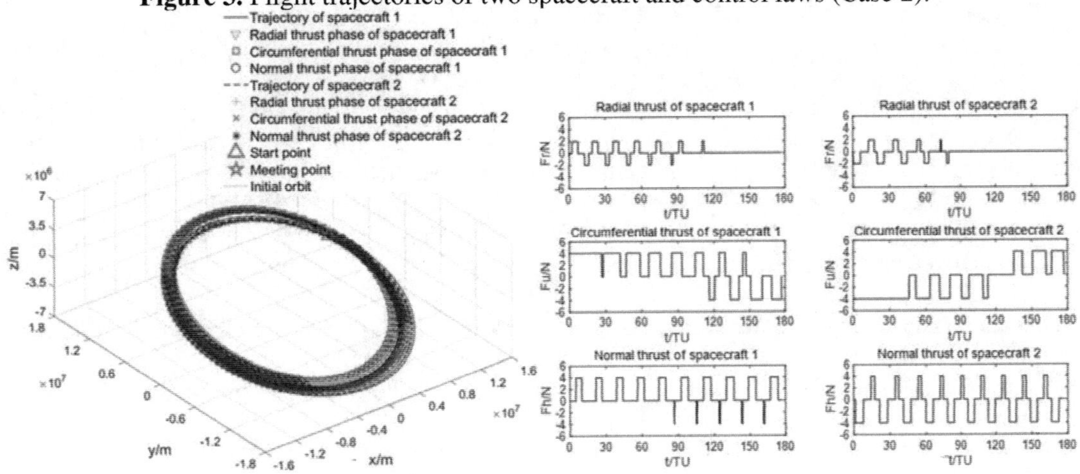

Figure 4. Flight trajectories of two spacecraft and control laws (Case 3).

The fuel consumptions of Case 1 and Case 2 are almost the same, however, the time consumption of Case 2 is much higher than that of Case 1. The reduced thrust magnitudes in Case 2 make the number of burning and coasting arcs increase greatly. The radial thrust is not necessary in Case 1 and Case 2. The rendezvous time is limited in Case 3, the radial thrusts appear to help the spacecraft complete the rendezvous quickly, meanwhile the fuel consumption greatly increases. From the results of the coplanar rendezvous cases, Case 4 and 5, it can also be found that the rendezvous can be accomplished without the radial thrusts, and the thrust magnitude has almost no effect on the fuel consumption.

The simulation results also show the advantages of the indirect optimization method to solve the continuous-thrust far-distance rendezvous problems. In a far-distance rendezvous, the engine sometimes needs to switch on & off frequently and many burning arcs will appear in the trajectories (such as Case 2 and 3). In such complex cases, the application of the direct optimization method is limited by the discretization scale, a very small discretization scale is required to assure the accuracy. The hybrid optimization method needs to introduce too many intermediate variables for the burning arcs, causing the number of variables of the optimization problem very large. Therefore, Case 2 and 3 can hardly be solved by using direct method or hybrid method.

5. Conclusion

A hybrid algorithm combined homotopy method and SQP algorithm is proposed for solving the continuous-thrust far-distance cooperative rendezvous problems. The homotopy-SQP hybrid algorithm used in this paper can overcome the nonsmooth integration problem. The proposed optimization method can effectively obtain feasible optimal solutions even though the indirect optimization method exhibits a narrow convergence domain, improving the feasibility of the indirect optimization method for solving complex trajectory optimization problems.

It is found from the simulation results that radial thrust is not necessary for the rendezvous if time is sufficient. But the radial thrust can accelerate the rendezvous process for the time-limited rendezvous missions, consuming relatively more fuel. The thrust magnitude has almost no effect on the fuel consumption if time is sufficient. High thrust is more efficient, making the rendezvous less time-consuming. For the time-limited rendezvous mission, a high thrust maneuver is desirable.

References

[1] Prussing J E and Conway B A 1989 Optimal terminal maneuver for a cooperative impulsive rendezvous *J Guid Control Dyn* vol 12(3) pp 433–435
[2] Mirfakhraie K and Conway B A 1994 Optimal cooperative time-fixed impulsive rendezvous *J Guid Control Dyn* vol 17(3) pp 607–613
[3] Crispin Y and Seo D 2011 Rendezvous between two active spacecraft with continuous low thrust *Advances spacecraft technologies* ed J Hall (IntechOpen) pp 585–596
[4] Kechichian J A 1997 Optimal low-earth-orbit-geostationary-earth-orbit intermediate acceleration orbit transfer *J Guid Control Dyn* vol 20(4) pp 803–811
[5] Ranieri C L and Ocampo C A 2008 Indirect optimization of two-dimensional finite burning interplanetary transfers including spiral dynamics *J Guid Control Dyn* vol 31(3) pp 720–728
[6] Pontryagin L S, Boltyanskii V G, Gamkrelize R V, Mishchenko E F 1962 *The mathematical theory of optimal processes* (New York: Wiley)
[7] Bertrand R and Epenoy R 2002 New smoothing techniques for solving bang-bang optimal control problems—numerical results and statistical interpretation *Optim Control Appl Meth* vol 23 pp 171–197
[8] Jiang F H and Tang G 2016 Systematic low-thrust trajectory optimization for a multi-rendezvous mission using adjoint scaling *Astrophys Space Sci* vol 361 p 117
[9] Tang G and Jiang F H 2016 Capture of near-earth objects with low-thrust propulsion and invariant manifolds *Astrophys Space Sci* vol 361 p 10
[10] Feng W M, Ren F, Shi L 2014 Optimal control for far-distance rapid cooperative rendezvous *Proc Inst Mech Eng, G J Aerosp Eng* vol 228(14) pp 2662–2673
[11] Feng W M, Zhao D, Shi L, Yang K, Zhao J F 2015 Optimization control for the far-distance rapid cooperative rendezvous of spacecraft with different masses *Aerosp Sci Technol* vol 45(Sep) pp 449–461
[12] Feng W M, Wang B, Yang K, Zhao D 2017 Cooperative rendezvous between two spacecraft under finite thrust *CEAS Space J* vol 9 pp 227–241
[13] Yang K, Feng W M, Liu G, Zhao J F, Su P Y 2018 Quantum-behaved particle swarm optimization for far-distance rapid cooperative rendezvous between two spacecraft *Adv Space Res* vol 62(11) pp 2998-3011
[14] dos Santos Coelho L, Mariani V C 2008 Particle swarm approach based on quantum mechanics and harmonic oscillator potential well for economic load dispatch with valve-point effects *Energy Convers Manage* vol 49(11) pp 3080–3085
[15] dos Santos Coelho L 2008 A quantum particle swarm optimizer with chaotic mutation operator *Chaos Solitons Fractals* vol 37(5) pp 1409–1418
[16] Bolte J and Pauwels E 2016 Majorization-minimization procedures and convergence of SQP methods for semi-algebraic and tame programs *Math Oper Res* vol 41 pp 442-465

A Novel Route Planning Scheme Based on Dynamic Time Variables

Jinghe Zhang[1,2,*], LeiLi[2], Yuechao Zhang[2], Shaofeng Cheng[2] and Weidong Fang[3]

[1]BUAA EMC Technology Institute, Beihang University, Beijing 100083, P. R. China.
[2]Beihang Beidou Application Research Institute of Jinhua, Jinhua 321037, P. R. China.
[3]Science and Technology on Micro-system Laboratory, Shanghai Institute of Micro-system and Information Technology, Chinese Academy of Sciences, Shanghai 200051, P. R. China.

[*]Corresponding author email: 1661395998@qq.com

Abstract. In this paper, the processing method of dynamic time variables in the route planning nodethat we designed involves the field of route planning, including determining the route time and stay time, combining the floating reserved time to determine the redundancy; after the entity has visited a node, the current value is calculated Redundancy: According to the calculated redundancy value, combined with the preset redundancy threshold, the entity's route and stay time are reminded. By converting the dynamic time variable into a fixed value, the model can complete the search for the optimal path within a limited time. At the same time, the concept of reserved floating time is introduced, thereby retaining the dynamic characteristics of time, so that visitors can have sufficient time stay, and can complete the stay according to the fixed time requirements, can effectively track the stay time and make corresponding processing in the corresponding situation to ensure the stay ability. The proposed scheme has better performance.
Keywords: Dynamic time variables, route planning, route time.

1. Introduction

Route planning is one of the main research contents of motion planning. Motion planning is composed of route planning and trajectory planning (time). The sequence of points or curves connecting the start and end positions is called the path, and the strategy that constitutes the path is called Route planning. The path is a certain sequence of the robot's pose, regardless of the factors that change the robot's pose parameters over time.

Route planning (generally referred to as location planning) is to find a series of path points to be passed. Path points are positions or joint angles in space, while trajectory planning is to give time information to the path. Motion planning (also known as motion interpolation) is to insert a sequence of intermediate points for control between the endpoints of a given path to achieve smooth motion along a given path. Motion control is mainly to solve the problem of how to control the target system to accurately track the command trajectory. That is, for a given command trajectory, select suitable control algorithms and parameters, generate output, and control the target in real time and accurately track the given command trajectory [1-5].

The goal of route planning is to make the distance between the path and obstacles as far as possible and the length of the path as short as possible; the purpose of trajectory planning is to make the robot's running time as short as possible or the energy as small as possible while the robot is moving in the joint space. Trajectory planning adds time series information on the basis of Route planning to plan

Content from this work may be used under the terms of the Creative Commons Attribution 3.0 licence. Any further distribution of this work must maintain attribution to the author(s) and the title of the work, journal citation and DOI.
Published under licence by IOP Publishing Ltd

the speed and acceleration of the robot when performing tasks to meet the requirements of smoothness and speed controllability.
With the vigorous development of smart tourism, scenic spots have made certain achievements in the construction of smart management and smart services, but there are still some problems, such as how scenic spots can assist in effective planning of tourist routes and improve service efficiency. In view of the problems of smart scenic tourism, we can make full use of new network technologies such as mobile Internet [6-9], Internet of Things [10, 11], and cloud computing, to solve these issues, and make the entire journey of scenic tourists more intelligent, humanized, specialized and refined.

2. Related Works

In order to solve the Route planning problem, many effective solutions have been explored. They are not mutually exclusive, and often combine to achieve a common path. The typical methods are as follows:
1) Geometric method
The geometric method extracts the geometric characteristics of the environment. Using its combination characteristics to map the environment space to a weighted (weight can make the geometric distance between two points) graph, so that the Route planning problem of avoiding obstacles can be transformed into a simple graph search problem. The Route planning method based on geometric method is generally divided into 3 steps:

- Find the starting point in the search graph;
- Find the target point in the search graph;
- Use these two points as a polyline that does not pass through obstacles in the graph or connect the curves to get a collision-free path.

Geometric methods include:
a. Visibility Graph
This method connects the vertices of all obstacles (set as V0), the starting point s, and the target point g with a straight line combination, and requires that the line between the three cannot pass through the obstacle, That is, the straight line is visible. Assign weights to the edges in the graph to construct the graph G (V, E) [12]. Then use a certain search method to plan the optimal path. This method can be used when the number of obstacles or the shape of the line is not complicated, but when the two are not satisfied, the constructed graph G (V, E) will increase the amount of calculation by hundreds of thousands, so it is somewhat clever People have made some improvements to it, so that it can complete Route planning without saving so much information. This method is called tangent method. Advantages: intuitive concept and simple implementation. Disadvantages: lack of flexibility, that is, once the starting point and target point are changed, the visible view must be reconstructed, and the path is not optimal
b. Voronoi diagram method
This method uses a series of nodes to define; these nodes are equidistant from the edges of two or more obstacles nearby. The Voronoi diagram divides the space into several areas, each area only contains the edge of an obstacle (similar to the brushfram method) Advantages: The path is very safe. Disadvantages [13]: The amount of calculation is very large, and the path is not optimal. A typical Voronoi diagram is shown in Fig. 1.

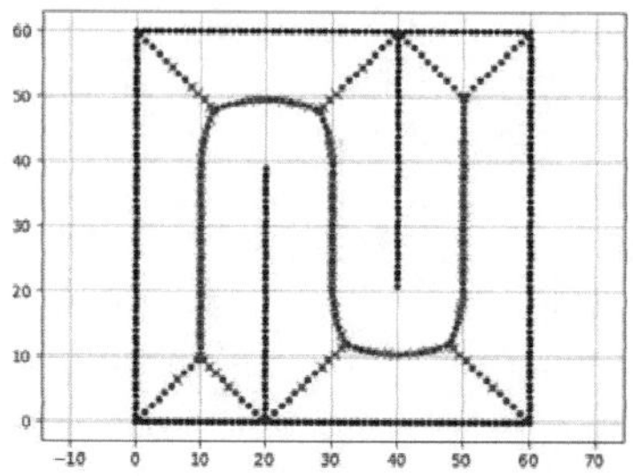

Figure 1. A typical Voronoi diagram.

c. Free space law
The environment is divided into two parts, namely obstacle space and free space. Use a certain search strategy to find a path in free space. According to the different methods of dividing free space, it can be divided into convex area method, triangle method, and generalized cone method. Advantages: The path has no collision and is more flexible [14]. The change of the starting point and the target point will not cause the reconstruction of the connected graph. Disadvantages: In some cases, the path deviates too far from the foreground target, and the shape of the planned path is more complicated and the accuracy is not high.
2) Unit division method
The typical feature of unit division is that the division of space and the search are cross-processed. This method divides the free space into a set of simple units, and the adjacency of the lines between the units is also calculated at the same time. First identify the unit where the starting point and the target point are located, and then connect the consecutive cells between the two points to obtain a safe path. The pollen of the unit can be dependent on obstacles or independent of obstacles. For the former, the boundary of the obstacle is used to generate the boundary of the cell, and the resulting set of free cells accurately defines the free space [15]. Its advantage is: the obstacle is effectively expressed, and the resulting cell is less, which is independent of obstacle cell division method searches for fewer nodes. The disadvantage is that the adjacency relationship between the unit decomposition and the calculation unit is expensive. For the unit decomposition independent of obstacles, the environment space is divided into some regular-shaped units, and the algorithm is simple. In all units, if there are no obstacles, it is called an empty unit: if it is full of obstacles, it is a full unit, and if part of it contains obstacles, it is called a mixed unit. Eight-segment elements are regarded as nodes, and the adjacent relations between them are connected by arcs to obtain a network connected graph, so the problem of finding a safe path becomes a graph search. Its advantages are: simple division and easy implementation. The disadvantages are: Obstacles are not necessarily represented accurately, and the way to improve is to increase the number of units to increase the accuracy of the points exchange.
3) Mathematical analysis method
This method transforms the problem of finding the optimal path from the start point to the end point into a problem of seeking a set of extreme values of objective functions with constraints, and transforms the problem of Route planning into a problem of function optimization. Because this kind of optimization is non-linear and has many kinds of restriction conditions, discretization methods are often used to find the optimal solution.

3. Route Planning Scheme Based on Dynamic Time Variables

In this section, after abstracting the route planning of scenic spots, we propose a route planning scheme based on dynamic time variables. It is described as follows:
According to the travel mode of the entities, we determine the route time t_1 spent by the entities on the journey, the stay time t_2 spent by the entities staying in the nodes according to the travel mode of the entities, and determine the floating reserved time $t = t_1 + t_2$ according to the route time and stay time;
Then, we determine the actual floating reserve time t_h, combined with the floating reserve time t to determine the redundancy η;
The moment when the entity finishes visiting a node is recorded as a critical moment node, and the calculation step of calculating the redundancy η corresponding to the critical moment node is triggered at each critical moment node to obtain the redundancy corresponding to each critical moment node;
According to the calculated redundancy value, combined with the preset redundancy threshold, the entity's tour route and stay time are reminded.
By converting the dynamic time variable into a fixed value, the model can complete the search for the optimal path within a limited time. At the same time, the concept of reserved floating time is introduced, thereby retaining the dynamic characteristics of time, so that visitors can have sufficient time stay, and can complete the stay according to a fixed time requirement. In order to ensure the completion of the route, we introduced the concept of time redundancy, which can effectively track

the stay time and make corresponding processing in the corresponding situation to ensure the stay ability.
Specifically, according to the difference in the route time of entities in each node, the redundancy corresponding to the actual route time is determined, and it is stipulated that the redundancy corresponding to the current moment is obtained every time the entity finishes visiting a node Values, and then remind visitors of the way of sightseeing based on redundancy, to achieve effective tracking of visitors' stay time, to remind the actual stay time, to effectively plan the visitor's stay time, and to avoid uneven distribution of stay time. As a result, the sights are missed.
In the actual implementation process, according to the different estimated time spent by entities at each attraction, the actual redundancy changes, and the dynamic time variables are converted into quantitative processing, which reduces the complexity of the model and facilitates the optimization. The search of the path, and the setting of time redundancy and key time nodes, allows us to monitor the whole process to ensure the smooth progress of the game.
1. Determining the route time spent by the entities on the journey according to the transportation mode of the entities, including:
Leave t_1^k, k=1, 2, ..., n, mark $t_1 = \sum_{k=1}^{n} t_1^k$, t_1 is the time that the entity will spend on the journey.
In the implementation, the journey to the node will be divided into several sections during the actual travel of the entities. Here, the time spent on each section of the journey is recorded as t_1^k, and the time spent by the entities during the journey is recorded as t_1 , The latter is the cumulative sum of the former, in this way, the time spent by entities in the course of the journey is determined by the cumulative sum.
2. According to the transportation mode of the entities, the staying time t_2 spent by the entities staying in the nodes includes:
Set t_2^k, k=1, 2, ..., n, mark $t_2 = \sum_{k=1}^{n} t_2^k$, as the time that entities will spend in the node.
Similar to the previous step, the time spent by entities in each node is recorded as t_2^k, and the time spent by entities during the journey is recorded as t_2, and the latter is the cumulative sum of the former. In this way, the sum is summed by accumulation. The way is determined the time spent by entities in the process of staying in the node.
3. Determine the actual floating reserve time th, combined with the floating reserve time t to determine the redundancy η, including:
The floating reserve time used is t_h, n=1 or 2, and time redundancy η is introduced. $\eta = 1 - \frac{t_h}{t}$ indicates the actual route's utilization of floating time. Among them, set t as the floating reserved time.
4. Determine the floating reserved time t_h according to the t_1 and t_2 obtained in the first two steps, and then determine the utilization degree of the floating time that represents the actual route time of the entitys in the node according to the floating reserved time.
The greater the redundancy, the higher the time utilization rate of the entity's current stay route, the less time spent on the road or other things unrelated to stay, and the more adequate stay; on the contrary, if the redundancy is less, Which shows that the time utilization rate of the entity's single-money stay route is low, and the time spent on the road or other things unrelated to the entity is more, and the stay is not sufficient.
5. According to the calculated redundancy η value, combined with the preset redundancy threshold, the entity's tour route and stay time will be reminded
In order to be able to remind visitors of the entity's stay time based on the calculated redundancy value, redundancy thresholds η_0, η_1 and η_2, are set, and the relationship among the three is in order of $0 < \eta_0 < \eta_1 < \eta_2 \leq 1$; when the actual calculated redundancy is close to 1. It shows that the current entity's stay time is basically used for the sightseeing of nodes in the node, and basically no time is spent on irrelevant things.

η_2 is the warning line, if $\eta_1 < \eta < \eta_2$ will trigger a warning, remind visitors that they must pay attention to the stay time, control the stay season to ensure the smooth completion of the pre-selected route;
η_1 is the termination line. When $\eta_0 < \eta < \eta_1$, a reminder "reserved time has been exhausted" will be issued. You must stay according to the set time to ensure the completion of the scheduled route;
When the time redundancy is lower than η_0, it shows that the current route cannot be completed in the remaining time, and the route search must be performed again so that the entitys can complete the game within the limited time.

4. Conclusion

When entities are currently visiting the node, it may be due to unreasonable route planning and excessive traffic at a certain node that the actual stay time may not match the expected stay time, or even vary greatly, which may cause the planned node to not be visited, which affects stay experience. In this article, we aim to provide the concept of time redundancy, so as to effectively track the stay time and make corresponding processing in the corresponding situation to ensure the processing method of dynamic time variables in the route planning.

References

[1] Fang W, Zhang W, Liu Y, et al. (2020) BTDS: Bayesian - based trust decision scheme for intelligent connected vehicles in VANETs. Wiley Transactions on Emerging Telecommunications Technologies. 2020; e3879. doi: 10.1002/ett.3879
[2] Tang C, Xia S, Li Q, Chen W and Fang W. (2021) Resource pooling in vehicular fog computing. Journal of Cloud Computing. 10, 19 (2021), https://doi.org/10.1186/s13677-021-00233-x
[3] Fang W, Chen W, Zhang W, et al. (2020) Digital signature scheme for information non-repudiation in blockchain: a state of the art review. EURASIP Journal on Wireless Communications and Networking. 2020, 56. doi: 10.1186/s13638-020-01665-w
[4] Fang W, Zhang W, Chen W, Liu Y, Tang C. (2020) TMSRS: trust management-based secure routing scheme in industrial wireless sensor network with fog computing. Wireless Networks, 26(5): 3169-3182. doi: 10.1007/s11276-019-02129-w.
[5] Ji X, Du J, Jia G, Fang W, (2021) Low Complexity Decoding Algorithm for Uplink SCMA Based on Aerial Spherical Decoding. Intelligent Automation and Soft Computing, Vol. 27, No. 3, 2021, pp: 737-746. doi: 10.32604/iasc.2021.013009
[6] Fang W, Cui N, Chen W, Zhang W, Chen Y. (2020) A Trust-based Security System for Data Collecting in Smart City. IEEE Transactions on Industrial Informatics, Vol. 17, No. 6, 2021, pp: 4131-4140. doi: 10.1109/TII.2020.3006137
[7] Ni Y, Chai J, Wang Y and Fang W. (2020) A Fast Radio Map Construction Method Merging Self-Adaptive Local Linear Embedding (LLE) and Graph-Based Label Propagation in WLAN Fingerprint Localization Systems. Sensors 2020, 20(3), 767.
[8] Fang W, Xu M, Zhu C, Han W, Zhang W, Rodrigues Joel J. P. C. (2019) FETMS: Fast and Efficient Trust Management Scheme for Information-Centric Networking in Internet of Things. IEEE Access. 7(1): 13476-13485
[9] Yi L, Fang W, Zhang W, Gao W, and Li B (2021). Game-Based Trust in Complex Networks: Past, Present, and Future. Complexity. Volume 2021, Article ID 6614941, 20 pages, https://doi.org/10.1155/2021/6614941
[10] Fang W, Zhang W, Zhao Q, et al. (2019) Comprehensive analysis of secure data aggregation scheme for industrial wireless sensor network. Computers, Materials and Continua, 61(2): 583-599.
[11] Chen W, Zhou K, Fang W, Wang K, Bi F, and Assefa B. (2020) Review on Blockchain Technology and Its Application to the Simple Analysis of Intellectual Property Protection, International Journal of Computational Science and Engineering, Vol. 22, No. 4, 2020, pp.437–444.
[12] Fang W, Zhang W, Yang Y, Liu Y, Chen W. (2017) A resilient trust management scheme for defending against reputation time-varying attacks based on BETA distribution. SCIENCE

CHINA Information Sciences, 2017, 60(4): 040305, doi: 10.1007/s11432-016-9028-0

[13] Zhang W, Fang W, Zhao Q, Ji X. and Jia G. Energy Efficiency in Internet of Things: An Overview. Computers, Materials and Continua, Vol. 63, No. 2, 2020, pp: 787-811. doi: 10.32604/cmc.2020.07620

[14] Huang L, Jia G, Fang W, Chen W, Zhang W. Towards Security Joint Trust and Game Theory for Maximizing Utility: Challenges and Countermeasures. Sensors 2020, 20, 221.

[15] Fang W, Zhang W, Yang W, Li Z, Gao W, Yang Y, (2021) Trust management-based and energy efficient hierarchical routing protocol in wireless sensor networks, Digital Communications and Networks, Available online 26 March 2021, In Press. https://doi.org/10.1016/j.dcan.2021.03.005

Spinneret Image Region of Interest Segmentation Algorithm

Dejun Li*, Guiyang Zhou, Kang Cheng, Cheng Wang, Yifan Wang and Guanglong Li

School of Electronic and Electrical, Wuhan Textile University, Wuhan, China

*Corresponding author email: 1986015@wtu.edu.cn

Abstract. In order to improve the accuracy of spinneret defect detection, a spinneret image region of interest segmentation algorithm is proposed for the problem that the complex background of spinneret image interferes seriously with the subsequent detection. The mask image is obtained by separating the fixed plate area and the spinneret wall area using the diffuse water filling method, and the minimum external circle and the maximum internal circle in the mask image are found using contour detection to obtain the mask image of the spinneret area, and then the spinneret area, i.e. the Region of Interest (ROI), is extracted. The experimental results show that this method can effectively separate the spinneret region and reduce the background interference.
Keywords: Spinneret, image processing, flooding filling method, contour detection.

1. Introduction

Spinneret, also known as spinning cap, is an indispensable precision part of the spinning machine in the chemical fiber industry, and its state is an important condition to ensure the quality of finished fiber and good spinning process.[1-2] The main process of polyester fiber production is the formation of raw materials through high temperature melt, the melt will be sent into the spinning equipment, accurate measurement and filtration, from the spinneret micro-hole spray, after cooling device to form a silk thread, and then through a series of spinning process processing to get the final product. [3] After the production of raw materials (polyethylene terephthalate and ethylene glycol) to generate the melt through the spinning device, the spinneret area is prone to polyester fiber melt accumulation phenomenon, after a long time to form yellow spots, crystalline lumps, black lumps and other different defective points, these defective points will move with the silk strip to the post-spinning production, which will eventually affect the quality of the finished product.

The spinneret image is shown in Figure 1, the spinneret image consists of three areas, respectively, the spinneret wall area, the spinneret area and the fixed plate area, the spinneret defect detection is for the detection of defects generated by the spinneret area, it can be seen by looking at the image, the spinneret wall area and the fixed plate area has a lot of stains, these stains will produce great interference with the accuracy of defect detection.

At present, there is no mature equipment and methods for the detection of defects on the surface of the spinneret due to melt accumulation without removing the spinneret, and most domestic companies still use the manual visual observation method, which can only roughly determine whether there is melt accumulation on the spinneret, which will inevitably lead to is judgement and low efficiency, and cannot meet the intelligent needs of enterprises. The main purpose of this paper is to discover the spinnerets that need to be cleaned, but the methods proposed so far are almost always to determine whether the holes are clean or not. We know that dirt in the hole can lead to abnormal indicator values. But the question of which spinnerets on the production line need cleaning remains an unsolved problem.[4-7]

Content from this work may be used under the terms of the Creative Commons Attribution 3.0 licence. Any further distribution of this work must maintain attribution to the author(s) and the title of the work, journal citation and DOI.
Published under licence by IOP Publishing Ltd

The literature [8] proposes the use of Gabor filter for spinneret defect detection, through the Gabor filter after edge detection to extract the edge of the defect to achieve defect detection, but the disadvantage is that for the overall image of the spinneret, the detection effect is not obvious, because the spinneret image background is complex, the detection of serious interference. In order to improve the accuracy of spinneret defect detection and solve the problem of spinneret image background interference, this paper makes improvements based on it and proposes a spinneret image region of interest segmentation method to segment the spinneret image and obtain the region of interest (ROI) with spinneret region in the spinneret image to eliminate the useless image background in order to improve the detection accuracy.

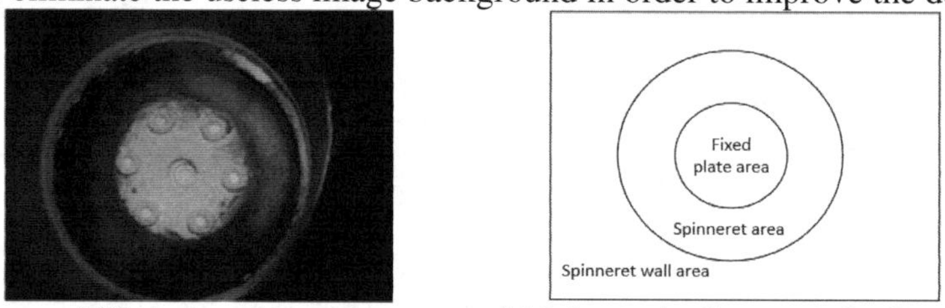

(a)Spinneret image (b)Schematic diagram of spinneret

Figure 1. Spinneret image.

2. Region Segmentation Algorithm

For the characteristics of the spinneret image, first Gaussian filtering of the original image to remove the random noise in the image, in preparation for the next step to extract the region of interest (ROI), and then use the diffuse water filling method for image segmentation, respectively, the fixed plate area and the spinneret area in the image, and then the mask image for contour detection, respectively, to find the minimum outer circle and maximum inner circle of the contour, and to get the mask image of the region of interest, and finally through the mask image of the complete segmentation of the region of interest image, the algorithm flow is shown in Figure 2:

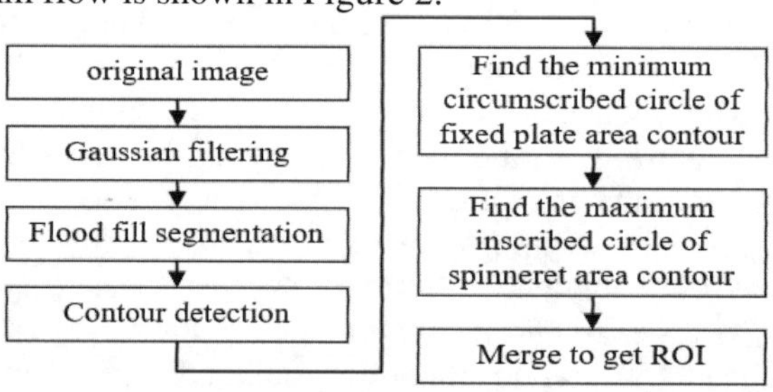

Figure 2. Algorithm flow.

2.1. Gaussian Filtering

Gaussian noise is a common noise that is easily generated during the numerous processes of image acquisition. In order to reduce the impact of random noise on subsequent image processing, Gaussian filtering is first applied to the image. A two-dimensional Gaussian filter is required to filter a two-dimensional image. A two-dimensional Gaussian filter can be obtained by multiplying a one-dimensional Gaussian filter in the X-direction and a one-dimensional Gaussian filter in the Y-direction:

$$G(x,y)=G(x)*G(y)=\frac{1}{2\pi\sigma^2}e^{-\frac{x^2+y^2}{2\sigma^2}} \tag{1}$$

Gaussian filtering has a good effect on Gaussian noise removal, but also causes blurring of the figure object, and the larger the filter size, the more blurred the filtered image becomes, so choose the minimum 3 × 3 size Gaussian filter for filtering, the results are shown in Figure 3.

(a) Details of original drawing of spinneret (b)Image details after filtering

Figure 3. Gaussian filter image.

2.2. Flood Fill

The flood fill algorithm is a very useful method that is often used to mark or isolate portions of an image for further processing or analysis. The diffuse fill algorithm can also be used to derive masks from the input image that can be used by subsequent algorithms. [9] The flood fill method is a kind of region fill method, which is essentially a seed fill method, and this method is applicable to the filling of internally defined regions. An internally defined region means that all pixels inside the region have the same color or luminance value, while pixels outside the region have another color or luminance value. The diffuse fill method can set all the pixels in the region to new values and determine a set of seed points (the seed points are intra-region points) by certain rules, and then decide whether to fill them by judging whether the neighboring pixels of the seed points and the seeds form a connected domain or not, until all the pixels in the region are found or the boundary of the contour line is reached.[10]

But the traditional flood filling method does not directly divide the spinneret area, by the light angle, the spinneret image brightness is not uniform, which will make the division of the spinneret area picture is incomplete or over-segmentation, in order to solve this problem, this paper first fixed plate area and spinneret cylinder area for segmentation.

2.2.1. Image segmentation. The flood filling method is divided into 3 main steps: selecting the seed point (x,y); using the seed point as the center, judging the difference between the pixel value of the 4-field or 8-field and the pixel value of the seed point, adding the pixel point whose difference is less than the threshold into the region; using the newly added pixel point as the new seed point, and executing the second step repeatedly until no pixel point is added into the region.[9]

According to the characteristics of the spinneret image, this paper selects the seed point at the center and edge position of the image, and fills the image with the pixel at that point, and this paper selects the diffuse water filling method in eight fields. In the segmentation of the image, the size of the threshold value directly affects the integrity and accuracy of the segmentation. The flood filling method has two thresholds, the lower bound difference value (loDiff) and the upper bound difference value (upDiff) added into the seed point region condition, which are abbreviated as L and U in this paper. when the L value is set too small, it will lead to incomplete image segmentation; when U is set too large, it will segment other regions. In order to achieve the best segmentation effect, this paper determines the optimal L and U by extensive experimental comparison. the results are shown in Figure 4:

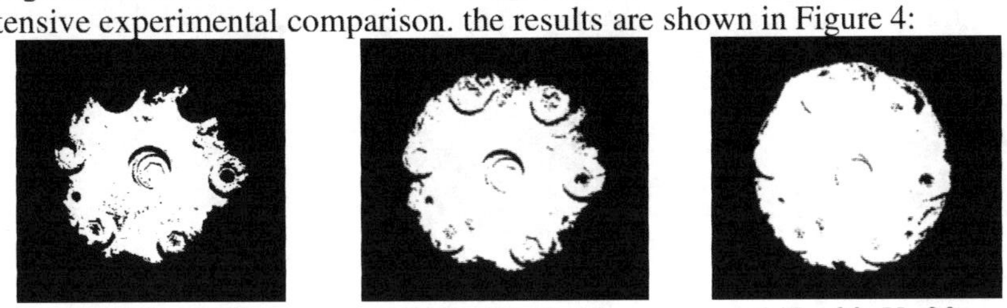

(a)L=20 U=30 (b)L=40 U=30 (c)L=80 U=30

(d)L=20 U=30 (e)L=30 U=30 (f)L=50 U=30

Figure 4. Mask image after segmentation.

Through the above experimental results can be found, when L is 80, U is 30, the fixed plate area segmentation effect is better; when L is 30, U is 30, the spinneret wall segmentation effect is better.

2.3. Contour Detection

The masked image after the flood filling method can be seen that the flood filling method does not completely segment the fixed plate area and the spinneret wall area, and the unsegmented part still interferes with the subsequent spinneret defect detection. In order to reduce the interference of the unsegmented part of the image and improve the accuracy of the subsequent spinneret defect detection, the mask image obtained in this paper is further processed. Specifically, firstly, the mask image of the separated fixed plate area and the mask image of the spinneret area are examined separately for contours, and the contour with the largest area is selected as the target contour, then the contour of the detected fixed area is searched for the smallest external circle of its contour, and the contour of the detected spinneret area is searched for the largest internal circle of its contour. The resulting two circles are combined to obtain the mask image of the spinneret region to lock the spinneret region and segment the ROI image with the spinneret region. The result is shown in Figure 5:

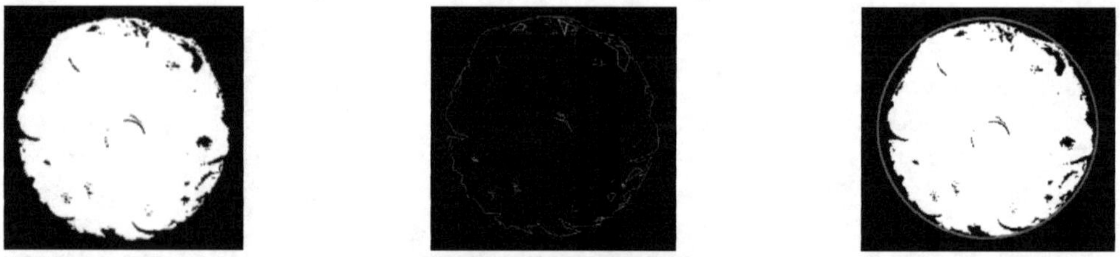

(a)Mask image of fixed plate area (b)Contour detection image (c)Minimum circumscribed circle

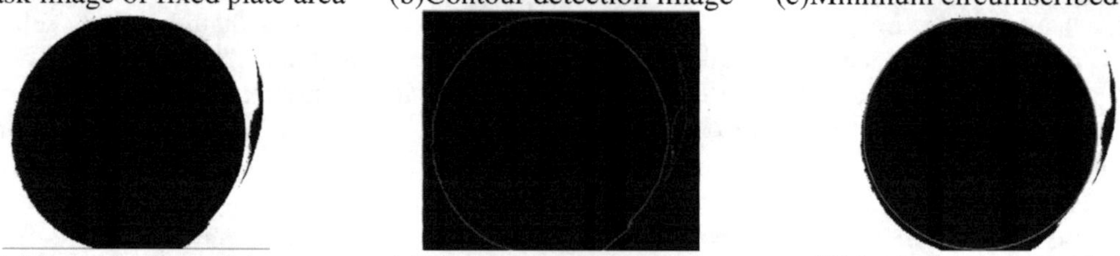

(d)Spinneret wall mask image (e)Contour detection image (f)Maximum inscribed circle

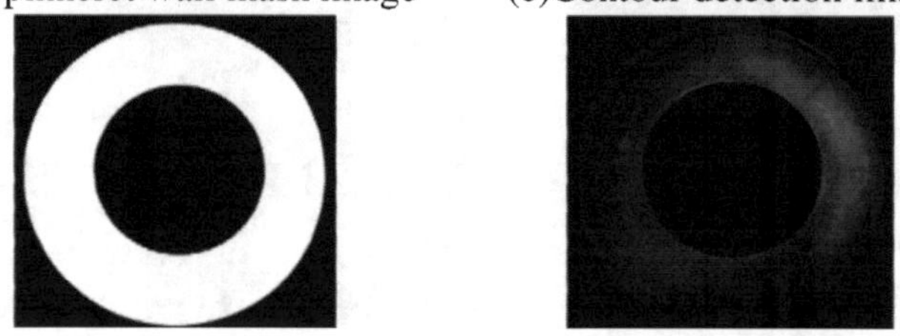

(g)mask image (h)Segmented ROI

Figure 5. Contour detection results.

The principle of finding the minimum external circle is as follows: after contour detection of the fixed plate area mask image, multiple contour point sets will be obtained, and the area of each contour will be found, and the maximum area contour will be found to get the corresponding point set, and the minimum

external circle of the contour will be obtained by using the minEnclosingCircle function.
The principle of finding the maximum inner connection circle is as follows: Invert the spinneret wall mask image, find the maximum connected domain in the inverted image and set it as a point set A. find the nearest distance R from each element in the point set A to the contour in turn, find the maximum value R_{max} of R and the corresponding pixel coordinates $M(x, y)$, and draw a circle on the original mask image with the point as the center and R_{max} as the radius to get the maximum inner connection circle of the contour. The maximum inline circle of the contour is obtained.
This paper locks the spinneret area of the mask image by finding the minimum external circle and the maximum internal circle of the mask image and segmenting it, which is an innovative point of this paper. It can completely segment the spinneret area image and avoid the interference of other areas in the image to the subsequent detection, which improves the accuracy of spinneret defect detection.

2.4. Defect Detection

After segmenting the image of the region of interest with the spinneret area, the image can be detected for defects. Without the complex background interference, only the filament interference in the image is considered for defect detection. In this paper, after grayscale processing, the image is filtered by wavelet mode maximization algorithm and wavelet thresholding algorithm to reduce the filament interference, and the filtered image is analyzed by SURF descriptor to determine the defect sector and identify the defect location. Since the defect detection algorithm is not the focus of this paper, it is not described in detail here. Since the generation of defects is random, a spinneret image with defects is selected for detection in this paper, and the results are shown in Figure 6 below.

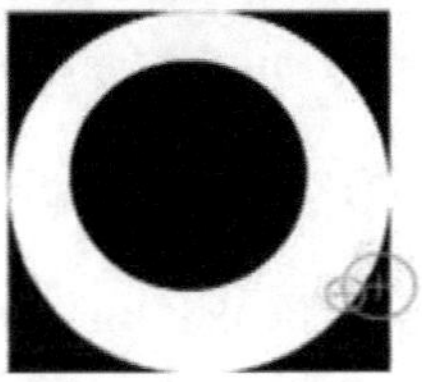

Figure 6. Defect detection results.

3. Experimental Results and Analysis

The computer configuration for this experiment: Intel(R) Core(TM) i5-10400F CPU @ 2.90 GHz 2.90 GHz, 16 GB RAM, experimental environment: Visual Studio 2017 and OpenCV 4.5.1. The selected sample images are a set of spinneret images collected at the factory site. Objective: To segment the spinneret region in the spinneret working image completely and effectively. Since there are few related studies and no related algorithms for comparison, in order to verify the effectiveness of the algorithm in this paper, the algorithm in this paper is compared with the commonly used flood filling method, watershed algorithm, and Grabcut algorithm, and the segmentation results are shown in Figure 7.
The experimental results in Figure 7 show that the spinneret area segmented by the flood filling algorithm and the watershed algorithm are incompletely segmented or over-segmented and the edges are rough, which makes it difficult to identify the defects in the image in the next step. The Grabcut algorithm can completely segment the spinneret region, but it needs to artificially frame the region to be segmented, and does not segment the fixed plate region still affects the subsequent detection results. The advantage of this algorithm is that it can accurately and completely segment the region of interest with the spinneret region, and the segmented target edge is smooth.

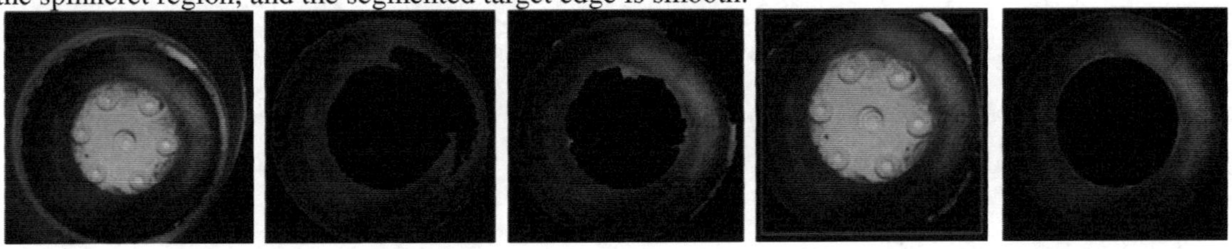

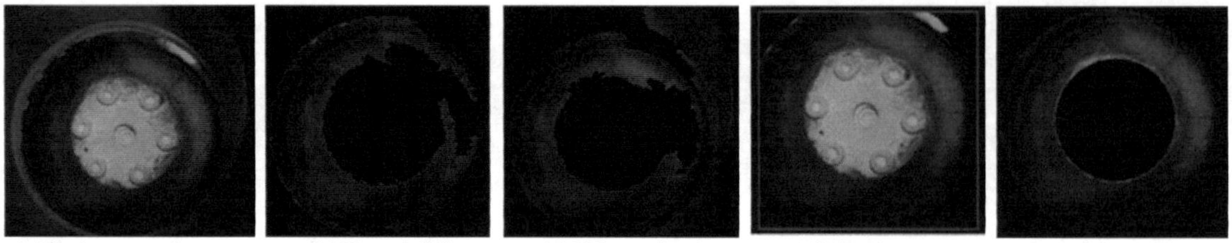

(a)Spinneret image (b)Flood fill (c)Watershed Algorithm (d)Grabcuts (e) this paper

Figure 7. Comparison of experimental results.

4. Conclusion

In this paper, by analyzing the acquired spinneret images and the characteristics of flood filling algorithm, we propose a spinneret image region of interest segmentation algorithm, which can better segment the spinneret region completely and accurately. The method effectively solves the problem of the complex background of the spinneret image on the subsequent defect detection, and improves the segmentation accuracy, so as to lay the foundation for the next step of identifying and analyzing the defects in the image.

References

[1] Yue Wang, Chongchang Yang, Huaping Wang, Leilin Xu, Xueli Wang, Min Zhang. Development status of spinneret [J]. textile machinery,2001(06):45-48.
[2] Rangtong Liu, Rui Li. Design and production of spinneret [J]. Textile industry and technology,2011,40(01):17-20.
[3] Caihong Ding, Shucheng Li, Xiru Wu. Design of automatic shovel robot system for spinneret [J]. Shanghai Textile Technology,2020,48(09):43-45+49.
[4] Pingping Yin, Chongjian Yang, Zhang Peng, Bingbing Chen. Automatic deviation correction based on machine vision in automatic detection of spinneret [J]. Journal of Donghua University (NATURAL SCIENCE EDITION),2007(01):117-119+124.
[5] Junxia Guo, Jianfeng Tao, Chengliang Liu. Detection path planning of concentric spinneret holes using improved ant colony algorithm [J]. Journal of textile,2017,38(11):124-130+136.
[6] Zengbo Xu,Cao GZ,Ruan SB.A method for inspecting the cleanliness of spinneret holes[J]. Journal of Engineered Fibers and Fabrics,2020,15.
[7] Xiaolong Lin, Tieqiang Zhang, Jiaquan Zhang, Liqiao He. Automatic detection system for hole plugging state of spinneret for viscose fiber [J]. Journal of Jilin University (Engineering Edition),2011,41(S1):293-296.
[8] Guanglong Li, Dejun Li, Huiqing Liu, Guiyang Zhou, Kang Cheng, Wang Cheng. Defect detection of spinneret based on Gabor filter [J]. Electronic test,2021(05):40-42+19.
[9] Gary Bradski, Adrian Kaebler. Learning OpenCV 3 [M]. California: O' Reilly Media, Inc.,2008
[10] Jiaxin Chen, Yingmin Jia. A real-time color target recognition method based on diffuse filling method [J]. computer simulation,2012,29(03):4-9.

Research on Modeling and Power Algorithm of Photovoltaic Cells for Solar Cars

Yan Zhong[1], Qing Yang[1,*] and Yanzhong Zhang[2]

[1]School of Mechanical Engineering, Beijing Institute of Technology, Beijing, China
[2]Department of Information, Beijing University of Technology, Beijing, China

[*]Corresponding author email: yangqing@bit.edu.cn

Abstract. Solar photovoltaic cells will be an effective supplement to electric vehicle batteries. In order to make the photovoltaic system of the solar car provide reliable power, this paper models the photovoltaic cell and designs the maximum power point (MPP) tracking algorithm based on three-point sampling with a fixed voltage. In addition, considering that the output power will change when the voltage and radiation intensity change under the peak power of the solar vehicle, the prediction of the peak output power of the photovoltaic array considering the influence of the radiation intensity and battery voltage is put forward for the first time. Experiments show that the designed power algorithm runs stably, and the power generation power prediction error is less than 1%. The peak output power prediction of the photovoltaic array takes into account the effects of real-time radiation intensity and battery voltage fluctuations, making the peak power prediction of the vehicle more accurate.
Keywords: Solar car, photovoltaic cells, modeling, maximum power generation, peak power prediction.

1. Introduction

Solar cars add solar energy as an energy source on the basis of electric cars, which can give full play to the environmental advantages of solar energy. At present, if the solar energy acquisition and utilization need to be accurately described in a mobile solar car, it can only be solved by component testing[1]. Because of cost considerations, solar cars do not have strong computing power. Therefore, based on the hardware components of solar cars, it is very important to establish a reliable photovoltaic cell model, and design an efficient and stable MPP tracking algorithm and peak output power prediction algorithm[2]. Based on the I-V characteristic curve of photovoltaic cells, many models have been established such as single diode model[3], double diode model[4] and modified double diode model. However, due to the large number of parameters of the dual-diode model and nonlinear coupling, the calculation of this type of model is complicated[5]. The model of a single diode is simple, and the amount of calculation is small, so the single-diode model is more widely used. Ali A N et al [6] and Pandey S V et al [7] both set up a photovoltaic simulator in consideration of irradiance and temperature, and explore the SOC and voltage changes of the battery under the state of charge. Singh B et al [8] uses the Lagrangian relaxation algorithm to optimize the hybrid solar cell system. Raj P J et al [9] explores the charging status of photovoltaic cells at the maximum power point based on fuzzy logic. The environment variables set by these schemes are relatively single, which does not correspond to the actual changed environment. Chtita S et al [10] considers the dynamic changes of weather conditions and load demand, and uses dual PI compensators to balance the power of the photovoltaic array and the battery. Baili Z et al [11] builds a photovoltaic power plant power prediction model based on genetic algorithm and BP neural network algorithm, and optimizes the

Content from this work may be used under the terms of the Creative Commons Attribution 3.0 licence. Any further distribution of this work must maintain attribution to the author(s) and the title of the work, journal citation and DOI.
Published under licence by IOP Publishing Ltd

initial weight and threshold of the neural network. These two schemes can predict photovoltaic power generation well, but require complex and powerful embedded systems and a lot of preliminary research. It does not meet the characteristics of solar cars that focus on practicability and low cost. So the purpose of the paper is to establish a photovoltaic cell model of solar vehicles, study its maximum power generation algorithm and peak power prediction algorithm, accurately estimate the power state of solar cars, and improve the performance of solar cars.

2. Modeling and Simulation

2.1. Photovoltaic Cell Type and Arrangement

The solar system of a solar car can independently drive the motor, the power of its solar power generation will also be the content that needs to be considered in battery management. As a mobile solar system, its power generation efficiency will be affected by changes in the external environment and restrictions on the shape of the body structure, so model selection and group arrays are particularly important. Therefore, SunPower's Me3 monocrystalline silicon solar panels are used for grouping, which has high conversion efficiency, small size, and is convenient for array design. With flexible processing technology, the solar panel array can be better attached to the roof of the car. Some parameters of Me3 are shown in Table 1.

Table 1. Parameters of single Si photovoltaic cell Me3.

Parameters	**Values**	**Parameters**	**Values**
Packing factor	82.9%	Open-circuit voltage	0.727V
Efficiency	25%	Maximum power	3.89W
Size	125mm*125mm	Voltage at the MPP	0.64V
Short-circuit current	6.45A	Current at the MPP	6.08A

In order to prevent the damage of local photovoltaic cells from affecting the overall solar system, the solar panels of the solar car are divided into multiple arrays. To study the power characteristics of photovoltaic cell systems, Maximum power point tracker (MPPT) is required to regulate the maximum power generation, and to predict the output power of the solar panel under the dynamic voltage output by the battery peak power duration. Therefore, this paper selects a set of 32 solar panels as the object to study the MPPT control algorithm and power prediction based on the influence of the external environment.

2.2. Modeling and Characteristic Simulation

The single diode model is used to model the Me3 solar cell, and according to the Kirchhoff's theorem, the output current of solar panels is expressed as below.

$$I = I_{ph} - I_{VD} - I_{sh} \tag{1}$$

$$s.t.\quad I_{VD} = I_0\left[\exp\left(\frac{q(U+IR_s)}{AkT}\right)-1\right] \tag{2}$$

$$I_{sh} = \frac{U+IR_s}{R_{sh}} \tag{3}$$

where I_0 denotes the reverse saturation current of a solar cell under the condition of zero radiation; q=1.6×10^{-19}C denotes the unit charge; A=1.2 is the correction factor; k=1.38×10^{-23}J/K is the Boltzmann constant; T denotes the surface temperature of the cell.

In the real-world solar-assisted EV, if a solar panel array is composed of N_s solar panels in series and N_p panels in parallel, the actual equivalent current is calculated as

$$I = I_{ph} - I_d - I_p \tag{4}$$

$$\textit{Series current} \quad I_p = \frac{UN_p / N_s + IR_s}{R_{sh}} \tag{5}$$

$$\textit{Parallel current} \quad I_d = \left(\exp\left[\frac{q(U / N_s + IR_s / N_p)}{AkT} \right] - 1 \right) I_0 N_p \tag{6}$$

Establish a photovoltaic cell array model, and set up a 32-cell module array according to the single-cell parameters in Table 1. The open-circuit voltage is set to 23.264V, the short-circuit current is 6.45A, and the series-parallel connection is 32S1P. Solar panels are greatly affected by temperature and radiation intensity, so setting different solar radiation intensity and temperature. And get the characteristic curve of the solar panel array as shown in Figure 1.

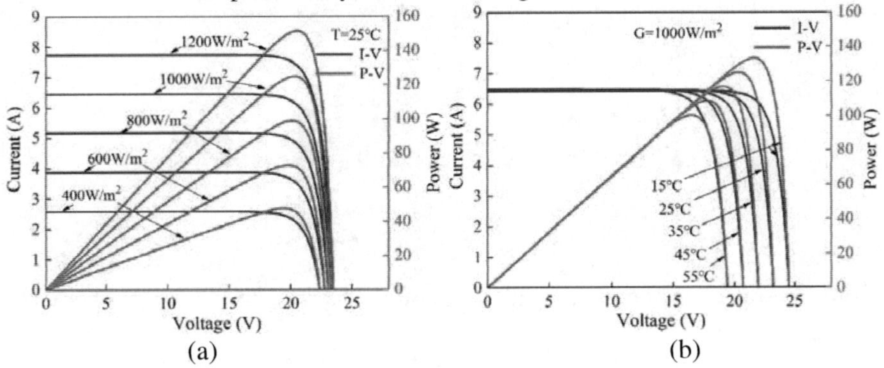

Figure 1. The characteristic curves of the photovoltaic cell.
(a) under different radiation intensities; (b) under different temperatures.

From Figure 1(a), with the increase of radiation intensity, the short-circuit current and maximum power of the solar panel array increase rapidly. In addition, as the radiation intensity increases, the voltage at the MPP increases slightly. Based on the simulation, the power generation of solar panels is greatly affected by the radiation intensity, but less affected by the temperature (only the temperature difference is larger than 10℃, the power generation is affected, which is not apparent in the practical application). Therefore, the short-term power generation prediction of solar vehicles can neglect the temperature, and only consider the influence of radiation intensity on the MPPT algorithm.

3. Power Algorithm Design

3.1. MPPT Tracking and Power Prediction

In this study, we adopt the adaptive fixed voltage numerical optimisation to track the MPP. The flowchart of this algorithm is depicted in Figure 2. The primary principle is to preliminarily determine the MPP, and then sample the power conditions at three points. Through simulating the characteristic curves, the MPP position can be solved. The above MPP can be modified by comparing the real-time power and the estimated power through the hysteresis comparison method. Finally, the real-time MPP of the solar panel array can be obtained cyclically. The details are expressed as follows.

(1) Before connecting the loads to the solar panels, open circuit voltage is measured, and the preliminary estimation of the MPP is calculated with $V_{set1} = k \times V_{oc}$. The MPPT devices regulate the power switches to set the output voltage V_{set1} and then sample the output current. Accordingly, the output power of the first point can be calculated. (2) The second and third points are determined by adding the disturbance variables. Specifically, two disturbance variables $-\Delta V$ and ΔV on the left and right sides of V_{set1} are determined, and then the MPPT devices would track these two voltages and calculate their output powers as the second and third points. (3) Use Newton interpolation method to find the MPP. (4) The MPPT device regulates the operative voltage to the calculated maximum voltage point, i.e. $V_{set} = V_{max}$, and the maximum power P_{max} can be determined.

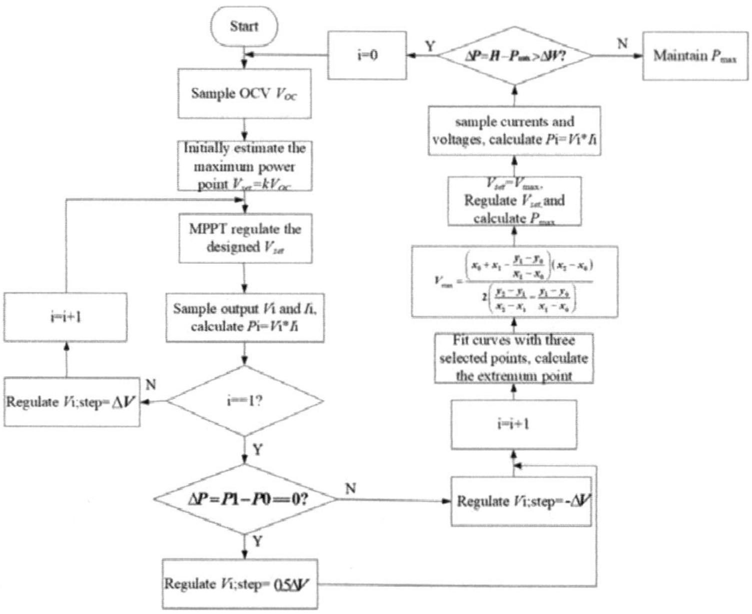

Figure 2. Flowchart of the designed the MPPT algorithm based on the adaptive fixed voltage points.

3.2. Correction of the SOP Based on the Real-time Generated Power

In the actual environment, the conversion efficiency of MPPT will be affected by the load and battery voltage, which would impact the accurate prediction of SOP state of batteries. To accurately predict the effective output power of MPPT, the following modification is adopted in this study. Since the duration of the peak power is relatively short, the temperature change can be regarded as a constant in the control period and only the radiation intensity is considered. The final predictive power selects the minimum predictive power within a continuous duration. Therefore, the power prediction correction for L sampling periods at current time k can be processed by the following formula.

$$P^{l}_{\max,k+L} = \min\left(\sum_{i=1}^{L} E^{i}_{k.k+L}\eta S\right) \tag{7}$$

where $E^i_{k.k+L}$ denotes the radiation intensity at timeslot *i*, and η denotes the generated efficiency of the solar panels, and *S* is the effective area of the solar panel.

Another important factor is the dramatical voltage drop. When the battery voltage is pulled down to the cut-off voltage, the generated voltage of solar panel would also be pulled down, and the boosting efficiency would be affected meanwhile. Therefore, the predictive power of solar panels under the condition of the cut-off voltage of batteries can be formulated as follows.

$$P^{sout}_{\max,k+L} = \min\left(P^{l}_{\max,k+L} - P_r\lambda_1\right)\lambda_2 \tag{8}$$

where $P^l_{max.k+L}$ denotes the peak generated power of solar panels during the timeslots $\Delta t = L \times T_s$ and P_r is the current generated power. λ_1 represents the factor on the generated power of solar panels when the battery voltage drops to the cut-off voltage. λ_2 represents the booster efficiency of MPPT.

4. Results and Discussions

In this section, experiments are carried out to verify the designed power algorithm. Design a 4P7S battery pack in parallel with the photovoltaic cell array, and the signal acquisition system collects the experimental results. The power loads, solar-generated power, and battery power are presented in Figure 3. As shown in Figure 3(a), the experiment uses a variable load. The cycle of a set of loads is 400s. This experiment has gone through 17 cycles. From the view of Figure 3(b), the power of panels is stable almost without the heavy fluctuations when the SOC is larger than the allowance value 0.1, which manifests that the generated power of panels is not affected by the variant loads. After then, the generated power shows a larger fluctuation, which is caused by the variant battery voltage with lower SOC value less than 0.1. Overall speaking, the average generated power of the array is 91.2 W and the average efficiency of MPPT device is up to 96.29%. To validate the effectiveness of the proposed predictive generated power algorithm, we make a comparison with the actual measured generated power and the results are presented in Figure 4. It can be seen that the predictive generated power (according to equations 3 and 4) can well follows the change of the actual generated power and the predictive error is lower than 2W, which is profited from that the proposed predictive algorithm is not significantly affected by the variant battery voltage.

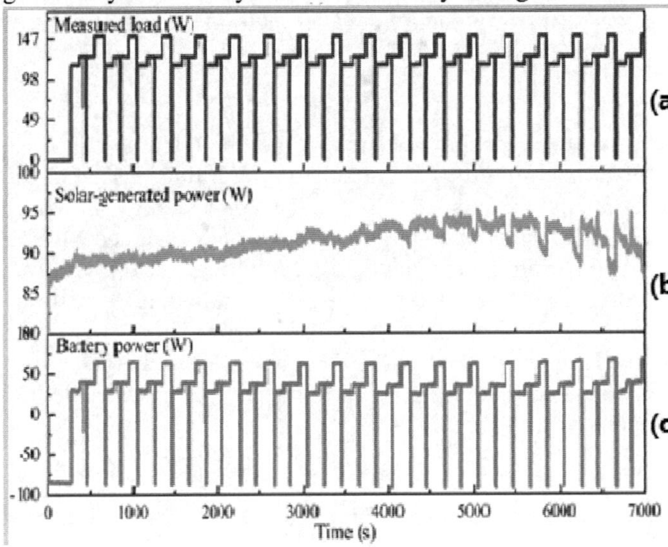

Figure 3. The experimental results of solar panel power: (a) Measured load; (b) solar-generated power; (c) battery power.

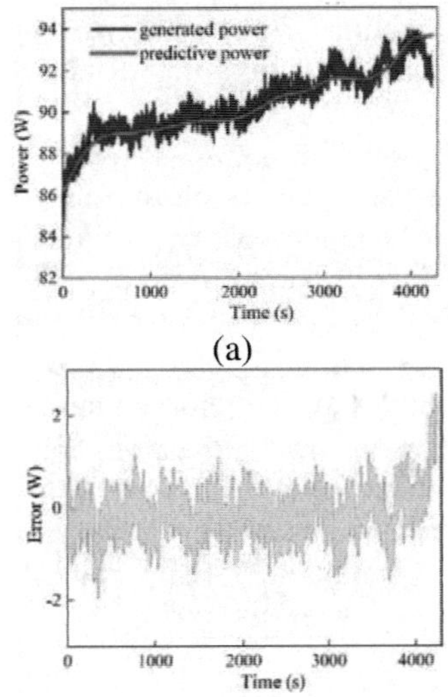

Figure 4. Comparison of the solar generated power and predictive power: (a)Power;(b)Error.

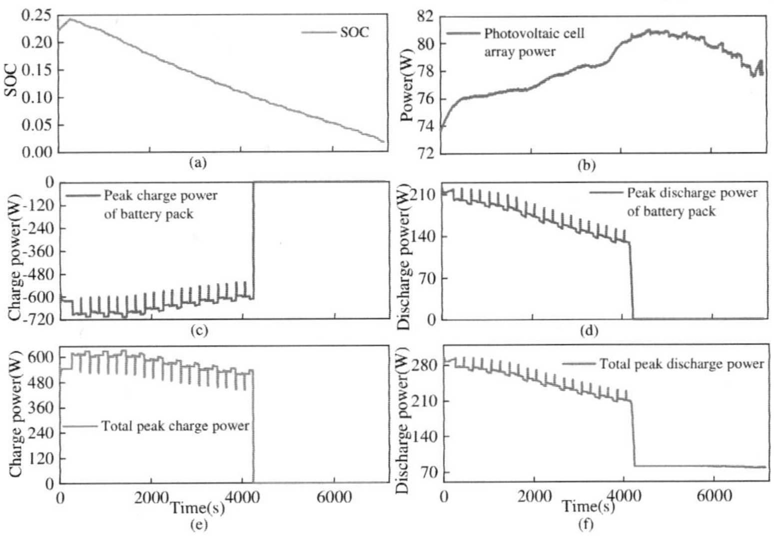

Figure 5. The predictive results of the battery power and the total BMS.

The change of SOC value and the resulting SOP value are illustrated in Figure 5. SOC estimation adopts extended Kalman filter algorithm. From the Figure 5(a), the initial SOC is sampled as 0.222, and after 17 operating cycles, the final SOC is lower than 0.017, which indicates that the battery pack is approaching to the full discharge. The peak generated power of the photovoltaic array shown in Figure 5(b) is lower than the predictive generated power in Figure 5(a) due to the impact of the peak voltage drop. Battery SOP prediction uses RC model algorithm based on SOC constraint. The predictive peak charging power of the battery pack is presented in Figure 5(c) and 5(d). Considering that excessive peak currents in the lower battery state would impair the battery pack, when the battery SOC decreases less than 0.1, the peak operating power would be constrained as zero in the SOP predictive algorithm. When the battery SOC value locates between 0.222 and 0.1, the predictive power in the SOP algorithm presents a decreasing tendency, which results in the increase of peak power of battery pack and the solar panels as shown in Figure 5(e) and 5(f). After comprehensively considering the maximal power, the increasing peak power of battery pack would promote the discharging limit, extending the travelling distance of the traditional electric vehicles.

5. Conclusion

This paper carries on the modeling and the power algorithm research to the photovoltaic cell of the solar car. Power algorithm includes MPPT maximum power tracking algorithm and real-time maximum output power prediction. Experimental results show that the solar energy utilization efficiency of the photovoltaic array reaches 23.4%, the conversion efficiency of the designed MPPT is higher than 96%, and the predicted power generation error is less than 1%. The peak output power prediction of the photovoltaic array takes into account the effects of real-time radiation intensity and battery voltage fluctuations, making the prediction data more accurate. At the same time, after adding the maximum power prediction of the photovoltaic array, the total peak charge and discharge capacity of the solar car battery system is improved. Designed models and algorithms provide a favorable basis for formulating more accurate energy management strategies.

References

[1] Taheri S H. Simple, fast and accurate two-diode model for photovoltaic modules[J]. Solar Energy Materials and Solar Cells, 2011.
[2] Gong W, Cai Z. Parameter extraction of solar cell models using repaired adaptive differential evolution[J]. Solar Energy, 2013,94(4):209-220.
[3] Bana S, Saini R P. A mathematical modeling framework to evaluate the performance of single diode and double diode based SPV systems[J]. Energy Reports, 2016.
[4] Kchaou A, Naamane A, Koubaa Y, et al. Second order sliding mode-based MPPT control for photovoltaic applications[J]. Solar Energy, 2017,155:758-769.
[5] Ranjhitha G, Rajkumar M V. Implementation of Genetic Algorithm based MPP Tracking for Photovoltaic System[J]. 2016.
[6] Ali A N, Premkumar K, Vishnupriya M, et al. Design and development of realistic PV emulator adaptable to the MPP tracking algorithm and battery charging controller[J]. Solar Energy, 2021, 220(2):473-490.
[7] Pandey S V, Patel J, Dhiman H S. Battery State of Charge Modeling for Solar PV Array using Polynomial Regression[J]. 2020.
[8] Singh B, Knueven B. Lagrangian relaxation based heuristics for a chance-constrained optimization model of a hybrid solar-battery storage system[J]. Journal of Global Optimization, 2021, 80(4):965-989.
[9] Raj P J, Prabhu V V, Premkumar D K. Fuzzy Logic Based Battery Management System for Solar Powered Li-Ion Battery in Electric Vehicle Applications[J]. Journal of Circuits, Systems and Computers, 2020.
[10] Chtita S, Derouich A, Ghzizal A E, et al. An improved control strategy for charging solar batteries in off-grid photovoltaic systems[J]. Solar Energy, 2021, 220(4):927-941.
[11] Baili Z, Dandan Y, Lianying X, et al. Distributed Photovoltaic Short-Term Power Prediction Based On Genetic Algorithm Optimized BP Neural Network[J]. IOP Conference Series Earth and Environmental Science, 2021, 692(2):022104.

Abnormal Detection to Big Data Using Deep Neural Networks

Jian Zheng[1,*], Zhaoni Li[2], Jiang Li[3] and Hongling Liu[4]

[1,3,4]Chongqing Aerospace Polytechnic, 401331, Chongqing, P.R.China
[2]College of Computer, Qinghai Normal University, Xining, 810008, P.R.China

[*]Corresponding author email: zhengjian.002@163.com

Abstract. It is difficult to detect the anomalies in big data using traditional methods due to big data has the characteristics of mass and disorder. For the common methods, they divide big data into several small samples, then analyze these divided small samples. However, this manner increases the complexity of segmentation algorithms, moreover, it is difficult to control the risk of data segmentation. To address this, here proposes a neural network approch based on Vapnik risk model. Firstly, the sample data is randomly divided into small data blocks. Then, a neural network learns these divided small sample data blocks. To reduce the risks in the process of data segmentation, the Vapnik risk model is used to supervise data segmentation. Finally, the proposed method is verify on the historical electricity price data of Mountain View, California. The results show that our method is effectiveness.
Keywords: Abnormal detection, big data, deep neural networks.

1. Introduction

It is a major challenge to detect abnormal behavior in a high-speed network environment and massive amounts of data in the field of big data security. Using traditional detection methods, such as misuse detection, is difficult to be used for the security detection of big data. Researchers integrate various machine learning methods to improve the detection accuracy, meanwhile, to reduce the dimensionality of the sample data to increase the detection speed. For instance, the fragmentation (segmentation) mechanism of data stream, whose core idea is to divide high-speed and massive data into different parts and assigned to distributed independent detection nodes for parallel processing.

Data segmentation is the key point to achieve distributed parallel detection of data. Since the quality of segmentation algorithms directly affect the detection results, the ideal segmentation algorithm should be consider as following factors: (1) The data block size is moderate and the amount of information is sufficient. The segmentation algorithm requires the amount of information contained in the data block as support. The division of the data block size can be determined by the amount of information. (2) The size of the data block obtained by each detection node is as the same as possible. Moreover, the segmentation algorithm is as simple and efficient as possible. There are many existing distributed parallel detection mechanisms, such as the fragmentation mechanism, the hash-based packet distribution mechanism, et al. The fragmentation mechanism is to preferentially allocate all data blocks connected at a time to a detection node with a shorter queue length. This mechanism does not consider whether the divided data blocks contain the amount of information to support node detection, when the amount of connected data is too large, this may cause congestion in some detection nodes, also increases the load of the detection nodes. Before the data blocks are distributed to each detection node, the data blocks are pre-treated by pre-screening and local caching methods to reduce the burden on the processing nodes. For instance, in [1], the advance screening and local caching are used to pre-treat data blocks. However, this not only increases the complexity of the segmentation algorithm, but also difficultly implement and requires a

Content from this work may be used under the terms of the Creative Commons Attribution 3.0 licence. Any further distribution of this work must maintain attribution to the author(s) and the title of the work, journal citation and DOI.
Published under licence by IOP Publishing Ltd

certain hardware environment to support. Using a hash-based packet distribution mechanism can reduce the burden of the load balancing module [2], but this manner does not give a basis for the division of data block sizes. The segmentation algorithm inevitably destroys the integrity of the data to a certain extent. Hence, the existing segmentation mechanisms need to be improved to reduce the error ratio caused by distributed parallel detection.

Neural network can establish a classification model with learning ability [3], and analyze a large amount of complex data [4], as well as have the ability to model complex problems and high tolerance to noise data [5]. In various neural network models, there are more applications and researches. The traditional network models have standard BP algorithms[6], the improved BP algorithm with adaptive adjustment of additional momentum and learning rate, Levenberg-Marquardt (LM) optimization method, etc. The disadvantages of the standard BP algorithm is slow convergence and falls into local minimum. The improved BP algorithm has slow training speed and low learning efficiency, but it converges faster than the standard BP algorithm. The LM method is also easy to fall into local minimum, moreover, it requires a lot of storage space to solve complex problems. In order to make up for the shortcomings of these above mentioned methods, it is necessary to proposed a new method, which requires the training target to be classified with high accuracy and not easy to fall into local minimization. Indeed, evolutionary strategies have good global search capabilities [7], which can improve search efficiency.

We summarize the main contributions of this work as follows:

(1) The two-hidden layers neural network is designed, which is used to detect anomalies in big data.

(2) This manner of consisting of neural networks and the Vapnik risk model can reduce the risk of data segmentation.

2. Method

2.1. Calculation Data Segmentation Risk

Vapnik-Chervonenkis proposed a machine learning method based on the theoretical framework of statistical learning [8], which not only considers the experience risk of training samples, but also takes into account the generalization ability of the algorithm. Empirical risk is a decision function, which is a function that minimizes empirical risk from a fixed set of functions. Vapnik proved that for any $\delta \in (0,1]$ and any function *f* in the set of candidate functions, the following inequality holds at least with a probability of $1-\delta$:

$$R(f) \le R_{emp}(f) + \sqrt{\frac{8}{\ell}(h(\ln\frac{2\ell}{h}+1)+\ln\frac{4}{\delta})} \tag{1}$$

where, *h* represents the VC dimension, and *R*(*f*) is the expected risk, which is the real risk that exists objectively. $R_{emp}(f) + \sqrt{\frac{8}{\ell}(h(\ln\frac{2\ell}{h}+1)+\ln\frac{4}{\delta})}$ is structural risk. $R_{emp}(f)$ is empirical risk. $\sqrt{\frac{8}{\ell}(h(\ln\frac{2\ell}{h}+1)+\ln\frac{4}{\delta})}$ is confidence range, which is the increasing function of *h*. The larger the $\sqrt{\frac{8}{\ell}(h(\ln\frac{2\ell}{h}+1)+\ln\frac{4}{\delta})}$ is, the worse the generalization ability is. Therefore, structural risk is an upper bound of expected risk. When *h* is larger, the experience risk is smaller, the confidence range becomes larger, and the generalization ability becomes worse. Vapnik proposes the principle of structural risk minimization: select the subset with the smallest structural risk, in this The smallest subset solves the function that minimizes the empirical risk. This smallest function is the optimal function. In order to prevent the loss of segmented data and ensure the integrity of the data, the principle of structural risk minimization is adopted in the distributed learning process.

2.2. Model Description

A two-hidden layer BP neural network is proposed, which is formulated in detail as follows.

Input layer. The layer receives the original input data.

Hidden layers. The layers include the encoding hidden-layers and the decoding hidden-layers, which are designed as following, respectively.

For the encoding hidden-layers, the input $H_i^l(in)$ and the output $h_i^l(out)$ of the i-th hidden layer in the l-th iteration are calculated using E.q(2) and E.q(3), respectively, having

$$h_i^l = act_i(\mathbf{w}_i(l)H_i^l(in) + \mathbf{b}_i(l)) \tag{2}$$

$$H_i^l(in) = h_{i-1}^l(out) \tag{3}$$

For the decoding hidden-layer, correspondingly, the corresponding output of the j-th hidden layer in the l-th iteration are calculated, as following

$$h_j^l = act_j(\mathbf{w}_j(l)H_j^l(in) + \mathbf{b}_j(l)) \tag{4}$$

$$H_j^l(in) = h_{j-1}^l(out) \tag{5}$$

where, the items act is activation functions. **w** and **b** are the weights and bias of each hidden layer.

Output layer. The layer is used to output results.

Cost function. The cost function, denoted as $J(\mathbf{w},\mathbf{b})$, consists of the mean square error (mse) and the error function $\phi(y, f_{tar}(x^*, w))$ in E.q (6). For the solution of the error function, the approximate value can be obtained by Taylor series expansion.

$$J(\mathbf{w},\mathbf{b}) = \text{mse} + \phi(y, f_{tar}(x^*, w)) \tag{6}$$

where, **w** and **b** are the weights and bias of each hidden layer.

3. Experimental Simulation

(1) Dataset description

The experimental test uses historical electricity price information from 2010 to 2015 in Mountain View, California [9]. Because electricity prices change in real time, the sample data is processed first. The sample data is roughly divided into five categories: industrial electricity (denoted as I), agricultural and commercial electricity (denoted as A), and residential electricity (denoted as R), Municipal electricity (denoted as G), other electricity (denoted as O). Each type of electricity price includes three attributes: peak period (denoted as pt), valley period (denoted as bt), and flat period (denoted as at). Each of these three attributes includes two sub-attributes: summer (denoted as sr).), winter (denoted as wr).

We set the one-hour electricity price as a data sample, and the one-hour electricity price of these five types of electricity is expressed as $I^{\gamma 1}, A^{\gamma 2}, R^{\gamma 3}, G^{\gamma 4}, O^{\gamma 5}$,where, $\gamma 1, \gamma 2, \gamma 3, \gamma 4, \gamma 5 \in [0, 23]$. The sample data contains 24 numeric attributes {0, 1, 2, ..., 23} and 5 symbol attributes {*pt*, *bt*, *at*, *sr*, *wr*}.

(2) Benchmark test

In order to verify the effectiveness of the algorithm, the algorithm is first benchmarked. The test sample is selected from the test data of the "3rd International Knowledge Discovery and Data Mining Tools Competition" [9]. 2000 data were randomly selected as training samples from the test data. The test data contained 23 target classifications. Each sample data contained 41 attributes, including 7 symbol attributes and 34 continuous attributes. In addition, we also compared the method in [6]. In the whole test process, when the accuracy of the two algorithms is the highest, the number of training times and the number of neurons are listed in Table 1.

Table 1. Benchmark test results.

Algorithm	Training steps	Accuracy	Neurons scale
Our method	380~525	81%~88%	117~207
Method in [7]	688~1003	73%~78%	73~165

(3) Testing results

In order to verify the performance of this algorithm, 200,000 pieces of data are selected as training samples. The learning results are shown in Fig.1. The learning results of the 16th to 37th blocks are relatively stable, and the stable knowledge will be stored in the corresponding neurons.

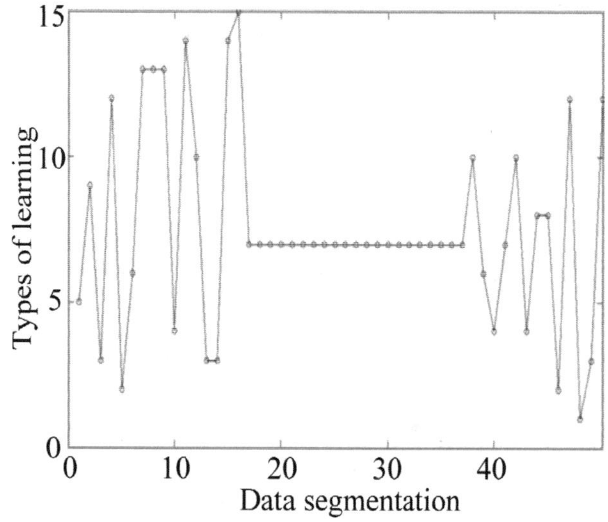

Figure 1. The number of types of learning behaviors for each data sample.

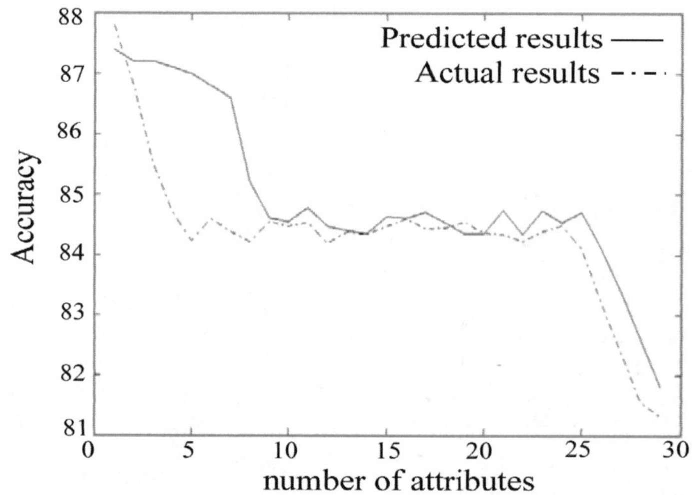

Figure 2. Comparison of prediction rate and actual detection rate.

The sample is used to predict the accuracy of the theoretical model before testing, and the prediction result is compared with the experimental simulation result. The comparison result is shown in Fig. 2. It can be seen from Fig. 2 that the sample attributes are 1-8, and the actual detection accuracy is lower than the predicted accuracy. The sample attributes are 9-25, and the predicted results are closer to the experimental results, and the error is relatively small. It can be seen from the learning result in Figure 2 that when the learning result is relatively stable, the knowledge is stored in the corresponding neuron. At this time, the accuracy of the experimental detection is close to the accuracy of the model prediction.

4. Conclusion

In this paper, we proposed a detection method of consisting of neural networks and the Vapnik-Chervonenkis model, to reduce the risk of segmentation data. Finally, the effectiveness of the proposed method is verified through benchmark tests and simulation experiments. Compared with competing methods, our method outperforms them in terms of detection precision. In the future work, we will continue to explore some new methods in regard to anomaly detection in big data.

Acknowledgement
This research was financially supported by Project of science and Technology Department of Qinghai Province (2016-NN-161).

References

[1] Lai H, Cai S, Huang H, et al. A parallel intrusion detection system for high-speed networks[C]. In: Jakobsson M, ed. Proc. of the Applied Cryptography and Network Security: 2nd Int'l Conf. Berlin, Heidelberg: Springer-Verlag, 2004.pp.439–451.

[2] Charitakis I, Anagnostakis KG, Markatos E. An active traffic splitter architecture for intrusion detection[C]. In: Kotsis G, ed. Proc. of the 11th IEEE/ACM Int'l Symp. on Modeling, Analysis and Simulation of Computer Telecommunications Systems. Washington: IEEE Computer Society, 2003. pp.238–241.

[3] Shang Lin, Wang Jin-Gen, Yao Wang-shu, et al. A Classification Approach Based on Evolutionary Neural Networks [J]. Journal of Software, 2005, 16(9). pp.1577–1583.

[4] Ahmad Javaid, Quamar Niyaz, Weiqing Sun, et al. A deep learning approach for network intrusion detection system[C]// In Proceedings of the 9th EAI International Conference on Bio-inspired Information and Communications Technologies (formerly BIONETICS), 2016:21-26.

[5] Ye Yuan, Guangxu Xun, Fenglong Ma, et al. Muvan: A multi-view attention network for multivariate temporal data[C]// In 2018 IEEE International Conference on Data Mining (ICDM), 2018:717-726.

[6] Guo Chong-hui, Tang Huan-wen. Global convergence properties of evolution strategies[J]. Mathematica Numerica Sinica, 2001, 23(1), pp.105–110.

[7] V.Vapnik, Statistical Learning Theory[C]. New York: John Wiley and Sons, 1998.

[8] California ISO[OL]. 2010. http://www. caiso.com/

[9] KDDCUP99 data [DB/OL].(1999). http://kdd.ics.uci.edu/databases/kddcup99/

Evaluations of An Algorithm for Large Multivariate Optimization

Yoel TENNE*

Ariel University, Israel

*Corresponding author email: y.tenne@post.com

Abstract. Many simulation-driven engineering and scientific problems require finding an optimum of a function with many variables. Such settings pose a challenge for standard algorithms due to the large search space which in turn can lead to poor final results. Therefore this paper proposes a new simplified approach in which the dimension of the problem is dynamically reduced during the search to formulate a problem of lower size (dimension) which is easier to solve. A main novelty of the algorithm is its simplicity. Numerical experiments show the potential of this approach.

1. Introduction

Engineering and scientific problems commonly use numerical codes to evaluate designs and are thus cast as optimization problems where the best system configuration is sought. Some examples include problems in structural, fluid dynamics, and heat transfer domains [1-4]. One main challenge is that the computer simulation is computationally intensive hence only a few simulation runs are allowed. Another challenge is the number of variables which defines a large search space.

Various optimization algorithms have been proposed but were developed for low-dimensional problems (namely, having a small number of variables). However, real-world problems often involve tens or more variables and in such settings the performance of such algorithms can degrade quickly. This can yield an inefficient search and an inadequate final outcome, particularly when the resources allocated to the search process are limited. As such this paper describes an approach in which the original problem with multiple variables is reformulated into a simplified problem with a reduced set of variables which is therefore easier to solve. A dedicated search algorithm is then used to seek an optimum and the obtained solution is mapped into the original high-dimensional space. Numerical experiments show the potential of the proposed approach and its applicability to different problems. The remainder of the paper is set as follows: Section 2 provides background information, Section 3 describes the proposed approach, and Section 4 the numerical experiments and their analysis . Section 5 then concludes the paper.

2. Background

As described in Section 1 optimization problems where objective values are derived from a numerical code introduce several optimization challenges where common techniques may struggle. This has motivated the development of new heuristic approaches which often mimic phenomena in nature as solution methods [6,7]. The extensive computer resources required by each simulation run restricts the number of possible evaluations and accordingly *metamodels* (alternatively *surrogate-models*), which are mathematical approximations trained by using sampled data, can be used to obtain approximate function values more efficiently [8], Metamodel types include polynomials, radial basis functions, and

Content from this work may be used under the terms of the Creative Commons Attribution 3.0 licence. Any further distribution of this work must maintain attribution to the author(s) and the title of the work, journal citation and DOI.
Published under licence by IOP Publishing Ltd

various neural networks. Due to the limited number of possible simulations runs the obtained metamodel will likely be inaccurate and to address this it is required to check this accuracy and to update the metamodel during the search to avoid convergence to a false solution [9]. To accomplish this the proposed method uses the trust-region (TR) approach which performs a series of bounded steps and based on the success or failure of each optimization trial various updates are performed, as explained in the following section.

3. The Proposed Algorithm

The proposed algorithm combines three elements: a hybrid metaheuristic search, a trust region framework, and a dimensionality reduction procedure, as follows.

The hybrid metaheuristic search begins by training a Kriging metamodel [10] based on previously evaluated vectors such that the metamodel prediction is

$$m(x) = \beta + r(x)^T R^{-1}(f - 1\beta) \tag{1}$$

where β is a global "drift" coefficient, $r(x)$ is the vector of correlation values between the new vector x and the sampled vectors, R is the matrix of correlations values between sampled vectors, and f is the vector of responses (observed function values). After the metamodel has been trained a metaheuristic search is used to find its optimum. The optimization search combined an evolutionary algorithm (EA) followed by a local gradient search to achieve a good global-local search balance. The EA used a population of 100 vectors, stochastic universal selection (75% probability), mutation (10% probability), 10% elitism, and 50 generations. The localized search was performed by a sequential quadratic programming (SQP) algorithm. As mentioned the search is performed based on a trust-region (TR) approach such that the search is confined to a region T of radius r around the best solution

$$T = \|x - x_b\|_2 < r \tag{2}$$

The solution found by the EA+SQP search is evaluated with the true objective function. If it is indeed better than the known best then the TR radius is enlarged. Otherwise either a new vector is added in the trust-region (in case the number of vectors was insufficient, to avoid premature termination [7]) or the TR radius is decreased.

With regards to the dimensionality reduction, the procedure involves a) selecting the reduced dimension d and b) selecting which d variables will be active out of the existing n. This accomplished by modifying d during the search: starting from the original dimension n the initial reduced dimension is set as $d^{(1)} = n/k, k > 1$ and a subset selection is performed to identify the $d^{(1)}$ most important variables (as explained in the following text) and then the metamodel based search is performed. Lastly, the reduced dimension is adjusted based on the progress in the recent s iterations: if a better solution has been found than the algorithm further reduces the dimension to attempt a more intensive reduction. Otherwise the dimension is increased, namely $d^{(i+1)} = d^{(i)} \cdot k$ so the reduced dimension is closer to the original one. Based on numerical tests the selected settings where a dimension is updated every $s = 5$ iterations and a dimensionality change factor of $k = 1.1$ which was selected to avoid too rapid changes in the dimension.

Regarding the subset selection step – given n variables and a reduced dimension $d < n$ then there are $\frac{n!}{d_r!(n-d)!}$ possible subsets which implies that the search quickly become too vast to systematically explore. Therefore to more effectively search for a good subset the algorithm uses a binary evolutionary search which is different from the real-coded EA used for search the metamodel. With this EA different combinations where represented as binary vectors (1 indicates an active variable and 0 and inactive variable) and the prediction accuracy of different combinations was assessed based on cross-validation, namely, for each perspective subset a metamodel is trained such that the sampled vectors are spilt in a 80–20 training-testing ratio, a metamodel is trained based on the training set and its accuracy is estimated based on the testing set. The binary evolutionary algorithm used stochastic universal selection, shuffle crossover with a probability of 0.75, discrete mutation with a probability of

0.1, a population size of 100, 10% elitism, and 50 generations. To finalize this section Algorithm 1 presents the pseudocode of the complete algorithm.

```
sample an initial set of vectors;
set an initial reduced dimension d;
use subset selection to select d active variables;
repeat
        train a metamodel based on the extracted data set (sampled vectors and d);
        perform a TR search based on a hybrid EA+SQP search;
        if the found optimum is better than the current best set it as the TR centre
        else if there are sufficient vectors in the TR then reduce the TR radius
        else sample a new vector in the TR;
        every s iterations
                if the optimum found is better than current best known reduce d
                else increase d;
                re-select the important variables with subset selection;
        until the prescribed number of function evaluations is reached;
```

Algorithm 1: The proposed algorithm

4. Evaluation

To test the proposed algorithm it was used with an established set of mathematical test functions [12] and was benchmarked against a baseline version which did not use dimensionality reduction (DR), namely identical in operation except that the dimensionality reduction step was disabled, which was done to emphasize the impact of the DR step. For each test the function dimension was $n = 50$, the limit of function evaluations was 100, and 30 tests were repeated with each function-algorithm combination to obtain statistically-valid results. The code was implemented in Gnu Octave.

Table 1 presents the results where lower results are better (all problems are minimization) and statistics are based on the best function values achieved over 30 tests. From analysis of the results the contribution of the DR step is evident (lower results are better as these are minimization problems). The proposed algorithm consistently achieved better final results than the baseline version across different test functions in 4 out 5 cases. The only function for which the baseline version was better was Rosenbrock which since its valley shape landscape allows to obtain the optimum effectively even in the original high-dimensional formulation. The consistent performance shows that the algorithm was effective across problems with significantly different features.

Table 1. Tests statistics.

Function	Statistic	Proposed	Baseline
Ackely	Mean	**1.578E+1**	1.729E+1
	Median	1.581E+1	1.725E+1
	SD	8.202E-1	7.351E-1
Rastrigin	Mean	**5.490E+2**	5.760E+2
	Median	5.375E+2	5.798E+2
	SD	6.105E+1	7.851E+1
Rosenbrock	Mean	2.621E+3	**1.259E+3**
	Median	2.472E+3	1.025E+3
	SD	9.517E+2	1.702E+2
Schwefel	Mean	**4.105E+6**	4.229E+6
	Median	3.955E+6	2.677E+6
	SD	1.093E+6	2.689E+6
Weierstarss	Mean	**-5.159E+1**	-4.721E+1
	Median	-5.181E+1	-4.882E+1
	SD	1.965E+0	1.337E+0

SD: Standard deviation

The best mean result statistic is emphasized

5. Conclusion

Real-world engineering and scientific problems often involve a large number of variables and in these cases classical optimization algorithms can underperform. Therefore this study proposed an approach where a lower dimension formulation is generated to simplify the search. The extent of dimensionality reduction is dynamically adjusted during the search. A main novelty of the proposed algorithm is its simplicity which requires minimal computational overhead. Evaluations based on test functions indicates that the algorithm is effective in varied settings. Future work could improve the computational efficiency and explore more elaborate adjustments of the reduction factor.

References

[1] Y. Tenne and C. K. Goh, Eds., Computational Intelligence in Expensive Optimization Problems, ser. Evolutionary Learning and Optimization. Berlin: Springer, 2010, vol. 2.

[2] X. Dinh and K. K. Anh, "Radial basis function neural network based adaptive fast nonsingular terminal sliding mode controller for piezo positioning stage," International Journal of Control, Automation, and Systems, vol. 15, no. 6, pp. 2892–2905, 2017.

[3] B. Talgorn, S. Le Digabel, and M. Kokkolaras, "Statistical surrogate formulations for simulation-based design optimization," Journal of Mechanical Design, vol. 137, no. 2, 2015. Paper number MD-14-1128 (Online).

[4] T. Wortmann, A. Costa, G. Nannicini, and T. Schroepfer, "Advantages of surrogate models for architectural design optimization," Artificial Intelligence for Engineering Design, Analysis and Manufacturing, vol. 29, no. 4, pp. 471–481, 2015.

[5] T. Santhanam and M. S. Padmavathi, "Application of K-Means and Genetic Algorithms for Dimension Reduction by Integrating SVM for Diabetes Diagnosis," Procedia Computer Science, vol. 47, pp. 76–83, 2015.

[6] C. Wang, J.-H. Wang, S.-S. Gu, X. Wang, and Y.-X. Zhang, "Elongation prediction of steel-strips in annealing furnace with deep learning via improved incremental extreme learning machine," International Journal of Control, Automation, and Systems, vol. 15, no. 3, pp. 1466–1477, 2017.

[7] D. A. Rodríguez, P. P. Oteiza, and N. B. Brignole, "Simulated annealing optimization for hydrocarbon pipeline networks," Journal of Industrial Engineering and Chemical Research, vol. 52, no. 25, pp. 8579–8588, 2013.

[8] F. A. C. Viana, R. T. Haftka, and L. T. Watson, "Efficient global optimization algorithm assisted by multiple surrogate technique," Journal of Global Optimization, vol. 56, no. 2, pp. 669–689, 2013.

[9] Y. Jin, M. Olhofer, and B. Sendhoff, "A framework for evolutionary optimization with approximate fitness functions," IEEE Transactions on evolutionary computation, vol. 6, no. 5, pp. 481–494, 2002.

[10] A. R. Conn, N. I. M. Gould, and P. L. Toint, Trust Region Methods. Philadelphia, Pennsylvania: SIAM, 2000.

[11] J. Muller and C. A. Shoemaker, "Influence of ensemble surrogate models and sampling strategy on the solution quality of algorithms for computationally expensive black-box global optimization problems," Journal of Global Optimization, vol. 60, no. 2, pp. 123–144, 2014.

[12] P. N. Suganthan, N. Hansen, J. J. Liang, K. Deb, Y. P. Chen, A. Auger, and S. Tiwari, "Problem definitions and evaluation criteria for the CEC 2005 special session on real-parameter optimization," Nanyang Technological University, Singapore and Kanpur Genetic Algorithms Laboratory, Indian Institute of Technology Kanpur, India, Technical Report KanGAL 2005005, 2005.

Research on Acoustic Transfer Performance of Power Cabin of Underwater Craft

Hanqin Li[1], Bin Fang[2] and Yongming Zhang[3,*]

[1,3]Department of Operational Research and Programming, Naval University of Engineering, China
[2]Department of Naval Architecture and Marine Engineering, Naval University of Engineering, China

[*]Corresponding author email: zym_xf@163.com

Abstract. In order to study the acoustic performance of underwater vehicle including vibration and noise, two kinds of structure models of an underwater craft cabin in the air medium and in the water medium were analyzed separately by finite element method. Through modal analysis, the impedance ratios of the two kinds of models were calculated. The influence of the thickness of the pressure hull on the impedance ratio of the structure was studied, and the acoustic transfer performance was analyzed and compared between the two kinds of models.
Keywords: Vibration and noise, finite element method, underwater vehicle.

1. Introduction

As people exploring the sea world deeper and further, underwater craft becomes an important craft under the sea. Acoustic vibration performance is one of the key research directions of underwater craft. According to different mechanism, the noise of underwater craft can be distinguished as mechanical noise, propeller noise and hydrodynamic noise [1]. When the underwater craft is sailing at a low speed, mechanical noise plays a leading role, which is propagated from the shell through the equipment base as the vibration generated by the mechanical equipment working inside the cabin.

At present, the numerical methods of acoustic problems at home and abroad include statistical energy method, boundary element method, finite element method and so on. With the development of various methods, it is more commonly to combine different kinds of numerical algorithms into use. Finite element method was applied in early time, which mainly divided structure and fluid into finite element for solution, but finite element method can't be used for infinite domain problems. Statistical energy analysis (SEA) can extract the research object from the whole by random parameters description. Instead of the specific details of the object, it only focus on the statistical mean value in time domain, frequency domain and space. In recent years, statistical energy method has made progress in optimization design, dynamic impact and so on[2][3]。

The boundary element method (BEM) is an efficient and accurate engineering numerical method developed after the finite element method (FEM). Its modeling is simple, and the forming units are less, while it only needs to model on the surface of the structure, which can reduce the dimension. However, the traditional boundary element method forms a non-symmetric fully populated matrix, which can be time-consuming and memory-consuming. In order to overcome the shortcoming, many fast methods have been proposed, such as fast multipole method, fast direct solution method and adaptive cross approximation method [4][5].

Content from this work may be used under the terms of the Creative Commons Attribution 3.0 licence. Any further distribution of this work must maintain attribution to the author(s) and the title of the work, journal citation and DOI.
Published under licence by IOP Publishing Ltd

It will consume a lot of memory and time to calculate and analyze the fluid structure interaction on the finite element model of underwater craft cabin structure. In contrast, structure in the air can be calculated faster without considering the fluid structure interaction. In this paper, the power cabin of a certain type of underwater craft was selected for research. The finite element method was adopted to establish the model to calculate the impedance ratio to reflect the match degree of vibration and noise transmission as well as the acoustic characteristics. And its parameters were changed to compare the change rules of acoustic characteristics of the cabin with the change of parameters. The impedance ratio of structure in the air and in the water was compared to reach a general conclusion of fluid structure interaction. The impedance ratio is the ratio of the impedance of two coupled structures, which can reflect the matching degree of vibration and noise transmission and the acoustic performance of underwater craft cabin.

2. Fundamental Principles of Finite Element Method and Calculating Impedance Ratio

2.1. Fundamental Principles of Finite Element Method

A set of solutions for finite element structure in infinite liquid field are introduced. Set ω as the vibration circular frequency of the structure, M as the structure mass, B as damping, especially the stiffness coefficient, u as the displacement response, F as the driving force, P as the fluid pressure, r as the radius of outer boundary of the artificial fluid, and R as the radius of the spherical shell. The dynamic equations is as follows:

$$\left[-\omega^2 M + j\omega B + K\right]u = F + GAP \tag{1}$$

$\{u\}$ is displacement complex amplitude vector, $\{p\}$ is the fluid pressure complex amplitude vector, which are unknown quantities to be solved. Because the amplitude of $\{p\}$ depends on $\{u\}$, while the magnitude of $\{u\}$ is affected by the amplitude of $\{p\}$, which is called fluid-solid coupling as they interact with each other. It is necessary to introduce a set of equations as follows to solve the problem. The acoustic equation is written as:

$$\nabla^2 P + K^2 P = 0 \tag{2}$$

The boundary condition equation of fluid solid coupling is as follows:

$$\frac{\partial P}{\partial n} = \omega^2 \rho u_n \tag{3}$$

The radiation condition equation is as follows:

$$\lim_{r\to\infty} r\left(\frac{\partial P}{\partial r} - iKP\right) = 0 \tag{4}$$

It is difficult to calculate and analyze due to the large amount of calculation when calculating the fluid unit of infinite fluid field by NASTRAN software. The finite element method is used. By using the finite element method, construct an outer boundary whose diameter is far greater than that of the spherical shell to approximate meet equation (4). The radiation condition equation is changed as follows:

$$\lim_{r\to r_0} r\left(\frac{\partial P}{\partial r} - iKP\right) = 0 \tag{5}$$

And the boundary impedance is as follows

$$Z_\rho = \rho C_0 \tag{6}$$

P is the fluid density and C_0 is the velocity of the fluid.

The rest of the equations keep the same. $\{u\}$ can be obtained by equation(1) to (5). From the equation that $v = i\omega u$, the average method velocity can be obtained.

2.2. Fundamental Principles of Calculation for Impedance Ratio of Single-mode Structural Vibration

Consider the single-mode structural vibration system as shown in the figure: structure A vibrates in the normal direction of its structural plane due to an additional excitation, and the structure is analyzed by isolation method.

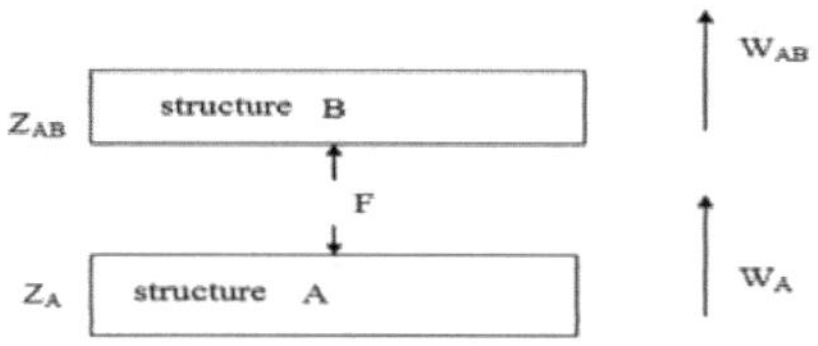

Figure 1. Single-mode structural vibration diagram.

When structure B is not considered, the normal vibration displacement response amplitude of structure A is w_A . While taking structure B in consideration, both structure A and B have the same normal vibration displacement response amplitude written as w_{AB} , and the interaction force amplitude at the interface is F ,

$$\begin{cases} w_{AB} = \dfrac{F}{Z_B} \\ w_{AB} - w_A = \dfrac{-F}{Z_A} \end{cases} \tag{7}$$

where Z_A and Z_B are the displacement impedance of structure A and B respectively. The following results are obtained by transforming equation (7) :

$$\frac{Z_B}{Z_A} = \frac{w_A}{w_{AB}} - 1 \tag{8}$$

It can be seen from equation (2) that displacement impedance ratio of the structure B and structure A can be directly calculated when w_A and w_{AB} are known in the single-mode vibration system. For that transfer of vibration energy between structure A and B is closely related to their impedance ratio, their impedance compatibility can be measured by their impedance ratio.

3. Establishing the Model and Determining the Mode

3.1. Establishing the Model

Part of the structural models of the underwater craft are given, and the finite element modeling is carried out, as shown in figure 2. The model includes pressure hull structure, non pressure hull structure, rib structure, brace structure, end cover structure, surface structure, fluid structure and outer boundary structure. The parameters are as follows: the length of cabin is 22 m, the diameter of pressure hull is 8 m, the diameter of non pressure hull is 10 m. The thickness of pressure shell plate is 30 mm, that of non pressure shell plate is 9 mm, that of the rib plate is 10 mm. The thickness of bracket plate is 7 mm, and that of bulkhead plate is 10 mm. The ribs are arranged with a spacing of 500 mm in pressure hull.

When only the dynamic cabin is studied, in order to consider the effect of the cabin at both ends and approximately simulate the structural vibration characteristics, the boundary condition is simplified as the simply supported at both ends. The bottom of the center of the cabin is subjected to a unit excitation force which is straight down. The frequency ranges from 2Hz to 100Hz, and the step size is 2Hz.

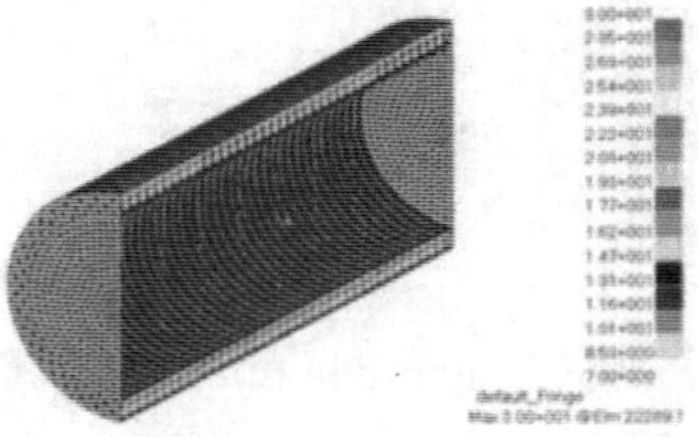

Figure 2. Longitudinal section of the finite element model of the cabin.

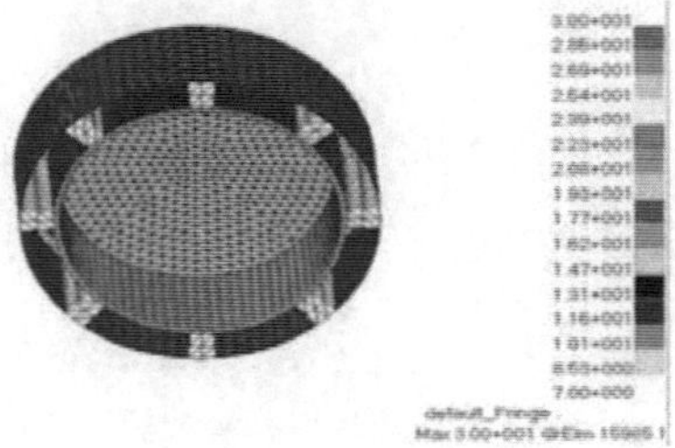

Figure 3. Middle section of the finite element model of the cabin.

3.2. Determining the Mode

Make modal analysis on the finite element model of the cabin structure in the air with NASTRAN software and figures out the mode shape. Then, the response of the cabin at different frequencies under the original given condition is obtained by calculating the model in the frequency range from 2Hz to 100Hz with the step of 2Hz. Through comparison, it can be seen in figure 4 and figure 5 that the modal shape of order 23 is similar to the response of the structure when the excitation frequency is 12Hz. The mode is selected as the main mode of the structural model and recorded as mode A.

For the structure in the water, take the isolation method to analyze the pressure hull and non pressure hull structure. So the vibration mode of pressure hull is not affected when considering the effect of fluid. For the case of parameter change, as long as the geometric characteristics do not change, such as length, diameter and so on, the mode shape does not change. Although the corresponding frequency of the corresponding complex mode will change, it does not affect the mode selection.

It can be seen that the mode A is the mode shape to be selected. The impedance ratio corresponds to the main part of the impedance ratio of complex structure.

Figure 4. Mode A at order 23.

Figure 5. Structural response when frequency is at 12 Hz.

4. Calculation Results of Impedance Ratio of the Structures

4.1. Characteristics of Impedance Ratio of the Structure in the Air

For the finite element model of the structure in the air, the impedance ratio of the typical mode, namely mode A, was calculated by changing the thickness of the pressure hull plate at different frequencies. The impedance ratio was compared with that of the original structure, and the change rules of the impedance ratio with thickness of the plate is given, which is represented in logarithmic, as shown in figure 6 and figure 7.

In the condition of mode A, the impedance ratio varies little as the thickness of pressure hull changes at low frequency; When the frequency increases, the logarithm peak value of impedance ratio decreases with the increase of plate thickness, and the logarithm peak value of impedance ratio moves backward; As the thickness of the plate decreases, the logarithm peak value of the impedance ratio increases and the logarithm peak value of the impedance ratio moves forward.

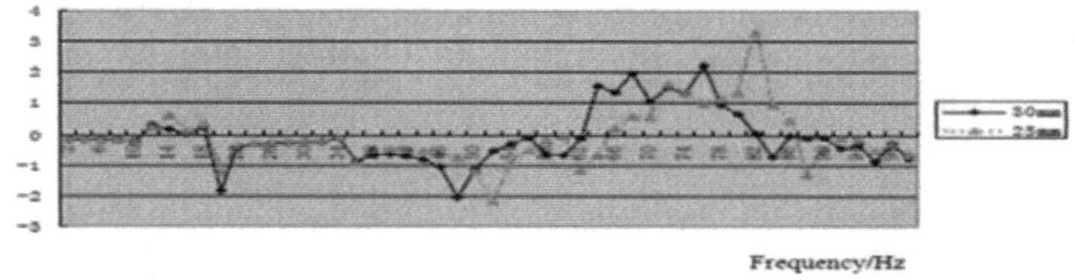

Figure 6. Comparison of impedance ratio of structure in the air in mode A with the thickness of 25mm and 30mm.

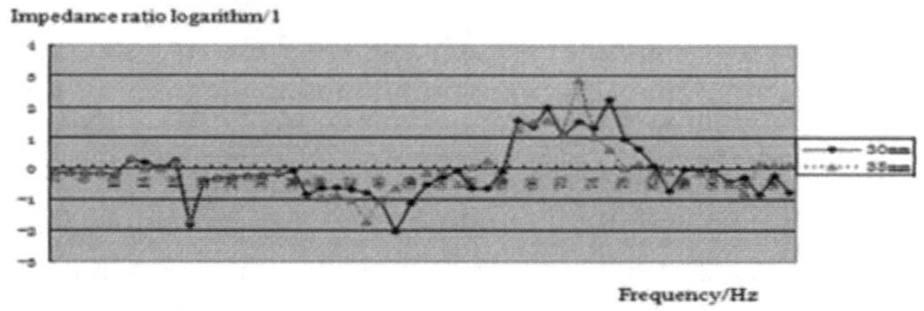

Figure 7. Comparison of impedance ratio of structure in the air in mode A with the thickness of 30mm and 35mm.

4.2. Characteristics of Impedance Ratio of the Structure in the Water

For the finite element model of the structure in the water, the impedance ratio of the mode A is calculated at different frequencies by changing the thickness of the pressure hull plate. The impedance ratio is compared with that of the original structure, and the changing rules of impedance ratio of the thickness of the plate is given in logarithm, as shown in figure 8 and figure 9.

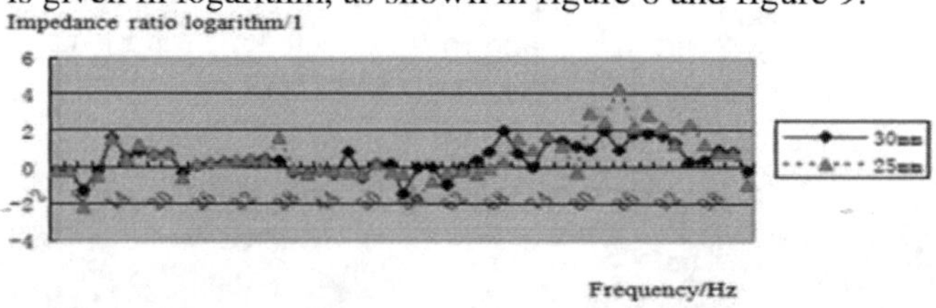

Figure 8. Comparison of impedance ratio of structure in the water in mode A with the thickness of 25mm and 30mm.

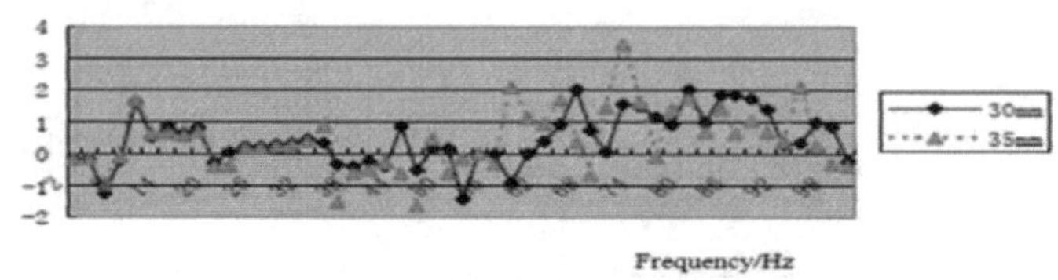

Figure 9. Comparison of impedance ratio of structure in the water in mode A with the thickness of 30mm and 35mm.

In the condition of mode A, it can be seen that, the logarithm of impedance ratio changes little at low frequency after the thickness changes, and the impedance ratio increases and the match degree increases when the thickness decreases. At high frequency, the increase of plate thickness causes the logarithmic peak value of impedance ratio to move backward, while the decrease of plate thickness causes the logarithmic peak value of impedance ratio to move forward.

4.3. Comparison Between Structures in the Air and in the Water

For mode A, the logarithm of the impedance ratio of the structure in the water is larger than that of the structure in the air. But when it is at high frequency, the impedance ratio difference between the two kinds of structures in different conditions increases rapidly, so the conclusion is given in different frequency bands. The conclusions are shown in the tables below.

Table 1. Comparison of impedance ratio between structure in the air and in the water at the frequency from 2Hz to 78Hz in mode A.

Range of frequency(Hz)	Thickness (mm)	Difference of impedance ratio between the two structures(1)
2-78	25	0.348
2-78	30	0.325
2-78	35	0.317

Table 2. Comparison of impedance ratio between structure in the air and in the water at the frequency from 78Hz to 100Hz in mode A.

Range of frequency(Hz)	Thickness (mm)	Difference of impedance ratio between the two structures(1)
78-100	25	1.463
78-100	30	0.889
78-100	35	0.317

5. Conclusion

The finite element method is used to calculate the cabin structure model of underwater craft. The result shows that: (1) The common changing rules of structures both in the air and in the water are that: the impedance ratio decreases and the impedance matching becomes better with the increase of pressure shell thickness. The peak value of impedance ratio moves forward. So in simulation and calculation for whether structure in the air or in the water, reducing the thickness of the pressure hull can decrease the transmission of mechanical noise so that decrease the vibration noise radiated by underwater craft. So it is beneficial to the acoustic performance to reduce the thickness of pressure shell properly. (2) The difference of changing rules between structures in the air and in the water is that: for the structure in the air and the other one in the water, the impedance ratio of the structure in the water is greater, and the impedance matching performance will be worse.

For the study of acoustic transmission performance of typical cabin structure of underwater craft, the following work should be further carried out: the acoustic transmission performance of the structure should be calculated by changing other parameters, such as thickness of non pressure hull, cabin density and so on. And the relationship of impedance ratio between the structure in the air and in the water should be further studied.

Acknowledgments

Thanks a lot to professor Fang and associate professor Zhang, who offered helpful advice for me. I would also like to appreciate my colleagues for their support. They all gave me a lot of encouragement.

References

[1] Wei Chen 2009 The Numerical Methods Research of Coupled Vibration and Sound Radiation of Underwater Structures
[2] Ruxin Gao and Yahui Zhang 2019 Reduction of Hybrid FE-SEA Model for the Mid-Frequency Vibration of Vibro-Acoustic Systems Using Dynamic Condensation Approach
[3] Richard D J and Gordon E 2017 Using the Modal Response of Window Vibrations to Validate SEA Wind Noise Models
[4] Leilei Chen and Chuang Lu 2020 Subdivision Surfaces-Boundary Element Accelerated by Fast Multipole for the Structural Acoustic Problem
[5] Leslie G and O'neil M 2021 multipole methods for evaluation of layer potentials with locally-corrected quadratures

Modeling and Analysis of Different Architecture for Civil Aircraft Hydraulic System

Xunan He*, YuanLi Kang and Yannian Hui

Department of More Electric System, Beijing Aircraft Technology Research Institute, China

*Corresponding author email: xunanhe2008@126.com

Abstract. For the evaluation of civil aircraft hydraulic system, Fuel Consumption is a very important index. However, there is no effective method to perform analysis for different architecture. This paper proposes that taking fuel weight penalty as a major mathmatical modeling method to analysis traditional architecture and more electric architecture. The result shows that the fuel consumption of more electric architecture is less than traditional architecture.
Keyword: Hydraulic system, modeling, analysis.

1. Introduction

Hydraulic system extracts mechanical energy and electric energy from engine and generator, and covert to hydraulic energy by pump, provide the hydraulic energy to its consumer. With the development of more electric technology, the hydraulic system also faces the option of traditional architecture and more electric architecture. For a traditional energy architecture aircraft, the hydraulic energy consumers include flight control system, landing gear system and so on. For a more electric architecture aircraft, take Airbus A380 and Boeing B787 as an example, Airbus A380 cancel a set of hydraulic system, flight control system partly uses electric energy instead of hydraulic energy, such as EHA, EMA end EBHA. For Boeing B787, the braking system uses electric brake instead of hydraulic brake. Therefore, it is very important and necessary to trade-off different architecture for hydraulic system that adopt more electric technology. However, there is no effective mathematical modeling method to perform analyze the fuel consumption of different architecture. It is very necessary to learn from evaluation methods of other systems.

2. System Overview

2.1. Traditional Architecture

The hydraulic system adopts three sets of central hydraulic energy network architecture, its pressure class is 5000psi, and it provides hydraulic energy to flight control system and landing gear system and so on. The architecture is shown in Figure 1.

2.2. More Electric Architecture

The hydraulic system adopts two sets of central hydraulic energy network architecture which is isolated from each other, its pressure class is also 5000psi, and it provides hydraulic energy to flight control system and landing gear system and so on. The architecture is shown in Figure 2.

Content from this work may be used under the terms of the Creative Commons Attribution 3.0 licence. Any further distribution of this work must maintain attribution to the author(s) and the title of the work, journal citation and DOI.
Published under licence by IOP Publishing Ltd

Journal of Physics: Conference Series **2068** (2021) 012028 doi:10.1088/1742-6596/2068/1/012028

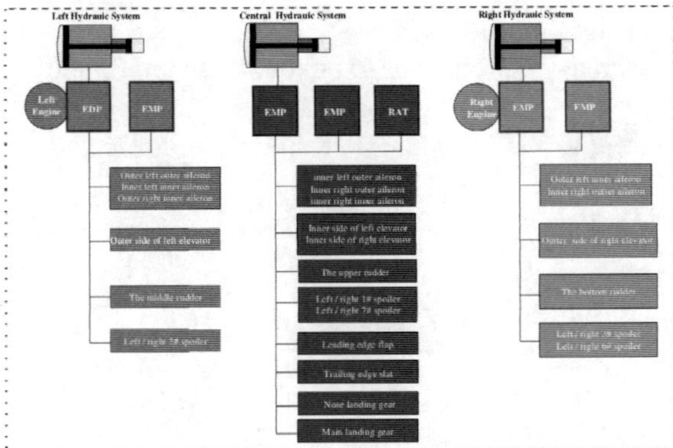

Figure 1. Traditional architecture of hydraulic system.

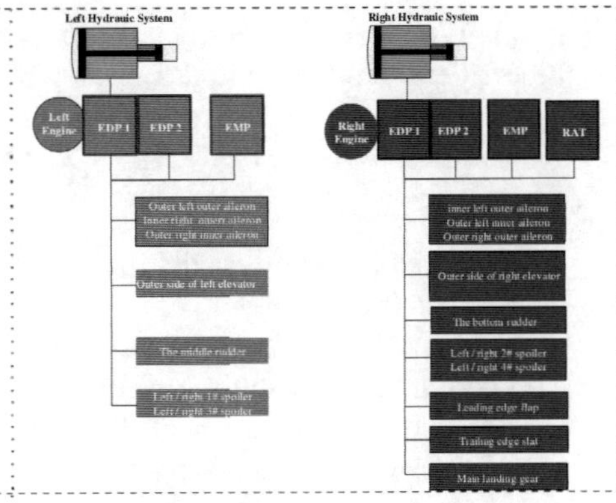

Figure 2. Traditional architecture of hydraulic system.

3. Modeling Method

There is no effective method to perform analysis the fuel consumption of different architecture for hydraulic system. However, it can take fuel weight penalty as a major mathematical modeling method like air conditioning system does. And the generally accepted methods of evaluation of fuel weight penalty include the equivalent weight method, the total take-off weight method and the equivalent resistance method。The transportation cost is a very important consideration for civil aircraft. It is suitable to take the total take-off weight method as the evaluation of hydraulic system. The fuel weight penalty for the total take-off weight method include fixed weight penalty, variable weight penalty, shaft power extraction penalty, bleed air penalty and ram air penalty.

3.1. Fixed Weight Penalty

Fuel is required to carry the system fixed weight and the dead weight of fuel up to the point where it is expended. The total take-off weight penalty for fixed weight consist of the system fixed weight plus the weight of the fuel required to carry it, and it is given by

$$M_{f,E} = M\left[exp\left(\frac{C_e \tau_0 g}{K}\right) - 1\right] \quad (1)$$

Where is:

g—acceleration of gravity, m/s^2

K—lift-drag ratio

C_e —specific fuel consumption for thrust, kg/(N•h)

τ_0—time of flight, h

3.2. Variable Weight Penalty

Expendable material (such as water for use in water boilers) may be carried aboard the aircraft. And it is assumed that the rate of consumption of the expendable material is constant. There are two cases in the following:

- Variable Weigh is used during any portion of the flight

The take-off fuel weight required for carrying the decreasing weight of expendable material, and also of fuel, is given by

$$M_{f,V} = \frac{W_V K}{C_e g}\left[exp\left(\frac{C_e \tau_0 g}{K}\right) - 1\right] - W_V \tau \quad (2)$$

Where is:

g—acceleration of gravity, m/s^2

K—lift-drag ratio

C_e —specific fuel consumption for thrust, kg/(N•h)

τ_0—time of flight, h

W_v —rate of consumption of expendable material, kg/s

- Variable Weigh is used during specific portion of the flight

The take-off fuel weight required for carrying the decreasing weight of expendable material, and also of fuel, is given by

$$M_{f,V} = \frac{W_V K}{C_e g}\left[exp\left(\frac{C_e(\tau_m - \tau_1)g}{K}\right) - 1\right] - W_V(\tau_m - \tau_1) \quad (3)$$

Where is:

g—acceleration of gravity, m/s²
K—lift-drag ratio
C_e —specific fuel consumption for thrust, kg/(N•h)
τ_1—start time of statistical flight, h
τ_m—end time of statistical flight, h
W_v —rate of consumption of expendable material, kg/s

3.3. Shaft Power Extraction Penalty

To maintain constant net thrust from the engine, the shaft power extracted must be compensated by an increase in fuel flow rate to the engine. Assuming power is consumed as a constant rate during the portion of mission being evaluated, the take-off fuel weight penalty is given by

$$M_{f,P} = \frac{P C_e' K}{C_e g}\left[exp\left(\frac{C_e \tau_0 g}{K}\right) - 1\right] \quad (4)$$

$$C_e' = \frac{C_{p,g} T_{in,t} \eta_c}{1004.5 * H_u \varepsilon_{kc} T_{in,c}(\pi_c^{0.286} - 1)} \quad (5)$$

Where is:

C_e'—specific fuel consumption for power, kg/(W•s)
P—Power being consumed, kW
$C_{p,g}$—The specific heat capacity of gas at constant pressure, J/(kg•KI)
H_u—The calorific value per unit of fuel combustion, J/kg
$T_{in,t}$—Inlet temperature of engine turbine, K
$T_{in,c}$—Inlet temperature of engine compressor, K
ε_{kc}—Complete combustion coefficient in the combustion chamber
π_c—Engine compressor boost ratio
η_c—Compressor efficiency

3.4. Bleed Air Penalty

If bleed air is extracted from the compressor of the turbine engine, the penalty can be evaluated on the basis of increase fuel flow required to maintain constant thrust. As a first approximation only, the increase in fuel flow rate due to bleed air extraction is given by:

$$M_{f,bl} = \frac{q_{m,f,bl} K}{C_e g}\left[exp\left(\frac{C_e \tau_0 g}{K}\right) - 1\right] \quad (6)$$

If bleed air from low pressure stage of Engine compressor,

$$q_{m,f,bl} = \frac{q_{m,bl} C_{p,g} T_{in,t}(\pi_0^{0.286} - 1)}{H_u \varepsilon_{kc}(\pi_c^{0.286} - 1)} + q_{m,bl} V C_e \quad (7)$$

If bleed air from high pressure stage of Engine compressor,

$$q_{m,f,bl} = \frac{q_{m,bl} C_{p,g} T_{in,t}}{H_u \varepsilon_{kc}} + q_{m,bl} V C_e \quad (8)$$

Where is:

$q_{m,f,bl}$—Increased fuel mass flow due to bleed air, kg/s
$q_{m,bl}$—Bleed air flow, kg/s
π_0—Engine compressor boost ratio of bleed air
π_c—Engine compressor boost ratio
ε_{kc}—Complete combustion coefficient in the combustion chamber
H_u—The calorific value per unit of fuel combustion, J/kg

3.5. Ram Air Penalty

Assuming complete momentum loss, the ram air penalty in term of take-off fuel weight is given by:

$$M_{f,D} = \frac{q_m V K}{g}\left[exp\left(\frac{C_e \tau_0 g}{K}\right) - 1\right] \tag{9}$$

Where is:

q_m—ram air flow, kg/s
V— Cruise velocity, m/s

3.6. Total fuel weight penalty

The hydraulic system is a fixed system, and there is no need to bleed air from engine and overcome ram air resistance. However, it need to extract mechanical energy from engine. Therefore, the fuel weight penalty of hydraulic system includes fixed weight penalty and shaft power extraction penalty. The total fuel weight penalty of hydraulic system is given by:

$$\Delta M_T = M_{f,E} + M_{f,P} \tag{10}$$

4. Modeling Analysis

4.1. Definition of Flight Envelope

To perform analysis for traditional architecture and more electric architecture of hydraulic system, there is a strongly necessary to define the flight envelope. The flight envelope should have the necessary information which contain the flight phase, the flight time, the flight altitude and the flight velocity. And the defined flight envelope is shown in table 1.

Table 1. Detailed definition of flight envelope.

Phase	Detailed phase	Time(min)	Altitude(m)	Velocity(Ma)
Start preheating	1	1~10	0	0
Taix out	2	11~20	0	0
Take off	3	21	10	0.22
	4	22~25	450	0.22
Climb	5	26~28	450	0.38
	6	29~38	3050	0.38
	7	39~41	3050	0.46
	8	42~55	10668	0.8
Cruise	9	26~60	10668	0.85
	10	61~65	13106	0.85
	11	66~220	13106	0.85
Descent	12	221~240	3050	0.46
	13	241~245	3050	0.38
	14	246~260	450	0.38
Approach	15	261~265	10	0.22
Landing	16	266	0	0.01

4.2. Modeling Analysis

4.2.1. The calculation of fuel weight penalty. According to the mathematical modeling method of fuel weight penalty, the different fuel weight penalty of traditional architecture and more electric architecture of hydraulic system in single flight envelope is shown in figure 3 and figure 4.

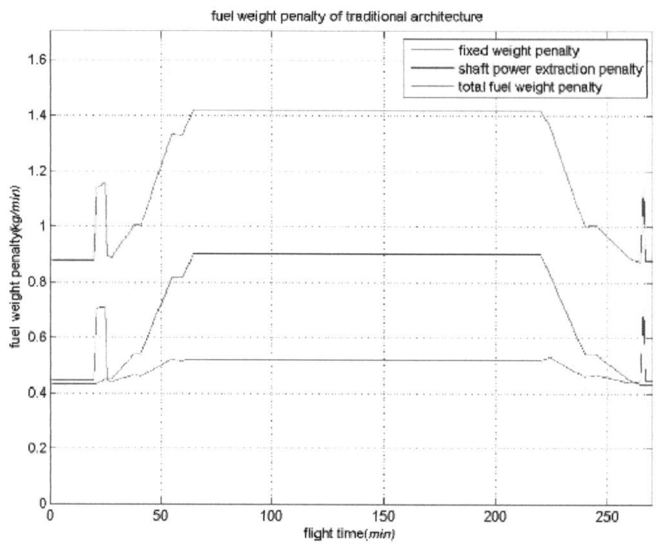

Figure 3. Fuel weight penalty for traditional architecture.

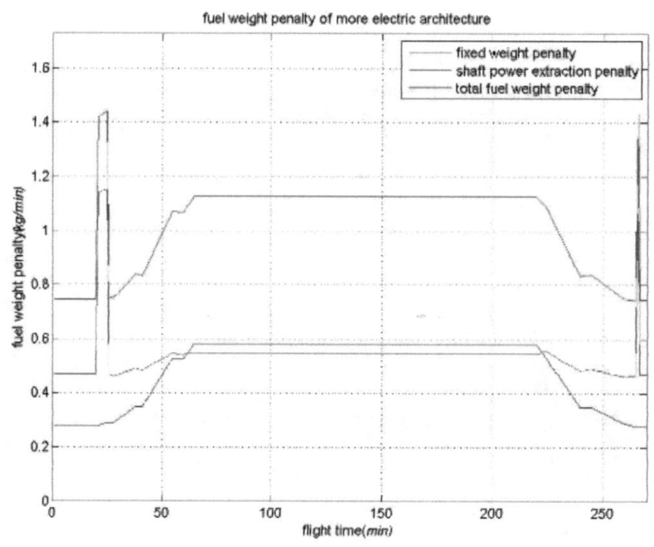

Figure 4. Fuel weight penalty for more electric architecture.

For a single flight mission, the total fuel weight penalty for traditional architecture is about 341.6kg. For the total fuel weight penalty of traditional architecture, the fixed Weight Penalty is about 205.47kg, which is about 60.1% of total fuel weight penalty. The shaft power extraction penalty is about 136.13kg, which is about 39.9% of total fuel weight penalty. the total fuel weight penalty for more electric architecture is about 277.52kg. For the total fuel weight penalty of more electric architecture, the fixed Weight Penalty is about 132.26kg, which is about 47.6% of total fuel weight penalty. The shaft power extraction penalty is about 145.26kg, which is about 52.4% of total fuel weight penalty.

4.2.2. The result of modeling analysis. The analysis of fuel weight penalty of traditional architecture and more electric architecture of hydraulic system in single flight envelope is shown in table 2.

Table 2. fuel weight penalty for different architecture.

Item	Traditional architecture	More electric architecture
Fixed Weight Penalty（kg）	205.47	132.26
Shaft power extraction penalty（kg）	136.13	145.26
Total fuel weight penalty（kg）	341.6	277.52

The statistical results which are from table 2 is shown in figure 5.

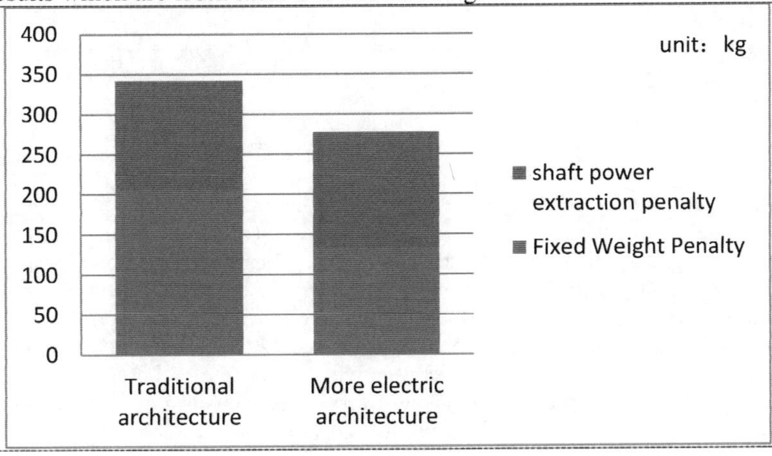

Figure 5 Fuel weight penalty statistical result.

It can be seen from Table 2, For a single flight mission, the total fuel weight penalty for More electric architecture is about 282.9kg, which is lower than traditional architecture about 13.5%. In other words, the fuel consumption of More electric architecture is less than traditional architecture.

5. Conclusion

The thesis proposes that taking fuel weight penalty as a major mathematical modeling method to analyze traditional architecture and more electric architecture. This method can help to calculate the fuel consumption of hydraulic system. However, it also has its limitations. It is only suitable for concept design of hydraulic system. And it does not adopt the maintenance cost as an evaluation index. In addition, the result of analysis is limited by the accuracy of some parameters for hydraulic system, such as weight and the power extracted from engine. In short, it still need to multi-dimensional evaluate the different architecture.

References

[1] Hui Yannian and Kang Yuanli 2018 OVERVIEW OF MEA KEY TECHNOLOGY AND SYSTEM INTEGRATION Journal of Engineerin p661-664.

[2] DENG Xishu, LI Ziguang 2003 Development status and trend of hydraulic system simulation technology[J]. Machine tool and hydraulics, no. 1, p20-22.

[3] M. E. Saleta, D. Tobia and S. Gil 2004 Experimental study of Bernoulli's equation with losses American Journal of Physic, vol. Volume 73, no. Issue 7, pp. 598-602.

[4] A. Queens, A. Korolev and A. Vaychulis 1996 Analysis of heat-hydraulic stability of multi-channel circulating thermal systems Automatic control of metallurgical processes, p89-103.

Modeling and Simulation of Single Battery Load Isolated Electric Vehicle Based on Simulink

Benchao Liu*
Department of Vehicle Engineering, Shandong Transport Vocational College, Weifang, China

*Corresponding author email: liubenchaoky@163.com

Abstract. The load isolation electric vehicle improves the range of electric vehicle, which separates the concept of power and energy, so that the engine can always work in the most economical operating area. Based on Simulink, a single battery load isolation electric vehicle model is established for a mini car. According to the requirements of vehicle performance indexes, the parameters of vehicle system model, gasoline generator set, battery pack and motor model were matched. In the process of accelerating the average speed to the set maximum speed, the battery voltage curve and the motor output power are analyzed. The analysis shows that the battery has no overvoltage danger and the motor does not exceed the peak power, thus verifying the feasibility and superiority of the model.

1. Introduction

In the face of severe resource and environment problems, developing electric vehicle industry has become a trend of automobile technology development. Pure electric vehicles abandon the complex mechanical structure of traditional cars. They use batteries as energy storage devices to provide electric energy for cars and motors as driving devices[1]. They have the advantages of simple structure, zero emission and low noise.

However, due to technical reasons such as low power density and energy density, the battery will only have a short range of electric vehicles [2].The load isolation vehicle is composed of gasoline generator set, battery pack, motor and so on. It liberates the engine from the energy drive device and keeps the engine working at the optimum steady-state operating point to charge the battery pack, thus improving the range of the electric vehicle. Two or more groups of batteries can be charged and discharged alternately to avoid the problems of overcharge, overdischarge and overtemperature [3].Based on Matlab/Simulink, a load isolation electric vehicle model is established in this paper. Since the number of battery packs has little influence on the modeling and simulation in this paper, in order to simplify the complexity of the model, only a load isolated EV model with a single battery group is established. In the model, the parameters of the vehicle power system and the main components are matched, and the performance of the vehicle is simulated.

2. Build the Body Model

The components needed by SIMULINK were rebuilt and encapsulated subsystems. This method can be represented by a group of modules or a single graphical subsystem module, which greatly enhances the readability of the system model. However, the parameters of each module in the subsystem need to be reset during the dynamic system simulation. Encapsulating a subsystem means encapsulating a subsystem that has already been built to have a certain function. The purpose is to generate user-defined modules that are completely consistent with the functions of the subsystem [4-5].

Content from this work may be used under the terms of the Creative Commons Attribution 3.0 licence. Any further distribution of this work must maintain attribution to the author(s) and the title of the work, journal citation and DOI.
Published under licence by IOP Publishing Ltd

To establish the model of load isolation electric vehicle, it is necessary to establish a vehicle longitudinal dynamics model first. Then the model is encapsulated into the vehicle longitudinal dynamics subsystem, and the parameters are set according to the requirements of the model[6].In this paper, CHANG AN BENBEN electric vehicle is selected. The parameters of the sample car are shown in Table 1.The initial longitudinal velocity was set at 0.11m /s.

Table1. Vehicle parameters.

number	parameter	Value
1	length/width/high（cm）	3730/1650/1560
2	Total mass（Kg）	1125
3	Coefficient of air resistance Cd	0.33
4	Windward area A（m^2）	1.9
5	Transmission efficiency η	0.93
6	The wheel radius（mm）	641
7	Rolling drag coefficient f	0.01

3. Parameter Matching and Model Building of Single Battery Group

The key to the power system of load isolation electric vehicle lies in its energy storage device matching. Battery is still the best choice of electric vehicle energy storage device because of the mature technology and reasonable price. Various factors should be considered comprehensively when selecting power batteries [7]. High energy density batteries are key to achieving the target range for electric vehicles. The integrated management system and thermal management system help display the battery SOC level and temperature in real time. These measures can prolong the life of the battery. Referring to the above factors, considering the battery usage of domestic and foreign electric vehicle manufacturers, existing equipment and safety, the model uses lithium iron phosphate battery.

Assuming that the model runs at an average speed of 42 km/h, the power output required by the motor is:

$$P_1 = \frac{1}{\eta} * \left(\frac{Mgf}{3600} u_1 + \frac{C_D A}{76140} u_1^{\ 3} \right) \tag{1}$$

Where, η is the efficiency of transmission system; M is the preparation mass; f is the rolling resistance coefficient; u_1 is the average speed; C_D is the air resistance coefficient; A is windward area.

Therefore, the output power of the battery pack is:

$$P_{out} = \frac{P_1}{\eta_{\mathrm{mo}}} \tag{2}$$

Where P_{out} is the average power; η_{mo} is the motor efficiency, and the value is 0.9.

The capacity of the battery is:

$$C_b \geq \frac{1000 P_{out}}{\lambda * U_1} \tag{3}$$

Where, λ is battery discharge rate; U_1 is the battery voltage.

By substituting the parameters of the sample vehicle into Equations (1), (2) and (3), it can be seen that $P_1 = 2.21$kW; $P_{out} = 2.42$ kW. $C_b \geq 28.7$ Ah, where k is the value 1, $C_b = 32$ Ah.

Model battery selection is based on average speed and average power. In order to ensure power and charge and discharge requirements, the load isolation electric vehicle selected 82 V, 100 Ah lithium iron phosphate battery. The internal resistance of a single battery is 30mΩ. Therefore, the internal resistance of a battery string is 0.069 mΩ.

4. Parameter Matching and Modeling Simulation of Engine and Generator

The output power of the generator is the first factor to be considered when selecting the generator. When the electric quantity reaches the alarm level, the generator needs to work to meet the steady-state driving power of the electric vehicle[8].The engine is completely separated from the load and only serves as an energy conversion device to provide electric energy to the battery in time. The model chooses the average speed and the average power as the reference index to select the engine[9].According to the average index, the volume and power of the engine can be greatly reduced, the weight of the car can be reduced, and the economy of the electric vehicle can be improved.

The power of the generator is roughly equal to the sum of the driving resistance of the vehicle at the average speed, namely:

$$P_t = \frac{1}{\eta \eta_{mo} \eta_{bo}} * \left(\frac{Mgf}{3600} u_1 + \frac{C_D A}{76140} u_1^3 \right) \tag{4}$$

Where, η is the efficiency of transmission system; η_{mo} is the motor efficiency; η_{bo} is the discharge efficiency of the battery, which is 0.96.

We can get Pg=2.6 kW, so let's take Pg=5 kW.

Taking all these factors into account, the four-stroke gasoline engine is still the most ideal engine for the load isolation electric vehicle generator set.

5. Motor Parameter Matching and Model Simulation

Motor has a long history of development, and motor technology has been very mature [10].At present, there are many kinds of electric motors used in automobiles. Induction motor, dc motor, AC motor and switched reluctance motor is the most widely used. Digital scoring algorithm is used to compare the performance of four kinds of motors. Considering all the factors, the model selects the low power AC induction motor as the driving motor of the load isolation electric vehicle. According to the power parameters of the battery pack and the relevant provisions on the power level of the motor, the low-speed motor with rated power of 10 kW and peak power of 16 kW is selected for the model.

The conversion relation between the maximum speed of the driving motor and the maximum speed of the vehicle is:

$$n = \frac{v_{\max} i_n i_m}{0.377 r} \tag{5}$$

Where, i_n is the speed ratio of the main reducer; i_m is transmission speed ratio; r is the radius of the car wheel.

The model eliminates the transmission system, so i_n =1. According to the set maximum speed of 75 km/h. The maximum speed of the motor was determined to be 4000 r/min, n =4500 r/min, and the coefficient β of motor extended constant power zone was 2.5. So the rated speed of the driving motor is 1800 r/min. According to formula (6), the rated torque of the motor is 56N·m.

$$T = \frac{9550P}{n} \tag{6}$$

6. Analysis of Simulation Results

The model is connected with an oscilloscope module to output speed at the speed output port. The model simulates ev acceleration from an average speed of 43 km/h to 75 km/h. Observe the voltage change curve of the battery string (Figure1) to check whether there is an overvoltage hazard. Whether the output power of the motor (Figure2) is in danger of exceeding the peak power. The results show that there is no overvoltage danger in the battery and the motor power does not exceed the peak power.

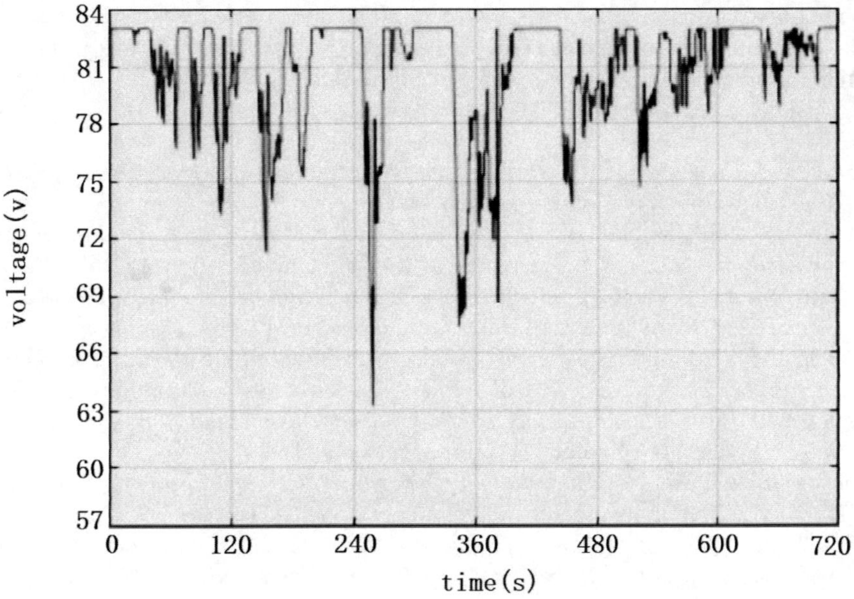

Figure 1. Voltage curve of the battery string.

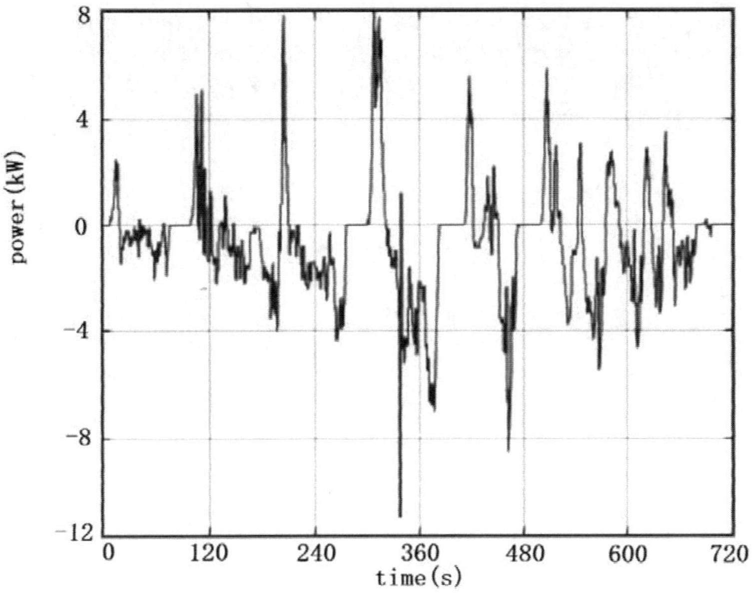

Figure 2. Voltage curve of the battery string.

7. Conclusion

Through MATLAB/SIMULINK software, the modeling and simulation of the load isolation electric vehicle are carried out. The simulation results meet the expected results. The feasibility of load isolation electric vehicle is verified. The motor and gasoline generator used in this model are slightly larger, so these components can be optimized in the future to achieve better results. The analysis of the simulation results in this paper provides the theoretical and experimental data basis for the future optimization design and the establishment and simulation of the load isolation electric vehicle model with two or more groups of batteries, so it has a very important guiding significance.

Acknowledgments
I would like to thank my team and friends for their Suggestions and opinions in the thesis research. Their opinion played an important role in the content research.

References

[1] Ke Song and Tong Zhang 2011*Research on power system of extended range pure electric drive vehicle* vol 7(Automotive Technology) pp 14-19
[2] Minrui Guo and Yingchun Gao 2016*Parameter matching and simulation of pure electric vehicle based on ADVISOR* vol 5(Journal of Chifeng University) pp 45-46
[3] Qinglong Wu and Zhongtu Liu 2010*Dynamic Performance simulation of pure electric vehicle based on ADVISOR* vol 10(Light vehicle technology) pp 13-16
[4] Qi Zhang and Xiaoling Fu 2016*Power System matching optimization and Regenerative Braking strategy for pure electric vehicles* vol 3(Journal of System Simulation) pp 600-609
[5] Su Zhou and Jigao Niu 2011*Study on Powertrain Design and Simulation of range extended electric vehicle* vol 11(Automotive engineering) pp 924-929
[6] Xiaohang Shi and Tiexiong Su 2012*Vehicle performance simulation of an electric vehicle based on Cruise* vol 6(Automobile practical technology) pp 80-83
[7] Liang Zhang, Qinzhong Song and Zuoxin Wang 2021*Braking Parameter Matching and Performance Simulation of Extended range Electric Vehicle* vol 3(Mechanical Engineering and Automation) pp 76-78
[8] Xinlei Liu, Hongwen He and Ximing Wang 2013*Research on fault tolerant control strategy of distributed drive electric vehicle* vol 30(Computer simulation) pp 152-155
[9] Lan Wei, Yan Yue and Qizhong Liu 2009*Design of control system for charging and discharging device of electric vehicle battery* vol 9(Power electronics) pp 69-71
[10] Xin Liu and Peng Ling 2013*Power parameter matching of pure electric vehicles based on ADVISOR* vol 3(Beijing automotive) pp 24-26

Research on Soft Sensor Modeling Method and Its Application in Wastewater Treatment

Mingzhu Li*, Shouxing Zhang and Yuanyuan Chen

Haikou University of Economics, Haikou 571127, China

*Corresponding author email: limingzhu110abc@163.com

Abstract. In order to solve the problem of real-time monitoring of water quality index which is difficult to measure in the process of wastewater treatment, a dynamic modeling method of soft sensor based on the VIP (Variable Importance in Projection) and JIT (Just-in-Time) methods was proposed. Firstly, the important variables were selected by the VIP method, and then the data were updated by the JIT method. Finally, the soft sensing model of PLS (partial least squares) was established. The model was used to predict BOD_5 which was an important water quality parameter in the process of wastewater treatment. The simulation results indicate that the proposed model achieves better mean square error and correlation coefficient.
Keywords: Soft sensor, variable selection, just in time, wastewater treatment.

1. Introduction

The problem of water pollution has been paid more and more attention by the state and relevant government departments. The state has formulated a series of laws and regulations to restrict the treatment and discharge of wastewater. In the process of wastewater treatment, there are many variables closely related to the safety of the production process, which are difficult to be measured directly online by hardware sensors [1]. In order to improve the qualified rate of effluent, save energy, reduce operating costs and improve system reliability, it is necessary to obtain the important process parameters and water quality parameters of wastewater treatment system in time [2]. Soft sensor is an effective way to solve this problem [3]. The soft sensor has the advantages of rapid response, low cost, relatively simple maintenance, etc. [4], which makes it applied in many industrial processes that are difficult to be measured by hardware sensors.

In this paper, the selection of auxiliary variables and online updating methods are studied to further improve the performance index of the model, so that the important water quality parameters in the process of wastewater treatment can be tracked in time, providing further guarantee for the safe production of sewage, and saving the cost of hardware instruments.

2. VIP Method

In the process of wastewater treatment, there are many measurable variables, which put forward the requirements for variable selection. Choosing appropriate auxiliary variables before establishing soft sensor model can not only reduce the sample data, but also strengthen the explanatory role of variables. Through the selection of auxiliary variables, the auxiliary variables unrelated to the output variables are eliminated, making the prediction effect of the soft-sensing model more accurate [5].

Many variable selection methods have been proposed, but it is difficult to propose a valid variable selection method for all objects [6].The common variable selection methods include VIP method[7], genetic algorithm [8], moving window [9], etc. In this thesis, the VIP method will be further studied.

Content from this work may be used under the terms of the Creative Commons Attribution 3.0 licence. Any further distribution of this work must maintain attribution to the author(s) and the title of the work, journal citation and DOI.
Published under licence by IOP Publishing Ltd

The VIP method is a method for variable selection, and it is based on the PLS method (partial least squares regression).The variable is selected by accumulating its importance in the PLS model. In this method, VIP index can be used to measure the importance of auxiliary variables to target variables, so as to select the auxiliary variables that contribute greatly to target variables. The VIP index can be calculated by formula (1).

$$VIP_j = \sqrt{p\sum_{k=1}^{l}\left[(q^2{}_k t'_k t_k)(w_{kj}/\|w_k\|)^2\right] / \sum_{k=1}^{l} q^2{}_k t'_k t_k} \tag{1}$$

Where p is the number of auxiliary variables, t_k is the k-th column element of the score vector, and q_k is the k-th column element of the regression coefficient. $w_{kj}/\|w_k\|^2$ shows the importance of the variable j. Setting the threshold TH, when $VIP_j<TH$, the variable j can be eliminated. It is generally believed that the screening effect is better when TH is 1. However, some literature [10] believed that TH between 0.83 and 1.21 could generate more relevant variables.

3. JIT Algorithm

Although soft sensor has achieved great development, but most of the researches are based on static model, that is, global model. With the passage of time, the static model will lead to the change of the operating point of the system, making the soft-sensing model is no longer suitable for the application of working conditions. So the model needs to be updated in time.

The JIT algorithm was proposed by Bontempi G in 1999[11], which has gained attention due to its good adaptability [12]. Based on the principle of "similar input produces similar output", the algorithm memorizes the sample data in the memory, finds the similar data in the sample data according to the input points, and establishes some local models of the system on-line rolling with the changes of the system working points, so as to meet the needs of real-time modeling and control of nonlinear systems. Similar data can be expressed by the Euclidean distance between the samples. If the distance is within the specified range, the two samples are considered to be similar. The Euclidean distance between two points is expressed as formula (2).

$$d(x_i, x') = \|x_i - x'\|_2 \tag{2}$$

If the local space composed of k sample points near the input point x' is Φ_k , then the data close to x' can be expressed as formula (3).

$$\Phi_k = \{x_i | d(x_i, x') < L\} \tag{3}$$

Where L is the window length and i ranges from 1 to k.

4. Soft Sensor Modeling Steps

The steps of soft sensor modeling for wastewater treatment are as follows:

Step 1: Requirement analysis. According to the characteristics and actual situation of the wastewater treatment system, the system needs are analyzed and the variables to be measured are defined.

Step 2: Build model. The principle, structure and characteristics of the soft sensing method are analyzed, and the dynamic model of soft sensing is established.

Step 3: Simulation experiment. After the singular value and missing data are processed, 70% of the data are used to train the model and 30% to verify the model. The system simulation is carried out by MATLAB to further analyze the prediction effect of the algorithm in the application of wastewater treatment.

5. Case Analysis

In this paper, the important water quality parameter BOD_5 which is difficult to measure in wastewater treatment is taken as the research object. The auxiliary variables are selected through the VIP method, combined with the on-line update method of JIT, a systematic PLS soft sensing dynamic model is established, in order to further improve the prediction effect of the algorithm.

Three models are established to predict BOD_5. The models are PLS model, VIP-PLS model (the model combining VIP variable selection), and JIT-VIP-PLS model (the model combining VIP variable selection and JIT method). The predictive effect is displayed in Figure 1, Figure 2 and Figure 3, respectively.

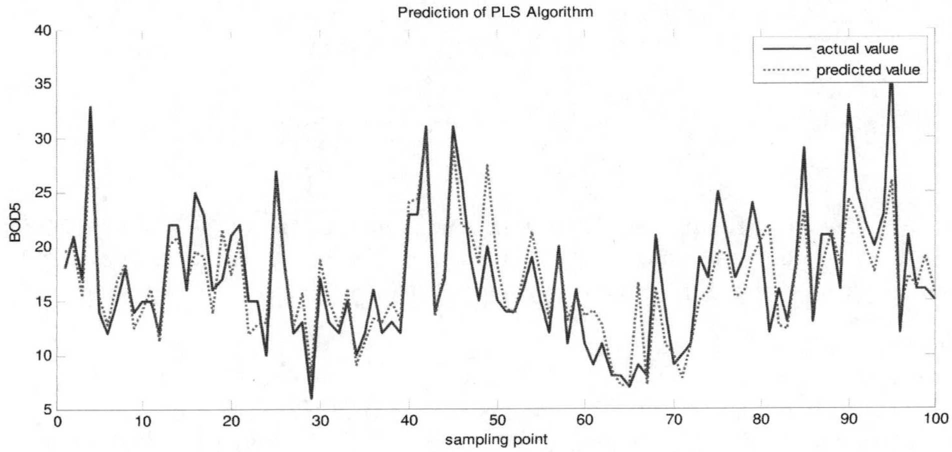

Figure 1. The prediction of PLS model.

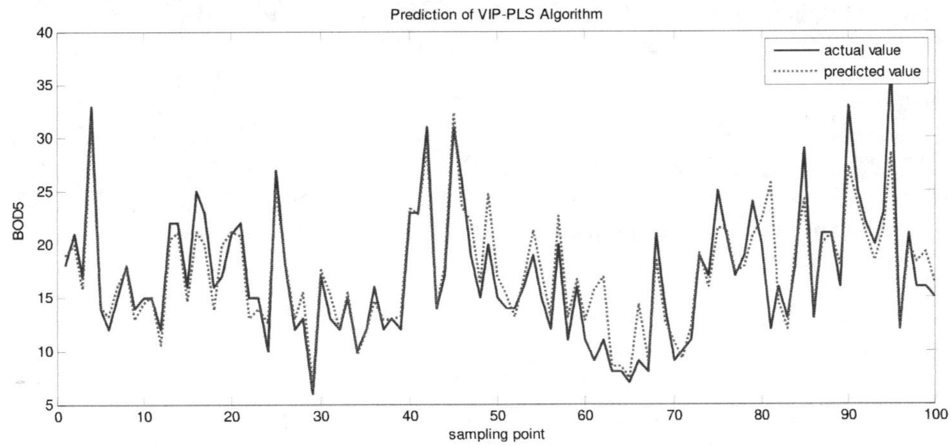

Figure 2. The prediction of VIP-PLS model.

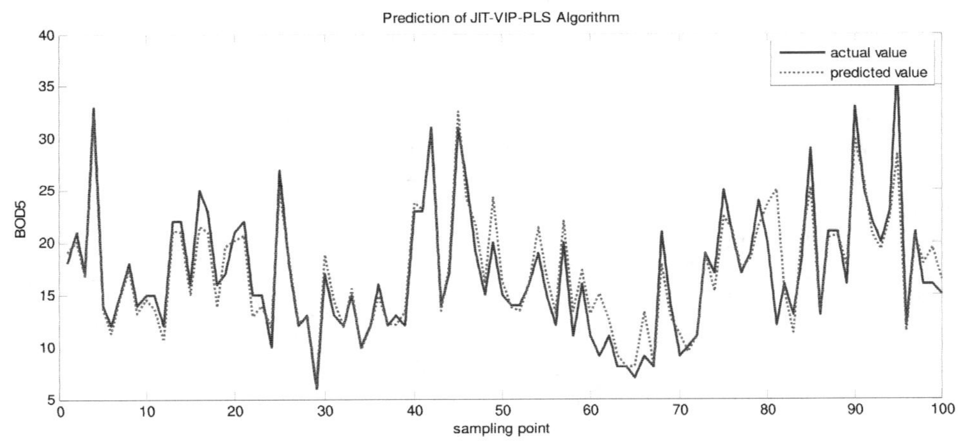

Figure 3. The prediction of JIT-VIP-PLS model.

The mean square errors (MSE) and the correlation coefficients (CC) of the three models are shown in Table 1.

Table 1. The results of three models.

Model	MSE	CC
PLS	9.2257	0.8676
VIP-PLS	6.7137	0.9064
JIT-VIP-PLS	5.3142	0.9253

From figure 1~3 and table1, it can be seen that the prediction accuracy of the system is improved after variable selection. At the same time, the prediction accuracy is further improved after online updating model, which illustrates the effectiveness of the proposed algorithm.

6. Conclusion

In this paper, the dynamic model of PLS soft sensing model is established by combining the VIP method and the online updating method of JIT algorithm, which provides an effective solution for the real-time measurement of BOD_5 in wastewater treatment. The research is helpful to improve the prediction accuracy of the model, make the important water quality index can be tracked in time, provide the basis for advanced control and optimization of production, and provide further guarantee for the safe production, and save the cost of hardware instrument. It has a certain engineering application value.

Acknowledgments

This research was supported by Hainan Provincial Natural Science Foundation of China (project number: 618QN254), Hainan Provincial Natural Science Foundation of China (project number: 620QN285), Haikou University of Economics Scientific Research Project (HJKY (ZD) 20-10), Haikou University of Economics Scientific Research Project (HJKY (ZD) 21-01).

References

[1] Chen Zhaobo, Ren Yueming. Wastewater Treatment Plant, Measurement, Automatic Control and Fault Diagnosis [M]. Beijing: Chemical Industry Press, 2008:1-20.
[2] Guan Qiu. Research on Soft-sensing Method Based on Neural Network for Wastewater Treatment [D]. Zhejiang University of Technology, 2005.
[3] Cao Pengfei, Luo Xionglin. Modeling of soft sensor for chemical process[J].CIESC Journal, 2013, 64(3): 788-800.
[4] Lai Huige, Zhu Xuejun,Yu Jinshou. Research on Intelligent Algorithm-based Soft-sensing Technology for Sewage Treatment Process [J]. Control and Instruments in Chemical Industry, 2011, 38(03): 241-244+278.
[5] Wang Jian, Yang Yaoquan, Ma Gaowei. Research on Method of Soft Measurement Secondary Variables Selection [J].Electric Power Science and Engineering, 2011, 27(7):37-40.
[6] Mehmood T, Liland K H, Snipen L, et al. A review of variable selection methods in Partial Least Squares Regression[J]. Chemometrics & Intelligent Laboratory Systems, 2012, 118(none): 62-69.
[7] Gosselin R, Rodrigue D, Duchesne C. A Bootstrap-VIP approach for selecting wavelength intervals in spectral imaging applications[J]. Chemometrics and Intelligent Laboratory Systems, 2010, 100(1):12-21.
[8] Hasegawa K, Miyashita Y, Funatsu K. GA strategy for variable selection in QSAR studies: GA-based PLS analysis of calcium channel antagonists[J]. Journal of Chemical Information & Computer Sciences, 1997, 37(2):306-310.
[9] Ypda D, Yzl B, Jhj C, et al. Spectral regions selection to improve prediction ability of PLS models by changeable size moving window partial least squares and searching combination moving window partial least squares[J]. Analytica Chimica Acta, 2004, 501(2):183-191.

[10] CHONG I G, JUN C H.Performance of some variable selection methods when multicollinearity is present[J]. CHEMOMETRICS AND INTELLIGENT LABORATORY SYSTEMS, 2005, 78(1): 103-112.
[11] Bontempi G, Birattari M, Bersini H. Recursive lazy learning for modeling and control[C]// European Conference on Machine Learning. Springer-Verlag, 1999.
[12] Kaneko H, Funatsu K. Moving Window and Just-In-Time Soft Sensor Model Based on Time Differences Considering a Small Number of Measurements[J]. Industrial & Engineering Chemistry Research, 2015, 54(2):150104102633007.

Dynamic Performance Simulation Analysis of Electric Forklift Truck Based on ADVISOR

Li Zhang*
Department of Vehicle Engineering, Shandong Transport Vocational College, Weifang, China

*Corresponding author email: zhangli.0424@foxmail.com

Abstract. According to the power system selection and technical parameters of the lithium titanate electric forklift that have been matched, the ADVISOR simulation software modeled the parameters of the drive motor, lithium titanate battery, reducer and other major components of the system.The acceleration ability, climbing ability and energy consumption of lithium titanate-electric forklift are simulated, and its technical parameters are obtained, which provides a basis for testing the rationality of system parameter matching.
Keywords: Dynamic performance, simulation analysis, forklift, ADVISOR.

1. Introduction

ADVISOR is a vehicle dynamics simulation software developed by NREL based on Matlab software. It can simulate and analyze various aspects of the performance of common vehicles, including electric vehicles. Software has the following advantages [1]: ① source code and open; ② Forward and backward hybrid simulation; ③ Modular simulation model; ④ Compatible with a variety of programming languages; ⑤ Combined with a variety of software simulation. The software can be used to simulate more than just traditional cars.

Lithium titanate electric forklift has the advantages of fast charging speed and long cycle life, which is of great significance in the field of pure electric forklift. In view of the particularity of the lithium titanate electric forklift, the power performance of the lithium titanate electric forklift was simulated by ADVISOR software to verify the rationality of the early power system design and selection. The simulation process is mainly divided into three steps [2-5].

(1)Build the forklift dynamics configuration. According to the design and selection parameters of the power system components of lithium titanate electric forklift truck, the dynamics models were constructed respectively.

(2) Set the simulation operation parameters. After completing the dynamic model of each part of the vehicle, the driving conditions and initial conditions of the forklift truck were set.

(3) Output simulation results. After completing the setting of technical parameters, the power performance curves of lithium titanate forklift truck are generated respectively, including: climbing capacity, mileage, etc.

2. Construction of Lithium Titanate Electric Forklift Dynamics Model

Power performance is an important index to evaluate the performance of a forklift. The power performance parameters of lithium titanate electric forklift determine its working efficiency. According to the selection and parameters of the main power system components in the matching design of the power system of the forklift truck, the vehicle dynamics model of the lithium titanate electric forklift truck was established by referring to the process and method of power system model

Content from this work may be used under the terms of the Creative Commons Attribution 3.0 licence. Any further distribution of this work must maintain attribution to the author(s) and the title of the work, journal citation and DOI.
Published under licence by IOP Publishing Ltd

construction in ADVISOR software. Key components driving motor, lithium titanate battery, reducer model construction is very important.

2.1 Drive Motor Model Construction

The parameter model of the moving motor of lithium titanate electric forklift truck is established according to the technical parameters of the driving motor after the vehicle selection and design. Motor parameters include rated voltage, rated speed, power curve, etc. Based on the heat exchange, operation characteristics and other performance indexes of the driving motor tested by experiments, the model is constructed. The specific parameters of the driving motor are shown in Table 1. Edit and input the corresponding data of each module shown in Table 1, define the characteristics, and get the walking motor ADVISOR model.

2.2 Lithium Titanate Battery Model

Lithium titanate battery is different from other lithium batteries. It has the characteristics of fast charging and long cycle life, which leads to the technical parameters different from the commonly used lithium iron phosphate battery and lead-acid battery.

Table 1. Model parameters of the motor.

number	variable	parameter
1	Speed(r/min)	0/700/1500/2200/2800
2	Torque(N·m)	0/50/90/130/180
3	Min voltage(V)	48
4	Max current(A)	330
5	Total mass(kg)	50

Through the test of lithium titanate battery, the current, voltage and State of Charge (SOC) of the battery during Charge and discharge are obtained. SOC are the ratio of the remaining capacity of the battery to the capacity of the fully charged State. The value range is 0~1, and when SOC=0, the battery is fully discharged. When SOC=1, it means that the battery is fully charged). The power limit, open circuit voltage, battery heat dissipation, calculation current and SOC algorithm required by the battery in the lithium titanate electric forklift truck are analyzed to determine the parameters required for model construction. According to the battery pack model parameters in Table 2, the battery model construction method in ADVISOR simulation software was applied to establish the parametric model of the power battery of lithium titanate electric forklift truck based on Matlab software.

Table 2. Model parameters of lithium titanate battery pack.

number	variable	parameter
1	Battery capacity(A·h)	98/98/98
2	Battery number	32
3	Single cell voltage(V)	1.6-2.8
4	The battery weight(kg)	13

2.3 Reducer Model

The basic performance parameters of the gearbox are obtained by testing the mechanical properties of the gearbox. The simulation test of straight and ramp conditions does not involve the performance of differential speed change, but mainly tests the transmission efficiency and the performance of torque increase and decrease. The performance parameters obtained after testing are shown in Table 3.According to the model parameters of the gearbox in Table 3, the battery model construction method in ADVISOR simulation software was applied to establish the parameterized model of the gearbox of lithium titanate electric forklift truck based on Matlab software.

Table 3. The model parameters of the gearbox.

number	variable	parameter
1	Gearbox transmission ratio	21.5
2	Total weight of reducer /kg	16.5

2.4 Other Control and Accessory Models

By referring to the process and method of ADVISOR simulation software, the parameter model construction of other components of lithium titanate electric forklift truck was completed [6-8].

(1) Forklift driving model. The driving model of lithium titanate electric forklift truck is built. In the software model, the input terminals ① and ② are the output outlets of decelerating torque and speed, and the output outlets of mechanical braking force. Combined with the weight of the whole machine, tire efficiency, rolling coefficient and moment of inertia and other factors, the actual acceleration of lithium titanate forklift truck was obtained, and its displacement and speed were calculated by formula.

(2) Forklift control model. The control model of lithium titanate electric forklift truck was constructed. In the software model, the input end ① is the target speed of the forklift truck.② is the throttle data; ③ is the brake data, and the output terminal ① is the speed and torque required by the driving motor;② is the data of its braking force. The calculation module is for the required forklift braking force, acceleration distribution, motor power, power feedback, etc.

(3) Driver model. In ADVISOR, the driver model is built based on the fuzzy PID control theory. The output is the accelerator or brake pedal, and the input is the required value of the speed. The driving condition of lithium titanate forklift truck is used to simulate the operation of the driver, so the simulation parameter flow is consistent with the actual operating condition.

2.5 Vehicle Parameter Model

In conclusion, this paper constructs the ADVISOR parameterized model of the lithium titanate electric forklift truck. The parameter model of lithium titanate electric forklift truck includes several sub-module models, The module model of transmission system includes the data of torque and moment of inertia. The data of the lithium titanate battery model includes temperature, SOC, battery resistance, voltage and thermal parameters. The driving motor model includes power, torque and other parameters. Axle and wheel are the data of speed and traction required in simulation. These parameters are defined by the type of the MATLAB software file. Finally lithium titanate electric forklift truck load parameters is added. Such as dashboard, lights and other accessories energy consumption data.

3. Simulation of Power Performance of Lithium Titanate Electric Forklift Truck

The acceleration ability and climbing ability of lithium titanate electric forklift are the main indicators of its dynamic performance[9-10], and its economic performance is also an important performance index. In this paper, The dynamic performance is simulated from the above three aspects. According to the design requirements, the dynamic performance should meet the following indexes: the acceleration time from 0 km/h to maximum speed is less than 12 s; Minimum climbing slope > 14%; Single continuous working time > 3.5h.

3.1 Simulation of Acceleration Capability

The acceleration capability of the lithium titanate electric forklift will be simulated by the acceleration performance test module in ADVISOR simulation software. First, the SOC value of the maximum current discharge in the module is set to 1, and the time required for the speed of the electric forklift from 0 km/h to the maximum speed of 18 km/h is calculated by simulation. First, the unit is converted and matched with the software. After parameter adjustment, the simulation program is run to calculate the time change curve required by the forklift to accelerate from 0 speed to maximum speed, as shown in Figure 1:

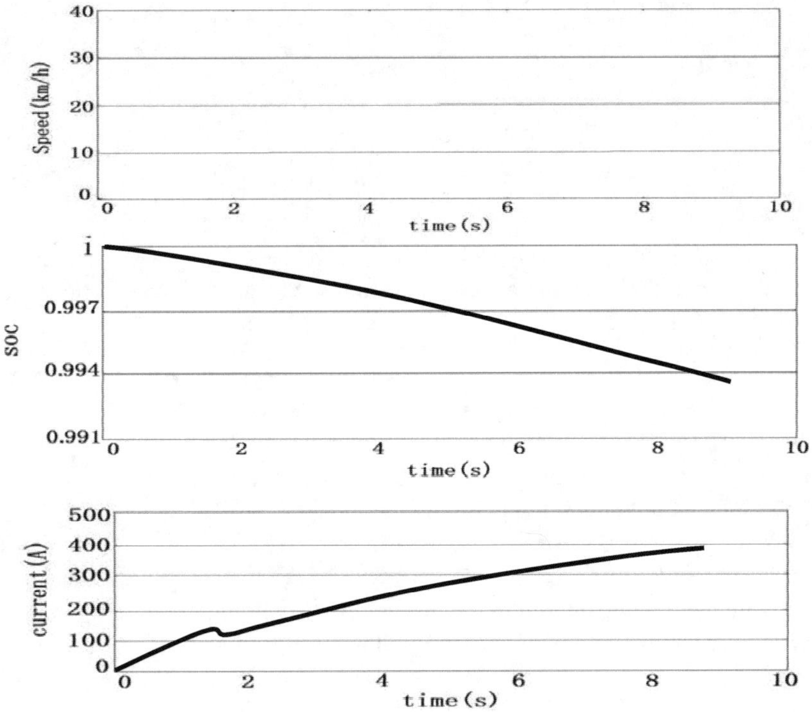

Figure 1. Battery discharge simulation results.

From the simulation results, it can be obtained that the distance of lithium titanate electric forklift accelerated to the maximum speed is 12 m. The maximum speed of 18.2 km/h was reached in 8.7s.With the output of power, the acceleration of lithium titanate electric forklift gradually decreases, while the speed increases. According to the SOC simulation curve of lithium titanate battery, the SOC value changes little during the acceleration process, and only decreases by 0.007 during the whole acceleration process. According to the simulated current curve, the lithium titanate electric forklift requires A large torque in the starting process, and the battery output current rises rapidly at first, and then the current eases after the startup. After that, due to the stable operation of the forklift, the current gradually rises and finally reaches the maximum current of 382A.

In the normal operation condition of lithium titanate electric forklift, the driving motor is not always running in the maximum current condition. Therefore, in order to ensure the thermal stability and safety of the motor, the discharge rate of the battery should be appropriately reduced. It is assumed that the discharge ratio is set to 0.6c and the time required for simulation calculation is 8.2s.The test results of lithium titanate battery SOC are basically the same as those in 1C condition. To meet the design requirements of forklift trucks, so as to ensure safety and reliability, the battery discharge current is set to 0.6C.

3.2 Simulation of Climbing Ability

According to the operating conditions of lithium titanate electric forklift, it should meet the requirements of normal climbing capacity: minimum climbing slope > 14%.In the ADVISOR, the climbing speed was set to 5.1 km/h, the climbing time was set to 15 s, and the SOC value was set to 0.5.The corresponding test results of the simulation test are shown in Figure 3, all of which are > 14%, reaching the design goal. At the same time, through the above simulation test, the variation curve of each parameter in the 5.1km /h climbing process of forklift truck can also be obtained, as shown in Figure 2 and Figure 3.

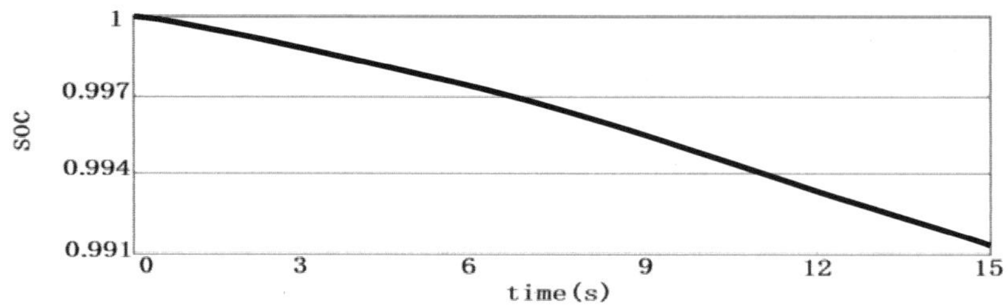

Figure 2. SOC change of 5.1km/h climbing lithium titanate battery.

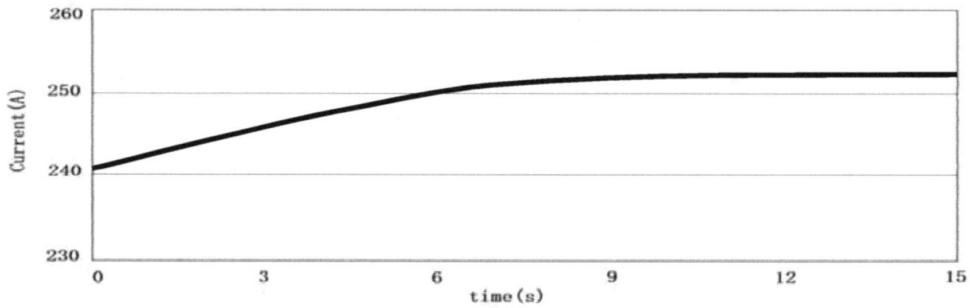

Figure 3. Current change of 5.1km /h climbing lithium titanate battery.

According to SOC curve and battery current curve, SOC changes little and current is large. In order to ensure safety and climbing ability, the maximum discharge current of lithium titanate battery is reduced appropriately.

3.3 Simulation of Energy Consumption

The energy consumption simulation of lithium titanate battery needs to first establish the equivalent cycle condition, then carry out simulation calculation, and finally achieve the requirements of the working condition. Charging a single continuous working time > 3 h. Firstly, according to the dynamic performance, working conditions, operating environment and national standard test requirements of electric forklift, the equivalent cycle simulation of energy consumption for forklift truck was established in ADVISOR, as shown in Figure 6.

Through the analysis of the forklift working conditions and the calculation of the running speed, the equivalent cycle of the operation times is about 85 times (running for 3 h). After setting the parameters in ADVISOR software, the SOC and current change curves of the lithium titanate battery during the energy consumption process were obtained, and the travel distance curve of the forklift was also obtained, as shown in Figure 4 and Figure 5.

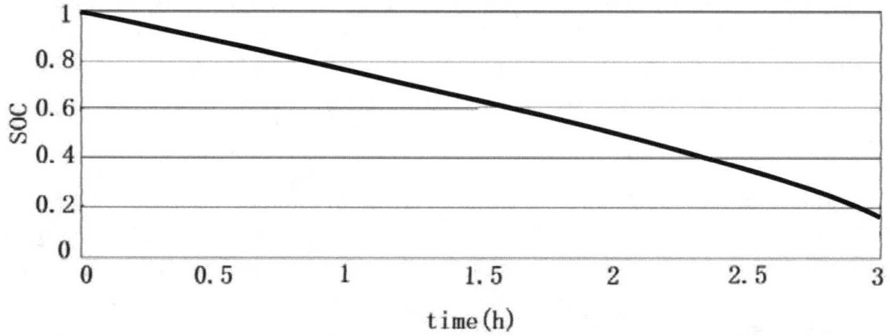

Figure 4. Changes in battery SOC during energy consumption cycle.

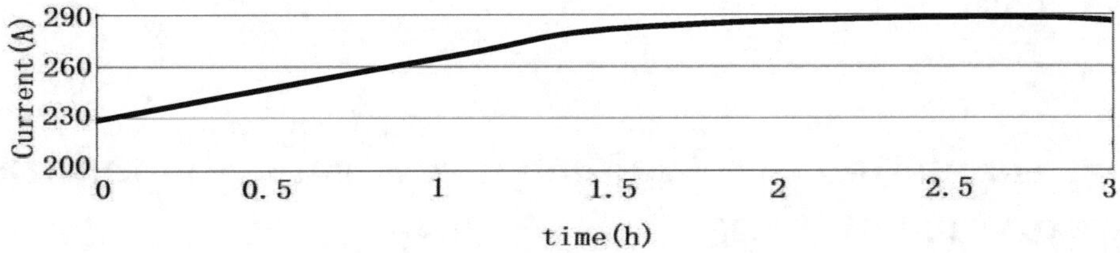

Figure 5. Battery current changes during the energy consumption cycle.

According to the above simulation curve, during the whole process of testing the electric forklift, the minimum power of the battery is not less than 15%, which meets the design technical requirements.

4. Conclusion

This paper introduces the ADVISOR simulation software and its test process, and builds the vehicle model in the software. In particular, the key components of driving motor, lithium titanate battery, reducer model construction are described in detail. Finally, the acceleration ability, climbing ability and energy consumption of lithium titanate-electric forklift are simulated to verify whether its performance indicators meet the design requirements.

References

[1] Zhaofeng Pan and Zhigang He 2012*Power System modeling and Simulation of hybrid electric vehicle* vol 37(Journal of Guangxi University) pp 284-290

[2] Yujian Jia and Da Xie 2011*Classification and characteristics of equivalent circuit models for electric vehicle batteries* vol 32(Electricity and Energy) pp 516-521

[3] Kai Bai and Shida Liu 2017*Dynamic performance simulation analysis of EV based on ADVISOR*vol 55(Agricultural equipment and vehicle engineering) pp 74-77

[4] Libiao Jiang and Bin Wu 2011*Simulation design and experimental verification of electric vehicle dynamic parameters* vol 12(Automotive engineering) pp 1013-1017

[5] Mingqiang Niu and Xingzhong Guo 2014*Power System Parameter matching and Simulation of pure electric vehicle based on CRUISE* vol 3(Journal of Anhui Engineering Universityg) pp 49-53

[6] Xiaohang Shi and Tiexiong Su 2012*Vehicle performance simulation of an electric vehicle based on Cruise* vol 6(Automobile practical technology) pp 80-83

[7] Xinlei Liu, Hongwen He and Ximing Wang 2013*Research on fault tolerant control strategy of distributed drive electric vehicle* vol 30(Computer simulation) pp 152-155

[8] Feng Liu and Shaopeng Tian 2011*Dynamic Performance calculation and Simulation of electric vehicle* vol 3(Beijing automotive) pp 44-46

[9] Guowei Dou and Fen Liu 2010*Development practice of pure electric vehicle drive control strategy* vol 5(Shanghai automotive) pp 8-10

[10] Yurong Liu 2018*Study on auxiliary transport System of mine double electric trolley* vol 2(Coal engineering) pp 32-36

Modeling Simulation and Maintenance Strategy Optimization of Multi-component System

Yujun Ma and Yan Zhou*

School of Economics and Management, Nanjing University of Science and Technology, Nanjing, China

*Corresponding author email: zhouyan77@njust.edu.cn

Abstract. As the system tends to be complicated and refined, the probability and cost of failure also increase. In the actual operation and production, the system will be affected by various factors such as temperature, humidity and other factors to change their original working environment, which will have greater impact on the reliability of the system. Based on the reliability theory, this paper mainly studies the *k-out-of-n: F* system, constructs the reliability model of the system in dynamic environment, gives the simulation algorithm of the system reliability. Finally, this paper puts forward two maintenance strategies which are optimized to achieve the purpose of minimizing the long-term average cost during system operation.
Keywords: Multi-component system, maintenance cost, preventive maintenance, dynamic environment, *k-out-of-n* system.

1. Introduction

The improvement of science and technology promotes the change of social production and manufacturing structure, and the product system structure produced by socialized mass production tends to be complicated, randomized and varied. The purpose of the system is to make the product have better performance and longer life. However, the more complex the system, the higher the probability of failure. Therefore, there is a great contradiction between complexity and reliability, so we need to find a balance between them. In the process of modernization in China, the proposal of "Made in China 2025" has aroused widespread concern in the industry about product reliability.
The reliability theory was born in the 1930s, and the first field studied was machine maintenance[1].The optimization of maintenance strategy is to maximize the reliability of the system under the condition of minimizing the cost and to ensure its high availability and security[2-5]; For reducing the probability of system or product failure, the method of update theory by Lotka[6] and Campbell[7] related to component replacement strategy has been discussed in the later part of the paper. Franceschini and Galetto[8] consider the potential efficiency of the system, propose a regular component replacement strategy, and finally establish and optimize the maintenance strategy model targeting cost minimization.
The innovation of this paper is that on the basis of regular replacement, this paper puts forward the strategy of preventive maintenance, and changes the research environment from simple environment to complex dynamic environment. The purpose of this paper is to find out a strategy to minimize the average cost of the system during operation.

Content from this work may be used under the terms of the Creative Commons Attribution 3.0 licence. Any further distribution of this work must maintain attribution to the author(s) and the title of the work, journal citation and DOI.
Published under licence by IOP Publishing Ltd

2. Reliability Modeling and Simulation of Multi-component System

2.1. Basic Assumptions and Reliability Modelling

The life of system and the performance of components are affected by different environments. The system operates in normal (*Environment 0*) and harsh environment (*Environment 1*), the change of environment is independent of the state of the system. Figure 1 shows an example of *k-out-of-n* system with *n=6* and *k=4*. The basic assumptions of the system are as follows:

n : The numbers of components

k : The minimum number of failed components that makes the system stop working

i : The environment where the system operates

Y_i: The duration of *Environment i, i=0,1*

λ_i: The parameter of the exponential distribution which Y_i submits to, *i=0, 1*

$H_i(t)$: Cumulative distribution function of Y_i, *i=0, 1*

$h_i(t)$: Probability density function of Y_i, *i=0, 1*

X_i: The life of the component working in the *Environment i, i=0,1*

Xmu_i: The parameter of the exponential distribution which X_i submits to, *i=0, 1*

$F_i(t)$: Cumulative distribution function of X_i, *i=0, 1*

$f_i(t)$: Probability density function of X_i, *i=0, 1*

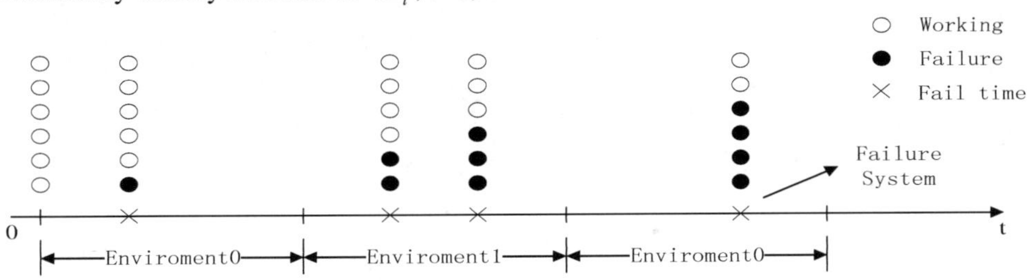

Figure 1. The *k/n*[*F*] system (*k=4*, *n=6*).

Based on the above discussion, we establish the system reliability model in two steps.

The first step is to model the environment. The system will be assumed to work alternately in two different environments, so we can study the reliability of the system in dynamic situations. The duration of each system in one environment is not fixed, but the duration of the ***Environment i (i=0,1)*** is subject to the specified exponential distribution. Therefore, it can be concluded that there is a need to introduce a two-state Markov Mechanism Transformation Model. If ***i*** represents two different states of the environment, the change of the environment can be represented by a two-state Markov process $\{Y(t), t \geq 0\}$. The transfer rate matrix of the system between different environments is

$$Q = \begin{bmatrix} -\lambda_0 & \lambda_0 \\ \lambda_1 & -\lambda_1 \end{bmatrix} \tag{1}$$

The second step is to establish a reliability model for the *k/n[F]* system. If the life of a single component is assumed ***X***, the reliability of the component in a stable environment is the probability that the system continues to work longer than the time ***t***, that is

$$R_0(t) = P\{X > t\} = 1 - F(t) = \overline{F}(t) \tag{2}$$

$R_0(t)$ is the probability of that the component ***X*** does not fail in time [***0, t***]. Now we study ***n*** components in the system, assuming that the life of each component is $X_1, X_2, X_3, ..., X_n$, and they are independent of each other, the reliability of each component is $R_0(t)$. At the beginning, all parts of the system start working at the same time ***t=0***, and they are all initially intact parts. So the reliability of the system above is

$$R(t) = \sum_{j=k}^{n} \binom{n}{j} P\{X_{j+1},...,X_n \le t < X_1,...,X_j\} = \sum_{j=k}^{n} \binom{n}{j} R_0^j(t)[1-R_0(t)]^{n-j}$$
$$= \frac{n!}{(n-k)!(k-1)!}\int_0^{R_0(t)} x^{k-1}(1-x)^{1-k}\,dx \tag{3}$$

Then, we take the factors of environmental change into account, and the system will work alternately in two different environments. After the environment changes (***i=i-1***), the system starts to work in the new environment, and judge whether the number of invalid components is more than ***k*** in the duration of the new environment.

2.2. The Simulation of System Reliability
In this section, the simulation results of the system reliability function are given by the MATLAB through simulation. The steps of system reliability simulation are as follows:

Step 1: Assign an initial value to the relevant parameters. Set ***n***, ***k***, the state of the environment $i = 0$, the number of failed components $S = 0$, and the life of the system $H = 0$.

Step 2: Randomly generate values for each component life and current environmental duration. The life of components $X_1, X_2, X_3,..., X_n$ in the same working environment are randomly generated by the MATLAB, which obey the exponential distribution of parameter Xmu_i and arranged in order from small to large. And then generates Y_i.

Step 3: Compare. After the initial ***n+1*** random numbers are generated, the life of each component is compared with the environmental duration: if $X_1, X_2,..., X_n > Y$, Then $H = H + Y$, $i = 1-i$ and returns to *Step 2*.

Step 4: Judge whether system fails. If $X_1, X_2,..., X_n > Y$, then count the number of failed components $SUM(X > Y)$ and judge them: if at least $k - S$ components have failed, the system will fail, Then and turn to *Step 5*; Otherwise count the number of failed components j at this stage. Then $S = S + j$, $H = H + X_{k-S}$, $i = 1-i$ and return to *Step 2*.

Step 5: End. Obtain the simulation result of system's life H.

3. Maintenance Modeling and Replacement Strategy Optimization

This section will continue to take the system in dynamic environment as the research object, and put forward two maintenance policies: *Strategy I* is Breakdown Maintenance, *Strategy II* is Preventive Maintenance, and the case is used to simulate the long-term average cost of the system under each Strategy.

3.1. Strategy I: Breakdown Maintenance Modeling and Simulation

3.1.1. Basic assumptions and modeling. The Breakdown Replacement Strategy means that when the system shuts down because of failure, we need to replace all the components. Every time the system is repaired, the system will start operating from the *Environment 0*. Figure 2 shows an example of *k-out-of-n* system with *n*=6 and *k*=4. The basic parameters of the system are as follows:

Y_i: The duration of *Environment i, i=0,1*

λ_i: The parameter of the exponential distribution which Y_i submits to, *i=0, 1*

X_i: The life of the component working in the *Environment i, i=0,1*

Xmu_i: The parameter of the exponential distribution which X_i submits to, *i=0, 1*

C_f : The cost of the replacing all components after shutdown each time

C_s : The daily loss cost caused by the shutdown

d : The maintenance time required after each shutdown of the system

dmu : The parameter of the exponential distribution which d submits to

C_T : The long-term cost of operation and maintenance

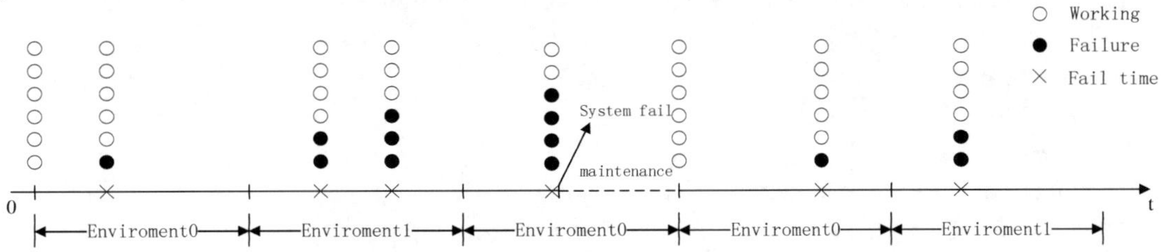

Figure 2. Strategy I: A sample path of *k/n*[*F*] system (*n=6, k=4*).

The long-term cost of operation and maintenance will be the sum of the replacement cost of the components and the cost of the system shutdown. So

$$C(T) = \frac{C_p * W(T) + C_s * D(T)}{T} \tag{4}$$

$W(T)$ is the expected number of system failures in time $(0,T]$ and $D(T)$ is the expected total number of days of system shutdown in time $(0,T]$.

When time T tends to infinity, the cost of long-term operating is

$$C(T) = \lim_{T\to\infty} \frac{\text{Expected cost in time } (0,T]}{T} \tag{5}$$

3.1.2. The Simulation of Cost. The simulation steps are as follows:

Step 1: Assign initial value to the relevant parameters. Set n, k, C_f, C_s; The number of days the system has run $R=0$; the total number of system failures $W=0$; the total number of days the system stops $D=0$; The initial *Environment* $i=0$; the total cost $C=0$, the average cost of long-term operation and maintenance $AC=0$; the number of failed parts $S=0$.

Step 2: Randomly generate d, and generate the system life H according to Section 2 algorithm.

Step 3: Define the total number of running days $LongT$ and compare with R. If $R < LongT$, then $R = R + H + d$, $D = D + d$ and the running times of MATLAB progress $W = W + 1$, then turn *Step 4*; if $R \geq LongT$,then $C = C_f * W + C_s * D$, turn to *Step 2*.

Step 4: End. Obtain the long-term average cost of operation and maintenance AC.

3.2. Strategy II: Preventive Maintenance Modeling and Simulation

3.2.1. Basic assumptions and modeling. Preventive replacement Strategy means that the system will be checked in a fixed time during the process of system operation. If it is found that the number of failure components reaches the pre-set threshold G, these invalid components would be replaced. Of course, if the system fails due to the excessive number of failure parts before checks, it would be replaced afterwards. Our aim is to find the optimal preventive maintenance threshold G to minimize the average cost of the system during a certain period of operation. Figure 3 shows a possible sample path of the system under Strategy II. When system failed, maintenance personnel will replace the all components. The basic parameters of the system are as follows:

Y_i : The duration of *Environment i, i=0,1*

λ_i : The parameter of the exponential distribution which Y_i submits to, *i=0, 1*

X_i : The life of the component working in the *Environment i, i=0,1*

Xmu_i : The parameter of the exponential distribution which X_i submits to, *i=0,1*

C_f : The cost of the replacing all components after shutdown each time

C_s : The daily loss cost caused by the shutdown

d : The maintenance time required after each shutdown of the system

dmu : The parameter of the exponential distribution which d submits to

C_T : The long-term cost of operation and maintenance

t : The interval for checking point

G : The preventive maintenance threshold

m : The number of invalid components

C_d : The cost of invalid component for replacement during the check

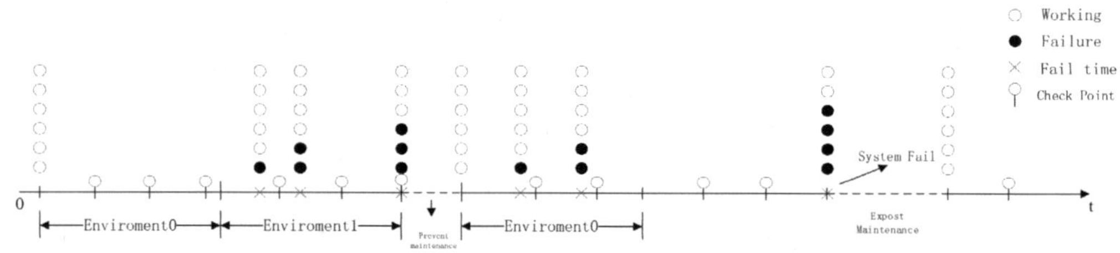

Figure 3. Strategy II: A sample path of *k/n*[*F*] system (*n=6, k=4, G=3*).

If $m < G$, the system will continue to work properly without any cost;

If in time $(0,T]$, $G \le m < k$, the system does not fail, and only the failed components need to be pre-replaced, the total cost is

$$C(T) = C_d * m(T) \tag{6}$$

$m(T)$ is the number of pre-replaced components in time $(0,T]$. The average cost is

$$AC(T) = \frac{C_d * m(T)}{T} \tag{7}$$

If in time $(0,T]$, the system has not only been pre-replaced processing, but also experienced Breakdown Replacement, then the total cost is

$$C(T) = C_d * m(T) + C_p * W(T) + C_s * D(T) \tag{8}$$

$W(T)$ is the number of system downtime in time $(0,T]$ and $D(T)$ is the total number of downtime in time $(0,T]$. The average cost is

$$AC(T) = \frac{C_d * m(T) + C_p * W(T) + C_s * D(T)}{T} \tag{9}$$

3.2.2. The Simulation of Cost. The simulation steps are as follows:

Step 1: Assign initial value to the relevant parameters. Set n, k, C_f, C_s, C_d, t, G; The number of days the system has run $R = 0$; The total number of system failures $W = 0$; The total days of the system stopped $D = 0$; The environmental status $i = 0$; The total cost $C = 0$; The average cost of long-term operation and maintenance $AC = 0$; The number of failed components $S = 0$, the number of failed parts at the time of check $m = 0$ and the cumulative detection time $T = 0$.

Step 2: Randomly generate values for each component life $X_1, X_2, ..., X_n$ and current environmental duration Y_i.

Step 3: Determine whether T is 0. If $T > 0$, turn to *Step 4*; Otherwise, $T = T + t$ and turn to *Step*

Step 4: Determine the relationship between the check point and the environmental change point. If $T < Y$, count the number of failed components in check $m = SUM(X < Y)$, then turn to *Step 5*; If $T \geq Y$, count the number of failed components in check $m = SUM(X < Y)$, and turn to *Step 6*.

Step 5: Determine whether to carry out preventive replacement Strategy. Set the value of dmu.

(1) If $m \geq k$, randomly generate d, then $H = H + X_{k-S}$, $C = C + C_f + C_s * d$, $R = R + H + d$ and turn to *Step 7*;

(2) If $m < G$, return to *Step 3*;

(3) If $G \leq m < k$, then $C = C + C_d * m$, $H = H + t$, $R = R + H$ and return to *Step 7*.

Step 6: Determine whether the system is invalid.

(1) If $m < k$, then $H = H + Y$, $i = 1 - i$, $S = m$. Then $T = t - \text{mod}(H, t)$ and return to *Step 2*;

(2) If $m \geq k$, randomly generate d, then $H = H + X_{k-S}$, $C = C + C_f + C_s * d$, $R = R + H + d$ and turn to *Step 7*.

Step 7: Define the total running days $LongT$ and compare it with R. If $R < LongT$, return to *Step 1*, $i = 0$, $T = 0$, $S = 0$, $H = 0$; If $R \geq LongT$, turn to *Step 8*.

Step 8: $AC = C / R$. Obtain the average cost of long-term operation and maintenance of system under Strategy II.

3.3. Case Study: 7-out-of-16 Bearing System

This section selects bearings to study the replacement Strategy for the system. Bearing is the core system of many mechanical equipment, and it is also the most vulnerable to environmental damage. The bearing will suffer from excessive temperature and fatigue damage under repeated loads, the temperature distribution of the bearing will also affect the life of bearing in varying degrees[9]. So we assume that a bearing contains 16 balls, and it will be considered to be invalid if the number of deformed balls is greater than or equal to 7. Therefore, the system studied in this section is a *k/n*[*F*] system with $n = 16$, $k = 7$, and the life of each component is exponentially distributed. The bearing will work alternately in normal *Environment 0* with moderate load and low temperature, or in *Environment 1* with heavy load and high temperature. Basic assumptions of this case are shown in Table 1.

Table 1. Basic assumption parameter of the bearing system.

Parameter	Value
λ_0	1/5
Xmu_0	1/20
λ_1	1/3
Xmu_1	1/10
C_f	24 yuan/time
C_s	60 yuan/time
C_d	1.5 yuan/time
dmu	1/5
$LongT$	10, 000

The probability distribution diagram of bearing's life can be obtained by running program in MATLAB for 100, 000 times. Figure 4 shows the result.

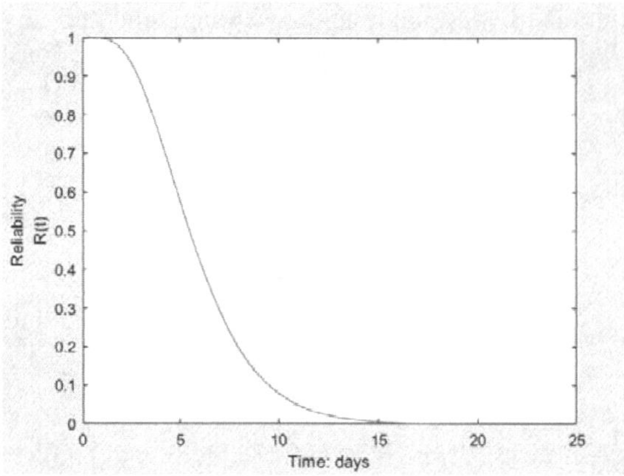

Figure 4. The life distribution of bearing system.

The specific optimization of the replacement policies are as follows:

Strategy I: The case is simulated according to the parameters, the average cost AC after simulation is 22.83 yuan / day.

Strategy II: Optimize the preventive replacement Strategy for the case. By enumerating different G, we can optimize the average cost of long-term operation and maintenance of the case. The enumeration interval of G is [1,6], and the cost distribution map is obtained as shown in Figure 5.

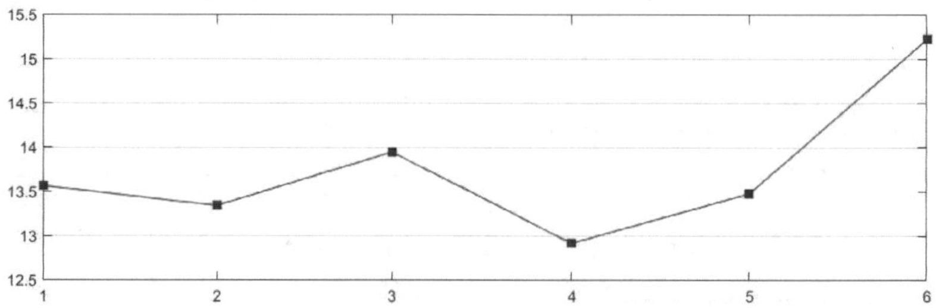

Figure 5. The cost distribution of preventive replacement strategy.

It can be seen that when the pre-maintenance threshold is set to 4, the average cost of long-term operation and maintenance of the system will reach the minimum cost, which is about 13.0 yuan / day. Compared with the Strategy I, the average cost of Strategy II is much lower. Therefore, when adopting Strategy II and set the pre-maintenance threshold $G = 4$, the purpose of maintenance strategy and cost optimization is achieved.

4. Conclusion

In this paper, the reliability of the *k-out-of-n: F* system in dynamic environment and the corresponding replacement Strategy optimization are studied. According to the state change caused by the number of parts failure, considering the change of the system in different working environment and the various costs during operation, the simulation and analysis of the bearing case are carried out with both replacement policies. The results show that the long-term cost of preventive maintenance Strategy is much lower than that of Breakdown replacement Strategy.

There are still many shortcomings in our study of multi-component systems. The research will consider the other possible aspects in reality, such as extension of the system state, extension of system structure and dependence between components.

References

[1] Palm C. Arbetskraftens fordelning vid betjaning av automatckiner[J]. Industritidningen Norden, 1947, 75: 90-94, 119-123.

[2] Coria V H, Maximov S, IVASDÁVALOSivasdavalo F, et al. Analytical method for optimization of maintenance Strategy based on available system failure data [J]. Reliability Engineering & System Safety, 2015, 135: 55-63.
[3] Jiawen Chen, Yuehua Cheng, Bin Jiang, et al. An In-orbit Reconfigurement Algorithm for Mission Constraint Oriented Spacecraft Attitude Control System [J]. Journal of Aerospace, 2017, 38(9): 989-997.
[4] Pham H, Wang H. Imperfect maintenance[J]. European Journal of Operational Research, 1996, 94(3): 425-438.
[5] Xiaokai GE, Jianbo Hu, Bofeng Zhang. Modelling of Degradation System Status Maintenance Strategy and Maintenance Activities [J]. Systems Engineering and Electronic Technology, 2013, 35(1): 102-108.
[6] Lotka A J. A contribution to the theory of self-renewing aggregates with special reference to industrial replacement [J]. Annals of Mathematical Statistics, 1939, 10: 110-130.
[7] Campbell N R. The replacement of perishable member of a continually operating system[J]. Journal of the Royal Statistical Society, 7(1941): 110-130.
[8] Franceschini F, Galetto M. A new approach for evaluation of risk priorities of failure models in FMEA[J]. International Journal of Production Research, 2001, 39(13): 2991-3002.

Design and Simulated Analysis of Low Heat Flux Cold Plate for High-power Power Electronics Equipment

Xunan He*, **YuanLi Kang and Peng Li**

Department of More Electric System, Beijing Aircraft Technology Research Institute, China

*Corresponding author email: xunanhe2008@126.com

Abstract. The thesis has carried out design and simulated analysis of low flux cold plate for high-power power electronics equipment. First, according to heat distribution and design requirements, two kinds of flow passage for low flux cold plate is designed. Then performance characteristics simulation of different flow passage for cold plate is performed. And the heat dissipation characteristics of two kinds of flow passage can both meet the design requirements. Finally, from the perspective of high temperature, temperature difference, pressure drop and so on, the simulated results are comparative analysis. And it is helpful to the design and optimizing of cold plate for high-power power electronics equipment.
Keyword: High-power power electronics equipment, low heat flux cold plate, low heat flux cold plate.

1. Introduction

With the development of more electric technology, there are more and more electrical equipment on civil aircraft, and electric power is also increasing. At the same time, its heat productivity is getting bigger and bigger, especially the calorific value of high-power power electronics equipment is also more and more[1,2]. And natural cooling or forced air cooling can't meet cooling demand, or it will affect the equipment's reliability and life. However, liquid cooling can effectively bring the heat produced by high-power power electronics equipment. And it can be achieved by cold plate. For civil aircraft, Boeing B787 adopts liquid cooling to cool the high-power power electronics equipment such as ATRU and CMSC. Therefore, the design of cold plate is the key to the heat dissipation of power electronics equipment. And it is very necessary to conduct performance characteristics simulation of cold plate during the phase of preliminary design. Meanwhile, it is also important to the optimization design of cold plate.

2. Cold plate for High-power Power Electronics Equipment Overview

Cold plate is a kind of heat exchanger that uses single cooling fluid. And it is a key cooling component for high-power power electronics equipment. The heat produced by power electronics equipment is brought by cooling fluid which flow through cold plate. The heat dissipation performance determines the power electronics equipment's working temperature, it directly affect the reliability and stability of power electronics equipment. And the design of flow passage for cold plate can also directly affect the heat dissipation performance.

Content from this work may be used under the terms of the Creative Commons Attribution 3.0 licence. Any further distribution of this work must maintain attribution to the author(s) and the title of the work, journal citation and DOI.
Published under licence by IOP Publishing Ltd

3. Design of Cold Plate for High-power Power Electronics Equipment

3.1. Design Requirement

The heat distribution of high-power power electronics equipment is shown as figure1.

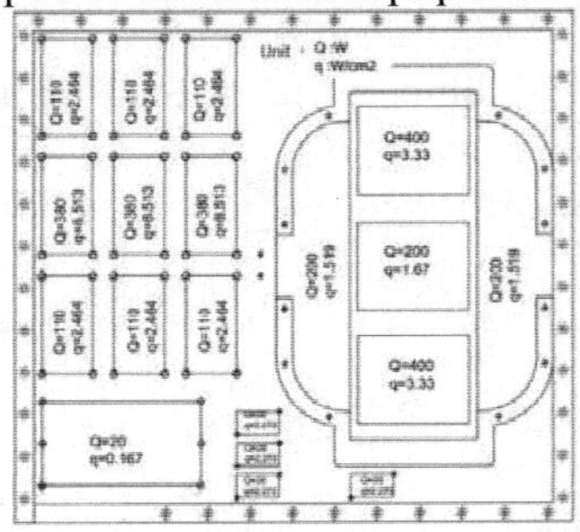

Figure 1. Heat distribution of high-power power electronics equipment.

The design requirements of low flux cold plate for high-power power electronics equipment are as follows:

- The heat loss for high-power power electronics equipment is about 3.3Kw. Among then, the heat loss of rectifying diode is about 1.8kW, the heat loss of transformer is about 1.4kW, the heat loss of resistance is about 800W, the heat loss of inductance is about 200W.
- The temperature of cooling fluid is from 9℃ to 30℃.
- The temperature should not more than 70℃ in a single cooling fluid cycle.
- The pressure drop of cooling fluid should not more than 10psi at rated flow.
- The highest temperature at normal working condition should not higher than 300℃.

3.2. Design of Flow Passage

According to the heat distribution and the design requirements of high-power power electronics equipment, A kind of serpentine flow passage and a kind of flow passage with tiny embedded fins are designed for the cold plate. The structures of flow passage are shown in figure 2 and figure 3. As is shown in figure 2, it is relatively sparse for ribs in the flow passage. In the right area of flow passage, the thickness is about 6mm, and the central distance for adjacent rib is about 12mm. In the left area of flow passage, the thickness is about 5mm, and the central distance for adjacent rib is about 15mm. the distribution of flow passage can cover all the heating components, and the height of the flow passage is about 14mm. As is shown in figure 2, the thickness of all fins is about 2mm, the central distance of adjacent fin is about 5mm, the height of fin is about 14mm. the arrangement of fin can form certain inclination, which can make the fluid velocity uniformly in the tiny flow passage. And the kind of flow passage have the outstanding superiority in heat dissipation. However, as the cooling fluid flow longer, the flow resistance is bigger and the pressure drop of import and export for cold plate is also bigger

Expendable material (such as water for use in water boilers) may be carried aboard the aircraft. And it is assumed that the rate of consumption of the expendable material is constant. There are two cases in the following:

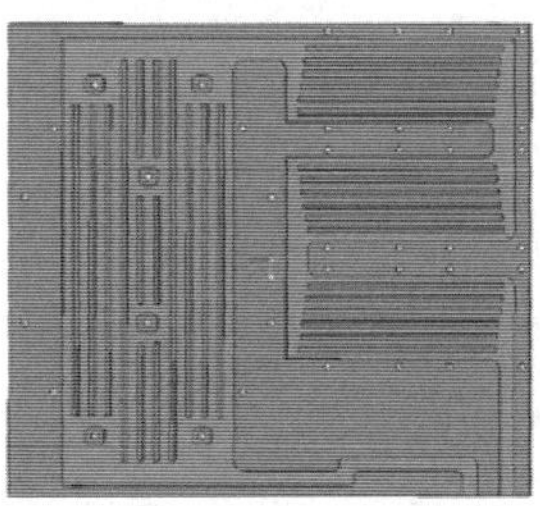

Figure 2. Flow passage 1.

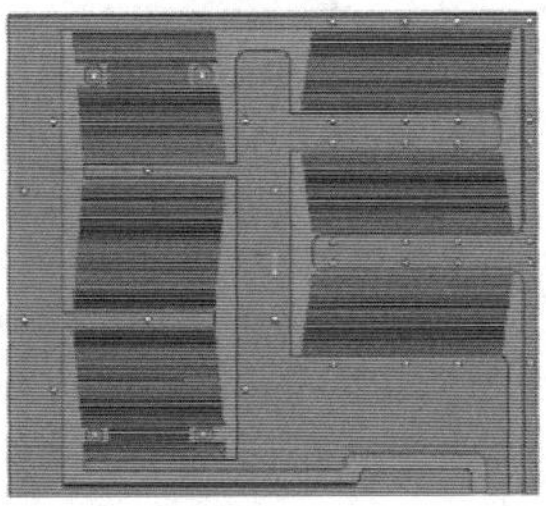

Figure 3. Flow passage 2.

4. Simulated Analysis

4.1. Simulated of Flow Passage 1 for Cold Plate

The heat dissipation characteristics have been simulated by the software Flow EFD. And the simulation conditions are as follow: the temperature of feed liquid is 9℃ and 30℃, and the flow of cooling fluid is 0.0004 m^3/s. when the temperature of feed liquid is 9℃, the simulated results are shown in figure 4 to figure 7. It can be obtained as follow:

- The highest temperature area of power electronics equipment is at rectifying diode, and the value is about 62.75℃.
- The highest temperature of cooling fluid is about 34.9℃, the temperature difference of import and export for cold plate is about 2.44℃.
- The highest velocity of cooling fluid is about 3.848m/s.
- The pressure drop of import and export for cold plate is about 3.12psi, which is less than 10psi.
- The total heat exchange amount is about 3.299kW, which is meet the heat dissipation requirement of 3.3kW.

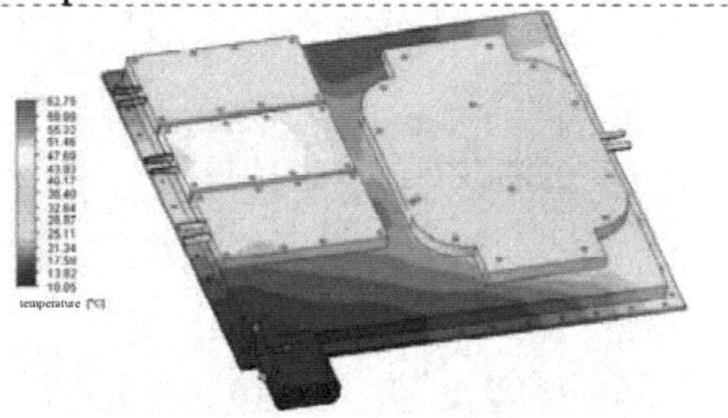

Figure 4. Temperature of cold plate.

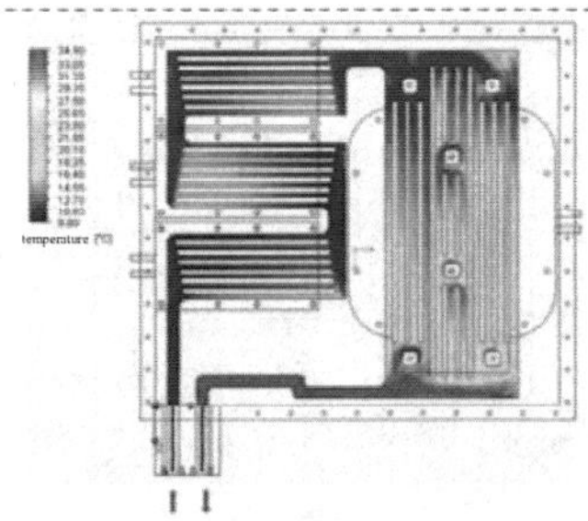

Figure 5. Temperature of cooling fluid.

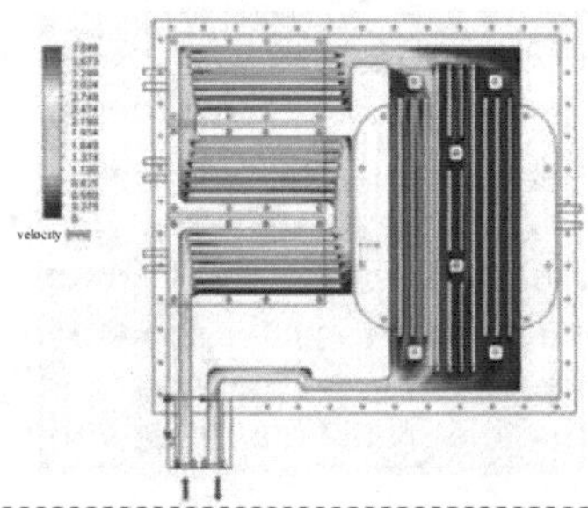

Figure 6. Velocity of cooling fluid.

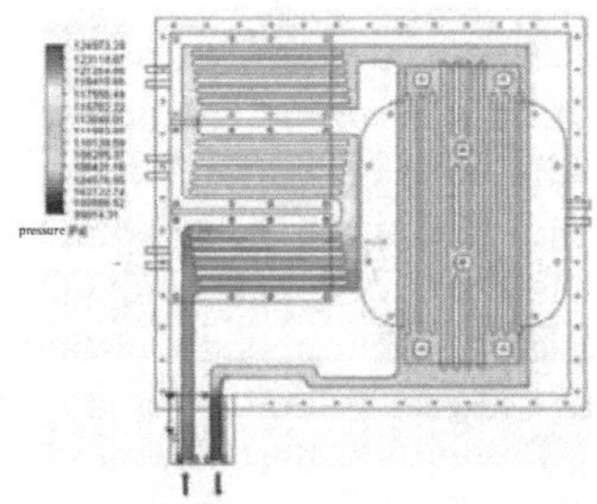

Figure 7. Pressure of cooling fluid.

when the temperature of feed liquid is 30℃, the simulated results are shown in figure 8 to figure 11. It can be obtained as follow:

- The highest temperature area of power electronics equipment is at rectifying diode, and the value is about 83.73℃.
- The highest temperature of cooling fluid is about 55.71℃, the temperature difference of import and export for cold plate is about 2.41℃.
- The highest velocity of cooling fluid is about 3.388m/s.
- The pressure drop of import and export for cold plate is about 2.19psi, which is less than 10psi.
- The total heat exchange amount is about 3.3kW, which is meet the heat dissipation requirement of 3.3kW.

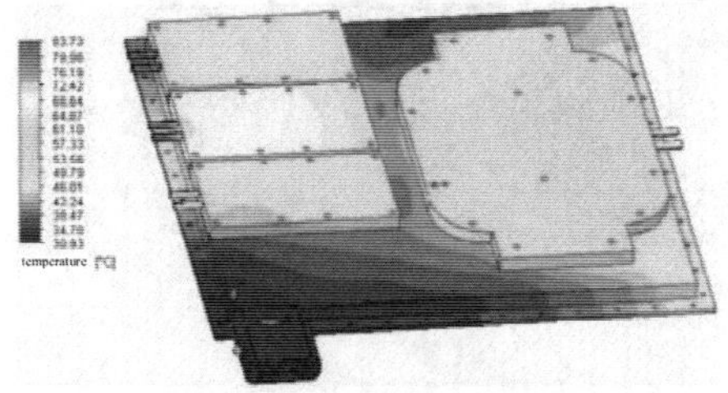

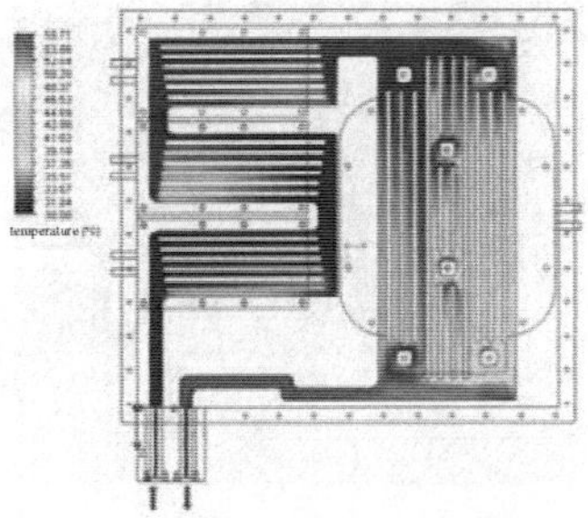

Figure 8. Temperature of cold plate. **Figure 9.** Temperature of cooling fluid.

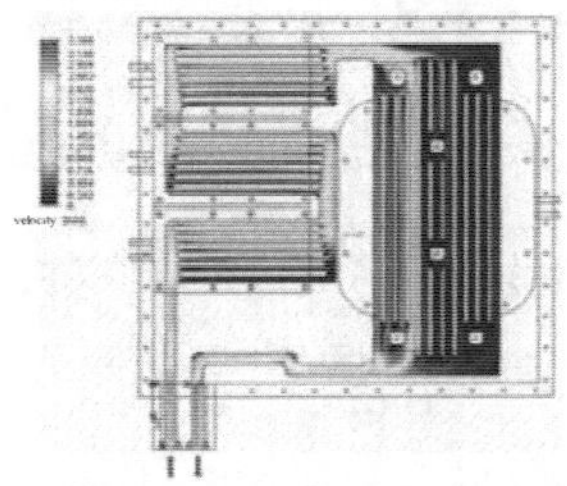

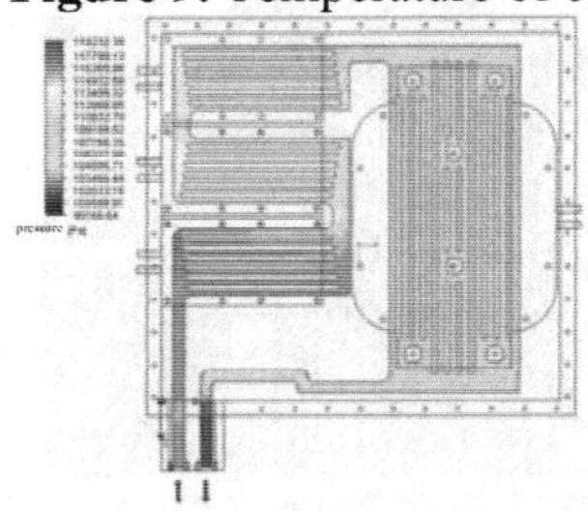

Figure 10. Velocity of cooling fluid. **Figure 11.** Pressure of cooling fluid.

The flow passage 1 can meet the design requirements, and when the temperature of cooling fluid at the import for cold plate increase from 9℃ to 30℃, the highest temperature area of power electronics equipment also increased from 62.75℃ to 83.73℃, which increased by 20.62℃, and the highest temperature of cooling fluid increased from 34.9℃ to 55.71℃, which increased by 20.81℃. Due to temperature increased and viscosity of cooling fluid decreased, the flow resistance also reduced from 3.12psi to 2.19psi, which decreased by 0.93psi.

4.2. Simulated of Flow Passage 2 for Cold Plate

The heat dissipation characteristics have been simulated by the software Flow EFD. And the simulation conditions are as follow: the temperature of feed liquid is 9℃ and 30℃, and the flow of cooling fluid is 0.0004 m^3/s. when the temperature of feed liquid is 9℃, the simulated results are shown in figure 12 to figure 15. It can be obtained as follow:

- The highest temperature area of power electronics equipment is at rectifying diode, and the value is about 55.62℃.
- The highest temperature of cooling fluid is about 28.27℃, the temperature difference of import and export for cold plate is about 2.41℃.
- The highest velocity of cooling fluid is about 3.934m/s.
- The pressure drop of import and export for cold plate is about 5.59psi, which is less than 10psi.
- The total heat exchange amount is about 3.29kW, which is meet the heat dissipation requirement of 3.3kW.

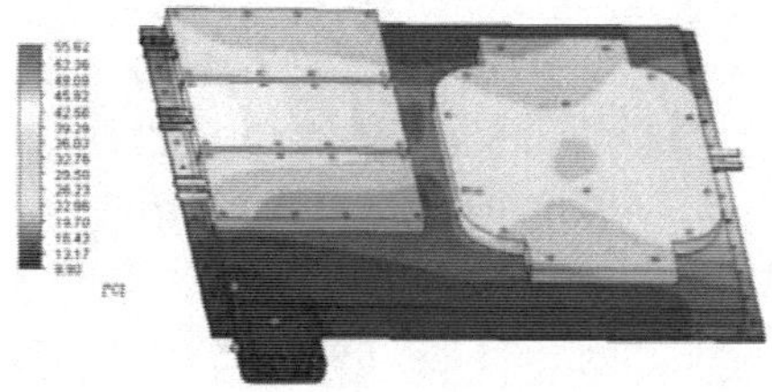

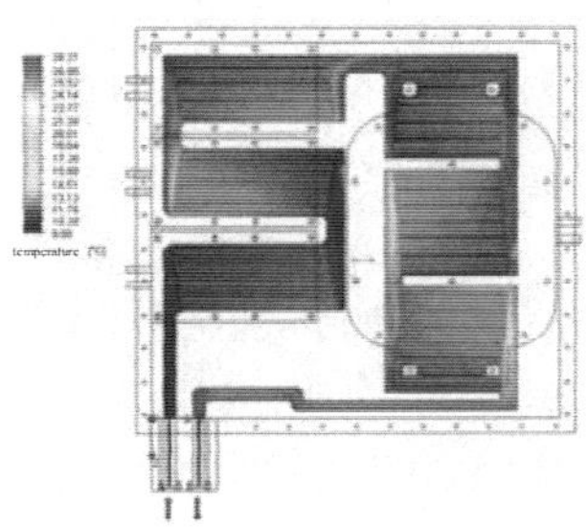

Figure 12. Temperature of cold plate. **Figure 13.** Temperature of cooling fluid.

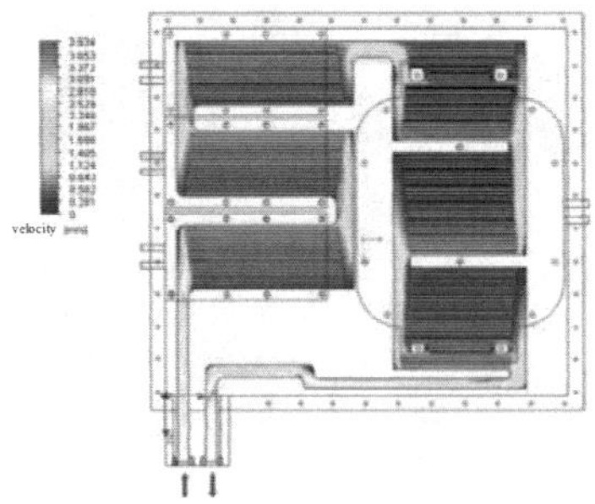

Figure 14. Velocity of cooling fluid.

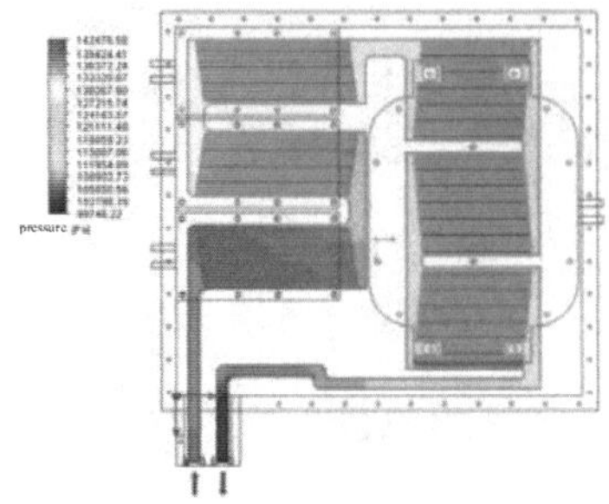

Figure 15. Pressure of cooling fluid.

when the temperature of feed liquid is 30℃, the simulated results are shown in figure 16 to figure 19. It can be obtained as follow:

- The highest temperature area of power electronics equipment is at rectifying diode, and the value is about 78.01℃.
- The highest temperature of cooling fluid is about 50.46℃, the temperature difference of import and export for cold plate is about 2.31℃.
- The highest velocity of cooling fluid is about 3.741m/s.
- The pressure drop of import and export for cold plate is about 4.01psi, which is less than 10psi.
- The total heat exchange amount is about 3.299kW, which is meet the heat dissipation requirement of 3.3kW.

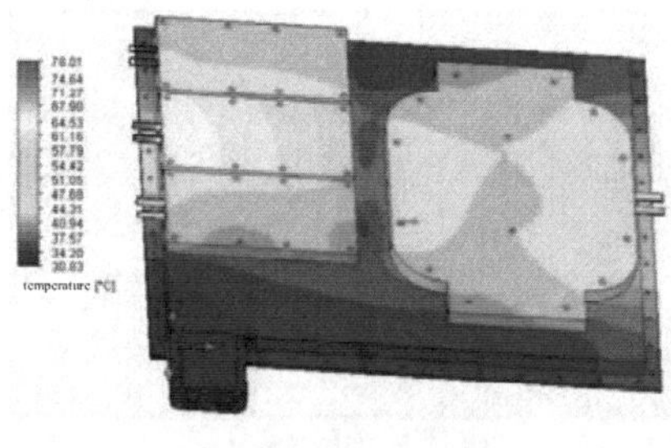

Figure 16. Temperature of cold plate.

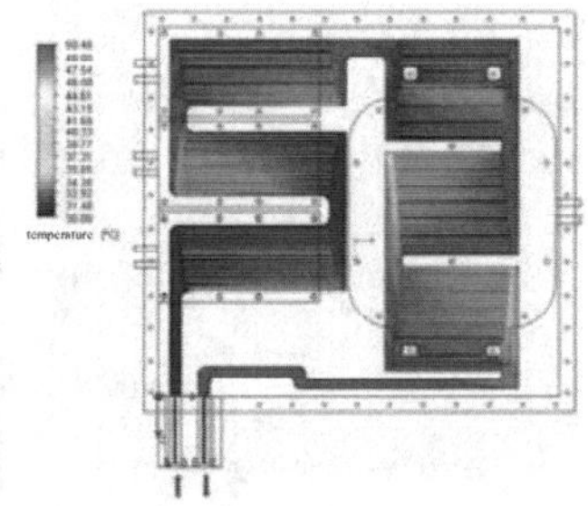

Figure 17. Temperature of cooling fluid.

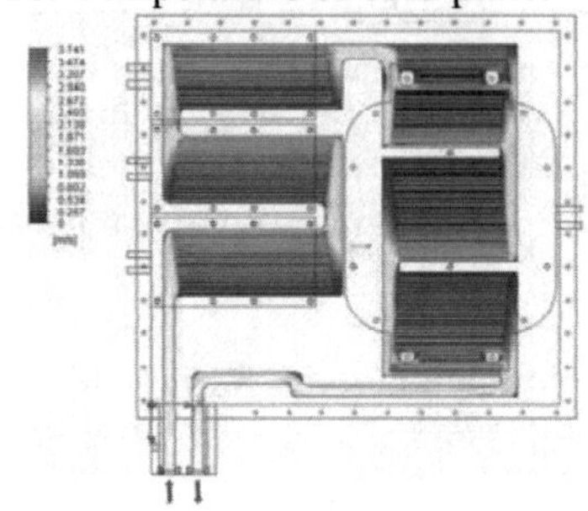

Figure 18. Velocity of cooling fluid.

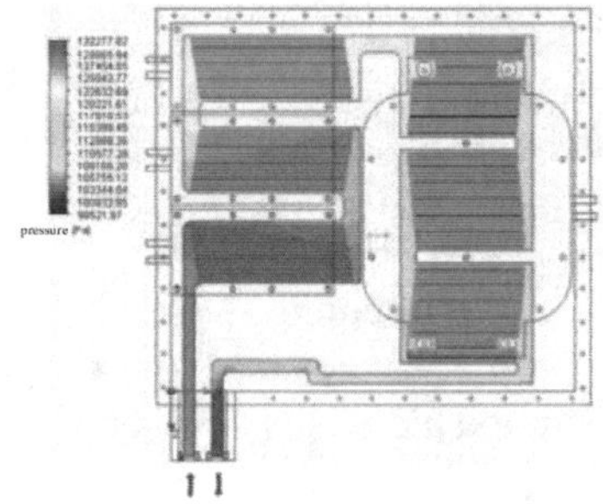

Figure 19. Pressure of cooling fluid.

The flow passage 2 can meet the design requirements, and when the temperature of cooling fluid at the import for cold plate increase from 9℃ to 30℃, the highest temperature area of power electronics equipment also increased from 55.62℃ to 78.01℃, which increased by 22.39℃, and the highest temperature of cooling fluid increased from 28.27℃ to 50.46℃, which increased by 22.19℃. Due to temperature increased and viscosity of cooling fluid decreased, the flow resistance also reduced from 5.59psi to 4.01psi, which decreased by 1.58psi.

5. Comparative Analysis

When the flow of cooling fluid is about $0.0004m^3/s$, and the temperature of cooling fluid at the import for cold plate is about 9℃ and 30℃ respectively, the simulated performance characteristics of two kinds of flow passage for cold plate are shown in table 1 and table 2. It can be seen from Table1 and Table 2, the heat dissipation capacity of flow passage 2 is stronger than flow passage 1. And the highest temperature of flow passage 2 is lower than flow passage 1. Although the pressure drop of import and export for flow passage 2 is higher than flow passage 1, the pressure drop of import and export for flow passage 2 can meet the design requirement of 10psi. Therefore, considering of high temperature, temperature difference, pressure drop, the flow passage 2 is better than flow passage1

Table 1. Comparative analysis of simulated performance for different flow passage.

	highest temperature of cold plate	highest temperature of cooling fluid	temperature difference of import and export for cold plate	highest velocity of cooling fluid m/s	pressure drop of import and export for cold plate psi
Flow passage 1	62.75	34.9	2.44	2.50	3.12
Flow passage 2	55.62	28.27	2.41	2.50	5.59

Table 2. Comparative analysis of simulated performance for different flow passage.

	highest temperature of cold plate	highest temperature of cooling fluid	temperature difference of import and export for cold plate	highest velocity of cooling fluid m/s	pressure drop of import and export for cold plate psi
Flow passage 1	83.73	55.71	2.41	2.5	2.19
Flow passage 2	78.01	50.46	2.31	2.5	4.01

6. Conclusion

The thesis has carried out design and simulated analysis of low flux cold plate for high-power power electronics equipment. The simulated performance results show that the serpentine flow passage and tiny embedded flow passage can both meet the heat dissipation requirement of cold plate for high-power power electronics equipment, but the heat dissipation performance of the tiny embedded flow passage is better than the serpentine flow passage. However, the simulated results also have its limitation, it can only be suitable for the preliminary and can not represent the actual heat dissipation capacity. In other words, the actual heat dissipation capacity can be obtained by performance test.

References

[1] G. Huang, Z. Min, L. Yang, P. X. Jiang, and M. Chyu 2018 Transpiration cooling for additive manufactured porous plates with partition walls, Int. J. Heat Mass Transf., vol. 124, pp. 1076–1087.

[2] Kandlikar S G and Hayner C N 2009 Liquid Cooled Cold Plates for Industrial High-Power Electronic Devices—Thermal Design and Manufacturing Considerations[J]. Heat Transfer Engineering, 30(12):918-930.

[3] David L Saums 2012 Liquid Cooling Systems for Electronics Thermal Management. IMAPS Chesapeake Chapter Symposium, 13-23.

[4] H. C. Chiu, R. H. Hsieh, K. Wang, J. H. Jang, and C. R. Yu 2017 The heat transfer characteristics of liquid cooling heat sink with micro pin fins," Int. Commun. Heat Mass Transf., vol. 86, no. June, pp. 174– 180.

Engine Fault Detection Approach Based on Angle Domain Signal Model

Hai Zeng, Ning Zeng, Jin Han and Yan Ding*

Chengdu Cigarette Factory, China Tobacco Sichuan Industry Co., Ltd, China

*Corresponding author email: dingyan830313@163.com

Abstract. Engine vibration signals include strong noise and non-stationary signals. By the time domain signal processing approach, it is hard to extract the failure features of engine vibration signals, so it is hard to identify engine failures. For improving the success rate of engine failure detection, an engine angle domain vibration signal model is established and an engine fault detection approach based on the signal model is proposed. The angle domain signal model reveals the modulation feature of the engine angular signal. The engine fault diagnosis approach based on the angle domain signal model involves equal angle sampling and envelope analysis of engine vibration signals. The engine bench test verifies the effectiveness of the engine fault diagnosis approach based on the angle domain signal model. In addition, this approach indicates a new path of engine fault diagnosis and detection.

1. Introduction

Engine is a kind of typical reciprocating and rotating machinery, featured by a complex structure, multiple excitation sources and unstable running. Therefore, engine failure diagnosis is always a difficulty in failure diagnosis. In 2003, Zunmin Geng put forward an engine vibration signal model, and extracted the failure feature of engine valve clearance by the wavelet packet decomposition approach [1]; A.P.Carlucci studied the relation between internal parameters and block surface vibration of IC engine by means of the conventional fourier transform approach and the time-frequency analysis approach [2]; S. Vulli extracted the combustion impact signal in multiple excitation sources of a three-cylinder engine by the short-time fourier transform approach [3]; Fuzhou Feng applied the bispectrum and genetic neural algorithm to diagnose piston pin failures in diesel engines [4]; Wang Chengdong classified the failure degree of engine valve clearance using WVD and the neural network [5]; Chi-man Vong diagnosed the engine ignition system failure using the wavelet packet transform and the K-means clustering algorithm [6]; Based on the statistical approach, S.BabuDevasenapati used the decision tree approach to diagnose and classify misfire failures of a four-stroke engine [7]; Liu Jianmin established a BP neural network model to locate and diagnose the misfire failure of a diesel engine [8]; Kamal Jafarian used linear and non-linear approaches to detect the misfire and valve clearance failures of a single-cylinder engine [9]; K.L.P Tharanga extracted the combustion impact signal from the multiple excitation sources of the diesel engine and uses the signal to detect the running state of the diesel engine [10].

Some progress has been made in engine failure diagnosis approaches for angle domain signals. Yujun Li proposed an empirical mode decomposition approach based on the angular domain signals and extracted the failure features of piston ring wear by this approach [11]; Yimin SHAO detected the failure of the connecting rod bearing clearance of a single-cylinder engine in the angular domain [12].

To sum up, engine vibration signals are mostly analyzed in the time domain or the frequency domain in conventional engine failure diagnosis approaches, while the conventional angle domain analysis approach is also susceptible to strong noise and multiple excitation sources. The failure features of the

Content from this work may be used under the terms of the Creative Commons Attribution 3.0 licence. Any further distribution of this work must maintain attribution to the author(s) and the title of the work, journal citation and DOI.
Published under licence by IOP Publishing Ltd

engine vibration signal, as a kind of non-stationary signal with strong noise, cannot be extracted accurately by the conventional analysis and processing approaches. For more accuracy of engine failure diagnosis, an engine angle domain vibration signal model is founded in this paper, based on which a new engine fault detection approach is proposed. The angle domain signal model reveals the modulation feature of the engine angle domain signal. The failure detection approach involves equal angle sampling and envelope analysis of engine vibration signals. The failure bench test of the connecting rod bearing clearance has verified the effectiveness of this algorithm, and some valuable theories are obtained, which provides a new path for the research of engine failure detection and diagnosis approach.

2. Engine Angle Domain Signal Model Based on Order Feature

When the engine runs normally, the vibration signals acquired by the sensor are composed of vibration signals of internal engine assemblies and vibration signals transmitted from other machinery or equipment. The components which are irrelevant to the vibration signal of internal assemblies can be regarded as noise signals. Therefore, the basic model of engine vibration signal can be represented as:

$$s(\theta) = x(\theta) + r(\theta), \theta = 1,2,3,\ldots,N \tag{1}$$

Where $s(\theta)$ are signals acquired by the sensor, $x(\theta)$ are signals of engine internal assemblies, $r(\theta)$ are noise signals.

Due to the speed fluctuation, the engine vibration signals in the time domain are non-stationary, and such signals are not periodic. In the equal angle sampling, the crankshaft angle position signal can be accurately repeated in each revolution, which is a deterministic signal. Therefore, engine vibration signals are transformed from non-stationary signals in the time domain to stationary signals in the angular domain, eliminating the influence of random fluctuations in crankshaft speed [12].

According to the principle of equal angle sampling, the angular domain signals of engine internal assemblies can be perceived as stationary signals, which can be expressed by the following function:

$$x(\theta) = \sum_{m=0}^{M} x_m \cos\left(\frac{2\pi\theta}{360} + \varphi_m\right) \tag{2}$$

Where x_m is vibration amplitude at point m of angle domain vibration signal, φ_m is vibration phase at point m of angle domain vibration signal, θ is crankshaft rotation angle.

When an engine operates, its internal assemblies (such as connecting rod bearing assembly and crankshaft assembly) will periodically collide with each other, thus generating periodic impact signals. Such impact signals often excite the natural frequency of the internal assemblies, which are modulated by the natural frequency of the system to generate modulated vibration signals of the engine. The vibration signals of the internal assemblies can be represented as:

$$x(\theta) = \sum_{n=0}^{N} A_n \cos\left(\frac{2\pi O_n n\theta}{360} + a_n\right)\sin\left(2\pi O_c \frac{\theta}{360}\right) \tag{3}$$

Where A_n is amplitude of order-n component of envelope signal, a_n is phase of order-n component of envelope signal, O_n is order component of envelope signal, O_c is order component of carrier signal.

Vibration signals of engine internal assemblies mainly include: gas impact signals, impact signals caused by reciprocating inertial force of piston, crankshaft rotation signals, piston knock signals, valve seating impact signals, impact signals generated by collision of lubricating oil with crankshaft case, impact signals of discharge valve fluid, etc. In this paper, gas impact signals and crankshaft rotation signals of engines are mainly considered, and other signals are regarded as noise signals. Thus, the angle domain vibration signal model in the normal state of an engine can be established as:

$$s(\theta)=\sum_{n=0}^{N}A_n\cos[\frac{2\pi(O_r+O_q)n\theta}{360}+a_n]\sin(2\pi O_c\frac{\theta}{360})+r(\theta) \tag{4}$$

Where O_r is order component of gas impact signal, O_q is order component of crankshaft rotation signal.

In case of an engine failure, the component of the vibration signal due to the failure should also be considered. Therefore, in the failure state, the engine angle domain vibration signal model can be represented as follows:

$$s(\theta)=\sum_{n=0}^{N}A_n\cos[\frac{2\pi(O_r+O_q+O_f)n\theta}{360}+a_n]\sin(2\pi O_c\frac{\theta}{360})+r(\theta) \tag{5}$$

Where O_f is order component of vibration signal caused by failure.

3. Fault Detection Approach Based on Angle Domain Signal Model

The engine fault detection approach based on angle domain signal model is shown in figure 1.

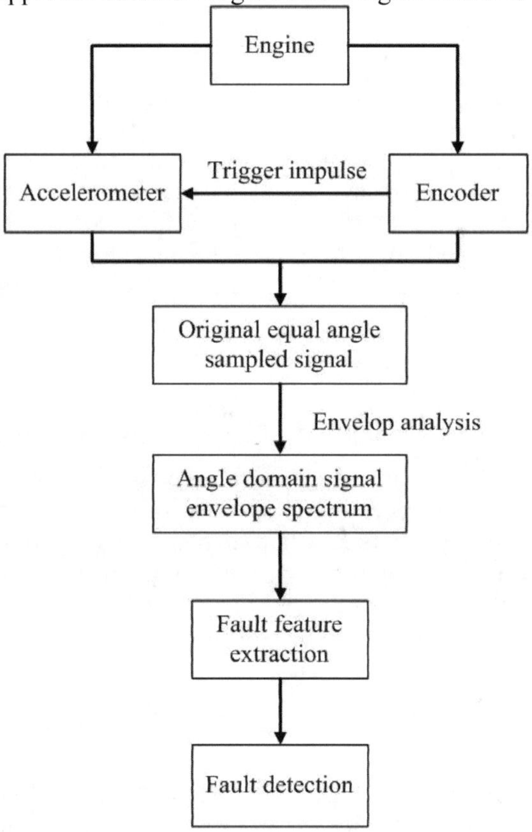

Figure 1. Engine fault detection approach based on angle domain signal model.

3.1. Equal Angle Sampling

According to the principle of equal angle sampling, as demonstrated in figure 2, the crankshaft rotation angle is taken as the sampling signal, the crankshaft angle position signal can be accurately repeated in each revolution. Therefore, engine vibration signals are transformed to stationary signals in the angular domain from non-stationary signals in the time domain, eliminating the influence of random fluctuations in crankshaft speed [12].

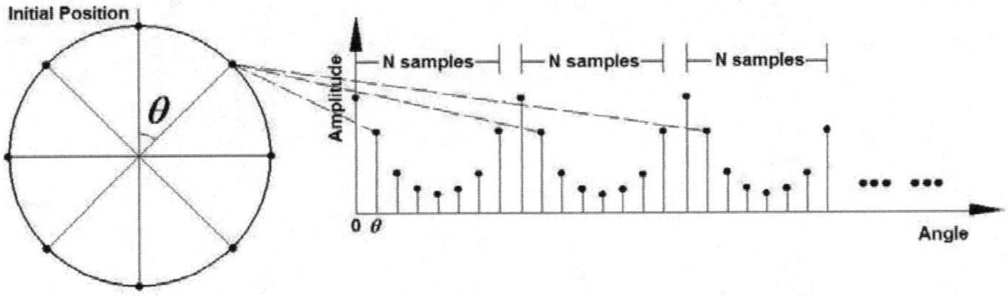

Figure 2. Schematic diagram of equal angle sampling.

3.2. Envelope Analysis

According to the engine angle domain vibration signal model, the engine vibration signals have obvious modulation features. The internal impact signals include gas impact signal, crankshaft rotation signal, vibration signal caused by failure, etc., which will be modulated by the natural frequency of the system, thus generating the modulated vibration signals of the engine, as shown in formula (5). Therefore, the envelope signals of engine vibration signals include gas impact signals, crankshaft rotation signals and vibration signals caused by failures, and the envelope analysis and the envelope spectrum can be used to extract and identify the vibration signals, including gas impact signals, crankshaft rotation signals and vibration signals caused by failures, so as to detect the operating state of the engine.

4. Test Results and Discussion

4.1. Engine Bench Test

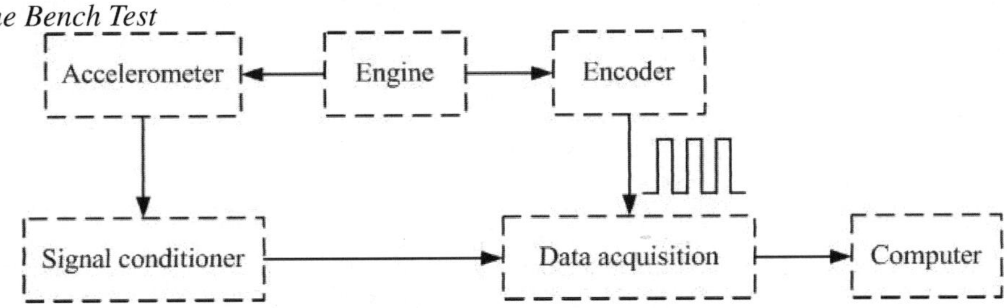

Figure 3. Signal acquisition system.

The data acquisition system is shown as figure 3. The system includes RT160 single-cylinder engine, accelerometer, optical encoder, signal conditioner, data acquisition card and computer. The optical encoder is set in the front of the crankshaft, together with the rotation of the crankshaft. The encoder can trigger 360 electrical impulses per cycle, which are regarded as the triggered signal, a sample is obtained when a pulse is triggered, and equal angle sampling will be realized. The monitoring point lies on the crankshaft bearing pedestal, with the piston moving in the same direction, and the failure setting is a fit clearance between the connecting rod and the crankshaft journal, as demonstrated in figure 4. The test conditions are detailed in table 1.

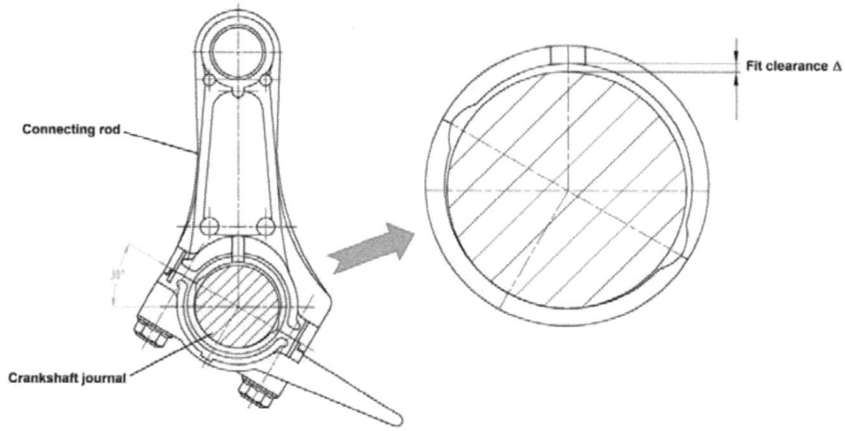

Figure 4. Drawing of connecting rod bearing with fit clearance.

Table 1. Test conditions.

Operating Condition	Δ (mm)	Rotational Speed (r/min)	Load ($N \cdot m$)
Normal	0.04	1800	0
Fault #1	0.07	1800	0
Fault #2	0.13	1800	0
Fault #3	0.19	1800	0
Fault #4	0.28	1800	0

4.2. Signal Analysis under Normal Operating Conditions

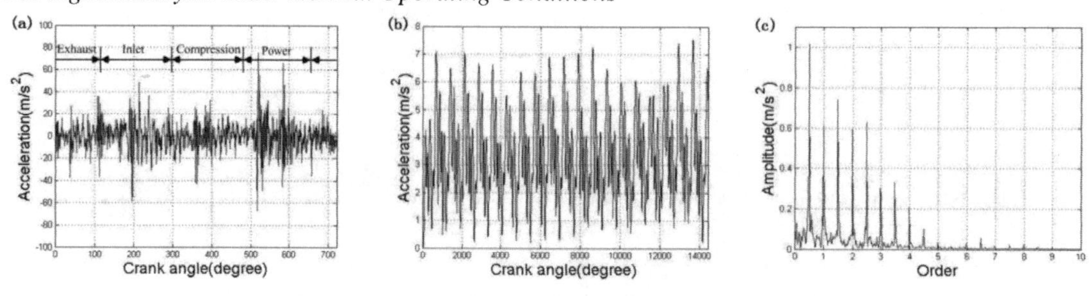

Figure 5. Analysis results under normal condition:
(a) Original signal; (b) Envelope Signal; (c) Envelope spectrum.

Figure 5 shows the results of engine angle domain vibration signals and envelope analysis under normal operating conditions. Figure 5 (a) shows the original signal in angle domain. Because the engine vibration signal is featured by strong noise and signal modulation, the impact signal caused by the fit clearance of connecting rod bearing cannot be accurately identified only from the signal itself. Figure 5 (b) shows the envelope signal of the angle domain signal, and figure 5 (b) shows that the envelope signal of the angle domain signal has a relatively obvious periodic component, and an order analysis can be performed on the signal. Figure 5 (c) shows the envelope spectrum of the angle domain signal, and figure 5 (c) indicates that the signal mainly contains two components of 0.5-order and 1-order. Among them, the 0.5-order corresponds to the order component of the gas impact signal, which characterizes once occurrence every two revolutions of the crankshaft; and the 1-order corresponds to the order component of the impact signal caused by the fit clearance of the connecting rod bearing, which characterizes once occurrence every revolution of the crankshaft. In figure 5 (c), the peak value of 1-order component is very clear, which can be used to extract the signal feature of the connecting rod bearing failure and detect the failure of the engine.

4.3. Comparison with Time domain Processing Approach

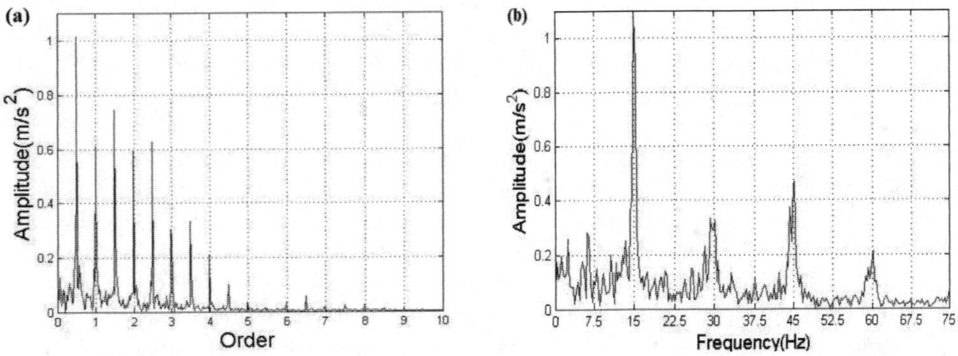

Figure 6. Comparison between the new approach and the time domain processing approach:
(a) Envelope spectrum of angle domain signal; (b) Envelope spectrum of time domain signal.

The comparison of analysis results between the new method and the traditional time domain processing method under normal working conditions is shown in figure 6. Figure 6 (a) shows the envelope spectrum of angle domain signal, and figure 6 (b) shows the envelope spectrum obtained by using the traditional time domain signal analysis method. According to figure 6 (b), this signal mainly contains frequency components of 15 Hz and 30 Hz, respectively corresponding to the gas impact signal and the impact signal caused by the connecting rod bearing clearance when the crankshaft rotates at 1800 r/min. However, due to the fluctuation of engine speed, obvious frequency smears appear in the envelope spectrum of time domain signal, especially at 30 Hz, which makes it difficult to identify engine failure features. Compared to traditional analysis methods, the 0.5-order and 1-order components in figure 6 (a) are very clear, which helps to marking up the success rate of engine failure monitoring. Therefore, compared to the traditional time domain processing approach, the engine failure monitoring approach based on angle domain signal model has more advantages in failure feature recognition and extraction.

4.4. Analysis Results under Different Operating Conditions

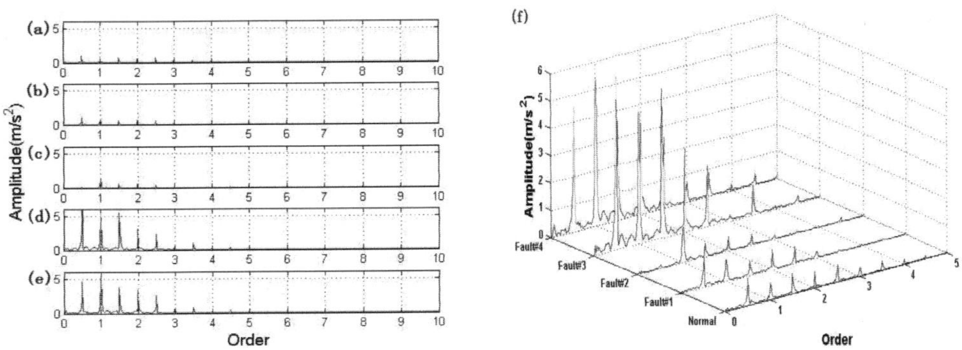

Figure 7. Results comparison under five conditions:
(a) Normal;(b) Fault #1;(c) Fault #2;(d) Fault #3;(e) Fault #4; (f) Waterfall spectrum.

The comparison of engine angular signals under five operating conditions is indicated in figure 7. Figure 7 indicates that the 1-order component caused by the fit clearance of the connecting rod bearing appear clearly in the analysis results of the five operating conditions, and with the increase of the fit clearance, the impact signal energy generated by the fit clearance increases gradually, and the amplitude of the 1-order component also gradually increases. Therefore, the signal processing approach based on the angle domain vibration signal model can effectively obtain the information of engine failure, and the 1-order component in the analysis results can also be regarded as the basis of failure detection. By extracting this order component, the failure of the connecting rod bearing of the engine can be monitored.

5. Conclusion

Engine vibration signals contain strong noise and non-stationary signals. With the time domain signal processing approach, it is hard to extract the failure features of engine vibration signals, so it is hard to detect and diagnose engine failures.

The engine fault detection approach based on angle domain signal model is presented in this paper. The engine angle domain signal model reveals the stationary signal feature of engine angle domain signals and expounds the obvious modulation feature of engine vibration signals. Angle domain signal model takes into consideration the engine failure information and abundant information generated during the operation and can be taken as the theoretical basis for failure monitoring.

With the new fault detection approach, the gas impact signal as well as the impact signal caused by the fit clearance of the connecting rod bearing can be effectively extracted. Compared to the traditional time domain signal processing approach, this new fault detection approach can effectively obtain the engine failure information, thus increasing the success rate of engine failure monitoring.

The engine bench test has proved the effectiveness of the engine fault detection approach based on the angle domain signal model.

References

[1] ZunminGeng, Jin Chen and J.Barry Hull 2003 Analysis of engine vibration and design of an applicable diagnosing approach International Journal of Mechanical Sciences (vol 45) pp 1391-1410.

[2] A.P.Carlucci, F.F.Chiara and D.Laforgia 2006 Analysis of the relation between injection parameter variation and block vibration of an internal combustion diesel engine Journal of Sound and Vibration (vol 295) pp 141-164.

[3] S.Vullia, J.F.Dunne and R.Potenza 2009 Time-frequency analysis of single-point engine-block vibration measurements for multiple excitation-event identification Journal of Sound and Vibration (vol321) pp 1129-1143.

[4] Fuzhou Feng, Aiwei Si and Hongxing Zhang 2011 Research on fault diagnosis of diesel engine based on bispectrum analysis and genetic neural network Procedia Engineering (vol 15) pp 2454-2458.

[5] Wang Chengdong, Zhang Youyun and Zhong Zhenyuan 2008 Fault diagnosis for diesel valve trains based on time–frequency images Mechanical Systems and Signal Processing (vol 22) pp 1981 -1993.

[6] Chi-man Vong, Pak-kin Wong and Weng-fai Ip 2011 Case-based expert system using wavelet packet transform and kernel-based feature manipulation for engine ignition system diagnosis Engineering Applications of Artificial Intelligence (vol 24) pp 1281-1294.

[7] S.BabuDevasenapati, V.Sugumaran and K.I.Ramachandran 2010 Misfire identification in a four-stroke four-cylinder petrol engine using decision tree Expert Systems with Applications (vol 37) pp 2150-2160.

[8] Liu Jianmina, Li Xiaolei and Zhang Xiaoming 2011 Misfire diagnosis of diesel Engine based on rough set and neural network Procedia Engineering (vol 16) pp 224-229.

[9] Kamal Jafarian, MohammadsadeghMobin and Ruholla Jafari-Marandi 2018 Misfire and valve clearance faults detection in the combustion engines based on a multi-sensor vibration signal monitoring Measurement (vol 128) pp 527–536.

[10] K.L.P Tharanga, Shuyong Liu and Shuai Zhang 2020 Diesel Engine Fault Diagnosis with Vibration Signal Journal of Applied Mathematics and Physics (vol 8) pp 2031-2042.

[11] Yujun Li, Peter W.Tse and Xin Yang 2010 EMD-based fault diagnosis for abnormal clearance between contacting components in a diesel engine Mechanical Systems and Signal Processing (vol 24) pp 193-210.

[12] Yimin SHAO, Yan Ding and Chris K. Mechefske 2013 Engine Fault Detection using Angle Domain Signal Envelope Algorithm Proceedings of the Institution of Mechanical Engineers, Part I: Journal of Systems and Control Engineering (vol 227) pp 541–551.

Cross-platform Simulation of Bipolar Junction Transistor Electrical Principle

Hongjun Wu*, Bangzheng Yin and Zetao Chen

Guangzhou Railway Polytechnic, Guangzhou, China

*Corresponding author email: wuhongjuntry@126.com

Abstract. This paper designs a set of simulation method about bipolar junction transistor (short for 'BJT') electrical principle, which can be used as the foundation of cross-platform simulation of semiconductor electronic technology. Firstly, a mathematical model is established to simulate the amplification effect of bipolar junction transistor current control, the influence of Base current and Collector-Emitter Voltage on the working area of bipolar junction transistor. Secondly, interactive multimedia such as graphics, images and animation are used for simulation presentation through cross-platform software design method which is the mainstream internet web page interaction technology. Finally, a logic structure of the whole simulation software is designed, which presents data transmission, control workflow, relationship between each class and each layer in the simulation software.

1. Background Introduction

Semiconductor technology is the cornerstone which supports the development of modern information technology industry in recent decades. Semiconductor devices have become an indispensable part of the electronic, electrical and chip industries. High integration, multi-function and low power consumption are the major goals of semiconductor technology devolopment, which is driven by the demand of modern information technology industries such as computer, internet and network communication. The application of simulation technology plays a more and more important role in the research and development of semiconductor technology and electronic products. For example, simulation is needed to verify electrical principle, reduce physical experiments and improve the efficiency of design[1] [2]. But there are the following problems in most semiconductor and electronic simulation software.

- It is difficult to share simulation data and files because there are barriers between different simulation systems;
- It is difficult to realize cross-platform simulation because high requirements of software and hardware for the running environment are necessary in most simulation systems;
- It is not easy for primary users to understand its principle becasue the human-machine interface(HMI) of most simulation softwares are too rigid and lack of interaction;
- The users can't directly observe the dynamic change trend of the simulation object becasuse the simulation results of most simulation softwares are displayed in tabular data or two-dimensional pictures and lack of dynamic effects;

BJT is the most basic component in semiconductor technology. This paper introduces a cross-platform simulation method for the electrical characteristics of BJT based on new generational multimedia technology. By the simulation method, the simulation system can be logged in and run easily as long as the users have the basic terminal equipment (such as PC, smartphone or tablet) on line. When using

Content from this work may be used under the terms of the Creative Commons Attribution 3.0 licence. Any further distribution of this work must maintain attribution to the author(s) and the title of the work, journal citation and DOI.
Published under licence by IOP Publishing Ltd

the simulation system, the users log in by website and do not have to install any software or plug-ins, which greatly facilitates users, improves the utilization efficiency and expands the application of the simulation system.

2. The Modeling of Electrical Principle about BJT

An accurate and appropriate model is a good foundation for a simulation system or software[3]. In order to make the simulation closer to the actual, this part will introduce the internal structure of BJT, the current-voltage relationship between electrodes and working areas in detail. And then present the mathematical model to discribe the electrical principle of BJT accurately.

2.1. Introdution of BJT

The dominated intrinsic semiconductor materials are silicon and germanium at present. Blend in intrinsic semiconductors with trivalent or pentavalent elements, which can produce impurity semiconductors with higher conductivity, i.e. N-type and P-type semiconductors. BJT is composed of three blocks with different types of impurity semiconductors, which lead out three wires and named Emitter, Base and Collector. It can be divided into NPN-type BJT and PNP-type BJT according to different composition structure. Each BJT contains two PN junctions and three blocks of imputity semiconductors [4] [5]. The structures are shown as the figure 1 and figure 2.

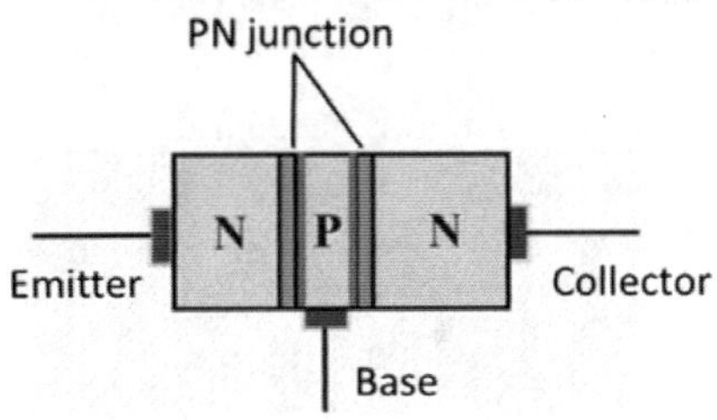

Figure 1. NPN-type BJT.

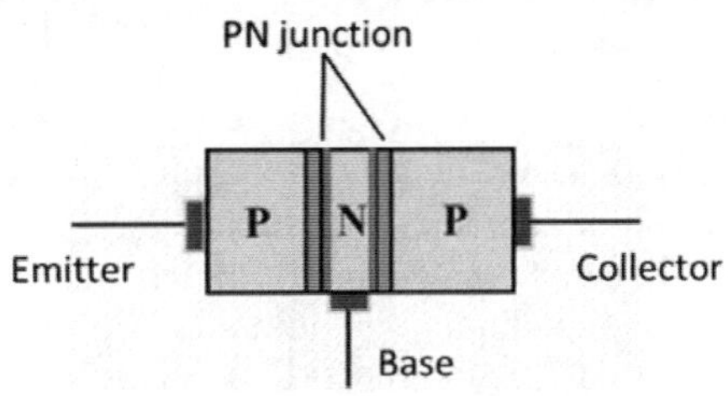

Figure 2. PNP-type BJT.

The main electrical characteristics of BJT are as following: when working in the amplified area, the relationship between Base current I_B and Collector current I_C is linear amplification; when working in the saturated area, Collector and Emitter are connected; when working in the cut-off area, Collector and Emitter are disconnected from each other. Figure 3 is a typical current-amplification circuit with NPN-type BJT and also is the simulation circuit in this paper.

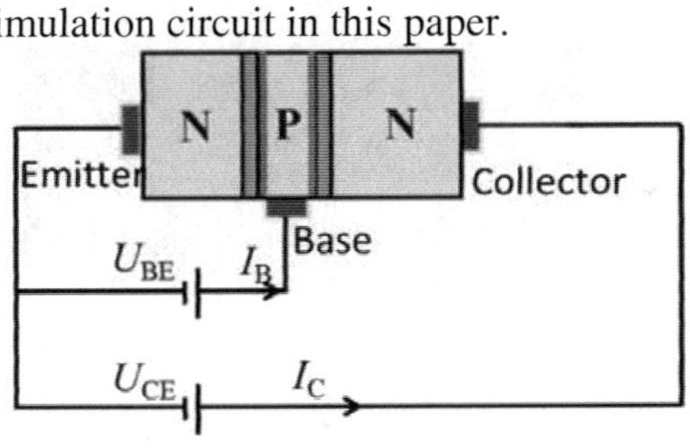

Figure 3. A typical current-amplification circuit with NPN-type BJT.

The current-voltage relationship between electrodes and working areas in detail are shown in table 1.

Table 1. The current-voltage relationship between electrodes.

Working areas	Electrical characteristics	Current-voltage conditions
amplified area	linear amplification between I_B and I_C	$U_{CE}>U_{BE}>U_{BE\text{-}D}$
saturated area	Collector and Emitter connected	$U_{BE}> U_{BE\text{-}D}$, $U_{BE}>U_{CE}$
cut-off area	Collector and Emitter disconnected	$I_{BE}=0$, $U_{BE}\leqslant 0$

2.2. *Modeling of BJT Electrical Priciple*

The voltage and current data of circuit in figure 3 are measured accurately in experiments to describe the electrical principle and characteristics of BJT, which generates the input-characteristic curve and output-characteristic curve with the voltage and current data. The input-characteristic curve is shown as figure 4 to descript the relationship between U_{BE} and I_B. The output-characteristic curve is shown as figure 5 to descript the relationship between U_{CE} and I_C.

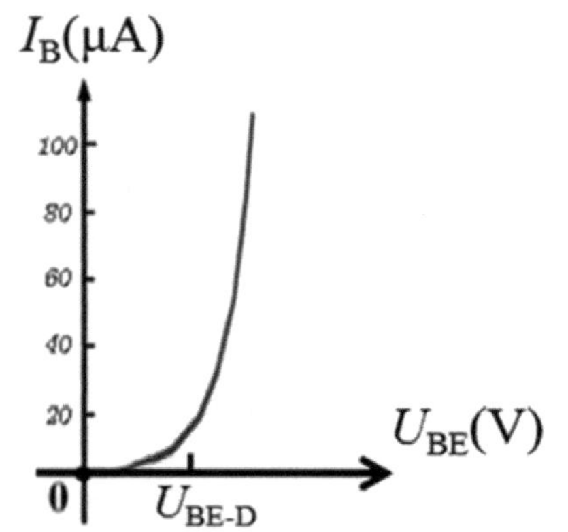

Figure 4. The input-characteristic curve.

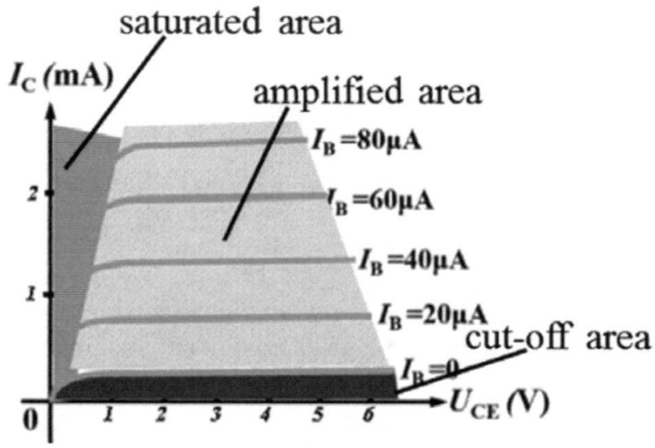

Figure 5. The output-characteristic curve.

It can be clearly observed that the relationship between U_{BE} and I_B can be described by exponential function in the input-characteristic curve through observation and data fitting. While the relationship between I_B and I_C can be described by linear function when BJT works in amplified area of the output-characteristic curve.

$$I_B = I_0 * \exp(U_{BE}/U_0) \quad (1)$$
$$I_C = \beta * I_B \quad (2)$$

The factors I_0 and U_0 are constants. And the factor β is a current amplification constant. All the constant factors are basically related to the specification of BJT, but also affected by temperature slightly.

3. The Cross-platform Simulation Software Design of BJT Electrical Principle

In this paper, the cross-platform simulation software designment adopts technical solutions of dynamic web technology with HTML5 and JavaScript in order to provide cross-platform and interactive features. The whole software system can be devided into five logic layers: basic pictures layer, graphic-image control layer, electrical principle layer, interaction-presentation layer and users layer. There are distinct functional boundaries and interfaces between each layer, and there are also corresponding interfaces between different class modules in the same layer. The purpose of special software designment is to ensure the low coupling between all class modules or all layers and improve each class module's universality and reusability.

The logic structure of the whole simulation software is shown as figure 6, which also presents the data transmission and workflow of the system. The HTML5 drawing tag <canvas> and specific object CanvasRenderingContext2D are the core of interaction-presentation layer, which receive command messages from the users' mouse and dynamically draw graphics and images to the users. The electrical principle layer and graphic-image control layer are primarily composed of six key class modules which are Base class, Collector class, Emitter class, Control buttons class, Current amplification class and Working areas class. The six key classes contain the core algorithm to simulate the BJT electrical principle, which compute the most important electrical parameters and transmit them to each other.

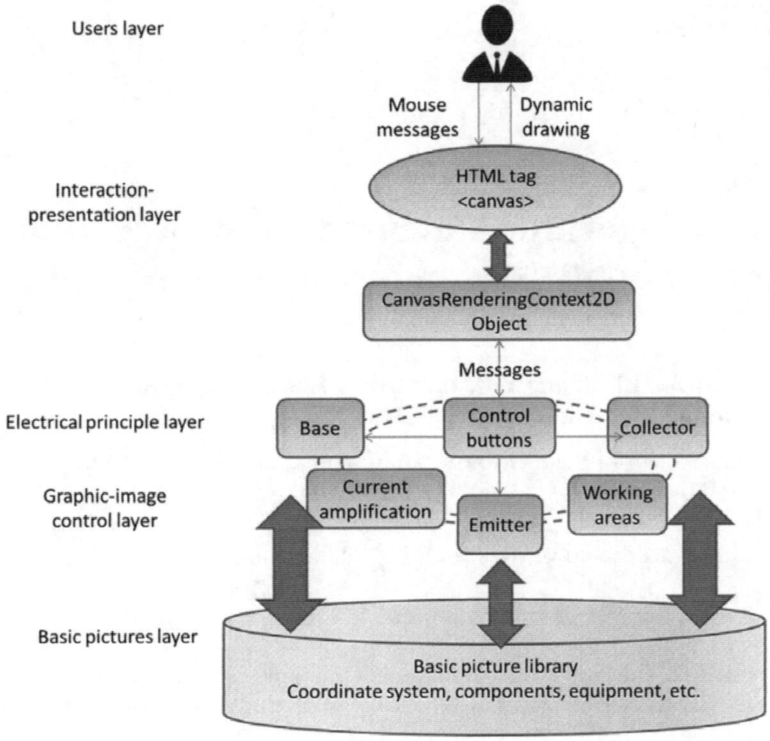

Figure 6. The logic structure of the whole simulation software.

4. Conclusion

This paper explores a set of method to simulate BJT electrical principle and designs a cross-platform software to provide simulation services which include current amplification, control of electrical parameters, visualization of electrical parameters and switching of BJT working area. It enables the users to access simulation services quickly and conveniently because the simulation software adopts cross-platform designment. The users can use simulation software just by browsing interactive websites through internet with basic hardware devices such as personal computers or smart phones. The simulation method and software are verified to match the BJT electrical principle and relative parameters accurately through experiment.

Acknowledgments

The work of this paper was supported by the 2021 Characteristic and innovative science research project of Guangdong Provincial Department of Education (Project No.2021KTSCX270).

References

[1] WU Yuanhao and TIAN Ling 2015 *Semantic-based simulation files cross-platform accessing method*(Computer Integrated Manufacturing Systems Vol.21)p1771-1780

[2] XUE L and TIAN I 2008 *Research on simulation process management for product development*(Proceedings of IEEE International Colloquium on Computing, Communication, Control and Management)p33-36

[3] LIU Hai-tao, GAN Ping, HUANG Yang-fan and WEN Zhi-yu 2014 *Method of Parameter Extraction and Experiment for Bipolar Junction Transistor EM Large-signal Model* (Research and Exploration in Laboratory vol 33)p62-65

[4] Hu Rong 2019 *Parameter analysis of inverted working state of bipolar type crystal transistor*(Electronic Test vol 19)p56-57

[5] Zeng Ling-qin and Chen Wei-ke 2019 *Semiconductor foundation and common devices*(electronic technology Fundamentals) chapter 1 pp 18–22

Modeling and Algorithm Research on Identification of Wrong Wiring Power Supply Region based on Classification Analysis

Zongpeng Li[1], Liang Yu[2], Jian Liu[3,*], Liangli Su[1], Huang Tan[4], Jiakui Zhao[1], Xiaojing Lin[1], Hui Cheng[1] and Hao Zhang[1]

[1]Data Center of State Grid Corporation of China, Beijing, China
[2]CHINA GRIDCOM CO. LTD, ShenZhen, China
[3]State Grid Information Telecommunication Group Co. Ltd., Beijing, China
[4]China Electric Power Research Institute, Beijing, China

[*]Corresponding author email: liujian3@sgitg.sgcc.com.cn

Abstract. The wrong wiring of the power supply region directly affects the accuracy of the metering data. This paper based on the existing data of the power supply regions, defined four indicators which include minimum negative line loss rate, maximum positive line loss rate, minimum power factor and maximum number of negative power values. Then by using the four indicators before and after the wrong wiring rectification of the historical wrong wiring regions and the decision tree algorithm, the wrong wiring regions identification model was built. The identification model was used to identify the wrong wiring regions in all regions of the power company, and the regions identified as wrong wiring by the model were inspected on-site. The inspection result shows that the precision of the model is satisfactory, which can effectively improve the inspection and rectification efficiency of the wrong wiring regions.

1. Introduction

Electric power data of the power supply region such as current, voltage, power, electrical energy supply and electricity consumption collected by the electric energy metering device lay the foundation for the power company to carry out the power supply quality monitoring and line loss assessment and other management work of the power supply region, which has strongly supported the service level improvement of the power company. With the rapid development of economy and technology, in order to meet the increasing demand of power supply service and power company management, more and more data indicators collected by power grid region, and the collection accuracy is constantly improved. As a result, there are more and more components and lines are more and more complex in the power supply region, and the lines of different components are interlaced together, which is prone to the wrong wiring of the power supply region. The wrong wiring of the power supply region will directly affect the accuracy of metering data of the power supply region, it causes property losses to the power company, and affect the power supply service level and management level of the power company[1-2].

As China has a vast territory and a large population, there are massive power supply regions and the regions are wide distribution. Therefore, it needs a lot of manpower and material resources to inspect the wiring of the electric energy metering device of the power supply region. At the same time, with the continuous development of the power supply region, the connection of the region may change, and the normal region may also become the wrong wiring region, so it is necessary to carry out the region wrong wiring inspect regularly.

Content from this work may be used under the terms of the Creative Commons Attribution 3.0 licence. Any further distribution of this work must maintain attribution to the author(s) and the title of the work, journal citation and DOI.
Published under licence by IOP Publishing Ltd

2. Related Work

Electric power companies and related scientific research institutions have studied the remote and automatic identification of wrong wiring of the electric energy metering devices. Chen Xiao et al. judge the wiring relationship of the metering device by rotating the phasor diagram between each metering element of the electricity meter, so as to realize the rapid and accurate identification of the wrong wiring of the electricity meter[3]. Zhang Qiang et al. identify the wrong wiring by calculating the probability distribution of the phase difference angle of voltage and current, and expound the construction of the wrong wiring identification software system[4]. By summarizing the data needed to generate the vector graph of metering device in the power supply region, Zhang Sheguo et al. put forward the method of remote diagnosis of the wrong wiring of the power supply region by drawing the vector graph according to the data of the power supply region[5]. In article 3-5, wrong wiring is identified mainly based on the index characteristics of metering devices when wrong wiring occurs in theoretical situations, and there are many indicators need to be collected and the calculation is complicated, but it is difficult to find out the wrong wiring situation outside the theory. Yang Pei et al. put forward a correlation analysis method of metrological abnormal events, which can identify the wrong wiring by analyzing the abnormal events related to the wrong wiring, but this method can only analyze the abnormal events that have occurred simultaneously with the wrong wiring in history[6]. Based on ADSP-BF537 logic control and pattern recognition, Xu Jinliang et al. designed an automatic judgment system for power grid wrong wiring on site, this automatic judgment system needs to be used on site in the power supply region and cannot be used for remote identification of wrong wiring[7]. Wang Hongxi et al. proposed a method to identify the wrong wiring of electricity meters based on HPLC technology, but HPLC electricity meters are still in the stage of promotion and the method requires additional data collection, so it is difficult to be applied in practice at present[8].

In view of the problems existing in the research of wrong wiring identification of the electric energy metering device, this paper proposes an identification method of wrong wiring region based on classification analysis. Firstly, a modeling data set was constructed based on the daily line loss rate, daily power factor and daily power curve data collected and calculated by the current power information acquisition system. Based on the data set, a decision tree classification algorithm was used to construct the identification model of the wrong wiring region. Then, the model is used to identify the wrong wiring region in a certain power company, and the on-site inspection is carried out according to the identification result. The on-site inspection result shows that the wrong wiring region identification method proposed in this paper has high precision and can effectively support the power company to carry out inspection and rectification work of the wrong wiring regions. Moreover, the method proposed in this paper is based on the existing data of the power information acquisition system, and the identification model constructed is easy to understand.

3. Algorithm Introduction

The identification of the wrong wiring regions is a typical dichotomy problem. At present, the commonly used classification algorithms include decision tree classification, Bayesian classification, support vector machine classification, logistic regression and so on. The decision tree classification is the most widely used classification algorithm, which is easy to understand, interpretable, easy to visualize, and requires small training samples. Due to the small number of the wrong wiring regions detected in the history, and the identification model should be easy to understand and use for basic business personnel, this paper uses the decision tree algorithm to build the identification model of the wrong wiring regions.

The concept of the decision tree first appeared in CLS (Concept learning system)[9]. In all decision tree classification algorithms, the most influential algorithm is ID3 presented by Quinlan in 1986[10], and based on the ID3 algorithm, Quinlan presented C4.5 algorithm in 1993[11], which becomes the most commonly used decision tree algorithm. The C4.5 algorithm users the information gain ratio to choose the node splitting attributes of decision tree, and overcomes the disadvantage of the ID3 tend to choose the attribute which has more values.

The detailed structure method of the C4.5 algorithm is as follows: 1. The given samples set is regarded as root node of the decision tree; 2. The information gain ratio of every attribute in the samples set is

calculated and choose the biggest as the partition attribute of current node; 3. The branches for every value of the partition attribute are established, partition the samples set into some subsets and establish new node for every subset. Step 2 and 3 above are repeated recursively for each new node until all nodes meet one of the following conditions:
1) All samples in the training sample set corresponding to the node belong to the same category;
2) The attribute values of all samples in the training sample set corresponding to the node are identical.

4. Data Sets and Model Building Processes

4.1. Data Sets Building Processes

The influence of wrong wiring on the power supply region indexes is mainly reflected in abnormal line loss, low power factor and negative power value in power curve. The influence of wrong wiring on line loss rate, power factor and power curve may not appear simultaneously on the same day, and the region may not be continuously in the state of wrong wiring due to the complex site environment and human factors, that is, the region is not in the state of wrong wiring every day. In order to avoid the effect of indexes influence of wrong wiring is not synchronized and the wrong wiring state fluctuation, in this paper, the line loss rate, power factor and the number of negative power values in the power curve of the region 10 days before the wrong wiring rectification (i.e. in the wrong wiring state) and 10 days after the wrong wiring rectification (i.e. in the non-wrong wiring state) are used to construct the modeling data set of the classified identification model.

Define the historical wrong wiring power supply regions set is $\mathrm{K} = \{a_1, a_2, a_3,..., a_k\}$, where k is the number of historical wrong wiring regions. Define the line loss rate, power factor and the number of negative power values in the power curve of the wrong wiring region ai 10 days before rectification is LL_{ai}^-, PF_{ai}^-, PC_{ai}^- as in equation (1),(2),(3) respectively.

$$\mathrm{LL}_{ai}^- = \left\{\mathrm{LL}_{ai}^{-1}, \mathrm{LL}_{ai}^{-2}, \mathrm{LL}_{ai}^{-3},..., \mathrm{LL}_{ai}^{-10}\right\} \tag{1}$$

$$\mathrm{PF}_{ai}^- = \left\{\mathrm{PF}_{ai}^{-1}, \mathrm{PF}_{ai}^{-2}, \mathrm{PF}_{ai}^{-3},..., \mathrm{PF}_{ai}^{-10}\right\} \tag{2}$$

$$\mathrm{PC}_{ai}^- = \left\{\mathrm{PC}_{ai}^{-1}, \mathrm{PC}_{ai}^{-2}, \mathrm{PC}_{ai}^{-3},..., \mathrm{PC}_{ai}^{-10}\right\} \tag{3}$$

Define the line loss rate, power factor and the number of negative power values in the power curve of the wrong wiring region ai 10 days after rectification is LL_{ai}^+, PF_{ai}^+, PC_{ai}^+ as in equation (4),(5),(6) respectively.

$$\mathrm{LL}_{ai}^+ = \left\{\mathrm{LL}_{ai}^{1}, \mathrm{LL}_{ai}^{2}, \mathrm{LL}_{ai}^{3},..., \mathrm{LL}_{ai}^{10}\right\} \tag{4}$$

$$\mathrm{PF}_{ai}^+ = \left\{\mathrm{PF}_{ai}^{1}, \mathrm{PF}_{ai}^{2}, \mathrm{PF}_{ai}^{3},..., \mathrm{PF}_{ai}^{10}\right\} \tag{5}$$

$$\mathrm{PC}_{ai}^+ = \left\{\mathrm{PC}_{ai}^{1}, \mathrm{PC}_{ai}^{2}, \mathrm{PC}_{ai}^{3},..., \mathrm{PC}_{ai}^{10}\right\} \tag{6}$$

According to the data situation and business meaning of the line loss rate, power factor and the number of negative power value in the power curve of the power supply region, line loss rate includes positive line loss rate and negative line loss rate, and greater the absolute value of line loss rate indicates that the region is more abnormal, smaller power factor indicates that the region is more abnormal, more negative power values indicates that the region is more abnormal. In this paper, the modeling data set of the classified identification model was constructed by using four indicators which include minimum negative line loss rate, maximum positive line loss rate, minimum power factor and maximum number of negative power values, and the four indicators were calculated by the data of the historical wrong wiring regions in 10 days before and 10 days after the rectification.

For the wiring region ai before rectification, the four indicators calculation equations are as follows:

1. the minimum negative line loss rate minLL_{ai}^- as in equation (7).

$$\mathrm{minLL}_{ai}^- = \begin{cases} \min(\mathrm{LL}_{ai}^-) & \min(\mathrm{LL}_{ai}^-) < 0 \\ 0 & \min(\mathrm{LL}_{ai}^-) \geq 0 \end{cases} \tag{7}$$

2. the maximum positive line loss rate maxLL_{ai}^- as in equation (8).

$$\mathrm{maxLL}_{ai}^{-} = \begin{cases} \max(\mathrm{LL}_{ai}^{-}) & \max(\mathrm{LL}_{ai}^{-}) > 0 \\ 0 & \max(\mathrm{LL}_{ai}^{-}) \le 0 \end{cases} \tag{8}$$

3. the minimum power factor minPF_{ai}^{-} as in equation (9).

$$\mathrm{minPF}_{ai}^{-} = \min(\mathrm{PF}_{ai}^{-}) \tag{9}$$

4. the maximum number of negative power values maxPC_{ai}^{-} as in equation (10).

$$\mathrm{maxPC}_{ai}^{-} = \max(\mathrm{PC}_{ai}^{-}) \tag{10}$$

For the wiring region *ai* after rectification, the four indicators calculation equations are as follows:

1. the minimum negative line loss rate minLL_{ai}^{+} as in equation (11).

$$\mathrm{minLL}_{ai}^{+} = \begin{cases} \min(\mathrm{LL}_{ai}^{+}) & \min(\mathrm{LL}_{ai}^{+}) < 0 \\ 0 & \min(\mathrm{LL}_{ai}^{+}) \ge 0 \end{cases} \tag{11}$$

2. the maximum positive line loss rate maxLL_{ai}^{+} as in equation (12).

$$\mathrm{maxLL}_{ai}^{+} = \begin{cases} \max(\mathrm{LL}_{ai}^{+}) & \max(\mathrm{LL}_{ai}^{+}) > 0 \\ 0 & \max(\mathrm{LL}_{ai}^{+}) \le 0 \end{cases} \tag{12}$$

3. the minimum power factor minPF_{ai}^{+} as in equation (13).

$$\mathrm{minPF}_{ai}^{+} = \min(\mathrm{PF}_{ai}^{+}) \tag{13}$$

4. the maximum number of negative power values maxPC_{ai}^{+} as in equation (14).

$$\mathrm{maxPC}_{ai}^{+} = \max(\mathrm{PC}_{ai}^{+}) \tag{14}$$

According to four indicators of the *k* historical wrong wiring regions before and after rectification, the modeling data set containing $2k$ records as shown in the Table 1 is obtained. When the state of the region is before the rectification, the label of the region is "Yes", i.e. the region is in the wrong wiring state. When the state of the region is after the rectification, the label of the region is "No", i.e. the region is in the non-wrong wiring state.

Table 1. The modeling data set of the wrong wiring region classified identification model.

region ID	*region before or after rectification*	*minimum negative line loss rate*	*maximum positive line loss rate*	*minimum power factor*	*maximum number of negative power values*	*label*
a1	before	minLL_{a1}^{-}	maxLL_{a1}^{-}	minPF_{a1}^{-}	maxPC_{a1}^{-}	Yes
a1	after	minLL_{a1}^{+}	maxLL_{a1}^{+}	minPF_{a1}^{+}	maxPC_{a1}^{+}	No
a2	before	minLL_{a2}^{-}	maxLL_{a2}^{-}	minPF_{a2}^{-}	maxPC_{a2}^{-}	Yes
a2	after	minLL_{a2}^{+}	maxLL_{a2}^{+}	minPF_{a2}^{+}	maxPC_{a2}^{+}	No
…	…	…	…	…	…	
ak	before	minLL_{ak}^{-}	maxLL_{ak}^{-}	minPF_{ak}^{-}	maxPC_{ak}^{-}	Yes
ak	after	minLL_{ak}^{+}	maxLL_{ak}^{+}	minPF_{ak}^{+}	maxPC_{ak}^{+}	No

4.2. Model Building Processes

In this paper, using the line loss rates, the power factors and the power curves of the historical wrong wiring regions in 10 days before the rectification and 10 days after the rectification, the four indicators before and after the rectification are calculated, which include minimum negative line loss rate, maximum positive line loss rate, minimum power factor and maximum number of negative power values. The modeling data set is constructed according to the four indicators of the regions before and after rectification, and the modeling data set is input into the C4.5 algorithm to obtain the decision tree classification model for the wrong wiring region identification. The construction process of the classified identification model for the wrong wiring region is shown in the Figure 1.

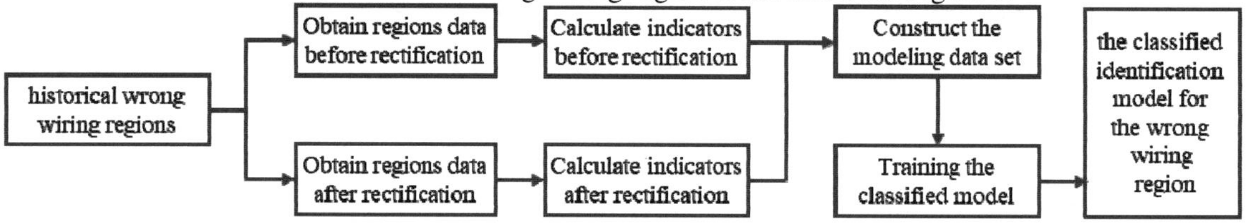

Figure 1. The construction process of the identification model for the wrong wiring region.

5. Model Building and Application

5.1. Model Building

In this paper, by using the data of 10 days before and 10 days after the rectification of the 1200 historical wrong wiring regions in a certain power company, the modeling data set with 2400 records for the classified identification model was constructed. And by using the modeling data set and the C4.5 algorithm, the decision tree classification model for the wrong wiring region identification was constructed. As the decision tree obtained is large and inconvenient to display, the identification rules of wrong wiring region are obtained after sorting out the decision tree, which are as shown in the Table 2.

From the Table 2, the line loss rate of region plays the greatest role in identifying whether a region is a wrong wiring region or not. When the minimum negative line loss rate is less than -100%, all regions are wrong wiring region, when the maximum positive line loss rate is more than 100%, the probability of the region being a wrong wiring region is 85.71%, and when the minimum negative line loss rate >=-7.16% and the maximum positive line loss rate <=9.13%, the region is basically not a wrong wiring region. In addition, when the maximum number of negative power values of the power curve of the region is more than 2, the region is more likely to be a wrong wiring region, and when the minimum power factor of the region is less than 0.8575, the region is more likely to be a wrong wiring region.

Table 2. The wrong wiring region identification rules based on historical wrong wiring regions.

rule ID	*minimum negative line loss rate*	*maximum positive line loss rate*	*maximum number of negative power values*	*minimum power factor*	*number of regions after rectification*	*number of regions before rectification*	*rate of regions before rectification*
1	<-100				0	169	100.00%
2	>=-100&<-7.16		<3	<0.8575	13	154	92.22%
3	>=-100&<-7.16		<3	>=0.8575	268	15	5.30%
4	>=-100&<-7.16		>=3		5	147	96.71%
5	>=-7.16	<9.13			672	23	3.31%
6	>=-7.16	>=9.13&<22.42	<3		68	31	31.31%
7	>=-7.16	>=9.13&<22.42	>=3	<0.8575	17	69	80.23%
8	>=-7.16	>=9.13&<22.42	>=3	>=0.8575	53	61	53.51%
9	>=-7.16	>=22.42&<100	<3		68	66	49.25%
10	>=-7.16	>=22.42&<100	>=3	<0.8575	11	372	97.13%
11	>=-7.16	>=22.42&<100	>=3	>=0.8575	19	57	75.00%
12	>=-7.16	>=100			6	36	85.71%
		total			1200	1200	50.00%

When the rate of regions before rectification of a rule exceeds 50%, the model classified the regions which meeting the rule as wrong wiring region, otherwise as non-wrong wiring region. In the model as Table 2, 1200 regions which before rectification were classified into two categories: 1065 wrong wiring regions (i.e. classified correctly) and 135 non-wrong wiring regions (i.e. classified incorrectly),

1200 regions which after rectification were classified into two categories: 124 wrong wiring regions (i.e. classified incorrectly) and 1076 non-wrong wiring regions (i.e. classified correctly).Therefore, the accuracy of the model is 89.21%, the precision is 89.57%, the recall is 88.75%, and the f1-Score is 89.16%, indicating that the model has good classification performance.

5.2. Model Application

Using the line loss rate, the active power curve and the power factor data of 55814 power supply regions in a power company from May 1 to 10, and according to the four indicators calculation equations of the minimum negative line loss rate, the maximum positive line loss rate, the minimum power factor and the maximum number of negative power values in Chapter 3, the input data set of the classified identification model containing 55,814 records were obtained. Then the wrong wiring regions identification was carried out by using the decision tree classification model which is obtained in 4.1, and the identification result of the 55,814 regions was obtained as shown in the Table 3. The wrong wiring probability in the Table 3 is the rate of regions before rectification in the Table 2.

According to the identification result in the Table 3, the power company conducted on-site inspection on 2709 regions whose wrong wiring probability are more than 75%, and the on-site inspection result is shown in the Table 4. From the Table 4, there are 314 regions are verified wrong wiring region in the 327 regions with the wrong wiring probability equal to 100%(i.e. precision is 96.02%), there are 1520 regions are verified wrong wiring region in the 1661 regions with the wrong wiring probability exceeding 90% but less than 100%(i.e. precision is 91.51%), there are 522 regions are verified wrong wiring region in the 721 regions with the wrong wiring probability greater than or equal to 75% but less than 90% (i.e. precision is 72.40%).

Table 3. The identification result of the 55,814 regions in a power company.

rule ID	*minimum negative line loss rate*	*maximum positive line loss rate*	*maximum number of negative power values*	*minimum power factor*	*region number*	*wrong wiring probability*
1	<-100				327	100.00%
2	>=-100&<-7.16		<3	<0.8575	776	92.22%
3	>=-100&<-7.16		<3	>=0.8575	6781	5.30%
4	>=-100&<-7.16		>=3		707	96.71%
5	>=-7.16	<9.13			34325	3.31%
6	>=-7.16	>=9.13&<22.42	<3		6232	31.31%
7	>=-7.16	>=9.13&<22.42	>=3	<0.8575	400	80.23%
8	>=-7.16	>=9.13&<22.42	>=3	>=0.8575	2651	53.51%
9	>=-7.16	>=22.42&<100	<3		3116	49.25%
10	>=-7.16	>=22.42&<100	>=3	<0.8575	178	97.13%
11	>=-7.16	>=22.42&<100	>=3	>=0.8575	176	75.00%
12	>=-7.16	>=100			145	85.71%

Table 4. The on-site inspection result of 2709 regions in a power company.

rule ID	*number of wrong wiring regions with wrong wiring probability >= 75%*	*wrong wiring probability output by the model*	*number of verified wrong wiring regions*	*precision*
1	327	100.00%	314	96.02%
2	776	92.22%	705	90.85%
4	707	96.71%	652	92.22%
7	400	80.23%	295	73.75%
10	178	97.13%	163	91.57%
11	176	75.00%	118	67.05%
12	145	85.71%	109	75.17%

According to the on-site inspection result, the identification method of wrong wring region proposed in this paper has high precision, which can improve the accuracy of on-site inspection of wrong wring region, so to be able to effectively improve the inspection and rectification efficiency of the wrong wiring regions and significantly reduce the manpower and material resources. In addition, the higher the wrong wiring probability output by the identification model, the higher the accuracy of on-site inspection is. Therefore, according to the wrong wiring probability given by the model and the manpower and material resources of the power company, the inspection and rectification plan for wrong wiring regions made in order of high probability followed by low probability would be scientifically and reasonable.

6. Conclusion

In this paper, based on the power supply region data of line loss rate, power factor and power curve collected and calculated by the power information acquisition system, four indicators of the region include minimum negative line loss rate, maximum positive line loss rate, minimum power factor and maximum number of negative power values are defined by analyzing the business meaning the region data, and the classified identification model for the wrong wiring region is built by using four indicators and decision tree algorithm. All the regions of a power company are identified by the model, and the regions with high wrong wiring probability are inspected on-site. The on-site inspection result shows that, the identification method of wrong wring region proposed in this paper has high precision, and can effectively improve the inspection and rectification efficiency of the wrong wiring regions, thus reducing the manpower and material resources of the power company. In addition, the region data used in this method is simple and new data collection is not needed, the identification model is easy to understand and easy to be deployed and popularized.

References

[1] Li Zhenna, Cai kun, Lin Rulan. Analysis of the Effects of Wrong Wiring on Three-phase Four-wire Electric Energy Meter. Metrology & Measurement Technique, vol.46, No.08. pp. 74-77, 2019.

[2] Li Mingzhu, Sun Weiting, Zhang Mengyao. Measurement Analysis of Several Wrong Wiring of Three-Phase Four-Wire Active Electric Energy Meter. Electric Engineering, No.22, pp.126-128, 2020.

[3] Chen Xiao, Zhou Yu, Fan Jie, et al. Study on the Rapid Error Connections Discriminating Method of the New Three-Phase and Three-Wire Measuring Device. Electrical Measurement & Instrumentation, vol.51, No.13. pp. 5-8, 2014.

[4] Zhang Qiang, Yu Ning, Cui Hui, et al. Design and implementation of fault wiring connection identification software for three-phase meters based on big data. Electrical Measurement & Instrumentation, vol.55, No.6. pp. 110-115, 2018.

[5] ZHANG Sheguo, LI Xinjia, WANG Rong. Research on wrong wiring diagnosis method of remote metering device in utility distribution transformer metering device. Power Demand Side Management, vol.16, No.4. pp. 44-47, 2014.

[6] Yang Pei, Li Jing, Chen Weimin, et al. A method of analyzing the fault causes based on the combination of metering abnormal events. Electrical Measurement & Instrumentation, vol.56, No.6. pp. 111-116, 2019.

[7] Xu Jinliang, Liu Pei, Chen Qijun. Design of high precision automatic judgement system for on-site miswiring of power network. Automation & Instrumentation, No.8. pp. 33-36, 2019.

[8] WANG Hongxi, TANG Ruyi, WU Yidi. Research and application of smart meter reading technology based on HPLC in customer side universal power internet of things. Power System Protection and Control, vol.48, No.3. pp. 92-98, 2020.

[9] Hunt E, Krivanek J. The effects of pentylenetetrazole and methylphenoxypropane on discrimination learning. Psychopharmacologia, vol.9, No.1. pp. 1-16, 1966.

[10] Ross Quinlan J. Induction of decision trees. Machine Learning, pp.81-106, 1986.

[11] Steven Salzberg. Book review: C4.5: programs for machine learning by J. Ross Quinlan, Morgan Kaufmann Publishers, Inc., 1993. Machine Learning, vol.16, No.3. pp. 235-240, 1994.

Hybrid Reconstruction and Evaluation of 3D Model of Equipment Parts for Virtual Maintenance System

Weijun Chu* **and Xiangyu Sun**

Army Engineering University of PLA, Nanjing, China

*Corresponding author email: chuwj2008@qq.com

Abstract. The research and application of reverse engineering technology are more and more extensive, but its process is still not fully mature. Aiming at the complexity of equipment parts and the diversity of accuracy requirements, the method of hybrid reconstruction and evaluation of 3D model of equipment parts for virtual maintenance system is illustrated. The general process is given. Taking the two-stage bevel gear worm reducer as the research object, the research on three-dimensional scanning, data processing and model reconstruction is carried out. This paper focuses on the reconstruction process and methods such as regional division, solid model construction based on maximum outer contour, creation of basic surface, patch fitting and so on by taking the bevel gear in the reducer as an example. Finally, 3D comparison, 2D comparison, cross-section accuracy and size of the reconstructed solid model are evaluated. The results show that the hybrid reconstruction method has strong adaptability and high efficiency. Reverse modeling and forward design are organically combined to give full play to their advantages, and the hybrid reconstruction method can effectively reverse the original design intention of the product and improve the parameterization ability of reverse modeling. It can meet the requirements of different precision models developed by virtual maintenance system, structural wall chart and atlas, multimedia and so on.

1. Introduction

Virtual maintenance system has the characteristics of high training efficiency, low cost and good safety. It effectively provides a realistic training environment for equipment operation training and maintenance training. It overcomes many adverse factors brought by the number of machinery and the training site environment, greatly reduces the repeated disassembly and assembly of equipment, and provides an efficient and economic maintenance support training means for new equipment. One of the bottlenecks of virtual training system is the construction of 3D model. In view of the complexity of equipment parts and the diversity of accuracy requirements for virtual maintenance system, structural wall charts and atlas, the 3D laser scanning technology is used to collect the 3D model data of parts efficiently and accurately, and then the large-scale point cloud data is processed. The 3D model hybrid reconstruction technology is used to carry out reverse reconstruction to obtain the parametric model, and the reconstructed model is evaluated and analyzed to provide necessary model resources for the development of virtual maintenance training system, structural wall chart and atlas.

At present, there are many reverse modeling software on the market. But the reverse modeling methods are different. Geomagic Design Direct is a forward and backward hybrid design software based on direct modeling technology. Through calculating and extracting the curvature and normal direction of different regions in the triangular mesh surface model, the corresponding 3D regular entity features are fitted. Geomagic Studio divides triangular mesh model according to the geometric features, then fits the corresponding 3D surface features, and finally reconstructs the surface model. Compared

Content from this work may be used under the terms of the Creative Commons Attribution 3.0 licence. Any further distribution of this work must maintain attribution to the author(s) and the title of the work, journal citation and DOI.
Published under licence by IOP Publishing Ltd

with the surface model, the solid model can express the 3D shape of the model more completely and strictly. Compared with them, this hybrid reconstruction method combines solid modeling and surface modeling, can quickly create the original model, and can guarantee the precision of the model, and can do the direct design on the basis of the reconstructed model.

2. Hybrid Reconstruction Process

2.1. General Process

The surfaces of a large number of parts in engineering equipment are relatively regular, there are few free-form surfaces, and there will be no large transition surfaces between faces. Basically, they transition with small fillets, showing obvious edges. The parts with this feature are suitable for the reverse modeling method based on solid features. For some free-form surface parts, such as outer cladding, the surface fitting reverse modeling method based on surface features can be used.

According to the structural characteristics and actual conditions of engineering equipment, the traditional method of machine measuring and drawing is actively used in reverse measurement to obtain some basic parameters of some parts like gears. The feature parameters extracted through advanced measurement and in the reverse modeling software are imported into the forward modeling software for editing, modification and solid modeling. The existing mature modeling technologies are made full use of by the organic combination of reverse modeling and forward design and cross multiple 3D modeling software platforms. They should be comprehensively applied, give full play to their modeling advantages, learn from each other, complement each other, and give full play to their respective advantages. Therefore, the hybrid reconstruction of engineering equipment 3D model not only adopts a variety of hybrid reverse modeling methods, but also adopts forward and reverse hybrid modeling. Its process is shown in figure 1.

Patch optimization can eliminate impurities by smoothing, reduce the surface roughness, so that the surface is more smooth. It can be optimized on the whole surface, but also on the local surface, according to the actual situation to improve the quality of the surface.

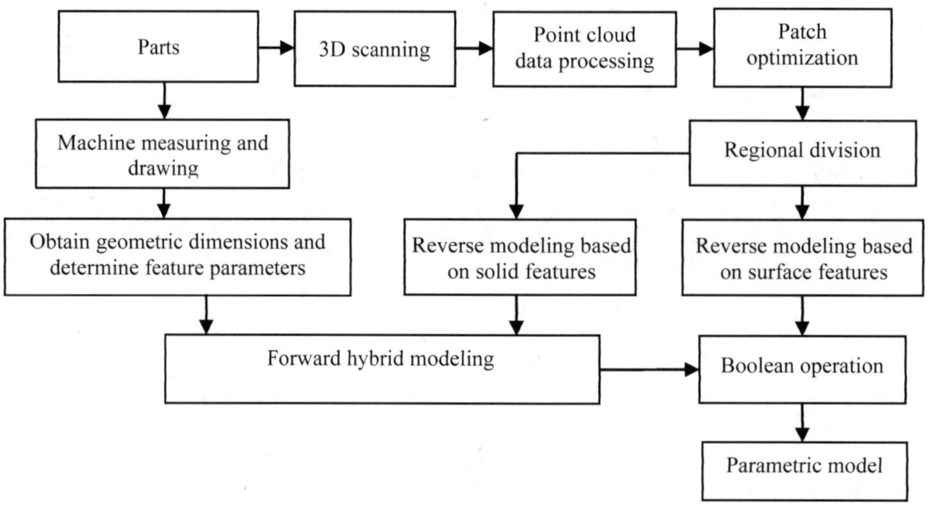

Figure 1. Hybrid reconstruction process of 3D model.

2.2. AXE-B11 Handheld Global 3D Scanner

This paper takes the two-stage bevel gear worm reducer as the research object. Two-stage reducer is shown in figure 2. The appearance of AXE-B11 scanner is shown in figure 3. Based on the decomposition and determination of basic parameters, AXE-B11 handheld global 3D scanner is selected to quickly scan the reducer to obtain the point cloud data of each part, and then carry out large-scale processing of point cloud data to restructure a regular and accurate model. The accuracy of reconstructed solid model data is required to reach part level, and the key model accuracy is to the millimeter level. Considering the reuse of existing models, it supports the transformation of a variety of standardized model formats. After importing the virtual system, the shape, structure, proportion and texture of the parts are basically consistent with the actual ones.

Figure 2. Two-stage reducer.

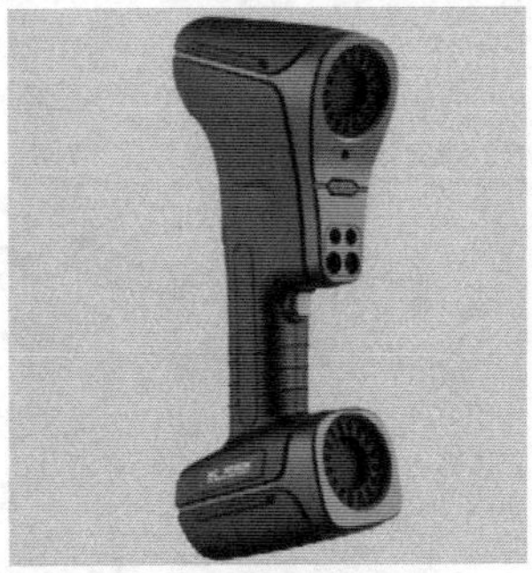

Figure 3. 3D scanner.

The scanner obtains spatial three-dimensional point cloud by using binocular vision principle. The space position of the scanner and the measured object is obtained by scanning the mark point of the current frame and matching the mark point library. The laser is emitted by the laser transmitter and irradiated on the surface of the scanned workpiece. Then the reflected light is captured by two calibrated industrial cameras, and the shape data of the workpiece is obtained through calculation. The light source is 22 crossed blue laser lines (plus 1 blue laser line), the blue light reference distance is 450mm and the depth of field is 500mm. Deep hole scanning supports an accuracy of 0.020 mm, and it has the scanning rate of 1300000 times per second.

3. Reverse Reconstruction of Bevel Gear

3.1. Determination of Basic Parameters

The picture of bevel gear part is shown in figure 4. After measurement, the diameter of the reference circle at the big end of the bevel gear is 58.13mm and the number of teeth is 30, the calculated value of the module at the big end is 1.937mm, then the standard value is 2mm. See Table 1 for some parameters.

Table 1. Some parameters of bevel gear.

Number of teeth	Module	Diameter of reference circle at the big end	Crossed axis angle	Outer cone distance	Addendum height	Root height	Apical gap	Root angle
30	2mm	60mm	90°	42.43mm	2mm	2.4mm	0.4mm	3.23°

3.2. Coordinate System Establishment and Model Alignment

Scanning picture and point cloud data are shown in figure 5. To import the scanned data to Geomagic Design X, coordinate alignment is required to make the model coordinates coincide with the system coordinates.

Figure 4. Bevel gear.

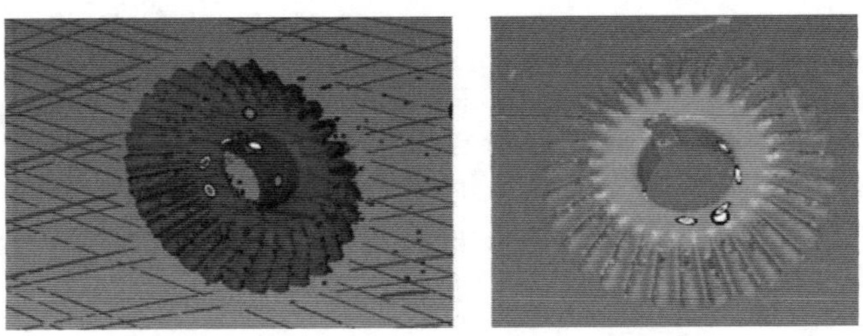

Figure 5. Scanning picture and point cloud data.

3.3. Solid Model Construction Based on Maximum Outer Contour

We can change the direction of the intercepted feature and adjust the view for inspection. As long as the intercepted range includes the tooth top, the maximum outer contour can be intercepted.

After the sketch is drawn (figure 6), the deviation of the line is analyzed. As shown in figure 7, the deviation is displayed in green within plus or minus 0.1mm and yellow within plus or minus 0.2mm. When the positive and negative deviation is close to or even more than 1mm, the line is red. Enlarge the figure, we observe and adjust the line to make it as green as possible to reduce the deviation.

Figure 6. Patch sketch fitting.

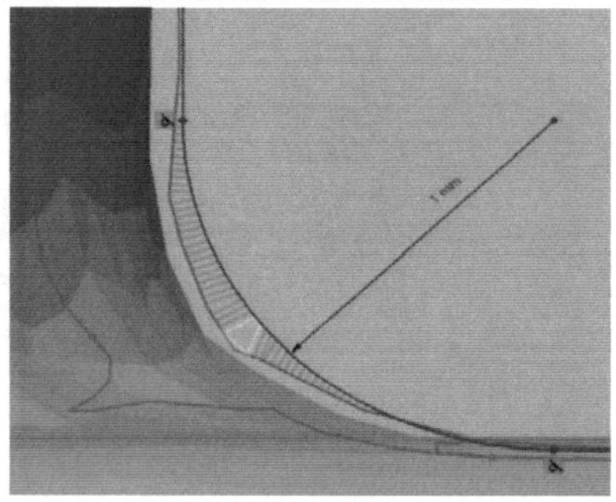

Figure 7. Line deviation analysis.

After checking the deviation, we select the separation end point and check whether the sketch is closed. If it is not closed, light green dots will appear, as shown in Figure 8. If it is not deleted, the entity creation operation cannot be realized.

3.4. Creation of Cutting Surface

3.4.1. Basic Surface Creation Method

- Select the field color block to create the base surface.

The region is an important basis for fitting patches. It uses different color blocks to split different surfaces according to the curvature changes of model surfaces, as shown in figure 9. The color block is attached to the STL triangle, which reflects the characteristics of the model, and is used to extract the shape and dimension information of the STL model. In the field color block of bevel gear, we select smooth color block to create cylindrical surface for the target field. The color block with complete smooth is helpful to the creation of our surface and reduce the deviation. The accuracy of regional division directly affects the accuracy of the fitting patch.

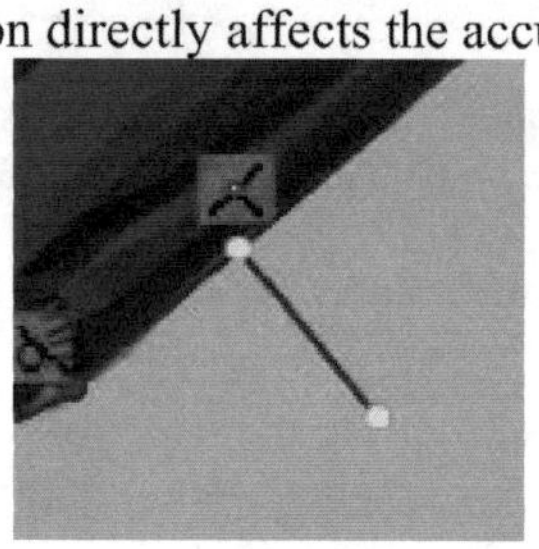

Figure 8. Line separation point inspection.

Figure 9. Regional division.

Firstly, the reverse parametric modeling method based on regional division is to divide the polygon model according to curvature, and express the features of the model through region groups. Then, the 2D section line and projection boundary contour are created according to the feature extraction of the region, and the parameter values are modified and the line constraint relationship is added to obtain more accurate 2D sketch. Then, stretch, rotate, scan the sketch, and create a parametric model. At the same time, surface fitting is carried out according to the region to create patches. Finally, the patch and parametric model are Boolean operation to get the accurate parametric model.

- Drawing UV curve.

UV curve provides a connection between the model surface and the texture image. UV curve is responsible for determining which vertex (pixel) on the texture image should be placed on the model surface, so that the whole texture can be covered on the model.

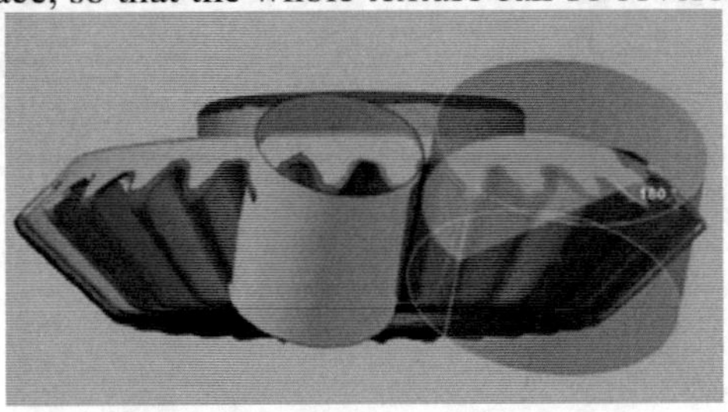

Figure 10. Drawing UV curve.

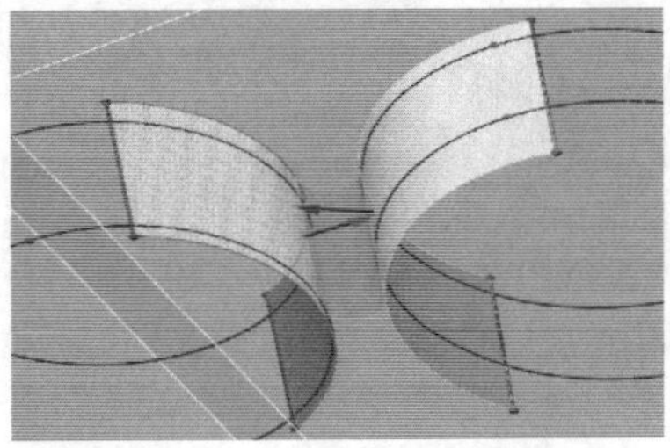

Figure 11. Fitting surfaces.

Select UV curve to loft to create a surface. Then, the remaining surface parts are obtained by cutting, and the groove parts of the two surfaces are fitted, as shown in figure 11.

3.4.2. Patch Fitting Method

- Select the fitting object and adjust fitting range.

Rotate the model to find the field color block with complete surface and good quality as the fitting object, as shown in figure 12. Adjustment of fitting range is shown in figure 13.

Figure 12. Selection of fitting object.

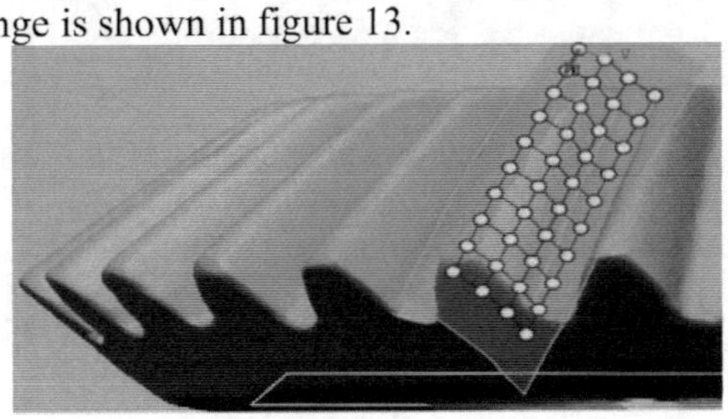

Figure13. Adjustment of fitting range.

- Check patch fitting effect.

A patch is a series of points, edges and faces (usually triangles) that are digitized in 3d based on polyhedra. Surface triangulation is the process of connecting three points and constructing a surface. There may be some wrong triangles in the process of Triangulation, such as cross triangles and redundant triangles, which should be deleted to improve the quality of the surface. In addition, defects such as holes and bulges may appear in the initial piece, so it needs to be repaired. The missing hole is filled with the element surface according to the shape curvature of the local patch, and the convex part of the patch is removed and repaired. The fitting quality of tooth surface needs to be checked to make the deviation analysis of fitting tooth surface as green as possible, as shown in figure 14.

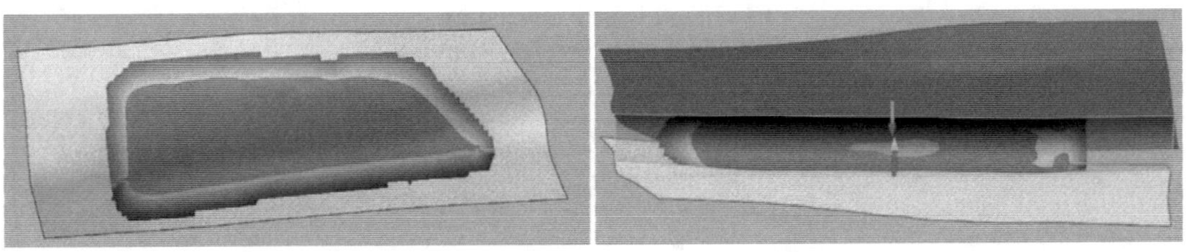

Figure 14. Tooth surface checking.

The resulting surface arraying is shown in figure 15. After solid cutting and keyway creation, the whole bevel gear model is reconstructed. Through deviation analysis and parameter adjustment, we can obtain the model meeting the accuracy requirement, as shown in figure 16. The the final model is shown in figure 17.

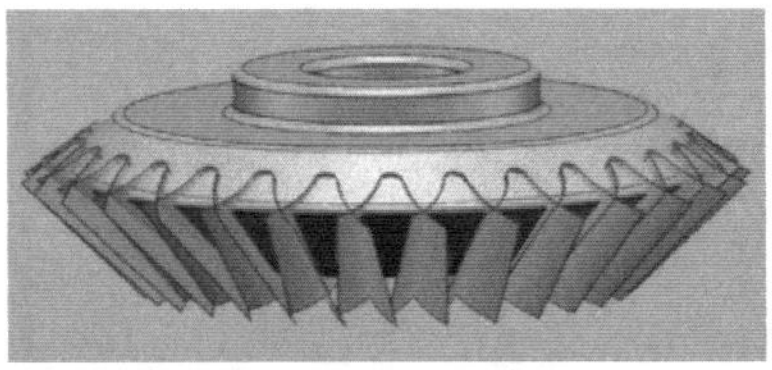

Figure 15. Surface array.

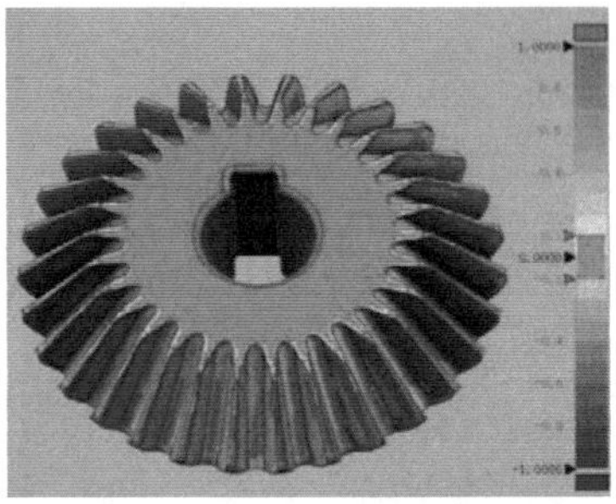

Figure 16. Deviation analysis.

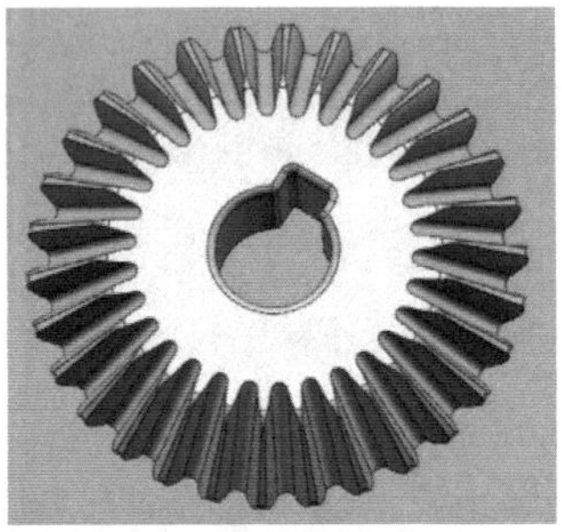

Figure 17. Final model.

4. Evaluation of Reconstruction Model

Geomagic Control platform provides the ability to simplify almost every process, reduce human interaction in measurement and recording, reduce measurement time and significantly enhance "GD&T" results. Geometric tolerance, hard measurement and azimuth inspection can not only speed up the measurement speed and improve the accuracy of parts.

After Importing the exported STL file and STP file into Geomagic Control software, the patch file and the model file are best fitted. We can analyze multiple aspects, such as 3D comparison (figure 18), 2D comparison (figure 19), cross section through the object; 2D dimensioning (figure 20), etc. Finally, we can also create PDF evaluation reports.

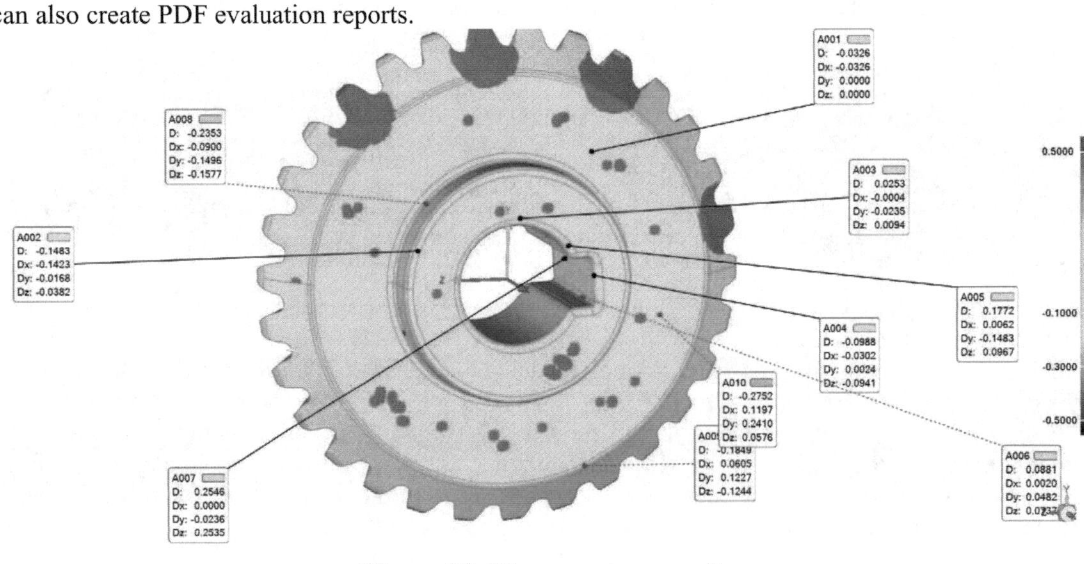

Figure 18. 3D comparison results.

Table 2. Results statistics.

≥Min (mm)	<Max (mm)	*Ni* (Number of points)	R_i (%)
-0.5000	-0.3000	*N*1=256	0.40
-0.3000	-0.1000	*N*2=13373	20.99
-0.1000	0.5000	*N*3=50076	78.60

$$R_i = N_i/(N1 + N2 + N3)100\% \quad (1)$$

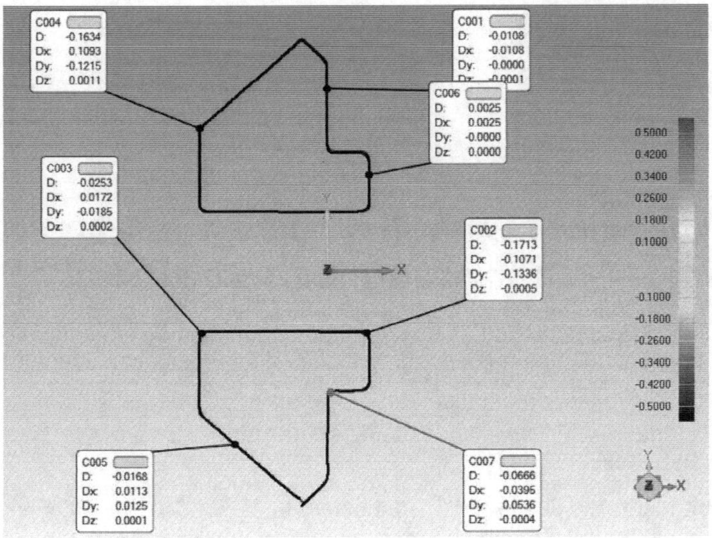

Figure 19. 2D comparison.

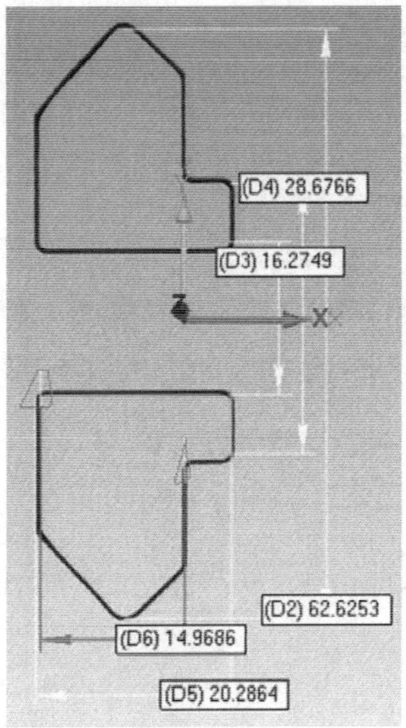

Figure 20. 2D dimensions.

The Statistics for 3D comparison are shown in Table 2. The deviation between -0.1 and 0.5mm accounted for 78.61%. The result indicates that the reconstruction model of bevel gear fully meet the requirements which the accuracy needs to be in millimeter for virtual maintenance system.

5. Conclusion

In the process of hybrid reconstruction, different features can be better divided according to the curvature division region, and the same feature can be edited manually. It is more conducive to the expression of features by dividing several regions. Therefore, it is important to divide the regions reasonably and accurately to construct the precise reverse parametric model effectively. The hybrid reconstruction method saves time compared with the traditional surface fitting method.

In addition, after obtaining the design parameters of the existing model, the relevant parameters and models can be easily adjusted to meet the accuracy requirements of the final model. Therefore, this method can meet the requirements of different precision models such as virtual maintenance system, structural wall chart and atlas, multimedia and so on. And if we want to improve or innovate the design on this basis, it also has great freedom and flexibility.

Reference

[1] Siyuan Cheng and Xuerong Yang 2019 *Geomagic Design X reverse design technology* (Beijing: Tsinghua University Press) pp 3–5

[2] Haichen Cong, Siyuan Cheng, Xuerong Yang, Xiangwei Zhang 2016 Reverse Parametric Modeling Based on Division of Region *Modular Machine Tool & Automatic Manufacturing Technique* No.6 pp 71-74

[3] Xie Yuan and Siguang Hou 2019 Reverse Modeling for Car Washer based on Geomagic Design X *Journal of Liaoning Institute of Science and Technology* Vol.21 pp 3–5

[4] Xuan Sun, Minjie Chai, Bo Shao 2018 Reverse Modeling of the Motorcycle Fender Based on Geomagic *Journal of Xingtai Polytechnic College* Vol.35 pp 61–63

[5] Chaochao Feng, Siyuan Cheng, Xuerong Yang, Shaoming Luo 2017 Forward and Reverse Hybrid Modeling Based on Geomagic Design X *Machine Tool & Hydraulics* Vol. 45 pp 3–5

[6] Haifei Qiu 2017 Self-replication design of crankshaft based on reverse technique *Modern Manufacturing Engineering* No. 7 pp 157-160

[7] Yong Liu, Xiao Lai, Sheng Guo, Ning Song 2017 Study of Reverse Design and Reverse Modeling Based on Geomagic for Impeller *Mechanical Engineering & Automation* No.6 pp 51–52

Modeling and Finite Analysis of Overall Structure Mechanical Behavior of Steel Frame—Cold-formed Thin-walled Steel Composite Wall

Huqing Liang[2,4], Mengxiong Tang[1,4], Ruicong Huang[3,*], Zhengrong Jiang[3], Junming Qiu[3] and Jian Cai[3]

[1]Guangzhou Research Institute of Construction Industry Co., Ltd., Guangzhou, China
[2]Guangzhou Construction Engineering Co., Ltd., Guangzhou, China
[3]School of Civil Engineering and Transportation, South China University of Technology, Guangzhou, China
[4]Guangzhou Municipal Construction Group Co., Ltd., Guangzhou, China

[*]Corresponding author email: cthuangruicong@mail.scut.edu.cn

Abstract. Based on the finite element method, the typical layout and parameters were selected to establish the overall structure model of steel frame—cold-formed thin-walled steel composite wall and their influence on the mechanical properties and the economic steel amount was analyzed. The results show that the screw spacing, purlin section and purlin spacing have little effect on the lateral stiffness and the amount of steel used in the overall structure, while the width of wall opening has obvious effect. It is suggested that the screw spacing should be consistent with the relevant provisions in the cold-formed thin-walled steel structure specification, the purlin section should be selected relatively small according to the wall thickness and the purlin spacing should be properly amplified. The applicable height of this structure system is higher than that of pure cold-formed thin-wall steel structure system in ordinary residential buildings. The amount of steel used in this structure system is obviously less than that of pure steel frame structure system.

1. Introduction

Cold-formed thin-walled steel structure housing system has the advantage of light deadweight, short construction period, simple manufacture, etc. Different from the beam-column system of ordinary steel frame structure, the force transmission system of this structure consists of composite floors and walls composed of cold-formed thin-walled steel rods and clad plates. Therefore, the structure is neat both inside and outside, and there is no beams and columns protruding from the interior which means a larger available space. However, due to the characteristics of elongated and small cross sections of cold-formed thin-walled steel bars, the bearing capacity and resistive overturning ability of this structure system are limited and it is mainly used for construction of low-rise residential and office. And the domestic technical specification [1] and standard [2] only stipulate the design and fabrication of cold-formed thin-walled steel structure within six floors and 20m high. There are some methods abroad to improve the cold-formed thin-walled steel structure system so as to increase its applicable height, such as replacing local load-bearing members with hot-rolled steel, or replacing the lower story of the structure with reinforced concrete frames [3]. Based on this strengthening concept, domestic research on the combination of steel frame and cold-formed thin-walled steel has begun. However, the

Content from this work may be used under the terms of the Creative Commons Attribution 3.0 licence. Any further distribution of this work must maintain attribution to the author(s) and the title of the work, journal citation and DOI.
Published under licence by IOP Publishing Ltd

existing tests and numerical simulations [4-8] mostly focus on the wall, and there are few analyses on the overall performance of steel frame—cold-formed thin-walled steel structure [9-10].
In this paper, the general finite element program SAP2000 is used to study the influence of different parameters of vertical structure system of steel frame—cold-formed thin-walled steel composite wall on the overall structure level and get the recommended parameters and applicable height of it.

2. Analytical Model

2.1. Modeling Method
The vertical lateral force members of this structure system are composed of steel frame and cold-formed thin-walled steel composite wall embedded in the steel frame. The general finite element program SAP2000 is used to model the overall structure. Frame beams and columns and cold-formed thin-walled steel purlins of the walls are simulated by beam elements, and wall panels are simulated by layered-shell elements. The two ends of the cold-formed thin-walled steel purlins in the wall are hinged to the frame beam, and the coupling between shell element and beam element at the screw joint is realized through the automatic division of finite element, without considering the slip between beam element and shell element [11]. The vertical displacement of upper purlin is released and the vertical in-plane axial stiffness of layered-shell element is set as inactive state, so that the embedded cold-formed thin-walled steel composite wall does not directly bear the vertical load but provides lateral stiffness. In addition, floor openings are not considered in all the models in this paper, and the command of partition constraint is used to realize the rigid floor assumption. As vertical structural system is mainly studied in this paper, the modelling of floors is not considered but taken as uniform-distributed plane load transmitted to frame beams.

2.2. Model Overview
The structure system studied is mainly applied to residential and office buildings, so the story height *H* is set as 3m and the column distance *A* is set as 4.5m or 6m. The openings of doors and windows are considered as the complete openings from the top of the lower beams to the bottom of the upper beams, and the opening width *K* is set as 1.5~2.7m (values are determined according to different analyses). The plane layout of the model is selected as horizontal and longitudinal three spans. The plane layout and the serial number of beams and columns are shown in figure 1. Z1~3 represent the serial number of frame column, L1 and L2 represent the serial number of frame beam, ☰ represents the composite wall composed of purlin and wall panel. DK represents the openings in the walls (the default opening position is located in the middle of the span). One Y-oriented secondary beam is arranged at the middle span of each X-oriented main beam and numbered as L*(omitted in figure 1).

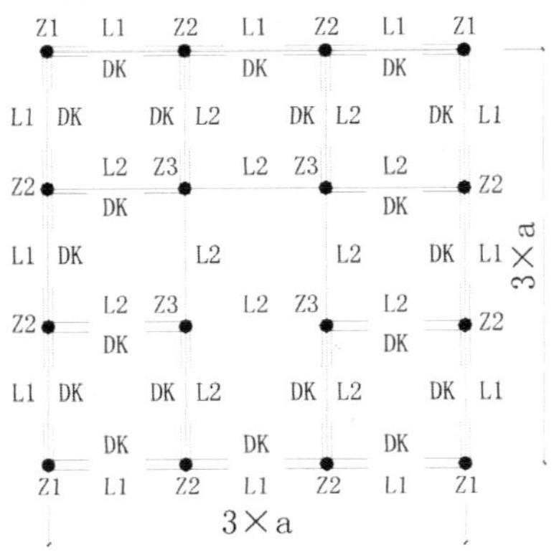

Figure 1. Model plane layout and beam and column serial number.

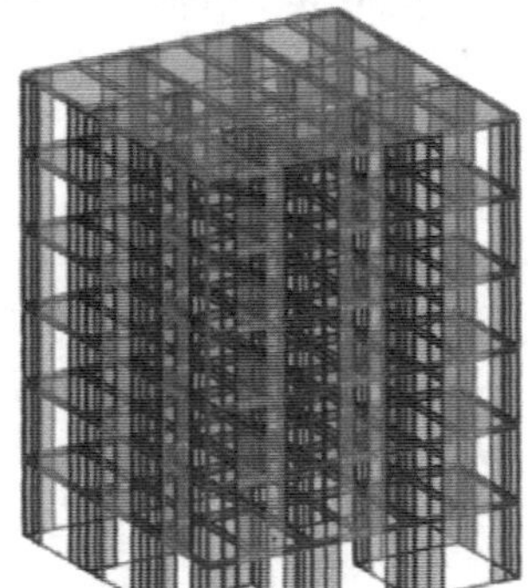

Figure 2. Model schematic diagram.

In practical engineering of cold-formed thin-walled steel structure system, the wall thickness is usually less than 150 mm and the beam and column sections made of cold-formed thin-walled steel sections do not protrude the wall. While improving the structural stiffness by adding steel frame, the advantages of original cold-formed thin-walled steel structure with neat interior and large usable space should be retained. Therefore, the section width of all frame beams and columns does not exceed the maximum wall thickness of 150mm in the models. Among them, the beam section is narrow flange H-beam or welded I-steel, and the column section is square steel pipe. Figure 2 is the schematic diagram of the model with 6 stories (N=6).

2.3. Materials and Loads

The beam element material of all models is Q355B and the layered-shell element material of all models is Oriented-strand board (OSB) and gypsum board. Among them, the external wall panel material is OSB with 9mm thickness on one side and gypsum board with 12mm thickness on the other side. The interior wall panel is gypsum board with 12mm thickness on both sides. Material parameters are shown in table 1.

Table 1. Material parameters.

	Elastic modulus (MPa)	Poisson ratio	Mass density (kg/m³)
Steel	2.06×10^5	0.30	7850
OSB	3500	0.30	600
Gypsum board	1125	0.23	900

The dead load of wall beam of all models in this paper is uniformly set as 3kN/m. The dead and live load of floor of all models in this paper are uniformly set as 2.0kN/m². The dead and live load of roof are set as 3.5kN/m² and 0.5kN/m² respectively. The basic wind pressure of all models in this paper is 0.5kN/m² and the surface roughness is B. Consider frequent horizontal earthquakes and the seismic fortification intensity is 7 degrees (0.10g) and the maximum horizontal earthquake influence coefficient is $\alpha_{\max}=0.08$. The design earthquake groups are all group 1 class II, corresponding characteristic period is $T_g=0.35s$. The damping ratio is 0.04. The default load combination is generated according to the load code.

3. Parameter Analysis

In this paper, the vertical structure system of steel frame—cold-formed thin-walled composite steel wall is studied on the overall structure level to analyze the influence of different parameters on the mechanical performance and economic steel amount. The parameters include: spacing of self-tapping screw on composite wall, section and spacing of cold-formed thin-walled steel purlin, width of wall opening. In this section, column spacing of all models is A=4.5m, story height is H=3m, number of floors is N=6. In the process of analysis and design, the strength, deformation and stability of all frame members are checked and optimized in accordance with the《Standard for design of steel structures》[12]. The limit of cross-section stress ratio is set at 0.9, and the design indexes such as structural period ratio and displacement ratio are ensured to meet the requirements of the code.

Three steel frame—cold-formed thin-walled steel overall structure models with screw spacing of 100mm, 150mm and 300mm are respectively named as Model 1, Model 2 and Model 3. The width of wall opening is 2.1m, the section of wall purlin is C109×50×15×2.0 and the spacing of purlin is 600mm. On the basis of model 3, the cross-sections of wall purlin are adjusted to C109×50×20×2.5 and C109×60×20×3.0, and the adjusted models are named as model 4 and model 5 respectively. On the basis of model 3, move the openings of all external walls to the right end of the wall and move those of all internal walls to the near external wall, and the width of opening K is still 2.1m. The purlin spacing is adjusted to 400mm, 600mm and 800mm, and name the adjusted models as Model 6 to 8 respectively. According to the result of design checking and optimization adjustment, the inter-story displacement angles of Model 1 to 8 are basically the same which are far less than the limit of the specification, and the stress ratio of beam-column member is also similar. The steel amount of Model 1 to 3 is all the same, and those of Model 4 and 5 increase with the increase of purlin section, and those of Model 6 to 8 decrease with the increase of purlin spacing. It can be seen that the screw spacing has almost no effect on the design section of the beam and column but plays a role in fixing the wall panel to make it deform in coordination with the purlin and frame. Therefore, it is suggested that the screw spacing in practical engineering should keep the requirements of " the ones at the edge of the panel should be within 150mm and that of in the middle of the panel should be within 300mm " in the code of cold-formed thin-walled steel structure [2]. Since the frame beams and columns in the model bear the main vertical load, the stress of the purlin section itself is very small, and it only acts to fix the wall panels and make them deform in coordination with the frame. To avoid waste, the wall purlin only needs to choose a relatively small section which meet the requirements of the wall thickness and the purlin spacing is recommended to be properly amplified to 600~800mm.

On the basis of Model 7, the width of the wall opening is adjusted to *K*=2.7m and *K*=1.5m and the adjusted models are named Model 9 and Model 10 respectively. Other parameters are the same as Model 7. After design checking and optimization adjustment, the beam and column sections of three models are all the same. The steel amount used increase from Model 9, 7 and 10, while the stress ratio of beam and column members decrease from Model 9, 10 and 7. Figure 3 shows the statistics of inter-story displacement angles of them. It can be seen that reducing the opening width of the wall will significantly improve the lateral stiffness of the structure, but too small opening width will lead to the increase of internal forces of the frame and the waste of steel used in the composite wall.

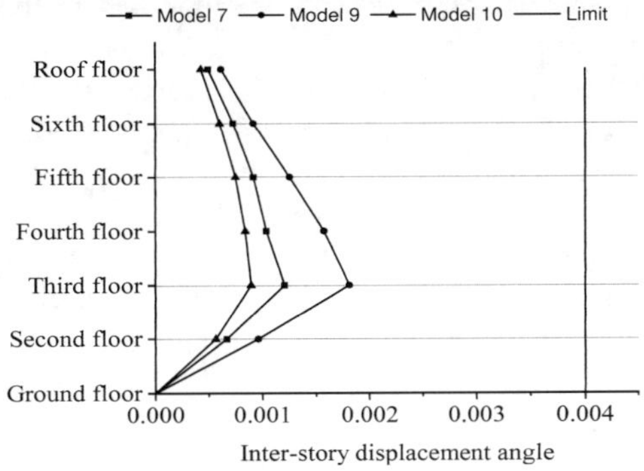

Figure 3. Inter-story displacement angle of Model 7、9、10.

4. Applicable Height

Based on the above analysis, a series of models are established according to Figure 1. The spacing of wall purlin of all models is 600mm, and the spacing of screws is 300mm. The section of the purlin is adjusted according to the width of the frame beam and column. When beam and column width is 130mm, 140mm and 150mm, the purlin section is C109×50×15×2.0 , C119×50×20×2.0 and C129×50×20×2.5 respectively. The two column distances are *A*=4.5m and *A*=6m, and the corresponding opening width is *K*=2.1m and *K*=2.4m respectively, which are located in the middle of the wall. The story height is *H*=3.0m. The seismic fortification intensity is 7 degrees (0.10g), and the wind load is 0.5kN/m^2. After design checking and optimization adjustment, table 2 and table 3 are the beam and column sections of models with two column distances and different stories.

Table 2. Beam and column sections of models with column distance *A*=4.5m.

Stories	6	7	8	9	10
Z1(n=1~3)	130×130×6	130×130×6	130×130×8	140×140×8	150×150×8
Z2(n=1~3)	130×130×6	130×130×6	130×130×8	140×140×8	150×150×10
Z3(n=1~3)	130×130×10	130×130×10	130×130×12	140×140×12	150×150×12
Z1(n=4~6)	130×130×6	130×130×6	130×130×6	140×140×6	150×150×6
Z2(n=4~6)	130×130×6	130×130×6	130×130×6	140×140×6	150×150×6
Z3(n=4~6)	130×130×6	130×130×6	130×130×8	140×140×8	150×150×8
Z1~3(n=7~10)		130×130×6	130×130×6	140×140×6	150×150×6
L1, L2	HN250×125×5×8	HN250×125×5×8	HN250×125×5×8	H250×140×5×8	H250×140×5×8
L*	HN250×125×5×8	HN250×125×5×8	HN250×125×5×8	HN250×125×5×8	HN250×125×5×8

Note: When *N* = 6, the column width is 130mm≈ the minimum width of beam is 125mm.
When *N* < 6, the column width should be kept equal to the beam width, resulting in steel waste.

When $N > 10$, the column width will exceed 150mm, which does not meet the section width limit.

Table 3. Beam and column sections of models with column distance A=6.0m.

Stories	4	5	6
Z1(n=1~3)	150×150×6	150×150×6	150×150×8
Z2(n=1~3)	150×150×6	150×150×6	150×150×10
Z3(n=1~3)	150×150×8	150×150×12	150×150×12
Z1~3(n=4~6)	150×150×6	150×150×6	150×150×6
L1, L2	HN360×150×7×11	HN360×150×7×11	HN360×150×7×11
L*	HN350×175×6×9	HN350×175×6×9	HN350×175×6×9

Note: When $N = 4$, the column width is 150mm= the beam width is 150mm= the section width limit.
When $N < 4$, the column width should be kept equal to the beam width, resulting in steel waste.
When $N > 6$, the column width will exceed 150mm, which does not meet the section width limit.

When the seismic fortification intensity is adjusted from 7 degrees (0.10g) to 8 degrees (0.20g), the design section of the beam and column is basically unchanged. According to the query, due to the light deadweight of structure and small load, the wind load is relatively large. The seismic action does not play a major control role, and the load combination corresponding to the design internal force of the beam and column section is mostly related to the wind load. At the same time, due to the large redundancy of the lateral stiffness of the structure, the increased inter-story displacement angles caused by the increase of seismic fortification intensity will not lead to the increase of the design section. With the further increase of seismic action and the adjustment of wind load, this characteristic may not be applicable. Therefore, when the wind load is 0.5kN/m², the seismic fortification intensity is less than 8 degrees, and the limit width of the section of beam and column is 150mm, the maximum applicable height of the steel frame—cold-formed thin-walled steel composite wall structure system with column spacing of 4.5m and 6.0m is about 30m and 18m respectively.

5. Contrast of Steel Amount

Figure 4 and figure 5 show the steel amount statistics for the models with column distance of 4.5m and 6.0m and the corresponding pure frame model. As can be seen from the figures, the steel amount of steel frame—cold-formed thin-walled steel structure system is about 30~40kg/m² when it is applied to residential and office buildings with column distance of 4.5m or 6m, which can significantly reduce the steel amount compared with pure steel frame structure. Among them, when the column distance is 4.5m, the steel amount is about 10-20% less than that of pure steel frame structure, and when the column distance is 6.0m, the reduction is about 10%.

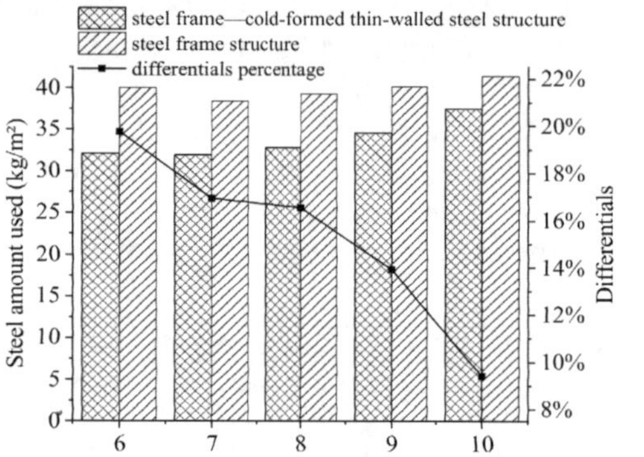

Figure 4. The steel amount statistics of models with column distance A=4.5m.

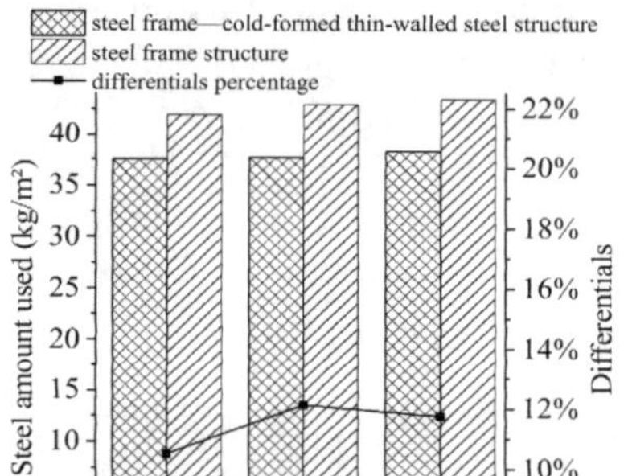

Figure 5. The steel amount statistics of models with column distance A=6.0m.

6. Conclusions

Based on the above test results, the following conclusions can be drawn: the screw spacing, purlin section and purlin spacing in the composite wall of the system have little effect on the elastic stage of the structure. It is suggested that the screw spacing in practical engineering should keep the requirements of " the ones at the edge of the panel should be within 150mm and that of in the middle of the panel should be within 300mm " in the code of cold-formed thin-walled steel structure, purlin section should be relatively small according to wall thickness, and purlin spacing should be 600~800mm. The opening width of the composite wall has a significant effect on the lateral stiffness and the steel amount of the structure. Reducing the opening width will significantly improve the lateral stiffness of the structure, but too small opening width will lead to the increase of the internal force of beam and column and the waste of steel. When the wind load is 0.5kN/m^2, the seismic fortification intensity is less than 8 degrees, and the limit width of the section of beam and column is 150mm, the maximum applicable height of the steel frame—cold-formed thin-walled steel composite wall structure system with column spacing of 4.5m and 6.0m is about 30m and 18m respectively. The steel amount of steel frame—cold-formed thin-walled steel structure system is about 30~40kg/m^2 when it is applied to residential and office buildings with column distance of 4.5m or 6m. When the column distance is 4.5m, the steel amount is about 10-20% less than that of pure steel frame structure, and when the column distance is 6.0m, the reduction is about 10%.

Acknowledgements
The research was financially supported by the Science and Technology Planning Project of Guangzhou Municipal Construction Group Co., Ltd. (Granted No [2019]-KJ304, [2019]-KJ018)

References
[1] JGJ 227-2011 2011 *Technical specification for low-rise cold-formed thin-walled steel buildings* (Beijing: China Architecture & Building Press)
[2] JGJT 421-2018 2018 *Technical standard for cold-formed thin-walled steel multi-storey residential buildings* (Beijing: China Architecture & Building Press)
[3] Dong Y Z, Lou N L 2005 A snapshot of multistory light steel housing in the United States *World Building* **1** pp 77
[4] Shi Y 2005 *Research on shear capacity of composite wall of low-rise cold-formed steel structure residential building* (Xian: Chang'an University) pp 13-52
[5] Li L 2012 *Research on hysteretic behavior of steel frame filled with cold-formed steel wall structure system* (Xian: Xi'an University of Architecture and Technology) pp 20-86
[6] Guo Y L 2018 Static lateral resistance performance and design method of steel frame-inlaid cold-formed steel thin-wall composite wall *Journal of Ningxia University (Natural Science Edition)* **39** pp 54-60
[7] Wu H H 2011 *Research on cooperative working performance of cold-formed steel composite wall-steel frame structure* (Xian: Chang'an University) pp 12-38
[8] Huang G Y 2013 *Study on the mechanical behavior and economic performance of thin-walled steel assembled-infilled wall—steel frame* (Xian: Chang'an University) pp 68-73
[9] Lu L, Lei Y D, Yu Y Q 2016 Integral behavior analysis of a new cold-formed thin-walled steel framing system of low-rise residential buildings *Building Structure* **46** pp 370-375
[10] Dan Dubina 2008 Behavior and performance of cold-formed steel-framed houses under seismic action *Journal of Constructional Steel Research* **64** pp 896-913
[11] Liu J B, Chen M, Liu X Q, et al 2008 Integral behavior analysis of cold-formed thin-walled steel framing system of low-rise residential buildings *Journal of Architecture and Civil Engineering* **25** pp 6-12
[12] GB 50017-2017 2017 *Standard for design of steel structures* (Beijing: China Architecture & Building Press)

Verification Simulation and Development of New Marine Transport Model

Jing Zhang*

Zhejiang Ocean University School of Information Engineering, China

*Corresponding author email: j.zhang.gr@zjou.edu.cn

Abstract. The structure of the global supply chain creates an imbalance in trade between regions of the world. It creates many empty containers on the ocean shipping route. Focusing on this problem, we will develop an empty container relocation plan for empty containers in a marine container logistics network connecting multiple bases and propose a model that minimizes the total cost of container transportation. The proposed model was verified by the simulation designed and developed in this study, and the effectiveness of the proposed model was shown by the data of the marine transportation company.
Keywords: Model, routing, algorithm, simulation, validation.

1. Introduction

With globalization of the world's economy, international maritime transport is becoming increasingly popular [1]. The global supply chain is configured to carry goods manufactured in East Asia to locations throughout the world. As a result, container transport has become active and there is an imbalance in the directionality of container travel. There are container shortages in East Asia, because exports exceed imports. Surplus containers accumulate in ports with an excess of imports. That is, while there are some ports with room to stack empty containers, there are other ports with no room. It therefore becomes necessary to circulate and rearrange empty stacked containers on surplus land.

As an existing approach to this problem, classified empty container relocation into three groups. The first focuses on empty container transport in a maritime container logistics network. The second considers inland container transport networks or empty container transport in integrated transport including inland transport. The third incorporates decision-making problems. We studied empty container transport as a constraint. Song et al also used dynamic programming to consider uncertain demands and to establish an optimal inventory management method that is responsive to acceptable shortage rates [2–3]. Zhang et al. found optimal combinations of ports with supply and demand for empty containers in a maritime container network by reducing the assignment problem in multiple ports [4–7].

In this research, we take a different approach from the existing research, focus on the current situation of shipping containers, and consider the relocation plan of the marine container network connecting multiple bases, the number of containers exchanged between ports and container inventory management. Propose a new model that minimizes the cost of moving empty containers. We consider both macro and micro perspectives. Macroscopically, we consider work from the perspective of product flows in the supply chain. In the network formed by container ship services connecting each port at the micro level, empty containers taken as stock at each port are interchanged. The aim is to minimize the cost of relocating empty containers by restricting empty container inventory shortage rates to below a certain level depending on the supply and demand of empty containers at each port. In addition, verification is

Content from this work may be used under the terms of the Creative Commons Attribution 3.0 licence. Any further distribution of this work must maintain attribution to the author(s) and the title of the work, journal citation and DOI.
Published under licence by IOP Publishing Ltd

performed using actual data obtained as part of industry-academia collaboration involving shipping companies, and the academic and practical values of the proposed method are shown.

2. Model Approach

2.1. Modeling and Entities

Under the constraint that the empty container shortage rate at each port remains below a certain level, let the objective function be cost minimization for empty container relocation. Furthermore, the optimal allocation of empty containers must consider port space and container vessels.
The following entities are in this model:
① Shipping company: A company engaged in the shipping business.
② Harbor: A company involved in port management.
③ Container leasing company: Consider containers owned by container leasing companies.
④ Container manufacturing company: Shippers tend to prefer new containers.
⑤ Cargo owner: Cargo owners are customers requesting ocean transportation.

2.2. Demand Forecasting for Empty Containers

First, we define empty container demand. As a premise, the occurrence of empty containers in ports is considered as "production," and "consumption" is when goods are loaded on empty containers. Empty containers are considered as stock until loading. Because empty containers are "consumed" when transport demand occurs, the number of inventory changes in each port is the number of empty containers returned minus the number of empty containers transported from the port. Here, the empty container demand is the same value as the decrease in empty container stock. In a given port (p), empty container demand $Demand_{p,d}$ in a certain period (*d*) is calculated using the transport demand $Transport_{p,d}$ and the number of empty container returns $Return_{p,d}$, expressed as a formula. As mentioned above, the manufacture and disposal of empty containers are not considered.

$$\boldsymbol{Demand_{p,d} = Transport_{p,d} - Return_{p,d}}$$

$\boldsymbol{Transport_{p,d}}$: Empty container consumed (transportation demand)

$\boldsymbol{Return_{p,d}}$: **Empty container refilled (stock replenishment)** (1)

2.3. Empty Container Inventory

Containers have economic value because they are used for marine logistics and generate profits. There are idle periods where they are not used for transport or at ports, and during that period they can be regarded as inventory that does not generate profits. The main objective of inventory management is to minimize total costs. There are various factors related to the cost of inventory control, but in this study, we mainly classify three.
① Inventory maintenance cost: In this study, we consider the cost per day of one empty container.
② Ordering cost: In this study, on the premise of regular container ships, container ships enter ports on specific days without a fixed number of empty containers.
③ Out-of-stock costs: We formulate a plan under the constraint that the shortage rate in each port must be kept below a certain level.

2.4. Empty Container Flexibility and Inventory

The ordering system mainly includes a quantitative ordering point system and a regular ordering point system. The former is for placing a fixed number of orders when the order interval is not fixed. The latter places periodic orders at different quantities. Considering the ordering method for empty container inventory, the interval of the ordering period of the empty container inventory at each port is determined by the schedule of the container ship calling on port.
In this study, on the premise of fixed-day service, the number of container ships determines the travel schedule. For example, assume it takes 21 days to complete a route and three container vessels will provide fixed-day service. In this case, one container ship will arrive each week.
The amount of accommodation between a port $p \in P$ and a container ship $l \in L$ is defined as $x_{l,p}$ as in

Equation 2, and takes a positive or negative integer value. $x_{l,p}$ assumes positive values when an empty container is transferred from a container ship to a port. As a result, it is possible to formulate the amount of interchange without distinguishing between supply or demand ports. Even in the same port, since $x_{l,p}$ can take a positive or negative value for each container ship, transshipment in a hub port can be reproduced. When link $l \in L$ does not call at port $p \in P$, $x_{l,p}$ becomes 0, so the corresponding $x_{l,p}$ is set to 0, thereby speeding up the calculation.

$$\mathbf{X} = \begin{pmatrix} x_{1,1} & \cdots & x_{1,p} & \cdots & x_{1,P} \\ \vdots & & \vdots & & \vdots \\ x_{l,1} & \cdots & x_{l,p} & \cdots & x_{l,P} \\ \vdots & & \vdots & & \vdots \\ x_{L,1} & \cdots & x_{L,p} & \cdots & x_{L,P} \end{pmatrix} \quad (2)$$

$x_{l,p}$: *Number of empty containers accommodated between link* $l \in L$ *and port* $p \in P$

In addition, inventory fluctuation can be calculated as the number of transferred empty containers minus the number of required empty containers. We therefore consider the total number of transferred empty containers and the total number of required empty containers. To handle the number of empty containers transferred at each port as a time series, we introduce a calendar constant $\mathbf{C_p}$ that creates a schedule from the number of transferred empty containers at ports (Equation 3). The element c^{p}_{ld} defines the case where container ship $l \in L$ calls at port $p \in P$ on day d as 1 and other cases as 0. According to this, $\mathbf{C_p}$ can be calculated for each port from the liner service schedule.

$$\mathbf{C_p} = \begin{pmatrix} c^{p}_{11} & \cdots & c^{p}_{1d} \\ \vdots & \ddots & \vdots \\ c^{p}_{l1} & \cdots & c^{p}_{ld} \end{pmatrix}$$

$$c^{p}_{ld} = \begin{cases} \mathbf{1} \ \ (\textit{if the Link } l(\in L) \textit{ calls at the Port } p(\in P) \textit{ on the Day } d(\in D)) \\ \mathbf{0} \ \ (\textit{if it does not}) \end{cases} \quad (3)$$

The number of orders $O_{p,d}$ on day d at port $p \in P$ can be obtained by Equation 4 for a planning period of D days.

$$O_{p,d} = \sum_{l \in L} x_{l,p} * c^{p}_{l,d} \ \textit{for each} p \in P, d \in D \quad (4)$$

The cumulative order quantity $TO_{p,d'}$ from $0 \in D$ to $d \in D$ can be obtained from the order placement schedule as Equation 5.

$$TO_{p,d'} = \sum_{d=1}^{d'} O_{p,d} = \sum_{d=1}^{d'} \sum_{l \in L} (x_{l,p} * c^{p}_{l,d}) \ \textit{for each}\ p \in P, d' \in D \quad (5)$$

To express the amount of stock transition using the flexible variable, regarding demand for empty containers at port $p \in P$, let the forecasted average cumulative demand from $0 \in D$ to $d \in D$ be $Demand_{p,d}$. Variation (standard deviation) in predictions uses the concept of "safety stock," that is, the optimal order quantity such that the shortage rate β remains below a certain level. Make a stock plan. Safety stock is calculated by multiplying a safety factor by the standard deviation σ of the accumulated demand in the planning period, and can be expressed as the second term on the right side of Equation 6. If the forecasted date is in the future, this safety stock will be larger than it is the future. A total demand forecast amount $Kdemand_{p,d}$ considering safety stock is defined as the sum of the above and the average cumulative demand forecast amount $Demand_{p,d}$. Here, "determining the optimal order quantity under the condition that the shortage rate β becomes lower than a certain level over the future" is "the cumulative order quantity minus the cumulative demand forecast quantity considering safety stock" "It is always positive during the planning period."

This condition can be expressed as Equation 7.

$$Kdemand_{p,d} = Demand_{p,d} + k * \sigma * \sqrt{d} \quad (6)$$

$$TO_{p,d} - Kdemand_{p,d} \geq 0 \quad (7)$$

We next consider how to represent shipboard stock using a flexible variable. First, according to the schedule of a given container ship $l \in L$, the port that calls at $d \in D$ is expressed as $PC_{l,d} \in P$ as in Equation 8. When no port is called, $PC_{l,d}$=0.

Denoting shipboard stock fluctuation for a container ship $l \in L$ in $d \in D$ as $SB_{l,d}$, the container ship $l \in L$ is the $PC_{l,d} \in P$, which is equal to the empty container accommodation amount $-x_{l,PC_{ld}}$. When $PC_{l,d}$=0 the container ship does not call at the port, so $SB_{l,d}$ becomes 0. Therefore, $SB_{l,d}$ is expressed as in Equation 9.
The onboard inventory $TSB_{l,d}$ at $d \in D$ is the sum of the cumulative amount of $SB_{l,d}$ at $d \in D$ and the initial onboard inventory (defined as ISB), as in Equation 10. This formula can represent the stock volume transition on a container ship using a flexible variable.

$$\boldsymbol{PC_{l,d}} = \begin{cases} \boldsymbol{p(if\ Link\ l\ calls\ Port\ p)} \\ \boldsymbol{0(if\ the\ Link\ does\ not\ call\ any\ Port)} \end{cases} \tag{8}$$

$$\boldsymbol{SB_{l,d}} = \begin{cases} \boldsymbol{-x_{l,PC_{l,d}}\ (if\ Link\ l\ calls\ Port\ p)} \\ \boldsymbol{0 \quad (if\ the\ Link\ does\ not\ call\ any\ Port)} \end{cases} \tag{9}$$

$$\boldsymbol{TSB_{l,d} = ISB + \textstyle\sum_{d=0}^{d} SB_{l,d}} \tag{10}$$

3. Modeling and Simulation

We formulate planning optimization methods. The following describes formulated objective functions and constraints.

3.1. Setting of Objective Function

The following constants are defined for the set P of ports to be designed, the set L of container ships, the set S of price stages in the stage charge system, and the period D for planning.

$l \in L$ Container ship
$p \in P$ Port
$s \in S$ Price stage
$d \in D$ Number of days until the forecast target date [days]
C_p Calendar constant [dimensionless]
IP_p Inventory management cost [yen/unit]
LE_p Lease expenses [yen/unit]
$SP_{p,s}$ Additional charge [yen/unit]
B Number range of each price stage
CHC_p Container handling cost [yen/unit]
FC_l Container ship fuel cost [yen/unit]
$LC_{l,d}$ Container ship loading limit [units]
μ_p Empty container average demand volume [units/day]
σ_p Standard deviation of empty container demand quantity

To facilitate description of the formulation, the following describes the storage cost, the empty container purchase cost, the empty container handling cost, and the stepwise charging cost.
The cost for storage is shown in Formula 12 as C_{Stock}. The port will require a maintenance storage fee according to the daily stock quantity. It is thus necessary to calculate the daily stock amount, which is obtained by subtracting the accumulated demand amount from the accumulated order amount.
Equation 13 shows the cost of handling empty containers at a port as $C_{Loading}$. The cargo handling volume at the port is considered to be proportional to the number of empty containers transferred between the port and the container ship. Therefore, the cargo handling amount $z_{l,p}$ is considered as the absolute value of the empty container interchange amount. The cargo handling cost per TEU is set as CHC_p for each port and multiplied by the cargo handling cost.
The stage charge cost is shown in Equation 14 as $C_{StageCharge}$. Formulation follows the concept of staged charges as described above. Restrictions on the staged charges will be described later. The cost for container shipping is shown in Formula 15 as $C_{Shipping}$, and is calculated for each container ship from the onboard inventory transition formulated as described above. In addition, the forecasted cumulative demand considering safety stock will be shown again. The forecasted cumulative demand is shown on a daily basis as a $Kdemand_{p,d}$ number sequence.

$$C_{Stock} = \sum_{p \in P} \left(W_p * \left(\sum_{d' \in D} TO_{p,d} - \sum_{d \in D} Kdemand_{p,d} + \sum_{d \in D} (D - d + 1) y_{p,d} \right) \right)$$

$$= \sum_{p \in P} \left(\boldsymbol{W_p} * \left(\sum_{d \in D} (\boldsymbol{D} - \boldsymbol{d} + \mathbf{1}) \sum_{l \in L} (\boldsymbol{x_{l,p}} * \boldsymbol{c^p_{l,d}}) - \sum_{d \in D} \boldsymbol{Kdemand_{p,d}} + \sum_{d \in D} (\boldsymbol{D} - \boldsymbol{d} + \mathbf{1}) \boldsymbol{y_{p,d}} \right) \right) \quad (11)$$

$$\boldsymbol{C_{Lease}} = \sum_{p \in P} \left(\boldsymbol{LP_p} \sum_{d \in D} \boldsymbol{y_{p,d}} \right) \quad (12)$$

$$\boldsymbol{C_{Loading}} = \sum_{p \in P} \left(\boldsymbol{CHC_p} \sum_{l \in L} \boldsymbol{z_{l,p}} \right) \quad (13)$$

$$\boldsymbol{C_{StageCharge}} = \sum_{p \in P} \sum_{s \in S} \left(\boldsymbol{SP_{p,s}} * \boldsymbol{w_{p,s}} \right) \quad (14)$$

$$\boldsymbol{C_{Shipping}} = \sum_{l \in L} \left(\boldsymbol{FC_l} \sum_{d \in D} \boldsymbol{TSB_{l,d}} \right) \quad (15)$$

The objective function is set as Equation 16. The optimal solution does not change even if terms that do not change due to variables in the objective function are eliminated. Therefore, the term related to the accumulated demand for the variable among the storage costs C_{Stock} is eliminated, and Equation 17 is the solution to the minimization problem. This is because the cumulative demand forecast amount considering the safety stock is a forecast value until boredom and is treated as a constant value in forecasts. Storage costs are thus determined by the cumulative order quantity and the cumulative lease quantity represented by the variables.

$$\min(\boldsymbol{C_{Stock}} + \boldsymbol{C_{Lease}} + \boldsymbol{C_{Loading}} + \boldsymbol{C_{StageCharge}} + \boldsymbol{C_{shipping}}) \quad (16)$$

$$\sum_{p \in P} \left(\boldsymbol{W_p} * \left(\sum_{d \in D} (\boldsymbol{D} - \boldsymbol{d} + \mathbf{1}) \sum_{l \in L} (\boldsymbol{x_{l,p}} * \boldsymbol{c^p_{l,d}}) + \sum_{d} (\boldsymbol{D} - \boldsymbol{d} + \mathbf{1}) \boldsymbol{y_{p,d}} \right) \right) \quad (17)$$
$$+ \sum_{p} \left(\boldsymbol{LP_p} \sum_{d} \boldsymbol{y_{p,d}} \right)$$
$$+ \sum_{p} \left(\boldsymbol{CHC_p} \sum_{l} \boldsymbol{z_{l,p}} \right)$$
$$+ \sum_{p} \sum_{s} \boldsymbol{SP_{p,s}} * \boldsymbol{w_{p,s}}$$
$$+ \sum_{l \in L} \left(\boldsymbol{FC_l} \sum_{d \in D} \boldsymbol{TSB_{l,d}} \right)$$

3.2. Setting Constraints

There are four constraints, for the stock shortage rate, ship size, cargo volume, and stage rate system. Equation 18 is a constraint that reduces the stock shortage rate to below a certain value during the planning period. The upper limit on the shortage rate is set using safety coefficient k, and formulation is performed according to the stock transition calculation method. Formula 19 sets lower and upper limits on the number of empty containers on a container ship. The amount of space on the container ship sets the upper limit value in chronological order. Although $x_{l,p}$ can take a negative value to indicate the amount of accommodation, the number of empty containers on a ship must be a nonnegative value, so the lower limit is 0. Equation 20 is set for the definition of $z_{i,j}$ that represents the amount of cargo handling. The cargo handling volume at the port is expressed as an absolute value of the porting volume $z_{i,j}$ of the container ship, so setting it as Equation 20 allows $z_{i,j}$ to be $x_{l,p}$ in linear programming. It can be treated as an absolute value of Equation 21 is an equation for a tiered billing system, set up to incorporate a system for staged charging.

$$\sum_{\mathbf{d=1}}^{\mathbf{d'}} \left(\sum_{l \in L} (\boldsymbol{x_{l,p}} * \boldsymbol{c_{l,d',p}}) + \boldsymbol{y_{p,d}} \right) + \boldsymbol{FirstStock_p} - \boldsymbol{Kdemand_{p,d'}} \geq \mathbf{0} \quad (18)$$
$$(\boldsymbol{for\ aDay\ d'} \in \boldsymbol{D}\ \boldsymbol{and\ Port\ p} \in \boldsymbol{P})$$

$$\mathbf{0} \leq \boldsymbol{TSB_{l,d}} \leq \boldsymbol{LC_{l,d}}\ (\boldsymbol{for\ Link\ l} \in \boldsymbol{L}\ \boldsymbol{and\ Day\ d} \in \boldsymbol{D} \quad (19)$$

$$-\boldsymbol{z_{l,p}} \leq \boldsymbol{x_{l,p}} \leq \boldsymbol{z_{l,p}} \quad (20)$$

$$\boldsymbol{predicted\ stock} - \sum_{s=n+1}^{S} \boldsymbol{w_{p,s}} \leq \boldsymbol{B} * \boldsymbol{n}\ (\boldsymbol{for\ Port\ p} \in \boldsymbol{P}\ \boldsymbol{and\ Stage\ s} \in \boldsymbol{S}) \quad (21)$$

The variables are $x_{l,p}, y_{p,d}, z_{l,p}, w_{p,s}$ for each port $p \in P$, each container ship $l \in L$, and each price stage $s \in S$. We set the variables at the time of planning formulation as follows.

$\boldsymbol{x_{l,p}}$ **Number of containers transferred from container ship** $l \in L$ **to port** $p \in P$
$0 \leq \boldsymbol{y_{p,d}}$ **Number of containers leased at port** $p \in P$
$0 \leq \boldsymbol{z_{l,p}}$ **Handling amount** (absolute value of $x_{i,j}$)
$0 \leq \boldsymbol{w_{p,s}} \leq \boldsymbol{B}$ Number of containers to which price stage s applies

3.3. Setting of Target Period and Simulation Design

This is a period of guarantee that the stock shortage rate β will be below a certain level at the time of planning. That is, the constraint is that during the planning period, the probability of shortage will be less than or equal to β. The following simulation is performed by the above-mentioned modeling.

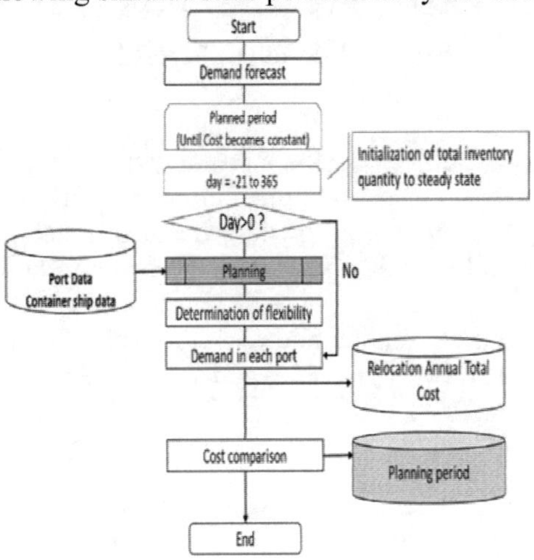

Figure 1. Simulation flowchart.

4. Verification with Actual Data

The following simulation verifies the proposed formulation model using actual data.

Table 1. Demand, inventory and handling costs at each port.

ID	Name	μ	σ	IP	CHC
1	Shanghai	100	10	800	30000
2	Singapore	100	10	0	7000
3	Tokyo	100	10	1500	30000
4	Yokohama	90	9	1500	30000
5	Nagoya	80	8	1500	30000
6	Manila	50	5	1500	30000
7	LA	-520	52	1200	30000

Figure 2 shows the results of a one-year simulation for the planning period. As the period considered in planning grows, it becomes possible to perform appropriate planning and reduce total costs. After 21 days, the fluctuation range of the total cost becomes smaller, and it becomes possible to formulate a container ship management plan.

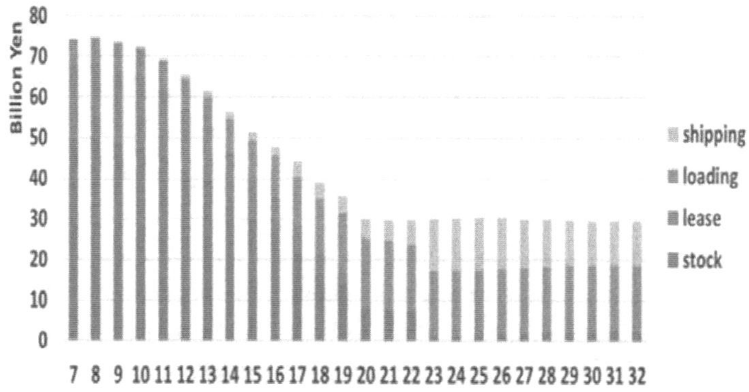

Figure 2. Simulation results.

5. Conclusion

In this study, we proposed a new empty container rearrangement model and formulation method in consideration of real companies such as Shipping company, Harbor, Container leasing company, Container manufacturing company, and Cargo owner. We also developed a simulation of the current state of container transportation and model validation. Furthermore, the validity of the proposed model was verified using actual data.

We proposed a new model and formulation method for relocation of empty containers, and developed a simulation that can be reproduced and verified. We were able to verify the effectiveness of the proposed method using actual data.

References

[1] Drewry (2011), Container market –annual review and forecast. Drewry Shipping Consultants, London

[2] Crinks, P. (2000) Container Usage Asset Management in the Global Container Logistics Chain, International Asset Systems.

[3] Song, Dong-Ping, and Jing-Xin Dong. "Cargo Routing and Empty Container Repositioning in Multiple Shipping Service Routes." Transportation Research Part B: Methodological 46, no. 10 (December 2012): 1556–75.

[4] Song, D., and Q. Zhang (2010) "A Fluid Flow Model for Empty Container Repositioning Policy with a Single Port and Stochastic Demand." SIAM Journal on Control and Optimization 48, no. 5: 3623–42.

[5] Furió, Salvador, Carlos Andrés, Belarmino Adenso-Díaz, and Sebastián Lozano (2013) "Optimization of Empty Container Movements Using Street-Turn: Application to Valencia Hinterland." Computers & Industrial Engineering 66, no. 4

[6] Wong, Eugene Y. C., Allen Tai, and Mardjuki Raman (2015) "A Maritime Container Repositioning Yield-Based Optimization Model with Uncertain Upsurge Demand." Transportation Research Part E: Logistics and Transportation Review 82

[7] Zheng, Jianfeng, Zhuo Sun, and Ziyou Gao (2015) "Empty Container Exchange among Liner Carriers." Transportation Research Part E: Logistics and Transportation Review 83

Research and Modeling Electronic Data Tracing Scheme Based on Blockchain

Zhengtao Jiang[1,2], Yi Chen[3], Zheng Liu[1,*], Hao Chen[3], Jianhong Zhang[4] and Yanchao Zhu[3]

[1]State Key Laboratory of Communication University of China, Beijing 100024, China
[2]Guangxi Key Laboratory of Cryptography and Information Security, Guilin, China
[3]Ele-Cloud Information Technology Co. Ltd., Beijing, China
[4]School of Information Sciences and Technology, North China University of Technology, Beijing, China

[*]Corresponding author email: liuzhengkeep@foxmail.com

Abstract. From the point of view that the traditional electronic data (documents in image, PDF, etc) is difficult to track and trace, taking image copyright as an example, this paper studies a new scheme to solve the problem of replication detection and traceability based on blockchain. With the decentralization and trust removing of blockchain, both copyright storage and evidence of infringement are stored in a tamper proof and transparent way. Based on the judgment of image similarity, a "resource reverse tracing strategy" is designed to control the infringement timely. The infringement evidence can reach the creator quickly, which provides convenience for the creator's rights protection. This method changes traditional tracing methods of image copyright. Also, it can be applied to the tracing problem of reuse of electronic invoice, short video, music, etc, and brings new ideas for the related problem.
Keywords: Blockchain, image copyright, P2P, cloud storage, image recognition, electronic invoice.

1. Introduction

In the era of "we media", all kinds of self-claimed original works are full of every corner of the Internet, but the original source of works is difficult to guarantee. As a centralized management mechanism, the traditional digital copyright protection technology is authorized by an authoritative management organization.

Since different copyright management platforms do not support interoperability, it's difficult to find the original copyright owner, which may harm the author's interests. For massive image infringement, it is impossible to realize accurate and rapid tracing of image copyright by manual means[1].

For providing better digital rights protection, the decentralization and traceability of blockchain are cocerned by researchers[2]. It realizes digital copyright registration, transaction, and tracing through its special mechanism which has no need of third party. Compared with the traditional way of image copyright protection, the image copyright tracking and tracing system based on blockchain improves the efficiency of image copyright protection, greatly saves the cost of image copyright protection, and reduces the risk of being infringed.

2. Related Work

Some researchers have studied copyright protection based on blockchain. In 2015, Kubesch et al anlysed the method to realize digital copyright protection by applying user access and the related code

Content from this work may be used under the terms of the Creative Commons Attribution 3.0 licence. Any further distribution of this work must maintain attribution to the author(s) and the title of the work, journal citation and DOI.
Published under licence by IOP Publishing Ltd

protocols. They also try to explore solutions to balance the interests and relationships among users. Wang et al designed a mechanism based on blockchain and the technique of watermark to reduce the need of a trusted third party[3]. In their paper, they proposed a blockchain-based image copyright protection framework and solved the blockchain data expansion problem by interplanetary file system. In addition, the image owner can authenticate the image and trace the copyright.
Zihao Lu et al introduced a blockchain-based digital rights management scheme to balance the problems of content protection, copyright protection and trading of works. According to the scheme, the off-chain design work and its copyright record are binded in the blockchain[4]. By using smart contracts and public key cryptography, their scheme has no need for active operation from participants in the trade.
D. Roman and K. Vu use smart contracts to solve the problem of single point failure for trusted third parties in data markets[5]. In their method, they proposed an architecture and then analysed the feasibility of the architecture for data markets which based on smart contracts and multi-party computation.
In [6], R. A. Dobre et al designed a system based on blockchain which aimed at detecting and avoiding copyright infringement. Photographers can register their photos with it. Photographers can use it to register their photos and the system can check the copyright condition of the image to be used. Their system also has the function to buy the right of photo for using. It is likely that recompressed versions of the same photo is stored in different places, since businesses can buy images on many sites.
We design an image copyright tracking and tracing system based blockchain technique. It mainly consists of two parts: the public chain composed of copyright owners and the alliance chain composed of "we media" platform. The copyright information of copyright owners and the image usage of we media can be traced through the system.

3. System Overall Design

The system uses virtual Ubuntu, and installs the open source application container engine Docker. It deploys the public chain hyperledger fabric1.2, builds a super ledger network (fabric), and compiles the smart contract of the blockchain using solidity. The copyright tracking and tracing system adopts B/S system architecture and utilizes the following techniques: public chain technology, private chain technology, MySQL database, massive image storage technology based on HBase, zookeeper, Kafka distributed cluster, image similarity recognition technology. It records user copyright data and infringement data based on blockchain technology.
Fabric is a distributed ledger based on blockchain. Each ledger has its own block, and the transaction data will be stored in the ledger block. The first block, genesis block, needs to be generated manually. Then the blockchain operation can be realized. Smart contract runs on a replicable and shared ledger program. It can process, receive, store and send information.
Massive images are stored by HBase's massive image storage technology. Image similarity is compared for infringement judgment. It realizes dynamic tracking of image copyright by returning copyright infringement warning to the copyright owner. The design framework of the system is illustrated in Fig. 1.

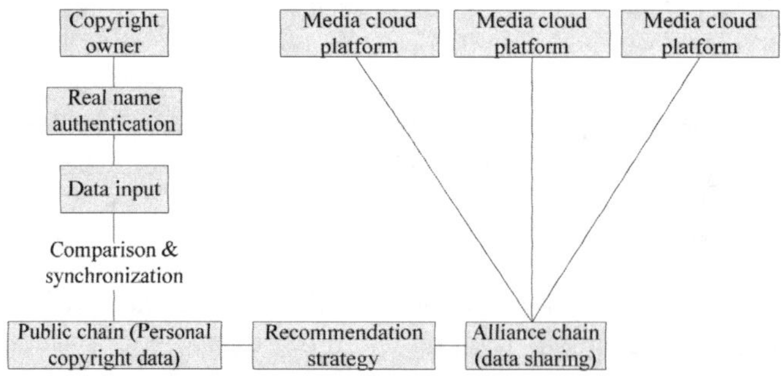

Figure 1. The framework of image copyright tracking and tracing system.

The recommendation strategy of the system in Fig.1 is based on machine learning and token voting mechanism.

Fig. 2 illustrates function composition of the image copyright tracking and tracing system based on blockchain. It is divided into two platforms: copyright user platform and media platform. The function of the copyright user platform is consist of real name authentication, copyright registration, copyright warning, infringement complaints, infringement information revocation, copyright inquiry, et al. The function of the media platform is consist of image usage record, copyright information query, infringement warning, infringement information management, infringement complaint and so on. Functions of the image copyright tracking and tracing system are constructed from blockchain.

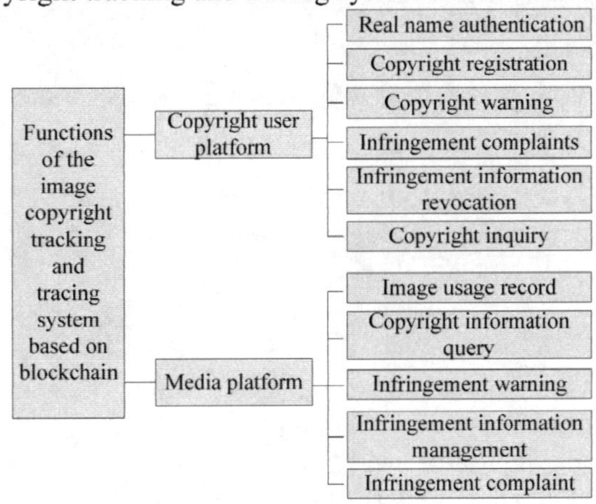

Figure 2. Functions of the image copyright tracking and tracing system based on blockchain.

The technological process of the blochchain based image copyright tracking and tracing system is shown in Fig. 3.

1) Images are uploaded to media platforms.

2) The uploaded date is deduplicated then is integrated into consortium blockchain. Also, it is encrypted and stored in the cloud.

On the other hand, 1') The copyright owner uploads copyright information. 2') This data is integrated into public blockchain. Also, it is stored in the cloud.

The system automatically compares the above two groups of data, and it identifies and selects the infringement data. Such data is stored in the infringement blockchain platform for evidence collection. This system helps copyright owners to track copyright in real time.

All copyright information is stored in the public chain and it is open and transparent.

When there is a need for image use, users can directly view the copyright notice of the images they use. This helps to prevent infringement.

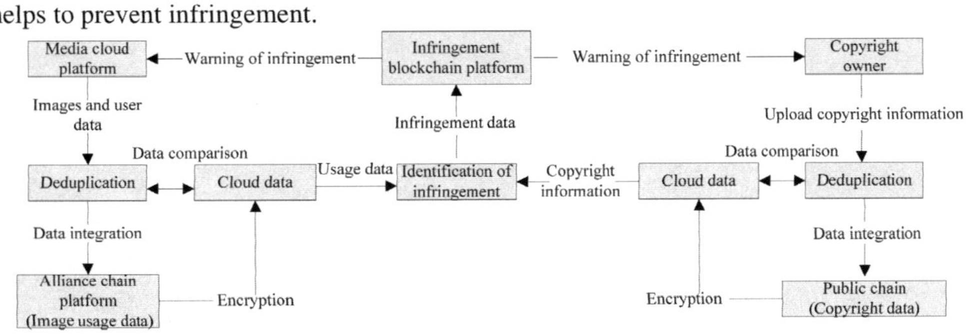

Figure 3. Process of the image copyright tracking and tracing system based on blockchain.

In the public chain of copyright information, the whole blockchain can be tracked and traced from any block to the creation block (that is, the first block of the blockchain) through the hash value of the previous block. If someone maliciously attacks, the data of the whole blockchain must be changed. However, computing the hash value is like a time-consuming operation, so it is almost impossible to change the data of the whole blockchain, which ensures the security of the blockchain. The picture copyright information tracking and traceability system model is shown in Fig. 4.

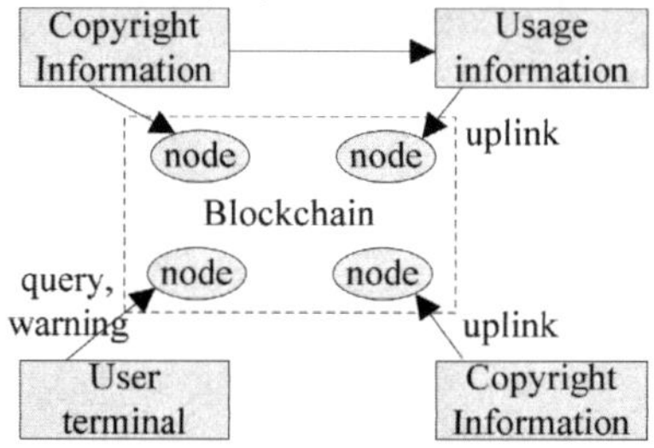

Figure 4. Image copyright information tracking and traceability system model.

4. System Implementation

The development is based on zookeeper, Kafka distributed cluster and PHP + MySQL database. MVC design pattern is used to realize the separatinforon of view, control and model code, simplify the workload, and accelerate the development speed.

4.1 Database Design of the System

The main entities of the system include: copyright user information table, "we media" platform information table, node information table, operation log table, image copyright information table, image usage information table, Merkletree leaf node information table, infringement information table. The main steps of image copyright registration are as follows:
1) Copyright users register first, and after completing the information registration, users can log in.
2) Users with real name authentication can upload their own image copyright information.
3) The system checks the duplicate of the copyright information.
4) Calculate hash value of copyright information.
5) The copyright owner signs the above hash value and broadcast to the network. Finally it returns the upload result to the copyright owner.
Brief process is illustrated in Fig. 5.

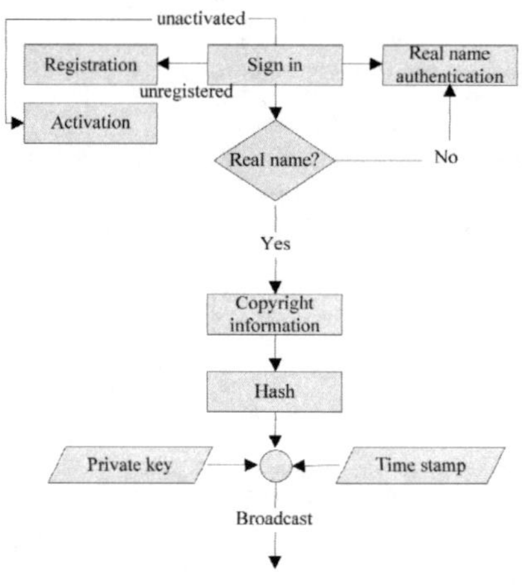

Figure 5. Copyright registration.

4.2 Image Similarity Comparison

Image similarity calculation is usually used to evaluate the similarity of content between two images. The common methods include pixel point contrast, center of gravity contrast, projection contrast, block comparison.

It cuts the image into several blocks, and then matches each block to calculate the similarity. And then the vector distance is calculated to get the similarity. The method of block comparison can improve the accuracy of the results significantly.

Euclidean distance algorithm is the most common algorithm for vector distance calculation:

$$d = \sqrt{(x_1 - x_2)^2 + (y_1 - y_2)^2 + \cdots + (y_{n_1} - y_{n_2})^2}$$

The above formula is used to match the image of the existing copyright database to get multiple similarity metrics. We multiply each similarity by a weight and add them to get the total similarity. The highest total similarity is the recognition result.

4.3 Image Copyright Warning

The data in the public chain is composed of image copyright, and the data in the alliance chain is composed of image usage information. Infringement can be detected by comparison of these two types of data. Unauthorized image information is fed back to the corresponding copyright owners and "we media" platform.

The system mainly links the picture copyright information, picture usage and other information to generate blocks through dynamic storage. In order to ensure the accuracy of the information on the chain, the module verification is added, and the Merkle tree structure is used to store the dynamic data. The block storage technology framework system is shown in Fig. 6.

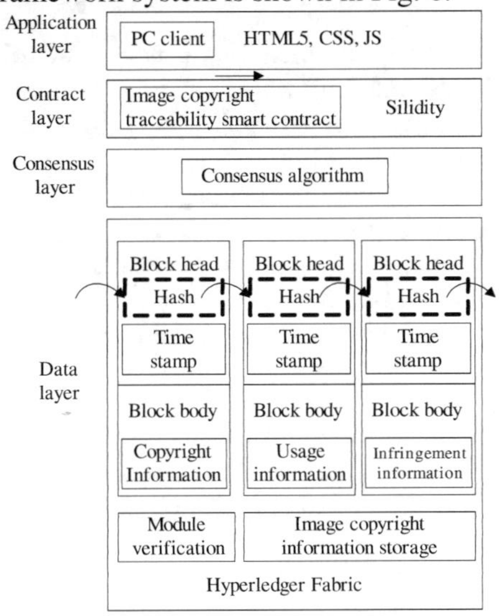

Figure 6. The block storage technology framework system.

When there is a query task in the image use database, PHP controllers (control layer) will determine the image similarity. If the calculated similarity less than 60%, it is determined that there is no data. If more than 60%, then it judges whether there is authorization information: if there is authorization information, it will be judged that there is no infringement; if there is no authorization information, a hash value of the infringement information will be calculated. Then a signature will be generated based on the user's private key and time stamp.

The signature will be added to the infringement blockchain, and is broadcasted to the whole network, which is the infringement data. Finally, the copyright owners and "we media" platform are informed to give copyright warning and realize timely tracking of the image infringement. The main steps are lustrated in Fig. 7.

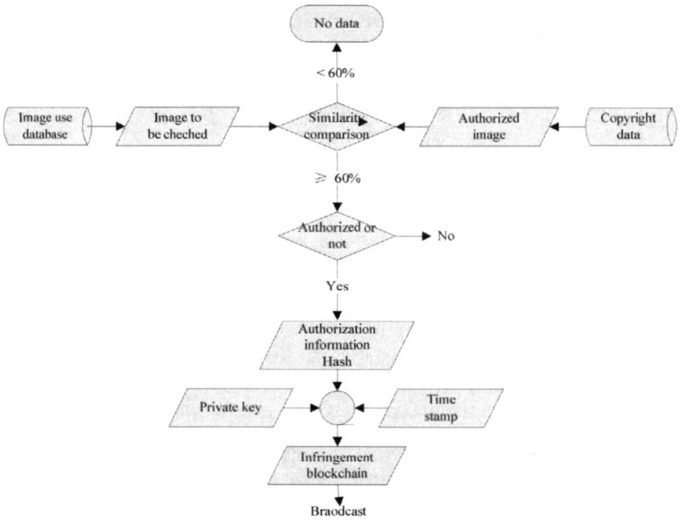

Figure 7. Copyright warning process.

5. Conclusion

Traditional image infringement is difficult to trace. In this paper, the following solutions are proposed:
1) Image copyright protection is integrated into blockchain, which enables the copyright information to be stored decentralized. It improves the efficiency of copyright authorization and tracing, and provides convenience for the creators and for users of images.
2) The system innovatively realized a recommendation strategy based on machine learning and token voting mechanism, namely "resource reverse tracing strategy". Copyright use information is timely fed back to copyright owners and "we media" platform to realize early warning of infringement.
This idea can also be used in other fields, such as electronic invoice (e-invoice) system, for detecting falsification of e-invoice, issuing invoice in time, and preventing the reuse of e-invoice.

Acknowledgments

Authors wishing to acknowledge financial support from National Key R&D Program of China(2018YFB0803900), Beijing Municipal Natural Science Foundation(4212019), Guangxi Key Laboratory of Cryptography and Information Security (GCIS201808).

References

[1] L. J. Xiao, W. H. Huang, Y. Xie, et al. A blockchain-based traceable IP copyright protection algorithm. *IEEE Access*, 2020, 49532-49542.
[2] G. C. Yuan, L. P. Fengng, J. Ning, et al. Survey on the application of blockchain in Digital Rights Protection. *International Conference on Intelligent Computing and Human-Computer Interaction (ICHCI)*, 2020, 183-187.
[3] B. W. Wang, J. W. Shi, W. S. Wang, et al. A blockchain-based system for secure image protection using zero-watermark. *IEEE 17th International Conference on Mobile Ad Hoc and Sensor Systems*, 2020, 62-70.
[4] Z. H. Lu, Y. Q. Shi, R. Tao, et al. Blockchain for digital rights management of design works. *IEEE 10th International Conference on Software Engineering and Service Science (ICSESS)*, 2019, 596-603.
[5] D. Roman, K. Vu. Enabling data markets using smart contracts and multi-party computation. *International Conference on Business Information Systems*, Berlin, Germany, 2018, 258-263.
[6] R. A. Dobre, R. O. Preda, R. A. Badea, et al. Blockchain-based image copyright protection system using JPEG resistant digital signature. *IEEE the 26th International Symposium for Design and Technology in Electronic Packaging (SIITME)*, 2020, 206-210.

Modeling Analysis and Comparision of Neural Network Simulation Based on ECM and LSTM

Lingyun Duan[1,2,3,4,5], Ziyuan Liu[1,2,3,4], Wen Yu[1,2,3,4,5,*], Wei Chen[1,2,3,4], Dongyan Jin[1,2,3,4], Denghua Li[1,2,3,4], Suhua Sun[1,2,3,4,5] and Ruixi Dai[1,2,3,4,5]

[1]Agricultural Information Institute, Chinese Academy of Agricultural Sciences, Beijing 100081, P.R.China
[2]Key Laboratory of Agricultural Information Service Technology, Ministry of Agriculture and Rural Affairs, Beijing 100081, P.R.China
[3]Key Laboratory of Intelligent Agricultural Early Warning Technology and System, Chinese Academy of Agricultural Sciences, Beijing 100081, P.R.China
[4]Beijing Engineering Research Center for Agricultural Monitoring and Early Warning, Beijing 100081, P.R.China
[5]Graduate School of Chinese Academy of Agricultural Sciences, Beijing 100081, P.R.China

[*]Corresponding author email: yuwen@caas.cn

Abstract. Comparing the prediction effects of traditional econometric algorithm model and deep learning algorithm model, taking regional GDP as an example, two prediction models of ARMA-ECM and LSTM-SVR are established for prediction, and the prediction results of different models are compared and analyzed. The results show that there are some deviations in the prediction results of the two models, but the prediction trends are the same. The prediction accuracy of LSTM-SVR model will decrease significantly with the reduction of time series data samples, while ARMA-ECM model is not so sensitive.
Keywords: ECM, ARIMA, LSTM, SVR.

1. Introduction

Firstly, this paper summarizes the prediction research of previous scholars using ARMA model, ECM model and LSTM model in various fields, and then introduces the algorithm principle of the model, such as cointegration test, error correction model, autoregressive moving average model and LSTM model. Finally, the prediction results obtained by the two algorithms are analyzed and the advantages and disadvantages of the two prediction methods are compared.

At present, the development of modeling algorithm is divided into traditional measurement algorithm model and emerging deep learning algorithm model. Traditional measurement models are mainly time series models, regression analysis models, grey system theory models and artificial neural network models. ARMA model and ECM model are classic time series forecasting methods, which can make good use of data rules for fitting and are widely used. Xia Junchang, Xie Kaigui, Cao Kan et al.[1] (2011) established an error correction model to predict short-term electricity prices, and established a two-variable autoregressive equation. The results show that the ECM model effectively describes the short-term fluctuations of the time series and helps predict the development trend of electricity prices. Lu Wanbo and Yang Dong[2] (2018) constructed a semi-parametric mixing data sampling error correction (SEMI-ECM-MIDAS) model for short-term prediction of China's CPI. The results show

Content from this work may be used under the terms of the Creative Commons Attribution 3.0 licence. Any further distribution of this work must maintain attribution to the author(s) and the title of the work, journal citation and DOI.
Published under licence by IOP Publishing Ltd

that the model has the best prediction accuracy when performing continuous predictions. , And the prediction result is not affected by the choice of mixing dynamic cointegration relationship. Zhang Jinping, Li Hongbin and Xiao Honglin [3](2020) established error correction models to predict the annual runoff in the source area of the Yellow River. The results show that the goodness of fit of the three variable error correction model is greater than that of the two variable error correction model, which can be used for the annual runoff prediction in the source area of the Yellow River. Fan Bo and song Wenbin [4](2021) established a sales volume prediction algorithm based on ARIMA model to predict the sales volume of the product. The results show that ARIMA model is applicable to the short-term prediction of the product. Shi Tingting [5](2021) constructed ARIMA (2,1,2) model to study the short-term change law of RMB exchange rate, and predicted the RMB exchange rate in the next three months. The results show that ARIMA model is suitable for RMB exchange rate prediction, and all pass the significance test. Ni Jie, Yu Li and Jin Xiaonan [6](2021) used the improved ARIMA model to predict the passenger flow of urban rail transit. The results show that the prediction effect is good and has strong applicability.

With the development of machine learning and deep learning and the popularization of algorithms, the fitting of models to complex systems is getting better and better, and the research on future prediction and analysis using deep learning algorithms has also begun to appear widely. Zhang Yang and Zhang Fuyi [7](2021) used LSTM-ARIMA model to predict the traffic changes of mobile communication base stations with a long time span, and achieved good prediction results. Zhu Qing and Zhou Shipeng[8] (2021) used LSTM neural network in deep learning to predict and analyze GDP growth, and compared the prediction results with traditional time series ARIMA and GARCH models. The experiments show that LSTM model has high application value in macroeconomic prediction. Li Dekui, Du Shubo and Zhang Peng[9] (2021) predicted and analyzed the delayed passenger flow of urban rail transit by using LSTM model. They believe that LSTM method has smaller root mean square error than Arima method and has better prediction effect. Zhang Lei, sun Shanghong and Wang Yue[10] (2021) predicted the RMB exchange rate based on the long-term and short-term memory deep learning network model LSTM. The results show that the LSTM model has higher prediction accuracy than the traditional volatility model. Hu Yuwen[11] (2021) proposed an LSTM prediction model based on PCA and LASSO to predict the stock closing price. The results show that PCA-LSTM model can greatly reduce data redundancy and obtain better prediction accuracy.

ARMA is an econometric model based on time series data. It selects a wide range of data types without too many restrictions. Aiming at the interference of sequence order, timeliness and uncertainty of random factors, ARMA model can still show good stationarity. Therefore, ARMA model can be applied in a wide range of fields, has strong application performance, and has a good effect in the application of short-term prediction of time series. LSTM model is developed from RNN. On the basis of inheriting the advantages of RNN, it also overcomes the influence of gradient disappearance and gradient explosion. It has long-term memory function and is simpler and more accurate in sequence prediction. Therefore, based on ARMA and LSTM time series models, this paper studies and analyzes the advantages and disadvantages of the model.

2. Modeling

2.1. Data Selection and Processing

Select the GDP of each region from 1990 to 2019, per capita education years of each region from 1990 to 2017, and the fixed asset investment data of each region from 1990 to 2017 are from the Data Center of the National Bureau of Statistics. Regional GDP, regional investment in fixed assets, and per capita years of education in the region are respectively expressed as gdp, inv, and edu.

2.2. Traditional Measurement Algorithm Model

2.2.1. Modeling Ideas

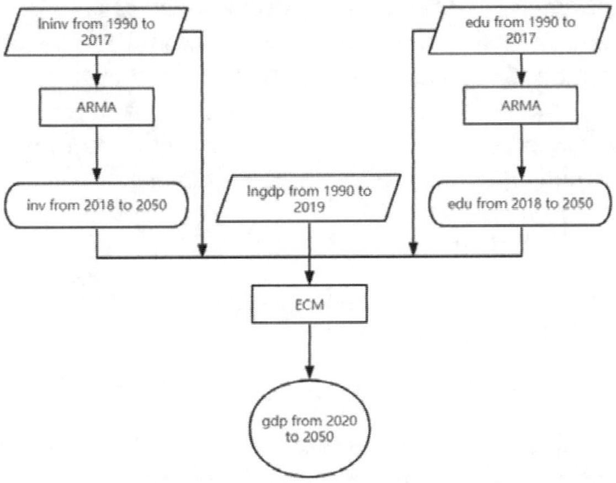

Figure 1. Modeling ideas of traditional measurement algorithm model.

In order to eliminate the heteroscedasticity between variables, the regional GDP and regional investment in fixed assets are processed in logarithm and named lngdp and lninv respectively. Firstly, the unit root test is used to test the stationarity of variables; Then, the cointegration test is carried out to determine the long-term relationship between variables and establish the error correction model. Finally, the autoregressive moving average model is used to calculate the prediction data of independent variables, and combined with the error correction model to predict the predicted value of regional GDP from 2020 to 2050.

2.2.2. Unit Root Test. Before testing whether there is a cointegration relationship between time series variables, it is necessary to use the ADF test to test the stationarity of the time series variables. Only when the variables are all single integrals of the same order can there be a cointegration relationship. The ADF test values of lngdp, lninv, and edu in each region are neglected. Its ADF test values of the first-order difference time series of lngdp, lninv, and edu are all less than 5%critical value ,-2.992, so the null hypothesis that there is a unit root is rejected at the 5% significance level, that is, the first-order difference time series of lnGDP, lninv and edu are stationary.

2.2.3. Cointegration Test. According to the ADF test, lngdp, lninv, and edu are all first-order single-integration time series, and the cointegration test can be performed to determine whether there is a cointegration relationship between variables. Using the EG two-step test method:
The first step is to use the least square method to perform cointegration regression on lngdp-lninv-edu. The cointegration regression equation of the three variables is:

$$Y_t = a_0 + a_1 X_t + a_2 Z_t \tag{1}$$

In the above formula, Y is lngdp, X is lninv, and Z is edu.
The second step is to test the stationarity of the residual sequence (et) of each regression equation. Using the ADF test, if the residual sequence is stationary, the cointegration relationship between the variables exists. The critical value of ADF test for cointegration test of residual series (et) is calculated by the method of cointegration test in *Econometrics* edited by Li Zinai. The critical value of ADF test for cointegration test with 3 variables and 30 sample sizes at 5% significance level is -3.461. After testing, the ADF test values are less than -3.461, so the original hypothesis of unit root is rejected at the significance level of 5%, Therefore, the residual series of each regression equation is stationary, so there is a co-integration relationship between lngdp-lninv-edu.

2.2.4. Error Correction Model (ECM). Through cointegration test between variables, it is proved that there is a long-term equilibrium relationship. In order to avoid short-term fluctuations in the long-term equilibrium relationship and reduce errors, Error Correction Model emerges as the times require. The stable residual series of regression equation with long-term equilibrium relationship is put into the model as error correction term, and the model is established together with other explanatory variables. Specific modeling steps:
The first step is to find the residual sequence as the error correction term (ecm) from the cointegration equation tested above. The residual formula of the three variables is:

$$ecm_{t-1} = Y_{t-1} - a_0 - a_1 X_{t-1} - a_2 Z_{t-1} \tag{2}$$

The second step is to establish the error correction model of lngdp-lninv-edu, and the first-order error correction model formula of three variables is:

$$\Delta Y_t = \beta_1 \Delta X_t + \gamma_1 \Delta Z_t - \lambda ecm_{t-1} + \mu_t \tag{3}$$

Substitute ecm as a new variable and the data corresponding to other independent variables into the above formula, and use ordinary least squares to estimate the corresponding parameters.

2.2.5. Autoregressive Moving Average Model (ARMA). The ARIMA model is a commonly used time series forecasting model at present, and it has a good fitting effect. The model first checks the stationarity of the time series, then uses the ACF and PACF functions to determine the order combined with the AIC and BIC criteria, and establishes the model prediction formula. The formula of ARMA (p, q) model is:

$$y_t = \beta_0 + \beta_1 y_{t-1} + \cdots + \beta_p y_{t-p} + \varepsilon_t + \theta_1 \varepsilon_{t-1} + \ldots + \theta_q \varepsilon_{t-q} \tag{4}$$

From the formula (3) of the Error Correction Model, the data of $\triangle Xt$, $\triangle Zt$ and ecm_{t-1} are needed to calculate the short-term fluctuation $\triangle Yt$. Therefore, to predict the short-term fluctuations of lngdp, we need to obtain the forecast data of lninv, edu and ecm. The ARIMA model is used to obtain the predicted values of lninv and edu in each region from 2018 to 2050.

2.2.6. Forecast Results. Substitute lninv and edu into formula (4) and ecm_{t-1} into formula (2) to calculate the forecast data. Substitute the forecast data corresponding to lninv, edu, ecm_{t-1} into formula (3) to find the annual growth value of lngdp, and finally, calculate the annual GDP. The forecast results are shown in figure 2.

Figure 2. GDP prediction results of ARMA-ECM prediction model.
Note: In the figure 2, the yellow part is the trend of GDP development from 1990 to 2019 and the red part is the trend of GDP development from 2020 to 2050.

2.3. Deep Learning Algorithm Model

2.3.1. Modeling Ideas

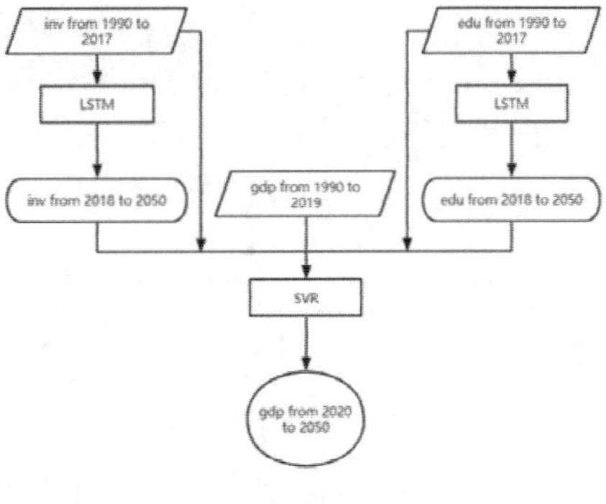

Figure 3. Modeling ideas of deep learning algorithm model.

Firstly, the long-term and short-term memory artificial neural network based on deep learning estimates the per capita years of education and fixed asset investment from 2018 to 2050; Secondly, the data of per capita years of education, fixed asset investment and GDP from 1990 to 2017 are used as the training set, and the support vector regression algorithm is used for supervised learning to establish the prediction model; Finally, the predicted value of regional GDP from 2020 to 2050 is estimated.

2.3.2. Short And Long Term Memory Artificial Neural Network (LSTM). In 1997, Hochreiter and Schmidhuber proposed short-term and long-term memory artificial neural network (LSTM). LSTM is an improved nonlinear recurrent neural network (RNN), which avoids the problems of gradient disappearance and gradient explosion caused by too many hidden layers and large amount of calculation in traditional RNN [12]. The traditional RNN model has the problem of low information persistence. If the useful information is far from the prediction point, RNN cannot correctly predict the results according to the useful information. Therefore, when the amount of information is too large, the use effect of RNN is poor. LSTM model is improved on the basis of traditional RNN. LSTM is composed of several basic repeating units that can realize hierarchical transmission of information. Each basic unit is formed by four neural network layers interacting with each other. Unlike the traditional RNN, there is only one simple Tanh module, which is also the main reason why LSTM model can realize "persistence".

LSTM model is composed of multiple memory cells. Each LSTM network memory cell contains three gating units. The memory ability of each cell is updated and deleted by input gate, output gate and forget gate. The functions of the three gates are as follows[13]:

1. Forget gate. It determines how much information is discarded from the previous cell state. Taking the input x_t of the current layer and the output h_{t-1} of the previous layer as the input, the cell state output at time t-1 is:

$$f_t = \sigma(W_f \cdot [h_{t-1}, x_t] + b_f) \tag{5}$$

2. Input gate. It determines how much information in the current newly entered information is saved in the memory cell and what value needs to be updated:

$$i_t = \sigma(W_i \cdot [h_{t-1}, x_t] + b_i) \tag{6}$$

$$\tilde{C}_t = \tanh(W_c \cdot [h_{t-1}, x_t] + b_c) \tag{7}$$

3. Output gate. It determines how much current information is output, and finally determines the output content:

$$o_t = \sigma(W_o \cdot [h_{t-1}, x_t] + b_o) \quad (8)$$
$$C_t = f_t * C_{t-1} + i_t * \tilde{C}_t \quad (9)$$
$$h_t = o_t * \tanh(C_t) \quad (10)$$

σ is the sigmoid activation function. The value range is [0,1]. 0 means to discard all, and 1 means to retain all; W_f, W_i, W_c and W_o are weight matrices; b_f, b_i, b_c and b_o are offset vectors, tanh is hyperbolic tangent activation function, and the value range is [-1,1].

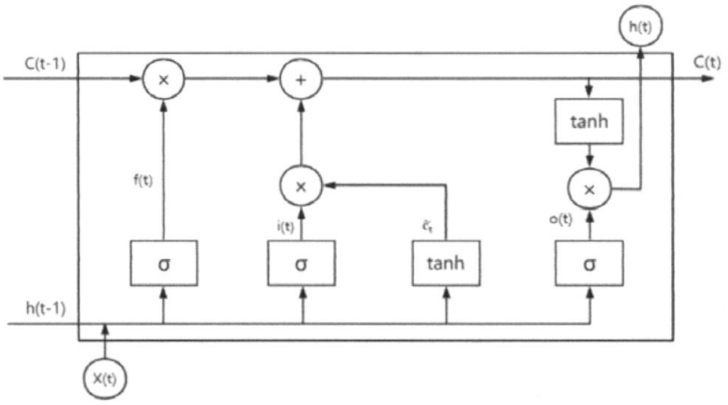

Figure 4. LSTM network model structure.

2.3.3. Support Vector Regression (SVR). SVR is a branch algorithm of support vector machine (SVM) and a linear regression model. SVM is a supervised learning algorithm, which mainly solves the classification problem. SVM determines the optimal decision surface by finding the maximum classification interval, and the sample nearest to the decision surface is included in the loss, so as to find the optimal classifier of the classification problem. SVR is the opposite of SVM. SVR aims to find the minimum classification interval and include the data outside the decision surface into the loss, so that most samples in a set are closest to the decision surface.
Firstly, the original data are normalized; Secondly, select the classifier as SVR, the data of per capita years of education, fixed asset investment and GDP from 1990 to 2017 are used as the training set to train the model and establish the prediction model.

2.3.4. Forecast Results. Taking the per capita years of education and fixed asset investment in 2020-2050 as independent variables and substituting them into the trained prediction model, the predicted value of regional GDP in 2020-2050 is estimated. The prediction results are shown in Figure 5.

Figure 5. GDP prediction results of LSTM-SVR prediction model.
Note: In the figure 5, the yellow part is the trend of GDP development from 1990 to 2019 and the red part is the trend of GDP development from 2020 to 2050.

3. Conclusions and Suggestions

Taking the regional GDP as an example, this paper uses the per capita years of education, fixed asset investment and regional GDP from 1990 to 2017, uses the traditional measurement algorithm model and the deep learning algorithm prediction model to predict, and establishes two prediction models of ARMA-ECM and LSTM-SVR. According to the empirical analysis results of the comparison of the two prediction models, the prediction results of the two models are biased, but the forecast trend is consistent. On the one hand, LSTM and SVR are data driven supervised machine learning approaches that do not take the statistical relations of data into account. In this scenario, the machine learning methods also suffer from the limited scale of training data. The prediction accuracy of LSTM-SVR model will decrease significantly with the reduction of time series data samples. On the other hand, the ECM model is a theoretically driven approach that can be helpful in the estimation of both short-term and long-term relations of time series data. ARMA-ECM model uses the long-term common random trend of basic variables to fit, and the response is not so sensitive.

Acknowledgements

This research was supported by the National Key R&D Program Project (2017YFE0104600).

Reference

[1] Xia Junchang, Xie Kaigui, Cao Kan, Li Chunyan. Error correction model for short-term electricity price forecasting in power market [J]. East China power, 2011,39 (02): 161-167

[2] Lu Wanbo, Yang Dong. Research on CPI prediction in China based on semi parametric mixing error correction model [J]. Statistical research, 2018,35 (10): 28-43

[3] Zhang Jinping, Li Hongbin, Xiao Honglin, Zhang Xin. Prediction of annual runoff in the source area of the Yellow River Based on error correction model [J]. People's Yellow River, 2020,42 (07): 5-8 + 66

[4] Fan Bo, Song Wenbin. Research on product sales forecast based on ARIMA model [J]. Industrial control computer, 2021,34 (05): 128-129 + 125

[5] Shi Tingting. Prediction of the fluctuation trend of COVID-19's RMB exchange rate based on ARIMA model [J]. modern business, 2021 (14): 108-110.

[6] Ni Jie, Yu Li, Jin Xiaonan. Passenger flow prediction and research of Urban Rail Transit Based on ARIMA model [J]. Intelligent computer and application, 2021,11 (04): 135-138

[7] Zhang Yang, Zhang Fuyi. Traffic prediction method of mobile communication base station based on LSTM and ARIMA model [J]. Information and computer (theoretical Edition), 2021,33 (07): 100-102

[8] Zhu Qing, Zhou Shipeng. Research on forecasting and modeling of national economic GDP growth based on LSTM model [J]. Economic Research Guide, 2021 (19): 5-9 + 37

[9] Li Dekui, Du Shubo, Zhang Peng. Comparison of delayed passenger flow prediction methods of Urban Rail Transit Based on ARIMA and LSTM [J]. Journal of Qingdao University of technology, 2021,42 (04): 135-142

[10] Zhang Lei, Sun Shanghong, Wang Yue. Exchange rate forecasting based on deep learning LSTM model [J]. Statistics and decision making, 2021,37 (13): 158-162

[11] Hu Yuwen. Stock forecasting based on optimized LSTM model [J]. Computer science, 2021,48 (S1): 151-157

[12] Chen Wei, Qi Weiyan, Yuan Fuxiang, Li Zhemin. Comparative analysis of rice yield prediction in Jilin Province Based on time series and cross-sectional data [J]. China agricultural information, 2018,30 (05): 91-101

[13] Duan Chuanming. Comparative analysis of traditional emotion classification methods and emotion classification methods based on deep learning [J]. Software guide, 2018,17 (01): 22-24

Modelling of Traffic Flows and Supply Chains Based on Geospatial Knowledge

A Kolesnikov[1,*], P Kikin[2] and E Panidi[2]

[1]Department of Cartography and Geoinformatics, Siberian State University of Geosystems and Technologies, Novosibirsk, Russian Federation
[2]Department of Cartography and Geoinformatics, Institute of Earth Sciences, Saint Petersburg State University, St. Petersburg, Russian Federation

[*]Corresponding author email: alexeykw@yandex.ru

Abstract. The field of logistics and transport operates with large amounts of data. The transformation of such arrays into knowledge and processing using machine learning methods will help to find additional reserves for optimizing transport and logistics processes and supply chains. This article analyses the possibilities and prospects for the application of machine learning and geospatial knowledge in the field of logistics and transport using specific examples. The long-term impact of geospatial-based artificial intelligence systems on such processes as procurement, delivery, inventory management, maintenance, customer interaction is considered.
Keywords: Geospatial knowledge, spatial modelling, machine learning, logistics.

1. Introduction

The modern stage of development of society is characterized by an increase in the accumulation and processing and analysis of various information resources. A characteristic feature of this stage is the tendency of an increase in the share of spatial data in the total volume of storage and processing [2]. At the same time, the importance of methods and technologies for obtaining, transferring and using knowledge and geospatial knowledge is increasing. Today, in many cases, geospatial knowledge does not have an established terminology and descriptions of the methods and specifics of their acquisition, transmission and use. The main feature of geospatial knowledge (geo-knowledges) is the trend towards the use of spatially coordinated objects. Based on this, first of all, the problem arises of formulating the concept of geospatial knowledge [4].

The main theoretical and practical basis for the formation of geo-knowledges is provided by geoinformatics / geomatics. But now, even more than before, geoinformatics is associated with related disciplines: geodesy, photogrammetry, cartography, remote sensing, etc. Therefore, one of the important features of geo-knowledges is solving the problems of storing, transferring, transforming, and using knowledge in these sciences, considering the spatial component.

Also, the issues to be formalized include the mechanisms of cataloging and searching for geo-knowledge, as well as the selection of the best metric for assessing and interpreting the result of their use. This study is part of a project for the development of Asian Russia, and specifically for the development of transport infrastructure. It represents the concept of extending a database with geo-knowledge. In this article, we will consider the methodological foundations of the formation and use of geo-knowledge in general, as well as their application for logistics purposes. The main point of the study was the formulation of the concept, definition, and components of geospatial knowledge.

Content from this work may be used under the terms of the Creative Commons Attribution 3.0 licence. Any further distribution of this work must maintain attribution to the author(s) and the title of the work, journal citation and DOI.
Published under licence by IOP Publishing Ltd

2. Materials and Methods

In addition to the direct formation of geosciences, the construction of a terminological base based on the scientific approaches used, the construction of ontologies and a semantic description is additionally carried out in order to realize the possibility of constructing a catalog. A working example of a geo-knowledge catalog is ESRI's Living Atlas, which contains pre-trained neural networks for semantic segmentation of satellite imagery. But, speaking of geo-knoledges, the existing catalogs are still far enough from catalogs of NLP models (for example, Huggingface).

The basic structure in most studies for representing geo-knowledge is a graph representation together with semantic information, mainly for finding the most similar objects [1,3]. In this case, the description format of one of the semantic markup languages is used, the most popular of which are RDF and JSON-LD.

It is the graph structure that is basic for solving logistics problems. These tasks can be divided into the following blocks:

- navigation. With the active development of satellite and inertial navigation systems, as well as SLAM methods, navigation problems have become even closer to geoinformatics. On the basis of knowledge, data is formed about the spatial position of the object relative to other spatial objects.
- transportation management. Geo-knowledges allows you to optimize transportation, improve traffic safety and predict logistics flows in the event of changes in transport infrastructure or resources at the end nodes.
- situational management. Allows to manage flows in any non-standard situations associated with emergencies, changes in the intensity of traffic flows, etc.

Next, we will consider the general methodology of geo-knowledge and the possible specifics when using them for logistics tasks.

3. Results

The authors propose a representation of geo-knowledge as a set of data (*S*) (including the necessary transformations), mechanisms for forming conclusions based on them (*P*) and methods of assessing (*M*) and interpreting the result (*I*):

$$GK = S \cup P \cup M \cup I \tag{1}$$

Data schema (S) is a description of data types (subdivided into mandatory and optional) (S_D) and required transformations (S_T). Conversions can include both standard format conversions, units of measure, and feature engineering, dimension change, etc.

$$S = S_D \cup S_T \tag{2}$$

Source data (D) is a collection of spatially coordinated data (*DG*) and statistics (*DS*) defined at one or more time points (1 ... n). It is understood that the statistics initially do not have coordinate values but can be associated with pre-existing spatial objects. Taking into account the already indicated interdisciplinary connections in geosciences, data can be presented in a variety of ways: vector cartographic data, satellite images, three-dimensional elevation models, point clouds, tabular statistical indicators, query results obtained from the protocols of map services, etc. Therefore, the necessary transformations (S_T) constitute a very important and no less voluminous part of the data schema (*S*) in order to form a single geospace from the entire available data set.

$$D = DG_{T1} \cup DS_{T1}, DG_{T2} \cup DS_{T2}, \cdots DG_{Tn} \cup DS_{Tn}, \tag{3}$$

The processing component can represent both simple conditions and complex mathematical models (including neural networks). At the same time, they can be ensembled, as in the case of processing data of any other types. Of all the elements of geo-knowledges, this component has the greatest variability, since it can be implemented by dozens of algorithms, methods and technologies. For some standardization, universal model formats can be applied (for example, ONNX, but it does not support all the necessary data preprocessing methods), but if they are applied to logistics, then there may be difficulties when working with graph models.

Since there is no direct data in geo-knowledges, but they come from outside, then for each new dataset, methods of assessing the quality and reliability of the result obtained (*M*) are needed. Also, in most cases, a set of spatial data is formed at the output, so specialized metrics are required that take into account both spatial and temporal components (for example, spatial autocorrelation, Gini Index, Getis-Ord, and others).

In addition to directly assessing the results of processing, it is important to interpret (*I*) that it is the process of forming conclusions (*P*) that occurs. The existing algorithms of artificial intelligence (machine learning) form the problem of the contradiction between accuracy and interpretability. Deep neural networks make it possible to solve more and more complex problems, but it is very difficult to understand how exactly this result came about. In the proposed concept of geoknowledge, it is assumed to use mainly model-independent (agnostic) approaches, implying operating with models presented in the form of a black box, and only their inputs and outputs are used for interpretation.

It should be noted that in the proposed version of geo-knowledge should not have a graphical representation in the form of a map. The map is obtained based on the results of processing using geo-knowledge, which are also spatial and statistical data, but with a greater degree of generalization.

$$GK: D \to D_{GK} \tag{4}$$

If we concretize the scheme of geo-knowledges related to logistics, then its elements should be economically significant objects (as a separate spatial element, or in generalized in the form of economic regions), connecting them with transport infrastructure of all types. Additional elements will be external factors affecting the integrity and sustainability of the transport infrastructure. These elements mean both negative factors that remove or reduce characteristics (for example, the amount of resource extraction, units of produced goods, the road capacity) of economically significant objects and connections between them, and positive ones that increase characteristics that add new transport links.

For the digital twin of the Asian Russia logistics system being developed, the authors propose the following initial components:

- description of the spatial structure based on maps of the territory;
- location of places for intermediate storage and transshipment of goods;
- structure of input and output streams;
- the structure of dynamic logistics objects that are moved and stored;
- the cost of performing operations in the nodes and arcs of the network;
- characteristics of transport channels;
- strategies for supply chain management;
- current and projected demand values;
- probabilities and characteristics of external factors affecting the nodes and arcs of the network;
- duration of technological operations.

For implementation, a multi-model DBMS is used that allows you to combine the formation of links between data and ontologies. The following types of links between ontologies are used.

- taxonomic relationships - for the implementation of the "is a child" relationship.
- compositional links - to implement the relationship "part-whole" in the form of one of the implementations "component-object", "ingredient-product", "set element".
- topological links - for the implementation of spatial relationships.
- relationships of entities with processes - for the implementation of relations of participation of the entity in the process, the execution of the process or its leadership.
- cause-and-effect relationships - for the implementation of relations of the influence of processes on entities, other processes (call, change, or termination).
- temporary connections - for the implementation of connections between entities flowing in time and between time intervals.

4. Conclusion

Thus, geo-knowledge will make it possible to combine the tools of a geographic information system, a traditional spatial database (for storing and most optimal processing of spatial data, which will be worked with using a geographic information system), a graph database (with extensions that allow storing and using semantic networks), technical solutions that implement algorithms and methods for constructing mathematical models obtained by both direct and reverse modeling. This approach will allow realizing both visualization and a comprehensive understanding of the internal processes and flows of the system and combine the two main approaches to the simulation of logistics networks - graphic and mathematical models. It will also allow the use of common modeling approaches - SCOR, EPCs, EPK, EOQ. Also, the developed concept of geodata will allow solving the tasks of strategic (assessment of the functioning of the production system at large time intervals), tactical (assessment of the current state and efficiency of the functioning of the existing production system), operational scheduling (building a production schedule for a short time - from hours to days, current planning , scheduling the loading of technological equipment, dispatching tasks) and simulation (a tool for playing scenarios of scenarios "what-if").

Acknowledgments

The research was prepared within the grant for carrying out major scientific projects in priority areas of scientific and technological development under the subprogram "Fundamental scientific research for the long-term development and ensuring the competitiveness of society and the state" of the state program of the Russian Federation "Scientific and technological development of the Russian Federation," project "Social and economic development of Asian Russia on the basis of synergy of transport accessibility, system knowledge of natural resource potential, and expanding area of interregional interactions", agreement with the Ministry of Science and Higher Education of the Russian Federation № 075-15-2020-804 (internal number 13.1902.21.0016).

References

[1] Ballatore A, Bertolotto M and Wilson D C 2015 A Structural-Lexical Measure of Semantic Similarity for Geo-Knowledge Graphs *ISPRS International Journal of Geo-Information* vol 4(2) (MDPI AG) pp 471–492 doi:10.3390/ijgi4020471.

[2] Lin H, Lan Y, Chuanbo H and Min C 2018 Prospect of Geo-Knowledge Engineering in the Era of Spatio-Temporal Big Data *Geomatics and Information Science of Wuhan University* vol 43(12) (Wuhan University) pp 2205–11. doi:10.13203/j.whugis20180318.

[3] Chen J, Shumin D and Huajun C 2017. CrowdGeoKG: Crowdsourced Geo-Knowledge Graph *Communications in Computer and Information Science* vol 784:165–72 (Springer Verlag) doi:10.1007/978-981-10-7359-5_17.

[4] Tsvetkov V Ia 2016 Geo-knowledge *European Journal of Technology and Design* vol 13(3) (Academic Publishing House Researcher)

[5] Lobanov A A 2014. Geo-knowledge as a new form of knowledge *Prospects of Science and education* vol 5 pp 37-44.

[6] Rudskoy A, Ilin I and Prokhorov A 2021 Digital Twins in the Intelligent Transport Systems *Transportation Research Procedia* vol 54:927–35. (Elsevier B.V.) doi:10.1016/j.trpro.2021.02.152.

[7] Katarzyna T and Kubik A 2020 Economic Aspects of Driving Various Types of Vehicles in Intelligent Urban Transport Systems, Including Car-Sharing Services and Autonomous Vehicles *Applied Sciences* vol 10(16) (MDPI AG, Switzerland) doi:10.3390/app10165580.

[8] Tripathi P S M, Kumar A and Chandra A 2021 An Overview of Intelligent Transport System (ITS) and Its Applications *Journal of Mobile Multimedia* vol 17 (River Publishers) doi:10.13052/jmm1550-4646.17134.

[9] Yujun D, Wang J, Rizos C and El-Mowafy A 2021 Vulnerabilities and Integrity of Precise Point Positioning for Intelligent Transport Systems: Overview and Analysis *Satellite Navigation* vol 2(1) (Springer Science and Business Media LLC) doi:10.1186/s43020-020-00034-8.

[10] Xingjuan C, Zhaoming H and Jinjun C 2020 A many-objective optimization recommendation algorithm based on knowledge mining, Information Sciences vol 537 pp 148-161 https://doi.org/10.1016/j.ins.2020.05.067.

Journal of Physics: Conference Series **2068** (2021) 012043 doi:10.1088/1742-6596/2068/1/012043

Simulation and Research of Urban Intelligent Traffic Light Control System

Chunling Sun* **and Fuhai Liu**

Department of Vehicle Engineering, Shandong Transport Vocational College, Weifang, China

*Corresponding author email: 16770604@qq.com

Abstract. Intelligent traffic light control system belongs to the field of transportation public safety control. So far, the traffic light control system on the market cannot change the working time of the traffic lights according to the actual traffic flow, which is not conducive to the easing of the traffic flow and easy to cause traffic jams. Therefore, it is necessary to solve the problem of the existing traffic light control system which cannot adjust the working hours of the traffic light according to the actual road conditions, so as to alleviate the traffic jam at the intersection to a certain extent. In this paper, the design of intelligent traffic light control system uses AT89C51 MCU as the core device. The system controls the traffic of different times and different conditions, and it can alleviate traffic jam to a certain extent. This paper analyzes the hardware and software designes of intelligent traffic light control system, and simulates traffic light control by simulation software. The system of this paper is simple in structure, economical and practical, reliable in operation and it can effectively dredge the traffic.

1. Development and Research Significance of Intelligent Traffic Light Control System

The earliest traffic lights appeared in London in 1858, they were mainly used in the streets by two colors signal lights of red and blue, and the work type of the signal lights was the mechanical wrench type. Then in London in 1868, the world's first traffic light came. Yet the yellow light was invented by Hu Ruding in China. In order to remind people to pay attention to the dangers at the intersection, he put forward a suggestion of adding a yellow light between the red and green lights. With a variety of transportation tools appearing in people's life more and more, in order to direct the traffic more conveniently, three colors lights of red, yellow, green appeared for the first time in 1918 in New York. With the deep development of automobile electronic technologies and computer control technologies, the control mode of traffic lights has developed from the original pure mechanical control to the current computer timing control, the control mode of traffic lights is becoming more scientific and automatic[1] [2].

With the continuous increase of traffic tools in China, the urban traffic flow is also increasing, and the urban traffic congestion has become an urgent problem to be solved. The problem of traffic congestion in large and medium-sized cities has become more and more serious. The serious traffic congestion has also seriously affected people's life. Among them, the problem of traffic congestion at urban intersections is getting worse. At present, the existing traffic light control systems on the market are only based on the time control to transform the different colors of the traffic lights, they cannot change the working time of the traffic lights according to the real time traffic flow of the actual intersections, so they are easy to cause the traffic jam at the intersections to a certain extent and they are not conducive to slow the traffic flow at the intersections. At present, many scholars are doing the

Content from this work may be used under the terms of the Creative Commons Attribution 3.0 licence. Any further distribution of this work must maintain attribution to the author(s) and the title of the work, journal citation and DOI.
Published under licence by IOP Publishing Ltd

simulation research and analysis of intelligent traffic light control system, but almost the existing traffic light control system works in a single fixed timing control, so it can not adjust traffic according to the actual conditions[3] [4].

The traffic light control system designed in this paper adjusts the intermittent time of the traffic lights according to the real-time road conditions, which can alleviate the traffic jam caused by more cars on one side and less cars on the other side. The system uses PROTUS simulation software to simulate the traffic light control, and verifies the feasibility of the system.

2. Research and Design of A Urban Intelligent Traffic Light Control System

Urban intelligent traffic light control system belongs to the field of transportation public safety control. At present, the existing traffic light control systems on the market can not change the working time of traffic lights according to the actual traffic flows which are not conducive to relieve the traffic flows to a certain extent, and they are easy to cause traffic jams[5]. Under this premise, how to achieve more reasonable traffic light control at the intersection and reasonable allocation working hours of traffic lights, so as to alleviate the intersection traffic congestion becomes particularly important. So we can design a reasonable intelligent traffic light control system which uses the control unit ECU to conduct perception of real-time road conditions and information processing, so the system has the advantages of low cost, high stability and reliability. This control system is more specifically related to the intersection of the traffic light conversion time control, so it can alleviate the intersection of traffic congestion to a certain extent, and it can bring greater convenience to people's life.

2.1. Hardware Circuit Design of the System

The intelligent traffic light control system is mainly composed of control module, traffic flow identification module, time display module and traffic light conversion module[6]. In addition, the control module is mainly composed of reset circuit, crystal oscillator circuit and power supply circuit. The system identifies the traffic flow at the intersection through the traffic flow identification module, passes the traffic flow information to the control module, and calculates the traffic flow by software programming on the control module, and adjusts the working hours of the traffic lights in time, so it can realize the reasonable time distribution and effective control of urban traffic[7]. The hardware circuit of intelligent traffic light control system is shown in Figure 1.

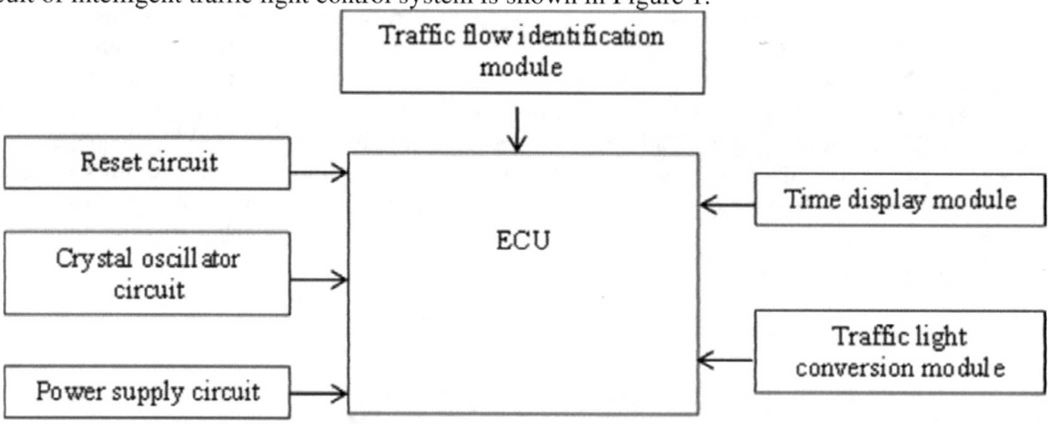

Figure 1. Hardware circuit of the system.

2.2. Hardware Circuit Design of the System

This intelligent traffic light control system software design process flow chart is shown in Figure 2.

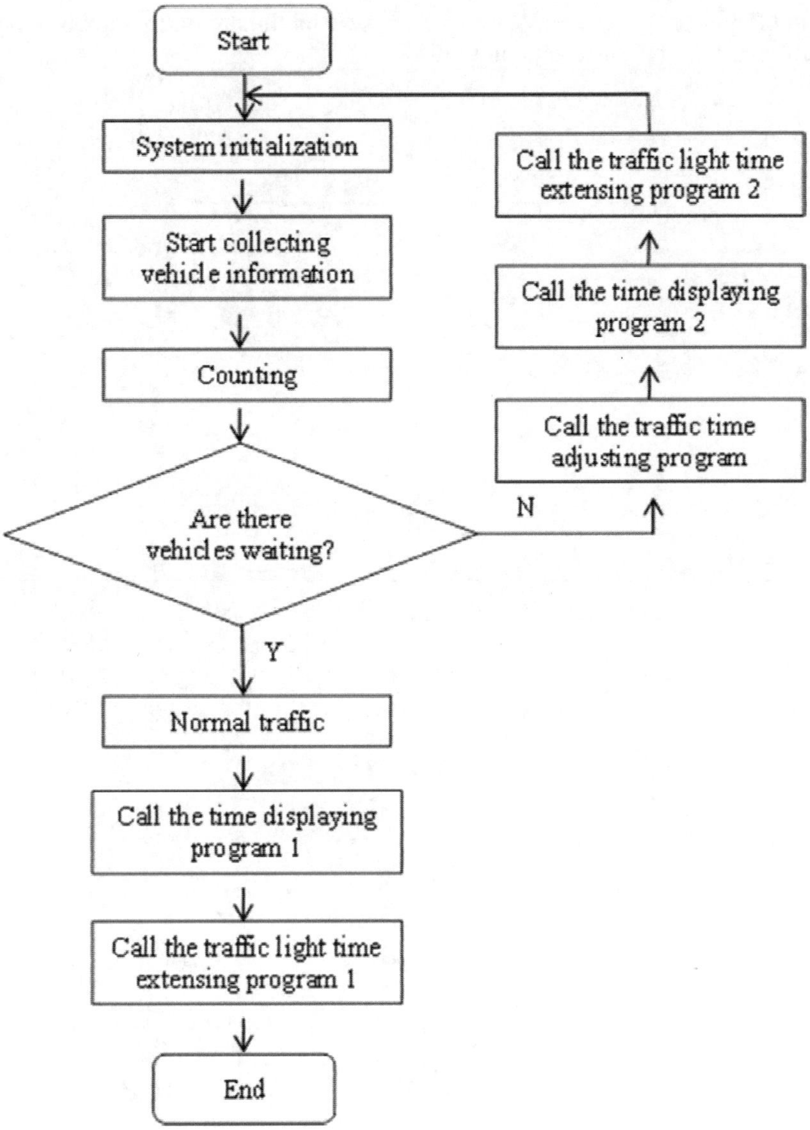

Figure 2. The flow chart of software design.

This traffic light control system aims to reasonably allocate the working hours of red light, yellow light and green light, alleviate traffic congestion, provide convenience to people's life, so it can produce certain benefits:(1)Quick real-time response: The vehicle information at the intersection can be monitored at any time through the traffic flow identification module, so the system can adjust the working time of the traffic lights in time.(2)High precision of the system: The system takes the control unit ECU as the main body, collects vehicle information, programs through software, accurately adjusts the working time of the traffic lights, takes real-time sampling and effectively alleviates traffic jams.(3)Small space: The system adopts MCU control system, reasonably configurates the input circuit and output circuit, takes up small space and it can be directly installed on the traffic lights.

3. Simulation Analysis of Intelligent Traffic Light Control System

The system uses PROTUS simulation software for simulation, after the completion of the drawing of hardware circuit by simulation software, we can add software applications to AT89C52, then we can carry out the simulation of the system. We can start system simulation by clicking the button at the bottom of the software interface as shown in Figure 3.

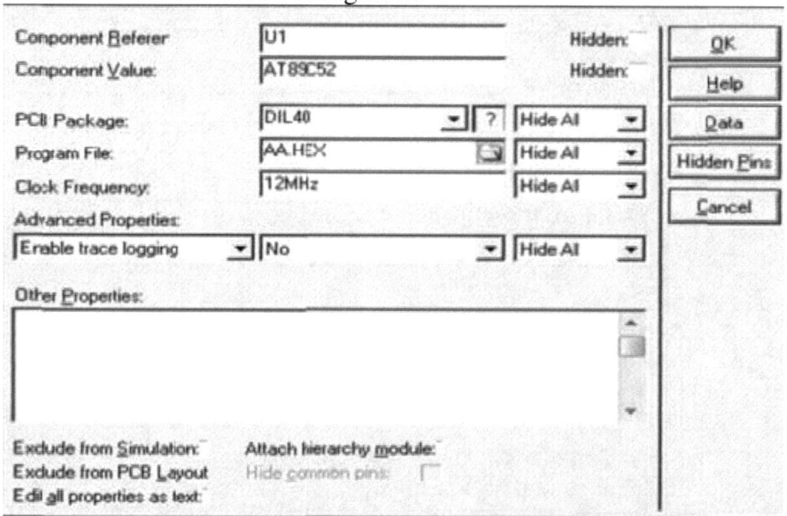

Figure 3. Add software program to AT89C52.

In the process of simulation, we can operate each control button and traffic flow simulation unit to simulate the traffic conditions of different road conditions, and change the working time of traffic lights by calling different time conversion programs, so as to alleviate the traffic jam problem to a certain extent. The simulation process is shown in Figure 4.

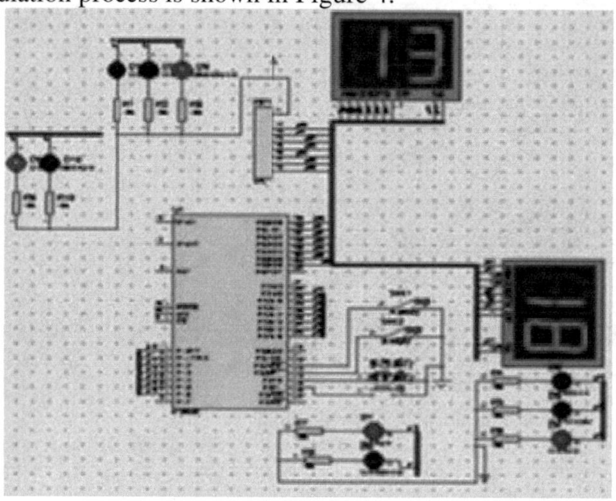

Figure 4. The simulation process.

4. Conclusion

The research object of this paper is the intelligent traffic light control system. The system uses ECU to complete real-time perception of real-time road conditions and information processing, which has the advantages of low cost, high stability and reliability. The intelligent traffic light control system is mainly composed of control module, traffic flow identification module, time display module and traffic light conversion module. The system uses the PROTUS simulation software to simulate the hardware and software of the system, and the feasibility of the system is verified.

The system of this paper identifies the traffic flow of the intersection through the traffic flow identification module, transmits the traffic flow information to the control module, and calculates the traffic flow through the software programming of the control module, adjusts the working time of the traffic light in time, so as to realize the reasonable time allocation and effective control of the urban traffic.

Acknowledgments
I would like to thank my team and friends for their Suggestions and opinions in this research. Their opinions played an important role in the content research.

References

[1] Zhanmiao Guo. Design of traffic lights based on STC89C52 MCU [J]. Industrial control computer, 2017, 30(06):138-139.
[2] Tianzhuo Lin. Research and Design of Traffic Light Control System Based on 51 MICROcontroller [J]. Electronics World, 2017(12):184.
[3] Haiying Tan, Cong Yang. Design and Analysis of Crossroad Traffic Lights Based on MCU Control [J]. Electronic Testing, 2017(13): 9+8.
[4] Limin Zheng. Industry & science & technology forum, 2017, 16(16):56-57.
[5] Xichen Yu. Science and technology information, 2017, 15(27):39-40.
[6] Qingni Hao, Wu Jixia, ZHANG Jiao. Design of Traffic Light Controller Based on MICROcontroller [J]. Electronic Manufacture, 2017(21): 69-70+82.
[7] Jun Chen. Design of intelligent Traffic Light based on AT89S51 microcontroller [J]. Electronic Technology and Software Engineering, 2016(01): 260-261.

Periodic Solution of A Delayed Intraguild Predation Impulsive System with Strong Allee Effect

Jiao Ai and Kaihua Wang*

School of Mathematics and Statistics, Hainan Normal University, Haikou, China

*Corresponding author email: kaihuawang@hainnu.edu.cn

Abstract. With periodic coefficients and strong Allee effects, we establish a delayed intraguild predation impulsive model. We obtain a set of sufficient conditions for the existence of positive periodic solution of the model using Mawhin's continuation theorem and analysis techniques. Finally, we identify the effectiveness of the theoretical results through some numerical simulations.
Keywords: Intraguild predation, delayed, impulses, periodic solution.

1. Introduction

Intraguild predation (IGP) model was first introduced by Polis et al. [1]. It is a combination of predation and competition because of eating the potential competitors. IGP is a kind of characteristics model of interaction between basal prey, IG prey and IG predators.

It has been paid more considerable attention by many authors because IGP system is an important community module to know the dynamics of complicated food webs [2-4]. For example, in literature [2], Anderson and Semlitsch studied the density dependence of IG predators in IGP system. The research shows that with the increase of IG predator population density, the potential encounter rate between IG predator and IG prey also increases nonlinearly, which will ultimately determine whether there is competition or predation relationship between them. In literature [3], Rabageo and Collera considered that under natural conditions, Predators can't catch prey and turn it into their own fecundity at the same time, so they established an IGP model with time delay. The dynamical properties such as nonnegative equilibrium points and Hopf bifurcation are studied. As another example, in reference [4], Kang and Wedekin considered the dynamical properties of the IGP system when the IG predator is a generalist predator or a specialist predator, and obtained the sufficient conditions for the extinction or persistence of the model.

Lately, Bai D et al. [5] considered an IGP model with strong Allee effect as follows:

$$\begin{cases} u_1' = u_1(u_1-\theta)(1-u_1) - \alpha u_1 u_2 - \beta u_1 u_3 \\ u_2' = \gamma_1 u_2 (u_1 - a_1 u_3 - d_1) \\ u_3' = \gamma_2 u_3 (u_1 + a_2 u_2 - d_2) \end{cases} \tag{1}$$

where u_1, u_2 and u_3 denote the basal prey, IG prey and IG predator, respectively. The significance of each parameter refers to [5]. However, in nature, some biological or environmental factors show periodic changes, such as climate and food supply, resulting in periodic coefficients in the model. In addition, if some biological factors have an instantaneous effect on the population at a certain moment, so that the population has a jump change, which needs to be described by impulse differential equations [6]. Also, after consuming the prey, the reproduction of predator is not instantaneous in real populations, and it is mediated by some time lag [7]. All of the above can induce complicated

Content from this work may be used under the terms of the Creative Commons Attribution 3.0 licence. Any further distribution of this work must maintain attribution to the author(s) and the title of the work, journal citation and DOI.
Published under licence by IOP Publishing Ltd

dynamical behaviors. So, on the basis of model (1), we take into account the following delayed IGP impulsive system with strong Allee effect:

$$\begin{cases} \left.\begin{aligned} x_1'(t) &= x_1(t)(x_1(t)-\theta(t))(1-x_1(t)) - \alpha(t)x_1(t)x_2(t) - \beta(t)x_1(t)x_3(t) \\ x_2'(t) &= \gamma_1(t)x_1(t-\tau)x_2(t-\tau) - \gamma_1(t)x_2(t)(\delta_1(t)x_2(t) + a_1(t)x_3(t) + d_1(t)) \\ x_3'(t) &= \gamma_2(t)x_3(t-\tau)(x_1(t-\tau) + a_2(t)x_2(t-\tau)) - \gamma_2(t)x_3(t)(\delta_2(t)x_3(t) + d_2(t)) \end{aligned}\right\} t \neq t_k, \\ \Delta x_i(t) = x_i(t_k^+) - x_i(t_k^-) = p_k^i x_i(t_k^-), \quad t = t_k, \quad k \in Z_+, \quad i = 1,2,3. \end{cases} \tag{2}$$

where x_1, x_2 and x_3 indicate the densities of basal prey, IG prey and IG predator respectively. τ is reaction-time lags by resource consumption into reproduction. All parameters in this model are positive. In addition, system (2) satisfies:

(H1) $\alpha(t)$, $\beta(t)$, $\theta(t)$, $\gamma_i(t)$, $a_i(t)$, $d_i(t)$ and $\delta_i(t)$ ($i = 1,2.$) are non-negative ω-periodic functions defined on real number field.

(H2) p_k^i, $t_k \in R$, $t_k < t_{k+1}$, $1 + p_k^i > 0$ and there is $q \in Z_+$ so that $p_k^i = p_{k+q}^i$ and $t_{k+q} = t_k + \omega$.

2. Existence of Positive Solutions

Let $U \subset R$, define $PC(U, R^3)$ the space of piecewise continuous functions $\phi: U \to R^3$ with the first kind discontinuities at $t = t_k \in U$. Let $PC'(U, R^3)$ denote the subspace of $PC(U, R^3)$ with $\phi' \in PC(U, R^3)$ for every $\phi \in PC'(U, R^3)$, $PC_\omega = \{\phi \in PC([0,\omega], R^3) \mid \phi(t+\omega) = \phi(t)\}$, $PC'_\omega = \{\phi \in PC'([0,\omega], R^3) \mid \phi(t+\omega) = \phi(t)\}$.

Lemma 1 ([8]). Suppose $\psi \in PC'_\omega$, then

$$\left| \sup_{s\in[0,\omega]} \psi(s) - \inf_{s\in[0,\omega]} \psi(s) \right| \le \frac{1}{2}\left(\int_0^\omega |\psi'(s)|\,\mathrm{d}s + \sum_{k=1}^q |\Delta\psi(t_k)| \right).$$

Let $u(t)$ is a periodic function with period ω, we introduce some notations for convenience:

$$\bar{u} = \frac{1}{\omega}\int_0^\omega u(s)\,\mathrm{d}s, \; u^L = \min_{s\in[0,\omega]} u(s), \; u^M = \max_{s\in[0,\omega]} u(s).$$

Lemma 2 ([9]). Suppose $L: \mathrm{Dom}\,L \subset X \to Y$ is a zero index Fredholm map, and $N: \bar{\Omega} \to Y$ is L-compact on $\bar{\Omega}$, where X, Y are Banach spaces and Ω is an open bounded subset of X. In addition, we assume that

(1) $Lx \neq \lambda Nx$, where $\lambda \in (0,1)$, $x \in \mathrm{Dom}\,L \cap \partial\Omega$;

(2) $QNx \neq 0$ for $x \in \partial\Omega \cap \mathrm{Ker}\,L$;

(3) Brouwer degree $\deg(JQN|_{\mathrm{Ker}\,L \cap \partial\Omega}) \neq 0$ (J is an isomorphism from $\mathrm{Im}\,Q$ to $\mathrm{Ker}\,L$);

then $Lx = Nx$ has at least one solution on $\bar{\Omega} \cap \mathrm{Dom}\,L$.

Theorem 1. Suppose that conditions (H1) and (H2) are hold. In addition, system (2) also satisfies:

$$\text{(H}_3\text{)}\ \frac{1}{4} + \frac{1}{\omega}\sum_{k=1}^q \ln(1+p_k^1) > 0; \ \text{(H}_4\text{)}\ \frac{\gamma_2^M \delta_2^M}{\beta^L \omega}\left(\frac{\omega}{4} + \sum_{k=1}^q \ln(1+p_k^1) \right) + \overline{\gamma_2 d_2} - \frac{1}{\omega}\sum_{k=1}^q \ln(1+p_k^3) > 0;$$

$$\text{(H}_5\text{)}\ \overline{\gamma_1\delta_1} e^{E_2} + \overline{\gamma_1 d_1} - \frac{1}{\omega}\sum_{k=1}^q \ln(1+p_k^2) > 0; \ \text{(H}_6\text{)}\ \frac{1}{\omega}\sum_{k=1}^q \ln(1+p_k^2) - \overline{\gamma_1 a_1} e^{C_3} - \overline{\gamma_1 d_1} > 0;$$

$$\text{(H}_7\text{)}\ \overline{\gamma_2^*} e^{E_1} + \overline{\gamma_2^* a_2} e^{E_2} + \frac{1}{\omega}\sum_{k=1}^q \ln(1+p_k^3) - \overline{\gamma_2 d_2} > 0;$$

then system (2) has at least one positive ω-periodic solution.

Poof. Let $x_i = e^{y_i(t)}$, $i = 1,2,3$, then the system (2) becomes

$$\begin{cases} y_1'(t) = (e^{y_1(t)} - \theta(t))(1 - e^{y_1(t)}) - \alpha(t)e^{y_2(t)} - \beta(t)e^{y_3(t)} := f_1(t) \\ y_2'(t) = \gamma_1(t)\left(\dfrac{e^{y_1(t-\tau)}e^{y_2(t-\tau)}}{e^{y_2(t)}} - a_1(t)e^{y_3(t)} - \delta_1(t)e^{y_2(t)} - d_1(t)\right) := f_2(t) \\ y_3'(t) = \gamma_2(t)\left(\dfrac{e^{y_1(t-\tau)}e^{y_3(t-\tau)} + a_2(t)e^{y_2(t-\tau)}e^{y_3(t-\tau)}}{e^{y_3(t)}} - \delta_2(t)e^{y_3(t)} - d_2(t)\right) := f_3(t) \end{cases} t \neq t_k \\ \Delta y_i(t_k^+) = \ln(1 + p_k^i) = \xi_i, \quad t = t_k, \quad k \in Z_+, \quad i = 1,2,3. \tag{3}$$

We set $X = \{x(t) = (x_1(t), x_2(t), x_3(t))^T \mid x_i(t) \in PC_\omega, i = 1,2,3\}$, $Y = \{\tilde{y} = (y', \xi_1, \ldots, \xi_q)\} = X \times R^{3q}$, where $y(t) = (y_1(t), y_2(t), y_3(t))^T$, $\xi_k = (p_k^1, p_k^2, p_k^3)^T = (\Delta y_1(t_k), \Delta y_2(t_k), \Delta y_3(t_k))^T$ ($k = 1,2,\ldots,q$) are constant vectors, then both X and Y are Banach space with $\|x\| = \sum_{i=1}^{3}\|x_i(t)\| = \sum_{i=1}^{3}\max_{0\le t\le\omega}|x_i(t)|$ and $\|\tilde{y}\| = \|y'(t)\| + \sum_{i=1}^{q}\|\xi_i\|$, $i = 1,2,3$. Set $L: \mathrm{Dom}L \subset X \to Y$ as $L(y) = (y'(t), \Delta y(t_1), \ldots, \Delta y(t_q))$, where $\mathrm{Dom}L = \{y \in X \mid y(t) \in PC'_\omega\}$; $N: X \to Y$ as $N(y(t)) = (f, u_1, \cdots, u_q)$, where $f = (f_1, f_2, f_3)^T$, $u_i = (\ln(1+p_i^1), \ln(1+p_i^2), \ln(1+p_i^3))^T$; $P: X \to X$ as $P(y(t)) = \frac{1}{\omega}\left(\int_0^\omega y_1(t)\mathrm{d}t, \int_0^\omega y_2(t)\mathrm{d}t, \int_0^\omega y_3(t)\mathrm{d}t\right)^T$; $Q: Y \to Y$ as $Q(f, \xi_1, \cdots, \xi_q) = (v_i, 0, \cdots, 0)$, where $v_i = \frac{1}{\omega}\int_0^\omega f_i(t) + \sum_{k=1}^{q}\xi_k^i$. Then L is a Fredholm operator of index zero, and that N is L-compact on $\overline{\Omega}$ for any given open bound subset Ω in X.
For equation $Ly = \lambda Ny$, $\lambda \in (0,1)$, we have

$$\begin{cases} y_i'(t) = \lambda f_i(t), \quad t \neq t_k, \\ \Delta y_i(t_k^+) = \lambda \ln(1 + \xi_k^i), \quad t = t_k, \quad k \in Z_+, \quad i = 1,2,3. \end{cases} \tag{4}$$

For certain $\lambda \in (0,1)$, suppose that $y(t) = (y_1(t), y_2(t), y_3(t))^T$ is a periodic solution of (4). By integrating (4) over the interval $[0,\omega]$, we can obtain

$$\int_0^\omega (e^{y_1(t)} - \theta(t))(1 - e^{y_1(t)})\mathrm{d}t + \sum_{k=1}^{q}\ln(1+p_k^1) = \int_0^\omega (\alpha(t)e^{y_2(t)} + \beta(t)e^{y_3(t)})\mathrm{d}t \tag{5}$$

$$\int_0^\omega \gamma_1^*(t)e^{y_1(t)}\mathrm{d}t + \sum_{k=1}^{q}\ln(1+p_k^2) = \int_0^\omega \gamma_1(t)(a_1(t)e^{y_3(t)} + \delta_1(t)e^{y_2(t)} + d_1(t))\mathrm{d}t \tag{6}$$

$$\int_0^\omega \gamma_2^*(t)(e^{y_1(t)} + a_2(t)e^{y_2(t)})\mathrm{d}t + \sum_{k=1}^{q}\ln(1+p_k^3) = \int_0^\omega \gamma_2(t)(\delta_2(t)e^{y_3(t)} + d_2(t))\mathrm{d}t \tag{7}$$

where $\gamma_1^*(t) = \frac{\gamma_1(\kappa^{-1}(t))}{e^{y_2(\tau)}}$, $\gamma_2^*(t) = \frac{\gamma_2(\kappa^{-1}(t))}{e^{y_3(\tau)}}$, $\kappa(t) = t - \tau$.
Furthermore, we can obtain

$$\int |y_1'(t)|\mathrm{d}t \le \frac{\omega}{2} + \sum_{k=1}^{q}\ln(1+p_k^1) + \sum_{k=1}^{q}|\ln(1+p_k^1)| := A_1 \tag{8}$$

$$\int_0^\omega |y_2'(t)|\mathrm{d}t \le 2\int_0^\omega \gamma_1^* e^{y_1(t)}\mathrm{d}t + \sum_{k=1}^{q}\ln(1+p_k^2) + \sum_{k=1}^{q}|\ln(1+p_k^2)| \tag{9}$$

$$\int_0^\omega |y_3'(t)|\mathrm{d}t \le 2\int_0^\omega \gamma_2^*(t)(e^{y_1(t)} + a_2(t)e^{y_2(t)})\mathrm{d}t + \sum_{k=1}^{q}\ln(1+p_k^3) + \sum_{k=1}^{q}|\ln(1+p_k^3)| \tag{10}$$

Since $y \in X$, there are ξ_i, $\eta_i \in (0,\omega)$, so that

$$y_i(\xi_i)=\inf_{t\in[0,\omega]} y_i(t),\quad y_i(\eta_i)=\sup_{t\in[0,\omega]} y_i(t). \tag{11}$$

By (5), (7) and (11), we get

$$\int_0^{\omega}\gamma_2^*(t)e^{y_1(\xi_1)}dt+\sum_{k=1}^{q}\ln(1+p_k^3)\le\frac{\gamma_2^M\delta_2^M}{\beta^L}\left(\frac{\omega}{4}+\sum_{k=1}^{q}\ln(1+p_k^1)\right)+\overline{\gamma_2 d_2}\omega$$

$$\int_0^{\omega}\alpha(t)e^{y_2(\xi_2)}\,\mathrm{d}t\le\frac{\omega}{4}+\sum_{k=1}^{q}\ln(1+p_k^1)$$

$$\int_0^{\omega}\beta(t)e^{y_3(\xi_3)}\,\mathrm{d}t\le\frac{\omega}{4}+\sum_{k=1}^{q}\ln(1+p_k^1)$$

which implies

$$y_1(\xi_1)\le\ln\frac{1}{\overline{\gamma_2^*}}+\ln\left(\frac{\gamma_2^M\delta_2^M}{\beta^L}\left(\frac{1}{4}+\frac{1}{\omega}\sum_{k=1}^{q}\ln(1+p_k^1)\right)+\overline{\gamma_2 d_2}-\frac{1}{\omega}\sum_{k=1}^{q}\ln(1+p_k^3)\right):=B_1 \tag{12}$$

$$y_2(\xi_2)\le\ln\frac{1}{\overline{\alpha}}+\ln\left(\frac{1}{4}+\frac{1}{\omega}\sum_{k=1}^{q}\ln(1+p_k^1)\right):=B_2 \tag{13}$$

$$y_3(\xi_3)\le\ln\frac{1}{\overline{\beta}}+\ln\left(\frac{1}{4}+\frac{1}{\omega}\sum_{k=1}^{q}\ln(1+p_k^1)\right):=B_3 \tag{14}$$

From Lemma 1, (8) and (12) , we have

$$y_1(t)\le y_1(\xi_1)+\frac{1}{2}\left(\int_0^{\omega}|y_1'(t)|\,\mathrm{d}t+\sum_{k=1}^{q}|\ln(1+p_k^1)|\right)\le B_1+\frac{A_1}{2}+\frac{1}{2}\sum_{k=1}^{q}|\ln(1+p_k^1)|:=C_1 \tag{15}$$

From (9) and (15), we have

$$\int_0^{\omega}|y_2'(t)|\,\mathrm{d}t\le 2\overline{\gamma_1^*}e^{C_1}\omega+\sum_{k=1}^{q}\ln(1+p_k^2)+\sum_{k=1}^{q}|\ln(1+p_k^2)|:=A_2 \tag{16}$$

Similar (15)'s derivation, we have

$$y_2(t)\le B_2+\frac{A_2}{2}+\frac{1}{2}\sum_{k=1}^{q}|\ln(1+p_k^2)|:=C_2 \tag{17}$$

From (10), (15) and (17), we have

$$\int_0^{\omega}|y_3'(t)|\,\mathrm{d}t\le 2(\overline{\gamma_2^*}e^{C_1}\omega+\overline{\gamma_2^* a_2}e^{C_2}\omega)+\sum_{k=1}^{q}\ln(1+p_k^3)+\sum_{k=1}^{q}|\ln(1+p_k^3)|:=A_3 \tag{18}$$

Similarly, from (14), (18) and Lemma 1, we get

$$y_3(t)\le B_3+\frac{A_3}{2}+\frac{1}{2}\sum_{k=1}^{q}|\ln(1+p_k^3)|:=C_3 \tag{19}$$

On the other hand, from (6), (11) and (19), we have

$$\int_0^{\omega}\gamma_1(t)\delta_1(t)e^{y_2(\eta_2)}\,\mathrm{d}t\ge\sum_{k=1}^{q}\ln(1+p_k^2)-\overline{\gamma_1 a_1}e^{C_3}\omega-\overline{\gamma_1 d_1}\omega$$

which implies

$$y_2(\eta_2)\ge\ln\frac{1}{\overline{\gamma_1\delta_1}}+\ln\left(\frac{1}{\omega}\sum_{k=1}^{q}\ln(1+p_k^2)-\overline{\gamma_1 a_1}e^{C_3}-\overline{\gamma_1 d_1}\right):=D_2 \tag{20}$$

From (16), (20) and Lemma 1, we get

$$y_2(t)\ge y_2(\eta_2)-\frac{1}{2}\left(\int_0^{\omega}|y_2'(t)|\,\mathrm{d}t+\sum_{k=1}^{q}|\ln(1+p_k^2)|\right)\ge D_2-\frac{A_2}{2}-\frac{1}{2}\sum_{k=1}^{q}|\ln(1+p_k^2)|:=E_2 \tag{21}$$

From (6), (11) and (21), we have

$$\int_0^{\omega}\gamma_1^*(t)e^{y_1(\eta_1)}\,\mathrm{d}t\ge\overline{\gamma_1\delta_1}e^{E_2}\omega+\overline{\gamma_1 d_1}\omega-\sum_{k=1}^{q}\ln(1+p_k^2)$$

which implies

$$y_1(\eta_1) \geq \ln\frac{1}{\overline{\gamma_1^*}} + \ln\left(\overline{\gamma_1\delta_1}e^{E_2} + \overline{\gamma_1 d_1} - \frac{1}{\omega}\sum_{k=1}^{q}\ln(1+p_k^2)\right) := D_1 \tag{22}$$

Similar (21)'s derivation, we have

$$y_1(t) \geq D_1 - \frac{A_1}{2} - \frac{1}{2}\sum_{k=1}^{q}|\ln(1+p_k^1)| := E_1 \tag{23}$$

From (7), (11), (21) and (23), we get

$$\int_0^{\omega}\gamma_2\delta_2(t)e^{y_3(\eta_3)}\,\mathrm{d}t \geq \overline{\gamma_2^*}e^{E_1}\omega + \overline{\gamma_2^* a_2}e^{E_2}\omega + \sum_{k=1}^{q}\ln(1+p_k^3) - \overline{\gamma_2 d_2}\omega$$

which implies

$$y_3(\eta_3) \geq \ln\frac{1}{\overline{\gamma_2\delta_2}} + \ln\left(\overline{\gamma_2^*}e^{E_1} + \overline{\gamma_2^* a_2}e^{E_2} + \frac{1}{\omega}\sum_{k=1}^{q}\ln(1+p_k^3) - \overline{\gamma_2 d_2}\right) := D_3 \tag{24}$$

Similar, from (18), (24) and Lemma 1, we get

$$y_3(t) \geq D_3 - \frac{A_3}{2} - \frac{1}{2}\sum_{k=1}^{q}|\ln(1+p_k^3)| := E_3 \tag{25}$$

It follows that $\max_{t\in[0,\omega]}|y_i(t)| \leq \max\{|C_i|,|E_i|\} := G_i$, $i=1,2,3$. So, there is a constant $G_0>0$, satisfied $\max\{G_i\} \leq G_0$. Let $r > G_0$, $\Omega = \{y \in X : \|y\| \leq r\}$, then Ω satisfies the requirement (1) of Lemma 2. For $y \in \partial\Omega \cap \operatorname{Ker} L$, obviously we have $QN(y) \neq 0$, which further implies $\deg(JQN|_{\operatorname{Ker} L\cap\partial\Omega}) \neq 0$. So far, we have verified that Ω meets the condition of all assumptions in Lemma 2. Thus, model (2) has a positive ω-periodic solution.

3. An Illustrative Example

In system (2), we take $t_k = k\pi$, $\theta = 0.001 + 0.0001\sin t$, $\alpha = 0.3 + 0.01\sin t$, $\beta = 0.2 + 0.01\sin t$, $\gamma_1 = 0.09 + 0.01\sin t$, $\gamma_2 = 0.08 + 0.01\sin t$, $a_1 = 0.4 + 0.05\sin t$, $a_2 = 0.3 + 0.05\sin t$, $d_1 = 0.5 + 0.1\sin t$, $d_2 = 0.6 + 0.1\sin t$, $p_k^1 = p_k^2 = p_k^3 = 0.1$, $\tau = 0.3$. Obviously, all the assumptions of theorem 1 are contented. Thus, the model has a 2π-periodic solution. The evolution figures of 2π-periodic solution of system (2) with the above parameters and initial value $(x_1(0), x_2(0), x_3(0)) = (0.6, 0.5, 0.4)$ are presented in figure 1-4. Figure 1-3 are time-series of $x_1(t)$, $x_2(t)$ and $x_3(t)$ respectively, and Figure 4 exhibits the phase portrait of 2π-periodic solution of system (2).

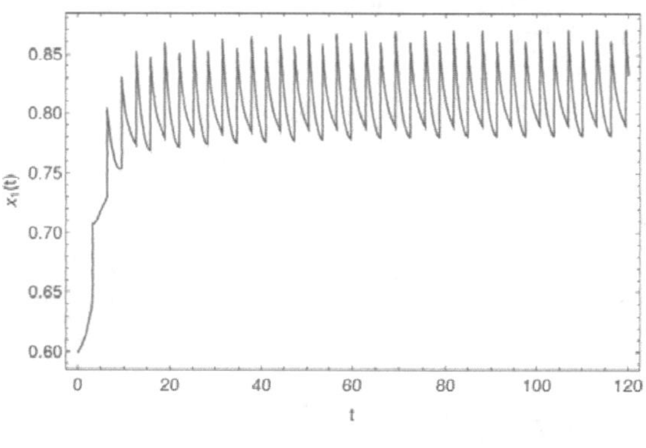

Figure 1. Time-series of $x_1(t)$ generated by system (2).

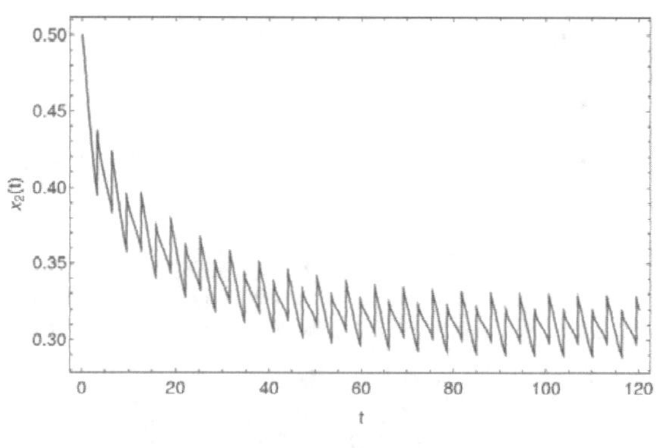

Figure 2. Time-series of $x_2(t)$ generated by system (2).

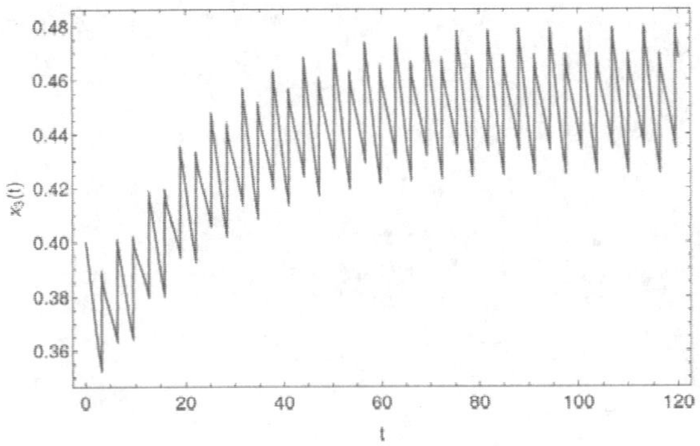

Figure 3. Time-series of $x_3(t)$ generated by system (2).

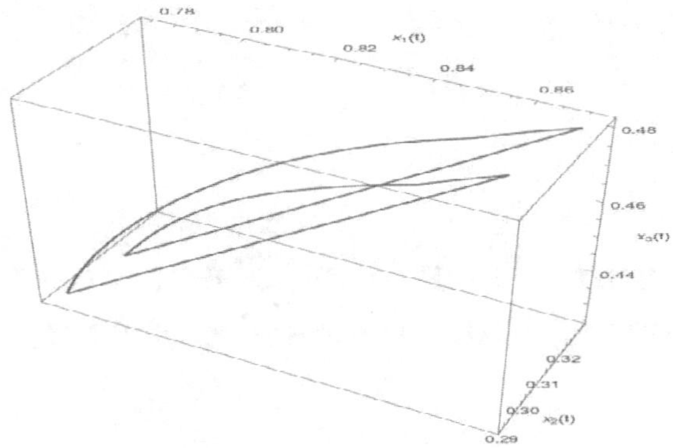

Figure 4. Phase portrait of 2π -periodic solution generated by system (2).

4. Conclusion

Intraguild predation (IGP) systems have been widely studied because of its predation and competition relations. In order to make the IGP model more applicable, we study a kind of delayed IGP impulsive system with strong Allee effect and periodic coefficients. Using Mawhin's method, the existence of positive periodic solutions is obtained. Finally, the effectiveness of the theoretical results is identified by some numerical simulations. The results show that under suitable conditions, the model can finally approach some kind of ecological balance.

References

[1] Polis G A, Myers C A and Holt R D 1989 The ecology and evolution of intraguild predation: potential competitors that eat each other *Annual review of ecology & systematics* 20(1):297-330

[2] Anderson T L and Semlitsch R D 2014 High intraguild predator density induces thinning effects on and increases temporal overlap with prey populations. *Population Ecology*, 56(2):265-273.

[3] Rabago J and Collera J 2018 Hopf Bifurcation in a Delayed Intraguild Predation Model *Southeast Asian Bulletin of Mathematics* 42(5):691-709

[4] Kang Y and Wedekin L 2013 Dynamics of an intraguild predation model with generalist or specialist predator *Journal of Mathematical Biology* 67(5):1227-1259

[5] Bai D and Yun K et al 2021 Dynamics of an intraguild predation food web model with strong Allee effect in the basal prey *Nonlinear Anal Real World Appl* 58:103206

[6] Wang Q and Ding M et al 2010 Existence and attractivity of a periodic solution for an N-species Gilpin-Ayala impulsive competition system. *Nonlinear Anal Real World Appl* 11(4):2675-2685

[7] Li Z and Dai B 2018 Global dynamics of delayed intraguild predation model with intraspecific competition. *International Journal of Biomathematics* 011(008):425-463

[8] Wang Q, Dai B and Chen Y 2009 Multiple periodic solutions of an impulsive predator-prey model with Holling-type IV functional response. *Math Comput Model* 49(9-10):1829-1836

[9] Gaines R E and Mawhin J L 1997 *Coincidence Degree and Nonlinear Differential Equations* (New York: Springer-Verlag)

Short-Term Wind Speed Forecasting Using Meta Learning-based Elman Neural Network

M. Madhiarasan*

Department of Computer Science and Engineering, Indian Institute of Technology Roorkee, Roorkee, Uttarakhand, India–247667, Orcid ID: 0000-0003-2552-0400

*Corresponding author email: mmadhiarasan.cse@sric.iitr.ac.in

Abstract. Adequate power provision to the customer and wind energy penetration into the electrical grid is necessitated for accurate wind speed forecasting in the short-term horizon to realize the scheduling, unit commitment, and control. According to the various meteorological parameters, the wind speed and energy production from wind energy are affected. Therefore, the author performs the multi-inputs associated Meta learning-based Elman Neural Network (MENN) forecasting model to overcome the uncertainty and generalization problem. The proposed forecasting approach applicability evaluated with real-time data concerning wind speed forecasting on a short-term time scale. Performance analysis reveals that the meta learning-based Elman neural network is robust and conscious than the existing methods, with a least mean square error of 0.0011.
Keywords: Meta-learning, elman neural network, generalization, forecasting, wind speed.

1. Introduction

The demand for global energy is increasing because of advancements in industry and technology. The use of fossil fuels causes environmental destruction; wind energy focuses on the power sector and stack holders to alleviate the above-said issue. Wind characteristics and climatic conditions vary by country, city, and region. Wind speed forecasting is essential in power system management and penetration of distributed energy systems **Madhiarasan M**[1] **Madhiarasan M and Deepa SN**[2]. During the past three decades, various forecasting approaches and methods have endeavored for wind speed forecasting **M Madhiarasan M and Deepa SN**[3, 4, 5, 6, 7, 8, 9, 10, & 11]. **Sun G, Chen Y, Wei Z, Li X & Cheung, K W**[12] a BPN (backpropagation neural network) with seasonal exponential adjustment based one-year ahead wind speed forecasting was pointed out. **Ul Haque A & Meng J**[13] suggested fuzzy logic and adaptive resonance theory (Fuzzy ARTMAP) based 12- and 24-hr forecasting of wind speed. Fractional-ARIMA model-based 24 and 48 hours ahead forecasting of wind speed presented by **Kavasseri R G & Seetharaman K**[14]. **Guo Z H, Wu, J, Lu, H Y, & Wang, J Z**[15] performed Relevance Vector Machine associated with the mixed kernel-based day-ahead wind speed forecasting. Less complex, high accurate forecasting of wind speed leverages the advanced power system operation and control. This research carried out meta-learning-based Elman neural networks by averaging four Elman Neural Networks (ENN) (2 inputs ENN, 3 inputs ENN, 4 inputs ENN, and 5 inputs ENN) based forecasting models to forecast the one-day and two days ahead wind speed. Neural network parameter selection is the prime factor that decides the forecast's stability and performance **Madhiarasan M**[16], **Madhiarasan M & Deepa SN**[17]. From the literature review, it is noted that a still generalized and precise forecasting model is required. Thus, in this paper, the author proposed new meta-learning-based Elman neural network averaging various inputs such as 2 inputs, 3 inputs, 4

Content from this work may be used under the terms of the Creative Commons Attribution 3.0 licence. Any further distribution of this work must maintain attribution to the author(s) and the title of the work, journal citation and DOI.
Published under licence by IOP Publishing Ltd

inputs, and 5 inputs based on individual Elman neural networks. The author considered inputs that highly influence wind speed forecasting. The proposed model addresses the issue of generalization and uncertainty in wind speed forecasting.

2. Highlights of the Proposed Forecasting Model

- Proposed new Meta learning-based Elman neural network.
- Carried out hidden neurons based on stability studies with various hidden neurons.
- Performed comparative analysis with an existing forecasting model.
- Achieve better robustness and generalization.

3. Proposed MENN Forecasting Model Modeling

Steps to modeling the proposed Meta learning-based Elman Neural Network are

i) Select the significant atmospheric parameters as inputs and acquire the data. The proposed MENN is designed with various inputs such as wind direction, temperature, relative humidity, pressure, and wind speed. The considering input parameters are acquired from the NOAA at location Tamil Nadu, India (latitude- 12.18 and longitude- 79.05). The dataset consists of 5 minutes averaged 2592 data of each input parameter.

ii) Perform the min-max scaling for range the considered inputs between 0-1 using the formula

$$\text{Scaled input}, I_i' = \left(\frac{I_i - I_{\min}}{I_{\max} - I_{\min}}\right)\left(I_{\max}' - I_{\min}'\right) + I_{\min}' \qquad (1)$$

Where I_i - input, $I_{\min}$ - minimum input value, $I_{\max}$ - maximum input value, $I_{\min}'$ - minimum target value, $I_{\max}'$ -maximum target value.

iii) Split the 2016 data for each input as training data and the remaining 576 data for each input as testing data.

iv) Modeling of the proposed meta-learning-based Elman neural network is done using the Table. 1 mentioned parameter, initialize the parameter. The proposed wind speed forecasting model architecture is depicted in Figure 1.

Table 1. Parameters of Meta-learning based Elman Neural Network.

Parameters of Meta-learning-based Elman Neural Network	
Parameters	**Values**
No. of the input neurons	2, 3, 4, & 5
No. of the hidden layer	1
No. of the hidden neurons	1-25
No. of the output neuron	1
No. of the epochs	2000
Threshold	1

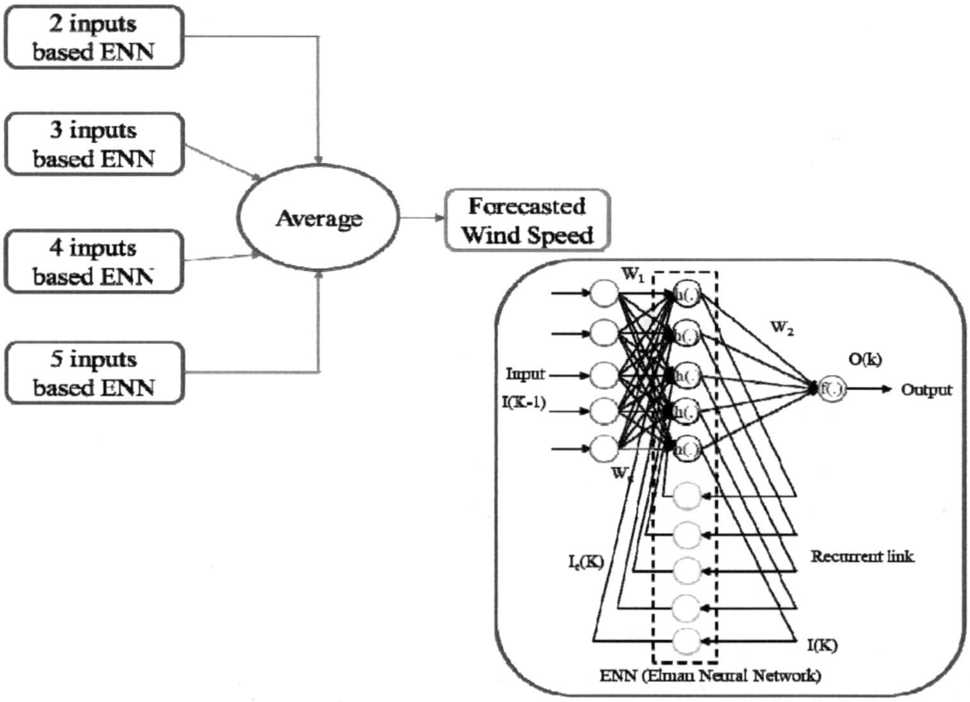

Figure 1. Proposed MENN architecture.

Individual ENN mathematical model

$$\text{ENN Input, } I(K) = h\left(W_c I_c(K) + W_1 U(K-1)\right) \tag{2}$$

$$\text{ENN Recurrent link layer inputs, } I_c(K) = I(K-1) \tag{3}$$

$$\text{ENN Output, } O(K) = f\left(W_2 I(K)\right) \tag{4}$$

Where, W_1 - weight between input and hidden layer, W_2 - weight between hidden and recurrent link layers, W_c - weight between the context and input layers, $h(\cdot)$ - hyperbolic tangent sigmoid activation function, and $f(\cdot)$ - purelin activation function.

MENN Mathematical Model:

$$\text{MENN output } = \frac{2inputs^{ENN} + 3inputs^{ENN} + 4inputs^{ENN} + 5inputs^{ENN}}{4} \tag{5}$$

Computation of MENN error measures:

$$MSE^{MENN} = Mean\left(2inputs_{MSE}^{ENN} + 3inputs_{MSE}^{ENN} + 4inputs_{MSE}^{ENN} + 5inputs_{MSE}^{ENN}\right) \tag{6}$$

$$MAE^{MENN} = Mean\left(2inputs_{MAE}^{ENN} + 3inputs_{MAE}^{ENN} + 4inputs_{MAE}^{ENN} + 5inputs_{MAE}^{ENN}\right) \tag{7}$$

$$MAPE^{MENN} = Mean\left(2inputs_{MAPE}^{ENN} + 3inputs_{MAPE}^{ENN} + 4inputs_{MAPE}^{ENN} + 5inputs_{MAPE}^{ENN}\right) \tag{8}$$

The recurrent link-layer retains the memory and stores the feedback. Hyperbolic tangent sigmoid activation function and purelin activation function adapted for the hidden layer and output layer, respectively.

v) Evaluate the proposed MENN model with training data a total of 2016 data (7 days, 5 minutes averaged) and testing data a total of 576 data (2 days, 5 minutes averaged). Calculate the error measures, namely, Mean Square Error (MSE), Mean Absolute Error (MAE), and Mean Absolute Percentage Error (MAPE). The mathematical formulas are

$$\text{MSE} = \frac{1}{N}\sum_{i=1}^{N}(O_i^{'} - O_i)^2 \quad (9)$$

$$\text{MAE} = \frac{1}{N}\sum_{i=1}^{N}(O_i^{'} - O_i) \quad (10)$$

$$\text{MAPE} = \frac{100}{N}\sum_{i=1}^{N}\left|(O_i^{'} - O_i)/\overline{O}_i\right| \quad (11)$$

Where N is the number of data, $O_i^{'}$-observed wind speed, O_i - forecasted wind speed, $\overline{O}_i$-average observed target wind speed.

vi) Using the computed error measures, select the optimal hidden neurons based on stability analysis with various hidden neurons 1-25.

vii) Print the output if the forecasting is satisfactory; else, do step (iv) fine-tune the parameters until reaching the stopping condition (max epoch or minimal error).

4. Discussion of Experimental Results

Proposed model meta-learning process training and averaging of four various inputs, namely, two inputs(wind direction, temperature), 3 inputs (wind direction, temperature, relative humidity), 4 inputs (wind direction, temperature, relative humidity, pressure), and 5 inputs (wind direction, temperature, relative humidity, pressure, and wind speed) based on individual Elman neural network. The proposed MENN wind speed forecasting model runs on an Acer laptop with a Pentium (R) Dual-Core processor (2GB of RAM and 2.30GHZ) and MATLAB environment used for simulation. The error measures are utilized to check the effectiveness of the suggested wind speed forecasting model.

The meta-learning process helps the proposed model generalize well and make reliable wind speed forecasting. Hidden neurons are essential in neural network stability and convergence **Madhiarasan M, Tipaldi M & Siano P**[18]. Proposed Meta learning-based Elman neural network model stability analyzed with various hidden neurons, and the obtained results are presented in Table 2. Based on the least error measure value, 15 hidden neurons are chosen as the optimal hidden neurons in the hidden layer. Figure 2 depicts the sample of the observed wind speed data. The obtained results based on the optimal hidden neurons associated with the proposed MENN are shown in Figures 3 and 4. Figure 4 infers that the forecasted wind speed values are very close to the wind speed observed values with improved forecasting accuracy; all computed error measures are minimal.

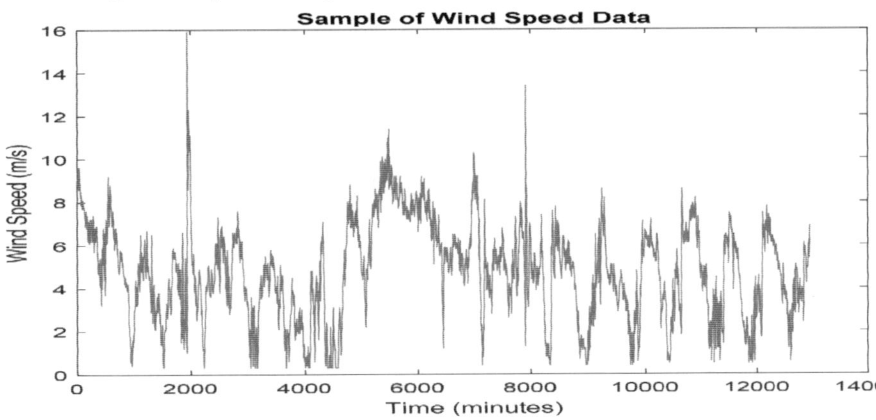

Figure 2. Samples of wind speed data.

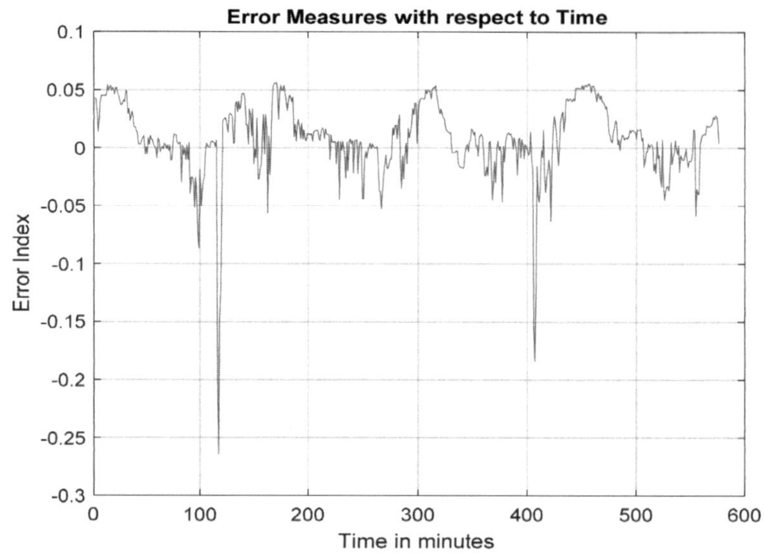

Figure 3. Error measures with respect to time.

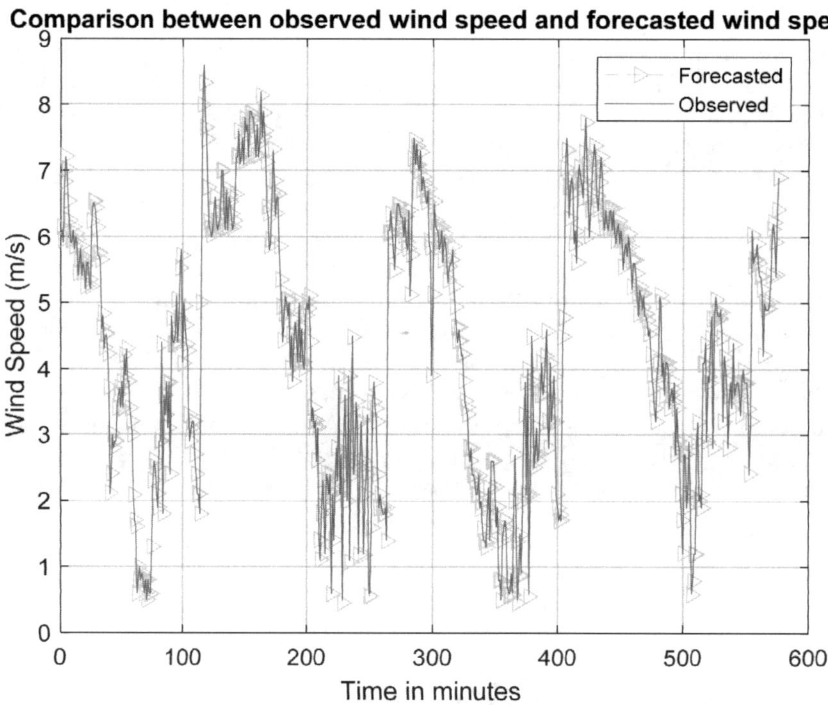

Figure 4. Observed wind speed compared with the forecasted wind speed.

Table 2. Stability analysis of MENN with various hidden neurons.

Number of Hidden Neurons in MENN	Error Measures		
	MSE	**MAE**	**MAPE**
1	0.0237	0.0191	1.9108
2	0.0037	0.0427	0.9812
3	0.0358	0.1034	2.3735
4	0.0075	0.0507	1.1638
5	0.0061	0.0526	1.2081
6	0.0055	0.5020	1.1948
7	0.0142	0.0760	1.7459
8	0.0506	0.1405	3.2274
9	0.0297	0.1019	2.3405
10	0.0178	0.0840	1.9278
11	0.0035	0.0405	0.9295
12	0.0027	0.0283	0.6507
13	0.0056	0.0556	1.2773
14	0.0111	0.0740	1.6984
15	**0.0011**	**0.0228**	**0.5235**
16	0.0023	0.0316	0.7254
17	0.0013	0.0232	0.5325
18	0.0053	0.0417	0.9584
19	0.0411	0.1210	2.7775
20	0.0328	0.1087	2.4957
21	0.0038	0.0352	0.8087
22	0.0162	0.0808	1.8565
23	0.0165	0.0878	2.0151
24	0.0039	0.0401	0.9214
25	0.0038	0.0436	1.0023

Table 3. Comparative analysis of existing different forecasting models with the proposed model.

S. No	Forecasting model	Error Measure (MSE)
1	Naïve model	1..2139
2	BPN model	0.2027
3	f-ARIMA model	0.1725
4	fuzzy ARTMAP model	0.0332
5	Relevance Vector Machine model	0.0132
6	ENN model	0.0054
7	**MENN model**	**0.0011**

The proposed model forecasts wind speed concerning one day and two days ahead; Figure 4 indicates that the proposed model correctly predicted wind speed concerns time horizon of short-term. The effectiveness of the proposed MENN is further quantified with a comparative analysis with other forecasting models tabulated in Table 3.

5. Conclusion

Forecasting of a wind speed avoids the encountering of grid stability and power quality-related issues. The proposed meta-learning ENN-based wind speed forecasting for short-term time scale results are very close to the observed wind speed values. This paper proposed a MENN forecasting model that handles the diversity of inputs, generalized to improved forecasting accuracy, which can be perceived from the experimental results. Based on the stability analysis, identified optimal hidden neurons as 15 for the proposed MENN. Compared with the existing forecasting models, the presented forecasting model can generalize well and offers outstanding short-term wind speed forecasting with a small Mean Square Error (MSE) of 0.0011. The proposed model is generic. It can handle the dynamic and uncertainty concern with various inputs and different time scale forecasting. In future research work, the author proved the validity of the proposed model with varying horizons of time and weather conditions.

Acknowledgments
The author conveys his gratitude to National Oceanic and Atmospheric Administration (NOAA), United States, for the dataset provision to carry out the simulation.

References

[1] Madhiarasan M & Deepa SN 2016 *Performance investigation of six artificial neural networks for different time scale wind speed forecasting in three wind farms of coimbatore region* International Journal of Innovation and Scientific Research vol 23(2) pp 380-411
[2] Madhiarasan M 2018 *Certain algebraic criteria for design of hybrid neural network models with applications in renewable energy forecasting* Ph. D. Thesis Anna University Chennai India
[3] Madhiarasan M & Deepa SN 2016 *Comprehensive study of various forecasting techniques for forecast of wind speed in the field of wind energy system* TERI Information Digest on Energy and Environment vol *15*(4) pp 439-457
[4] Madhiarasan M and Deepa SN 2016 *A novel criterion to select hidden neuron numbers in improved back propagation networks for wind speed forecasting* Applied intelligence 44(4) pp 878-893
[5] Madhiarasan M & Deepa SN 2016 *Deep Neural Network Using New Training Strategy Based Forecasting Method for Wind Speed and Solar Irradiance Forecast* Middle-East Journal of Scientific Research vol 24(12) pp 3730-3747
[6] Madhiarasan M & Deepa SN 2017 *A New Hybridized Optimization Algorithm to Optimize Echo State Network for Application in Solar Irradiance and Wind Speed Forecasting* World Applied Sciences Journal vol 35(4) pp 596-614
[7] Madhiarasan M & Deepa SN 2017 *Comparative analysis on hidden neurons estimation in multi layer perceptron neural networks for wind speed forecasting* Artificial Intelligence Review vol 48(4) pp 449-471
[8] Madhiarasan M & Deepa SN 2016 *ELMAN neural network with modified grey wolf optimizer for enhanced wind speed forecasting* Circuits and Systems vol 7(10) pp 2975
[9] Madhiarasan, M., & Deepa, S. N 2018 *A novel method to select hidden neurons in ELMAN neural network for wind speed prediction application WSEAS Transactions on Power Systems vol 13 pp* 13-30
[10] Madhiarasan M & Deepa SN 2018 *Determination of Adequate Hidden Neurons in Combo Neural Network Using New Formulation and Fine Tuning with IMGWOA for Enrich Wind-Speed Forecasting* International Journal of Applied Research on Information Technology and Computing vol 9 (1) pp 89-101
[11] Madhiarasan M & Deepa SN 2016 *Application of Ensemble Neural Networks for Different Time Scale Wind Speed Prediction* International Journal of Innovative Research in Computer and Communication Engineering vol 4 (5) pp 9610-9617
[12] Sun G, Chen Y, Wei Z, Li X & Cheung KW 2014 *Day-ahead wind speed forecasting using relevance vector machine* Journal of Applied Mathematics vol 2014 pp1-6
[13] Guo, ZH, Wu J, Lu HY & Wang JZ 2011 *A case study on a hybrid wind speed forecasting method using BP neural network* Knowledge-based systems vol 24(7) pp 1048-1056
[14] Kavasseri RG & Seetharaman K 2009 *Day-ahead wind speed forecasting using f-ARIMA models* Renewable Energy vol 34(5) pp 1388-1393
[15] Ul Haque A & Meng J 2011 *Short-term wind speed forecasting based on fuzzy ARTMAP* International Journal of Green Energy vol *8*(1) pp 65-80

[16] Madhiarasan M 2020 *Accurate prediction of different forecast horizons wind speed using a recursive radial basis function neural network* Protection and Control of Modern Power Systems vol 5(1) pp 1-9
[17] Madhiarasan M and Deepa SN 2016 *Precisious Estimation of Solar Irradiance by Innovative Neural Network and Identify Exact Hidden Layer Nodes through Novel Deciding Standard* Asian Journal of Research in Social Sciences and Humanities vol 6(12) pp 951-974
[18] Madhiarasan M, Tipaldi M and Siano P 2020 *Analysis of Artificial Neural Network Performance Based on Influencing Factors for Temperature Forecasting Applications* Journal of High Speed Networks vol 26(3) pp 209-223

A Failure Propagation Analysis Method Based on 3 × 3 Hierarchical Model

Shujie Pang and Jian Jiao*

School of Reliability and Systems Engineering, Beihang University, Beijing, China

*Corresponding author email: jiaojian@buaa.edu.cn

Abstract. In recent years, with the increasing complexity of equipment, system design and safety analysis for the equipment are becoming more and more difficult. In this paper, we make a more comprehensive analysis of the hierarchical relationship of the system from the vertical and horizontal angles and put forward the 3 × 3 Hierarchical Model. Then, we propose a failure propagation analysis method based on 3 × 3 Hierarchical Model. At the same time, SysML is used to build the models, which is convenient for different people to learn and modify the models. Finally, a hydraulic transmission system is taken as an example to illustrate the feasibility of the method.
Keywords: Safety analysis, 3 × 3 hierarchical model, failure propagation analysis, SysML.

1. Introduction

To ensure the safety and reliability of the system, the safety analysis should run through the whole life cycle of the system. Therefore, we should carry out the corresponding safety analysis from the early design stage. In this paper, we mainly study the hierarchical design method and failure analysis method of the system.

With the increasing complexity of the system, the system design needs to be divided into hierarchies and carried out in an orderly manner. The design work based on conceptual information in system design began in 1966. With the continuous development of system design, there are two main modeling frameworks in the industrial field: function-structure and function-behaviour-structure. In addition, there are other methods derived from them, such as task-function-resource, mission-capability-resource, function-behaviour-structure-state, etc. The existing methods can describe the design requirements of each hierarchy of the system. However, there is a lack of consideration in the modeling of the whole life cycle of the system.

At present, there are many failure propagation analysis methods. The state of art methods include the Systems-Theoretic Process Analysis (STPA), Functional Resonance Analysis (FRAM), Hierarchical Performed Hazard Origin and Propagation Studies (HIP-HOPS), Functional Failure Identification and Propagation (FFIP), etc. The STPA [1] establishes models under the framework of system structure and behaviour but lacks analysis of system function. The FRAM [2] involves the analysis of functions, but it is mainly used for social technology system and safety accident analysis. The HIP-HOPS method [3] uses if-FMEA to analyze the failure logic, and uses tables and words to describe failure causes, which leads to the described results are ambiguous. FFIP method [4] is a graphical modeling method, which is suitable for the early design stage of the system, so its scope of application is limited. Therefore, the shortcomings of current failure propagation analysis methods can be summarized into three aspects: (i) the analysis of system structure, function and behaviour are not comprehensive; (ii) there is ambiguity

Content from this work may be used under the terms of the Creative Commons Attribution 3.0 licence. Any further distribution of this work must maintain attribution to the author(s) and the title of the work, journal citation and DOI.
Published under licence by IOP Publishing Ltd

due to the description of natural language; (iii) there is a lack of analysis methods suitable for the whole life cycle of the system.
To sum up, aiming at the shortcomings of the current hierarchical design method and failure analysis method, we have conducted the following research. Firstly, we analyzed the hierarchical relationship of the system from a two-dimensional perspective and proposed the 3 × 3 Hierarchical Model. Then, we proposed a failure propagation analysis method based on the 3 × 3 Hierarchical Model. Moreover, we took a hydraulic transmission subsystem as an example, used the above method to analyze the failure propagation path. At the same time, we used SysML to describe the models, which not only avoided the ambiguity caused by natural language, but also facilitates different people to learn and modify these models.

2. 3 × 3 Hierarchical Model

In this part, we make a comprehensive analysis of the hierarchical relationship of the system from a two-dimensional perspective: vertical direction and horizontal direction. Then, we put forward the 3 × 3 Hierarchical Model, as shown in Figure 1. The advantage of the 3 × 3 Hierarchical Model is to help the system build a unified model. With the advance of system life cycle, different hierarchies can be selected for analysis.

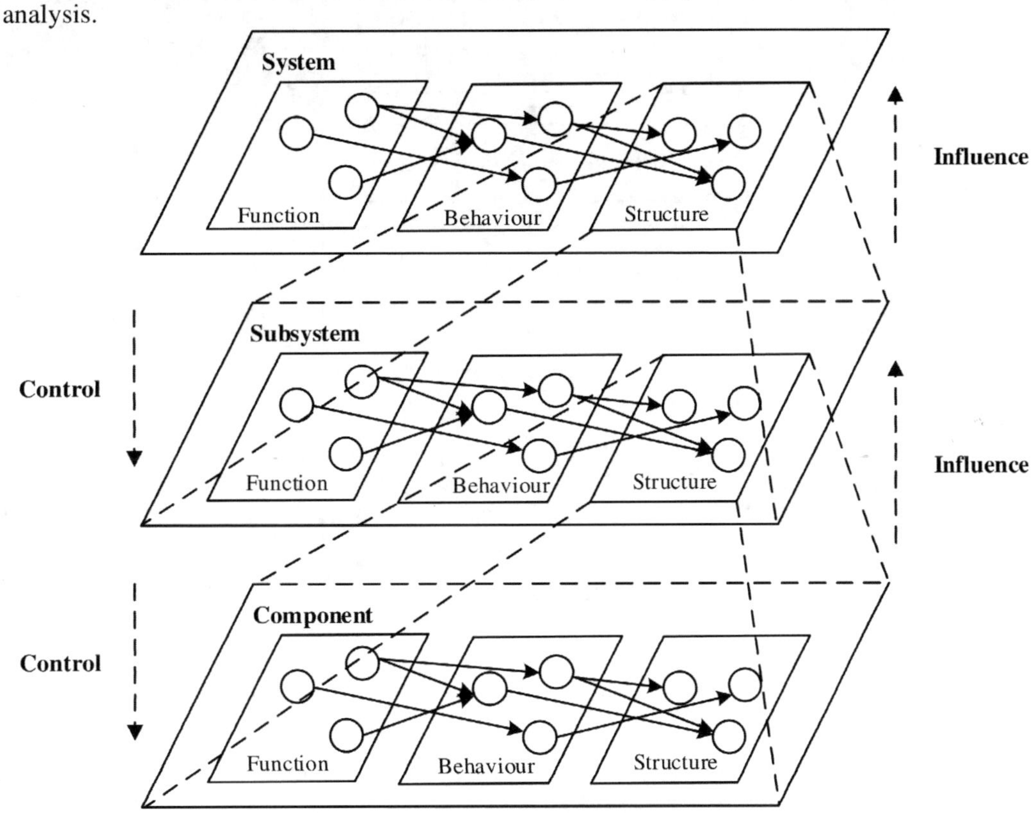

Figure 1. The 3 × 3 hierarchical model.

The system model can be built in both vertical and horizontal directions. Vertically, it can be divided into three layers: system-subsystem-component. Horizontally, the three layers can be divided into other three layers: function-behaviour-structure. So, the model is called 3 × 3 Hierarchical Model. Among them, function refers to an attribute that can meet a certain demand. Behaviour is the characteristic of variables in a system over time. Structure is a collection of interrelated system elements organized according to certain laws.
When studying the system hierarchical relationship, we first make a horizontal analysis from the system. The function of system layer is the beginning of design. It is the description of the tasks that the system finally needs to complete. Due to the realization of function needs to rely on the implementation of

behaviour, the next hierarchy is the behaviour layer. Behaviour needs a specific structure to complete. For the system layer, the system is composed of subsystems, which expand vertically to the subsystem layer. The subsystem layer can also be analyzed according to the function-behaviour-structure. The subsystem is composed of various components. Therefore, the structure layer of the subsystem can be extended vertically downward to the function-behaviour-structure of the components.

3. The Failure Propagation Analysis Method Based on 3 × 3 Hierarchical Model

In this part, we proposed a new safety analysis method. The method mainly includes six steps (as shown in Figure 2), they are determining analysis objectives and safety requirements, establishing hierarchical model, establishing function model, establishing behaviour model, failure logic reasoning and failure propagation path analysis. The analysis process of the whole framework needs to select the appropriate hierarchy vertically according to the purpose of the analysis. The specific analysis process is carried out in strict accordance with the horizontal three hierarchies of structure-function- behaviour. In addition, the purpose of this method is to analyze the failure propagation path of the system, so this method is named the Failure Propagation Analysis Method Based on 3 × 3 Hierarchical Model (the 3 × 3 Analysis Model for short).

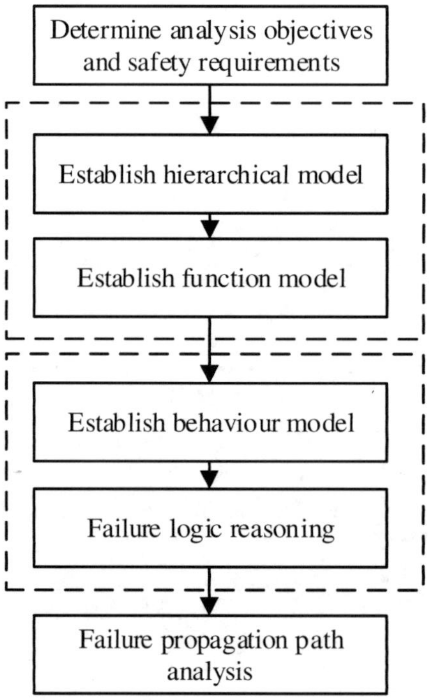

Figure 2. The 3 × 3 analysis model.

- Determine analysis objectives and safety requirements

Firstly, determine the analysis object and the analysis purpose according to the life cycle stage of the system. Next, select the appropriate system hierarchy. Then, analyze the possible risk factors and determine the safety requirements of the system.

- Establish hierarchical model

In this step, the hierarchical structure of the model is systematically established. The model can describe the interaction between the internal structures of the system.

- Establish function model

According to the hierarchical model, the functional model of the system can be established. Function model is a collection of functions that must be included in the system, reflects the transfer relationship of functions between components. It describes the information and data flow between component functions.

- Establish behaviour model

The behaviour model describes the time-varying characteristics of variables in the system. It uses the changing relationship between process variables to describe the behaviour state of components.

- Failure logic reasoning

Failure logic refers to the specific form of behaviour deviation caused by the system. Failure logic describes the changes of variable relationship which can lead to component failure.

- Failure propagation path analysis

By traversing the causality defined in the failure logic of each component, the failure propagation path in the whole system can be inferred and analyzed.

4. Case Study

Taking a hydraulic transmission system as an example, the specific implementation process of the above proposed method is described. According to the 3 × 3 Analysis Method, we establish model and analyze the failure propagation of hydraulic transmission system. In this paper, we describe the model with SysML which can improve the comprehensiveness, consistency and traceability of design.

The main source of risk factors of hydraulic transmission subsystem is component failure. The safety requirement is to ensure the load moves when needed to realize the hydraulic transmission function.

Firstly, we divide the analysis object into two hierarchies: hydraulic transmission subsystem hierarchy and component hierarchy. The component layer includes four categories: power components, control components, execution components and auxiliary components. Then, we mark the function of each component and the output variables.

In this paper, we use the block definition diagram in SysML to establish the hierarchical model and the function model of 3 × 3 analysis method. In this way, the structure, function and data flow can be shown at the same time, making the functional design of components clearer. The hierarchical model and functional model of hydraulic transmission subsystem are shown in Figure 3.

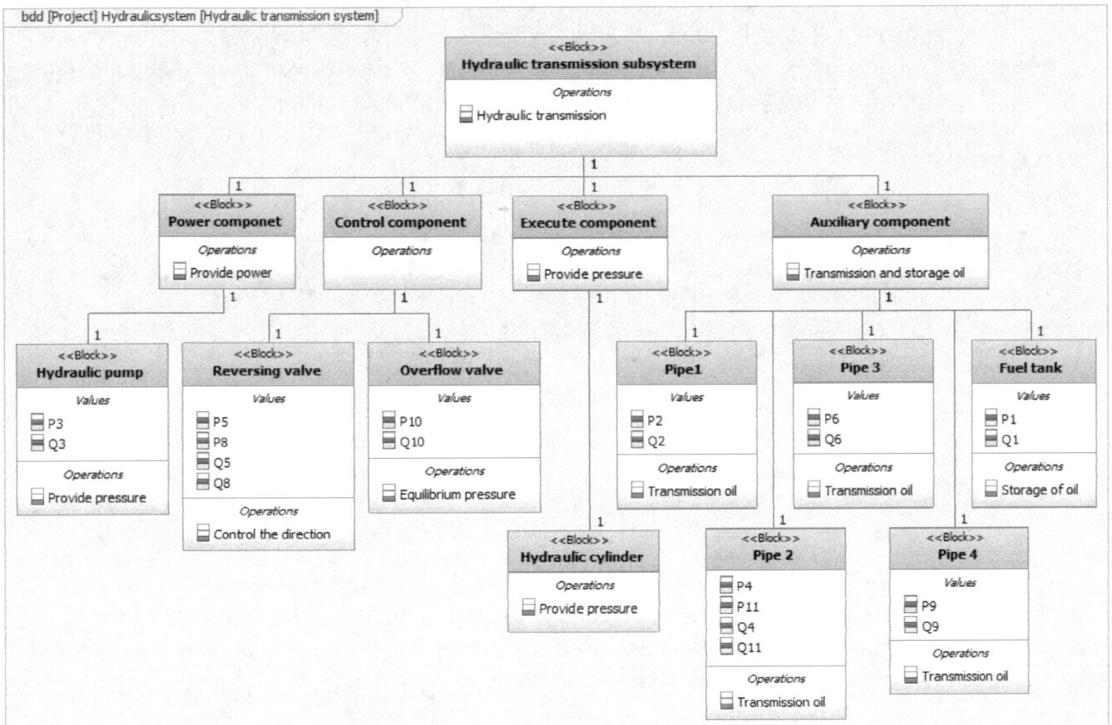

Figure 3. The hierarchical model and function model of hydraulic transmission subsystem.

Then, we establish the behaviour model using the sequence diagram in SysML to describe the information transfer relationship between modules, as shown in Figure 4. Figure 4 depicts the behaviour of the hydraulic transmission subsystem under normal conditions.

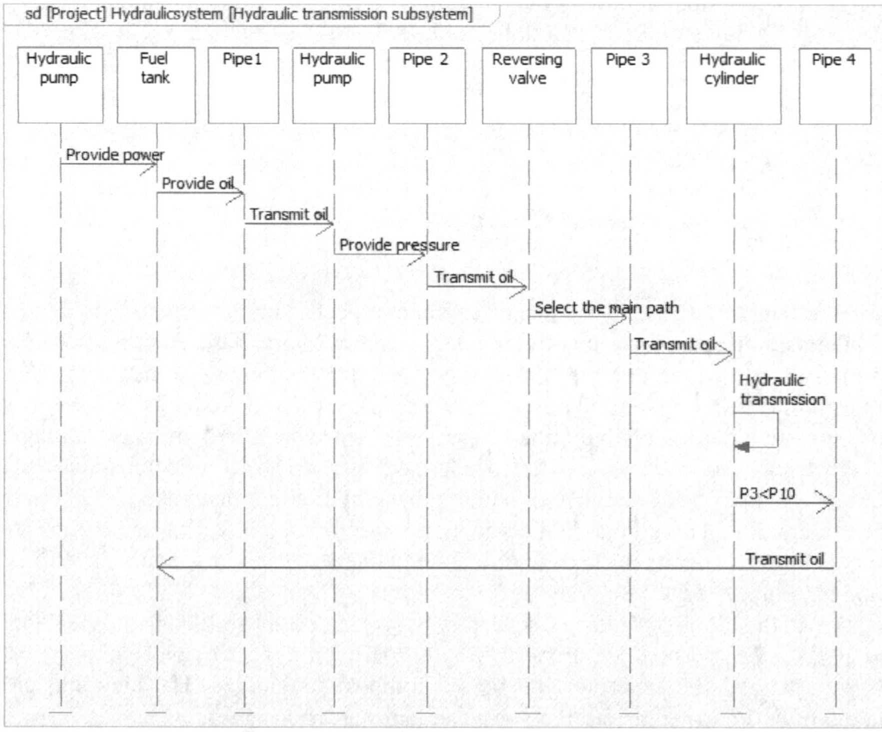

Figure 4. The behaviour model of hydraulic transmission subsystem.

Next, the failure logic is deduced. The state machine diagram of SysML can show different states of components and the transformation relationship between states. This paper uses the state machine diagram to represent the failure logic. Figure 5 describes the normal state and failure states of the hydraulic pump.

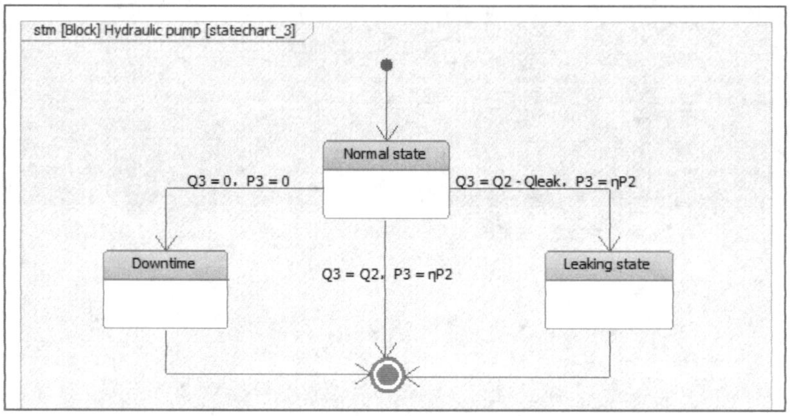

Figure 5. The failure logic relationship of hydraulic pump.

Finally, the failure propagation path is analyzed. The fault tree is used to represent the failure propagation path of the hydraulic transmission subsystem, as shown in Figure 6.

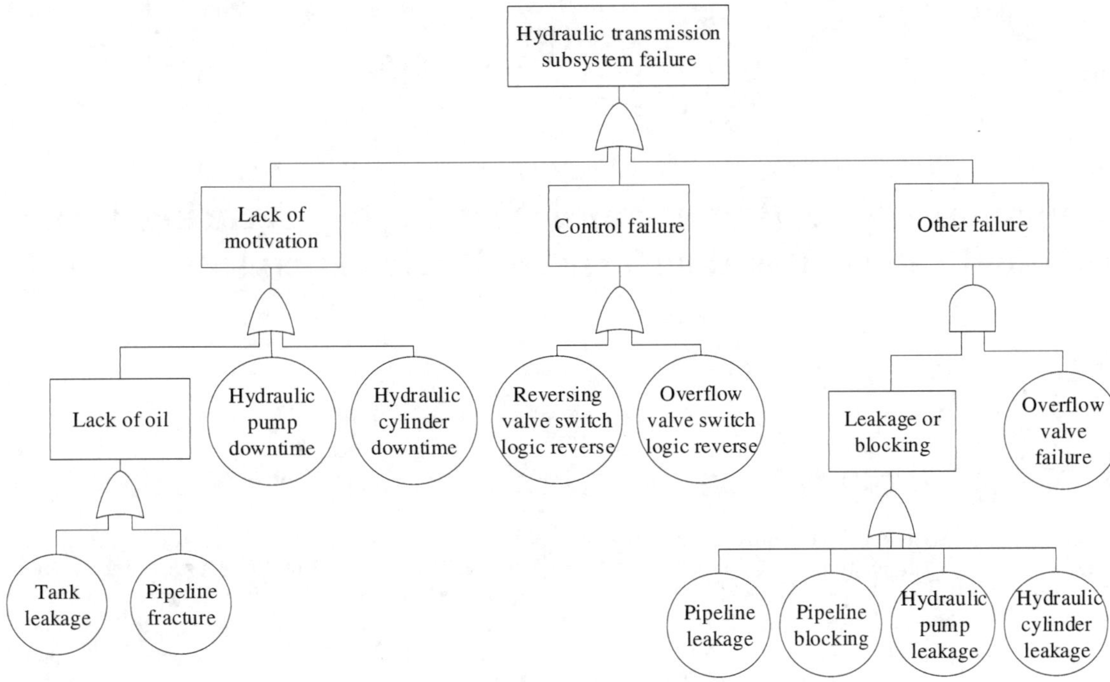

Figure 6. Fault tree of hydraulic transmission subsystem.

5. Conclusion

In this paper, we proposed the 3 × 3 Hierarchical Model by decomposing the system horizontally and vertically, which considered of the design and modeling of each stage of the system. Then, we proposed a failure propagation analysis method which made up for the shortcomings of the existing methods. In addition, we used SysML to build the system models, which improved the consistency and traceability of the model. Finally, we used the method to analyze the failure propagation path of the hydraulic transmission subsystem and obtained the fault tree.

References

[1] D A Longji; V Emilia. System safety assessment based on STPA and model checking. Safety Sci 2018, 109, 130-143.

[2] V Salehi; B Veitch; D Smith. Modeling complex socio-technical systems using the FRAM: A literature review. Human Factors and Ergonomics in Manufacturing & Service Industries 2021, 31, 118-142.

[3] S Sharvia; S Kabir; M Walker; Y Papadopoulos. Model-based dependability analysis. Software Quality Assurance 2016, 251-278.

[4] S J, Pang; J Jiao. Simulation and Risk Assessment of Sand Mixing Equipment Based on Function Failure Propagation. In 2020 4th International Conference on Modelling, Simulation and Applied Mathematics (MSAM 2021), Wuhan, China, 2021

Simulation of the Performance Metrics for Securing DSDV Routing Protocol Based on Trusted Environments

Chiang Ling Feng*

Department of Electric Engineering, ChienKuo Technology University, Taiwan

*Corresponding author email: acclaim0629v@gmail.com

Abstract. To organize the network in an efficient way to minimize the risk of illegal node and to safeguard protected information, a security mechanism is required to secure communication. In addition, a security mechanism is also required to ensure that received information have not been tampered with. In this paper, a more efficient mechanism for Securing the Destination Sequenced Distance Vector Routing Protocol (SDSDV) is proposed. This paper comprehensively investigates the performance impacts by varying the number of maximum connections and mobility on securing the ad hoc network with Destination Sequenced Distance Vector Routing Protocol (DSDV) Routing Protocol. From simulation results, we explore the causes for performance degradation. Based on the investigation, we indicate that we have to impose restrictions on the maximum connections to acquire an expected performance. These results also reveal that the performance decreases with the increment of the mobility and maximum connections that is unequal to 60. If we want to obtain the optimal performance, the number of nodes in a network should be constrained to be 60 if the maximum connections are 60.
Keywords: Mobility, maximum connections, mobile Ad hoc networks, modified SDSDV, hash function.

1. Introduction

Ad hoc network is a network with dynamic topology and may change randomly and rapidly at unpredictable times. Due to the fact of the vulnerability of channels, the possibility of denial-of-service attacks will increase due to the physical security treats in mobile wireless networks. The significant objective of ad hoc networks is to provide robust communication in a variation of environments. When a malicious node hijacks the network bandwidth by generating frequently unnecessary route requests, the network resources will be unavailable to other nodes or non-existent nodes and the network no longer operates correctly or crashes. In mobile Ad hoc networks, the system performance relies on cooperative routing. However, a selfish node [1-5] is defined as whom does only forward the control packets to gain his routing information about the networks, which refuses to forward data packets. But, the issue of selfish nodes is beyond the scope of this work. Hence, DSDV routing protocol is assumed that all nodes are trustworthy and cooperative here.

Since any attacker can simply look at the contents of packet, effectual algorithm and secret keys are two fundamental components in a cryptographic system. Specifically, a secret key should be not apt to be found and decrypted. One of the primary challenges in cryptography is how to implement the cryptography in wireless mobile networks. In general, secret-key algorithm is often used to protect the actual data. To decrypt the public-key encryption algorithm is more difficult due to the characteristic of key pair that one key used by the sender (private key) and one key used by the receiver (public key).

Content from this work may be used under the terms of the Creative Commons Attribution 3.0 licence. Any further distribution of this work must maintain attribution to the author(s) and the title of the work, journal citation and DOI.
Published under licence by IOP Publishing Ltd

Even if we know the public key of the peer, it is extremely difficult to figure out its private key unless the transaction of key's negotiation is eavesdropped. However, mobility transparency is an essential requirement in mobile wireless system to provide seamless mobility in any heterogeneous networks. This paper first depicts the design's considerations and the Literature survey is given in section 2. The implementations of security in our proposed MDSDV routing protocol in section 3 to enhance the security property of existing routing protocol. We present the system performance due to mobility and maximum connections in section 4. And the simulation results and detailed discussions are also given. Finally, we draw our conclusions in section 5.

2. Literature Survey

To safeguard protected information and ensure that the received information has not been tamped with, secure efficient distance vector (SEAD) has been proposed in [6-7] based on trusted environments. This proposed protocol can prevent a malicious node from modifying a sequence number or cost metric of an advertised route by using one-way hash function and does not use asymmetric cryptographic operation to guard against denial-of-service attacks. SDSDV [8] protects all messages combining with message authentication using pairwise shared keys. The message authentication code of a message is immediately sent after transmission of the message is finished.
Any nodes in SEAD network cannot successfully re-advertise the route with a distance shorter than 2. It also cannot prevent a malicious node from advertising a distance of 2 or longer. SDSDV also incurs a significant network overhead than DSDV because it has to check the consistency of an updated route due to the persistently changes of the sequence number, distance, or the next hop [8]. Based on the balance between the robustness of security and system performance, we are willing to have a similar risk on the security but produce a significant enhancement of the system performance as possible. Since any attacker can simply look at the contents of packet, effectual algorithm and secret keys are two fundamental components in a cryptographic system. Specifically, a secret key should be not apt to be found and decrypted. Mobility transparency is an essential requirement in mobile wireless system to provide seamless mobility in any heterogeneous networks. Therefore, we modify the algorithm of SDSDV routing [8, 9] using the encryption of pairwise shared secret key [10-14] to achieve the optimal performance as possible as that of original DSDV routing protocol.

3. MSDSDV Routing Protocol

In DSDV, each mobile node maintains a completely routing table containing routing information to every node in the network. The routing information includes all available destinations, the metric and next hop to each destination, and a sequence number generated by the destination node. Each node attempts to periodically update the routing table or when significantly available routing information is acquainted to maintain the consistency of the routing table with the dynamically changing topology of the ad hoc network. Sequence numbers and metric values containing in each update play a vital role in DSDV operation. However, a malicious node can easily disrupt the routing protocol by arbitrarily tempering the sequence numbers or the metrics. To enhance the robustness of DSDV routing protocol, we will review the security mechanism of SDSDV and characterize our MSDSDV (modified SDSDV) security mechanism in this section.
To provide a more realistic application in mobile wireless ad hoc network is not an easy task. SDSDV [8, 9] based on the regular DSDV protocol is particularly designed for wireless Ad hoc network. In order to establish a multi-hop route, SDSDV provide the voluminous information. Due to this reason, a hijacker can analyze the network topology to perform the attack of service of deny focusing on some critical nodes. Therefore, we modify the concept of SDSDV and implement it on the practical wireless mobile Ad hoc system based on the realistic security roaming. The major goal of our proposed SDSDV, denoted as MSDSDV, is to protect the sequence numbers and metrics in each entry of an update from being arbitrarily changed. To achieve this goal, hash chain is implemented to prevent the attack from arbitrarily tampered. We also adopt an authentication tag in the transmitted route to verify the integrity and authenticity of transmitted information. Authentication tag is a function of the node's identification (such as sequence number and cost metric) and the pairwise shared secret key. Authentication tag is

computed by the simple hash function. Let V be a set of nodes. For v_i, $v_j \in V$, assume $r_{v_i \to v_j}$ to be the route from node v_i to node v_j. Denotes $seq(v_i, v_j)$ and $cst(v_i, v_j)$ are the current sequence number and metric of $r_{v_i \to v_j}$, respectively. For $\forall v_i \in V$, v_i constructs a hash chain, denoted as $h^c(x_i)$, from a secret x_i. Therefore, we can replace the notation $h^c(x_i)$ in SDSDV routing protocol by the notation $h^c(seq(v_i, v_j), cst(v_i, v_j) | E_{K_{i,j}}[r_{v_i \to v_j}])$ in our proposed security scheme. Where the notation $E_{K_{i,j}}[X]$ denotes that a plaintext X is encrypted by the pairwise shared secret key $K_{i,j}$ which node v_i shares with node $v_j \in V$, $i \neq j$. The symbol | represents concatenation of the information.. But, instead of the statement in SDSDV routing protocol, $K_{i,j}$ is transmitted only at the phases of routing discovery and updated route, and is not sent combining with the message to node v_j to avoid the same-distance fraud attack to prevent the attack from arbitrarily tampered.

We assume some mechanisms are available for a node to deploy the authentication protocols and the authentic elements from its generated hash chains. Let the notation $D_{Ki,j}$ [X] represent the decrypting procedure of received message X using $K_{i,j}$. Therefore, when a neighbour node receives the forwarded packet,

$$r_{v_i \to v_k}(seq, cst+1) = D_{K_{i,j}}[E_{K_{i,j}}[r_{v_i \to v_j} \mid h^c(seq(v_i, v_j), cst(v_i, v_j))]] \quad (1)$$

Equation (1) is used to authenticate the adjacent node by computing the received sequence number and metric, which are received from the neighbour. Since security algorithm is designed for the protection of the network infrastructure, the packet is discarded and no action is executed if the constrained condition is mismatched. Otherwise, the receiving node encrypts its own routing table with the pairwise shared secret key related to the next hop if one of the following conditions is satisfied.

$$\begin{aligned} &(1).\ seq > seq' \\ &(2)\ seq = seq' \ and\ cst < cst' \end{aligned} \quad (2)$$

4. Simulation and Environment and Parameters

To comprehensively measure the performances among DSDV, SDSDV, and MSDSDV, the impact of the performance metrics must be evaluated. When a node is out of the communication range with the neighbour, a transmitted packet will be dropped. Therefore, the packet delivery fraction which is defined as the fraction of the data packets, generated by the source node, that are successfully delivered to the final destination node along the discovered routes can be used to evaluate the ability of the protocol to discover routes and to reflect the maximum throughput that the network can support

The simulation was done with ns2 simulator [15]. First, we assume that no means can be used to obtain the spatial information of nodes with respect to other nodes in this network. Each node is initially placed at a random location and moves according to the random waypoint mobility model. The moving speed ranges from 5 and 25 m/s. The destination node in the simulation environment is randomly chosen by each node at the beginning of the simulation. When a node reaches pre-selected location, we assume that this node stays for a period of time (assumed to be zero here), and then chooses a new location at random and moves with a randomly and uniformly chosen velocity. This process of pausing and selecting a new destination to which to move is repeated.

In order to address mobile limitations in IEEE 802.11 network, the physical channel bit rate is set to 2 Mbps CBR traffic for IEEE 802.11 DCF and the nominal radio range is 250 m communication range to satisfy the requirement of this standard. The size of application data payload size is 512 bytes for each packet. The number of active stations, denoted as nn, which have some data to be sent, in our simulated scenario is varied from 30 to 90. The number of maximum connections, denoted as mc, in our simulated scenario is varied from 20 to 80. For the routing zone radius for our simulation, the nodes are assumed to be uniformly distributed in 600 m * 600 m two-dimensional square spaces. The simulation time is selected to be 1000 seconds to stabilize the simulation results. All the mobile nodes are assumed to be in the communication range and move with the same constant speed.

Maximum connections, packet delivery fraction, mobility, and mean delay metrics are used to evaluate the security protocol. We change the only one parameter and maintain the others unchanged to clarify the effect of changed parameter as much as possible for each comparison. Altering the value of mc makes the first comparison. Each connection is specified as a randomly selected source-destination pair. Figure 1 and figure 2 show the packet delivery ratio versus maximum connections with respect to the DSDV, SDSDV, and MDSDV routing protocol for number of nodes being 30 and 40, respectively. From all of these figures, we see that packet delivery fraction gets worse when *mc* increases. The packet delivery fraction does not have much impact on the performance between DSDV routing protocol and DSDV routing protocol. But, MDSDV has better performance than SDSDV. We also see that increasing the number of active stations will decrease the packet delivery fraction. It means that the ratio of dropped packets increases. However, the position at *mc=60* seems to be worthy of discussion. Increasing the value of *mc* for each node, it represents that each node nearly has more direct links to all others in the network. Two or more transmitted packets to a same link are considered as collided packets at the same time. Therefore, the packet delivery ratio decreases because the probability of colliding packets increases.

To ensure the correction of our inference for the distinctive characteristic of mc related to each routing protocol in advance, we illustrate simulation result with the example of MDSDV. Varying the value of nn makes second comparison, In a mobile network, we allow mobile node to roam within the specific network. Therefore, the moving speed of nodes and the number of connections should be considered to evaluate the performance of MDSDV routing protocol and compare it with DSDV and SDSDV routing protocol. From figure 3, we can infer no obvious change in the throughput performance among DSDV, SDSDV, and MDSDV routing protocols for each moving speed. However, under the same traffic loads, we see that the increment of mobility of nodes will create the amount of update of the network topology and result in the degradation in network performance.

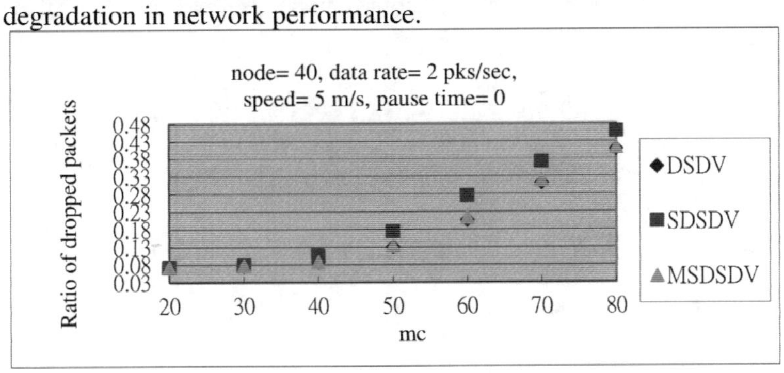

Figure 1. The packet delivery ratio versus maximum connections (mc) at $nn = 30$.

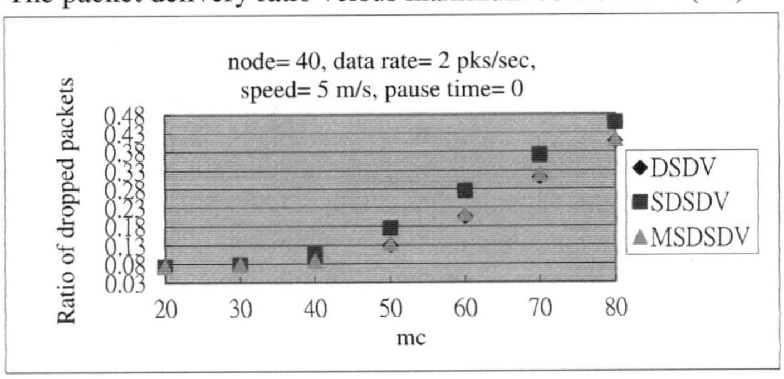

Figure 2. The packet delivery ratio versus maximum connections at $nn = 40$.

We also examined the dependence of the delay on each routing protocol at various maximum connections and mobility. We plot the results of average delay of a packet incurred by these three securing routing protocols at speed=5 m/sec and speed=20 m/s based on $mc = 20$, respectively. From figure 4, we see that MDSDV routing protocol offers lower average delay time than SDSDV. The higher system's processing loads incurred because every node spends extra time to achieve encryption and decryption mechanism. An extra delay of route maintenance incurred when the mobility increases. These protocols suffer from higher average delay time resulting from both higher probabilities to repair the broken links and the time to report link failure if these broken links cannot be repaired.

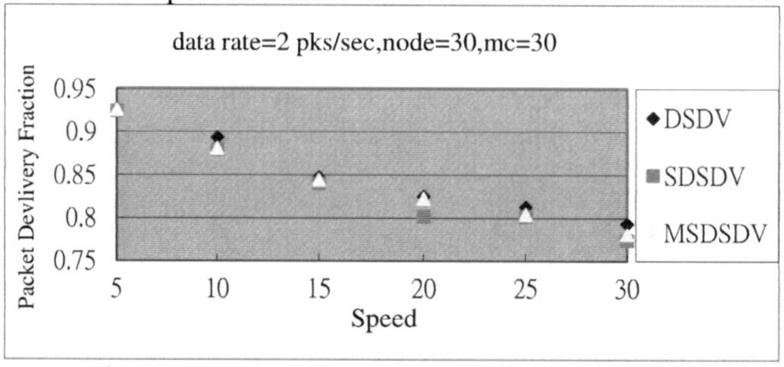

Figure 3. The performance impact of mobility.

When a node moves in or out of range with respect to each other, nodes first try to communicate with other nodes to maintain the connectivity information related to a changing topology. If possible connections are not found, the host will attempt to discovery a new connection for the destination. In this section, we plot the routing overhead incurred by these three securing routing protocols at speed=5m/sec and speed=20 m/s based on $mc = 20$, respectively. Figure 5 illustrates that the mobility seems to be not affected by the routing overhead at same moving speed (mobility) among these three routing protocols. This reason is that periodic updates are executed. However, the performance of MDSDV is better than that of SDSDV. In addition, when the mobility increases, unnecessary route reconstructions (i.e., to discover and maintain the communication paths) can be caused from frequently periodic and triggered updates.

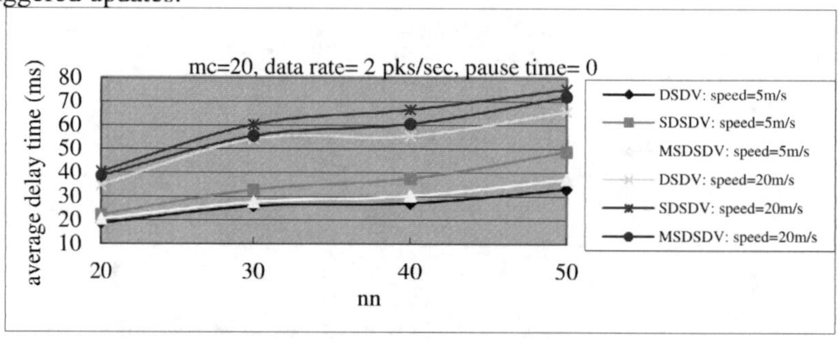

Figure 4. The impact of mobility on the delay performance.

5. Conclusions

In this paper, the vulnerabilities of DSDV and SDSDV routing protocols have been extensively studied. The solution has been carried out based on our proposed MDSD routing protocol in order to prevent attacks. Our proposed routing protocol does not require any additional security functionality to obtain enhanced performance. Our analysis results indicate that the performance is obviously decayed in network if the number of maximum connections is more required. However, the packet delivery fraction is independent of the use of DSDV, SDSDV, and MDSDV routing protocols at a fixed mobility. Suitable parameter selection obviously enhances the performance in the network. For the mobility, we observe that the packet delivery fraction significantly varies at the environments with mobility and nodes. In addition, reducing the number of maximum connections obviously enhanced the system performance in a network.

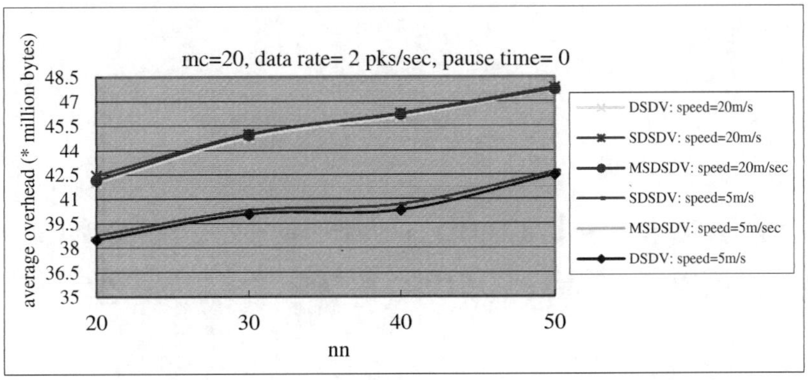

Figure 5. The impact of mobility on the routing overhead.

References

[1] K. Paul and D. Westhoff, "Context Aware Detection of Selfish Nodes in DSR based Ad-hoc Networks", *Proc. of IEEE GLOBECOM*, Taiwan, China, November, 2002

[2] L. Anderegg and S. Eidenbenz, "Ad hoc-VCG: A Trustful and Cost-Efficient Routing Protocol for Mobile Ad hoc Networks with Selfish Agents", *Proc. of ACM MobiCom*, San Diego, September, 2003

[3] S. Zhong, J. Chen, and Y.R. Yang, "Sprite: A Simple, Cheat-Proof, Credit-Based System for Mobile Ad-Hoc Networks", *Proc. of IEEE Infocom*. San Francisco, April 2003

[4] Yanchao Zhang, Wenjing Lou, and Yuguang Fang, "SIP: A Secure Incentive Protocol against Selfishness in Mobile Ad Hoc Networks", *WCNC 2004*, pp. 1679-1684, 2004

[5] Jyu-Wei Wang, and Yi-Ping Lin, "A Secure DSDV Routing for Ad hoc Mobile Networks", *Mobile Computing Workshop 2004*, pp. 413-418, 2004

[6] L. Buttyan and J.P. Hubaux, "Stimulating cooperation in self-organizing mobile ad hoc networks," *ACM Journal for Mobile Networks (MONET)*, Vol. 8, No. 5, October 2003

[7] Yih-Chun Hu, David B. Johnson, and Adrian Perrig, "SEAD: secure efficient distance vector routing for mobile wireless ad hoc networks", *Ad hoc Networks*, Vol. 1, pp. 175-192, 2003.

[8] Tao Wan, E. Kranakis, and P.C. Van Oorschot, "Securing the Destination Sequenced Distance Vector Routing Protocol", T*he Proceedings of the 6th International Conference on Information and Communications Security. LNCS,* Vol. 3269, pp. 358-374, October, 2004

[9] J. Cardinal, G. Van Assche, "Construction of a shared secret key using continuous variables", *Proceedings. 2003 IEEE Information Theory Workshop*. pp.135-138, March-April 2003

[10] B. Preneel, R. Govaerts, and J. Vandewalle, "Hash functions for information authentication", *Proceedings in Computer Systems and Software Engineering*, pp. 475 – 480, May 1992

[11] Xun Yi and Kwok Yan Lam, "Hash function based on block cipher", *Electronics Letters*, Vol. 33, Issue 23, pp. 1938-1940, November 1997

[12] G. Tsudik, "Message authentication with one-way hash functions", *INFOCOM '92*, Vol. 3, pp. 2055 – 2059, May 1992

[13] R. Bird, I. Gopal, A. Herzberg, P. Janson, S. Kutten, R. Molva, M. Yung, "The KryptoKnight family of light-weight protocols for authentication and key distribution", *IEEE/ACM Transactions on Networking*, Volume 3, Issue 1, pp. 31 – 41, February 1995.

[14] U.M. Maurer, "Secret key agreement by public discussion from common information", *IEEE Transactions on Information Theory*, Vol. 39, Issue 3, pp. 733-742, May 1993.

[15] Kevin Fall and Kannan Varadhan, The ns Manual (formerly ns Notes and Documentation). The VINT Project, 2001.

Modeling of Integrated Spectral Energy Distributions of Stellar Populations with Various Binary Fractions

Zhongmu Li* and Chen Yan

Institute of Astronomy and History of Science and Technology, Dali University, Dali 671003, China

*Corresponding author email: zhongmuli@126.com

Abstract. Binary stars are common in the universe, but binary fractions are various in different star clusters and galaxies. Studies have shown that binary fraction affects the integrated spectral energy distributions obviously, in particular in the UV band. It affects spectral fitting of many star clusters and galaxies significantly. However, previous works usually take a fixed binary fraction, i.e., 0.5, and this is far from getting accurate results. Therefore, it is important to model the integrated spectral energy distributions of stellar populations with various binary fractions. This work presents a modeling of spectral energy distributions of simple stellar populations with binary fractions of 0.3, 0.7, and 1.0. The results are useful for different kinds of spectral studies.
Keywords: Stellar population, spectral energy distribution, binary.

1. Introduction

Most stars in the universe are possibly binary stars. The observation shows that the binary fraction of the Milky Way is at least 50 per cent, and it may be as large as 80 per cent. According to the studies of stellar evolution, binary stars with two close components evolve differently from single stars like the sun, because two components can exchange mass and energy in their evolution. There are many binary interactions between two components, and these interactions can lead to the formation of special stars such as blue stragglers, red stragglers, and hot sub-dwarfs. They also relate to the formation of Type Ia supernovae and gravitational waves. In particular, binary stars are confirmed to have significant effects on the formation of UV-upturn integrated spectral energy distributions (SEDs) of old stellar populations, e.g., some globular clusters and early-type galaxies. This finally leads to many changes of SED fitting. For example, the determination of stellar metallicity and age will change obviously when binary stars are taken into account, **in SED, line index** and color-magnitude diagram (CMD) studies. One can read many papers, e.g., Li et al. (2006, 2008a, 2008b, 2008c, 2012a, 2012b, 2013, 2016, 2018, 2020a, 2020b)[1--11], Conroy et al. (2014)[12], Han et al. (2007)[13], and Eldridge et al. (2017)[14] for more details.

Although binary stars have been taken into account by some stellar population studies, a binary fraction of 0.5 is usually used. For instance, it is used by the two most important binary stellar population synthesis groups, Yunnan and Brussels (see the review of Eldridge et al. 2017)[14]. In the SED fitting works of Zhang et al. (2004, 2005)[15,16], Li et al. (2006, 2008a, 2008b, 2008c, 2013)[1,2,3,4,7], half stars of a population are considered as binary stars. In fact, the binary fraction varies in star clusters and galaxies. This can be seen from the detailed studies of binary fractions of some nearby star clusters. The binary fraction range is very large from 0 to 100 per cent. In this case, it is not a good choice to take only a fixed binary fraction for various SED studies.

Content from this work may be used under the terms of the Creative Commons Attribution 3.0 licence. Any further distribution of this work must maintain attribution to the author(s) and the title of the work, journal citation and DOI.
Published under licence by IOP Publishing Ltd

This work attempts to model the SEDs of simple stellar populations (SSPs) with different binary fractions. The results can be widely used for future SED studies, e.g., SED fittings and absorption index studies of star clusters and galaxies.

The second section introduces the parameter setting of SSP model. The third section analyzes the computation of stellar evolution and SEDs. The fourth section shows the main results of SED modeling of SSPs with various parameters. The last section summarizes the work done in this paper.

2. Inputs of SSP Models

Each SSP consists of 100, 000 stars that have the same metallicity and age. The initial masses of stars are generated randomly following the initial mass function of Charbrier (2003)[17], which is a piecewise function. The minimum and maximum masses of single stars are set to 0.1 and 100 $M_{\odot}$. For binary stars, the masses of primary stars (M_1) are given by the same way as single stars, and the masses of secondary stars (M_2) are calculated by multiplying M_1 by a mass ratio (q). Mass ratio q and orbital eccentricity (e) of binary stars are randomly generated in the range of 0--1. In order to check how metallicity and age affect the SEDs of SSPs, four metallicities and four ages are taken in the modeling. The detailed input parameters are listed in table 1.

Table 1. Input parameters of modeling.

	Font	Spacing
Primary mass	M_1	0.1 – 100.0 $M_{\odot}$
Secondary mass	M_2	< 100.0 $M_{\odot}$
Binary mass ratio	q	0 – 1
Orbital eccentricity	e	0 – 1
metallicity	Z	0.0003, 0.001, 0.02, 0.03
Bianry fraction	f_{bs}	0.3, 0.7, 1.0
Age	t	0.1, 5, 10, 13 *Gyr*

3. Computation of Stellar Evolution and SEDs

3.1. Computation of Stellar Evolution

When building SSPs, both single and binary stars are evolved using a rapid stellar evolution code of Hurley et al. (2000, 2002)[18,19]. This code calculates the stellar evolutionary parameters based on some fitting formulae of accurate evolutionary tracks. There are two advantages to take this code. On one side, the computation is very fast. In fact, this is the fastest code to compute stellar evolution. It is therefore suitable to model stellar populations consisting of a large amount of stars. On another side, both single and binary stars can be evolved by this code. Because all galaxies and star clusters possibly contain some binary stars, it is necessary to include binary star evolution. Thus this code is suitable for our study.

3.2. Computation of SEDs

The SEDs of stellar populations are computed from the evolutionary parameters of stars. The main physics is that the spectrum of a star directly corresponds to its gravity and effective temperature. The correlation between spectrum and evolutionary parameters are given by spectral libraries. In this work, the BaSeL 3.1 spectral library (Lejeune et al. 1997, 1998)[20,21] is used for computing the SEDs of stellar populations. The reason for choosing this library is that it covers enough wide wavelength range from UV to infrared. In addition, the uncertainty in the library is small.

4. SEDs of SSPs with Various Binary Fractions

This section shows the main results of SED modeling of SSPs with various parameters. The results of three binary fractions are presented in three sub-sections, respectively. The fluxes are plotted in the same range to make it easy to compare the results.

4.1. SSPs with 30% Binary Stars

Figure 1 shows the SEDs of SSPs with a binary fraction of 0.3. As some examples, four metallicities (Z = 0.0003, 0.001, 0.02, 0.03) and four ages (0.1, 5, 10, 13 Gyrs) are taken for the examples. As we see, when binary fraction and metallicity are fixed, SSPs become fainter when age increases. When binary fraction and age are fixed, SSPs become fainter with increasing metallicity.

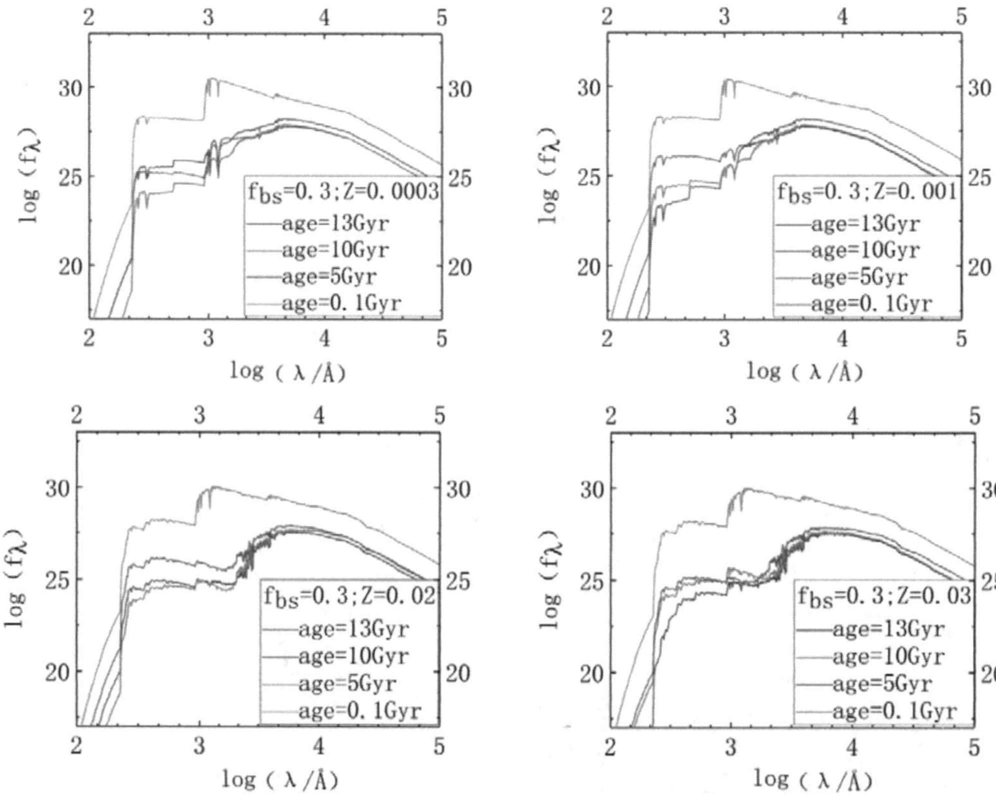

Figure 1. SEDs of SSPs with binary fraction of 0.3, four metallcities and four ages.

4.2. SSPs with 70% Binary Stars

We find that SEDs changes with similar trends as the SSPs with a binary fraction of 0.3. Figure 2 shows the SEDs of SSPs with a binary fraction of 0.7.

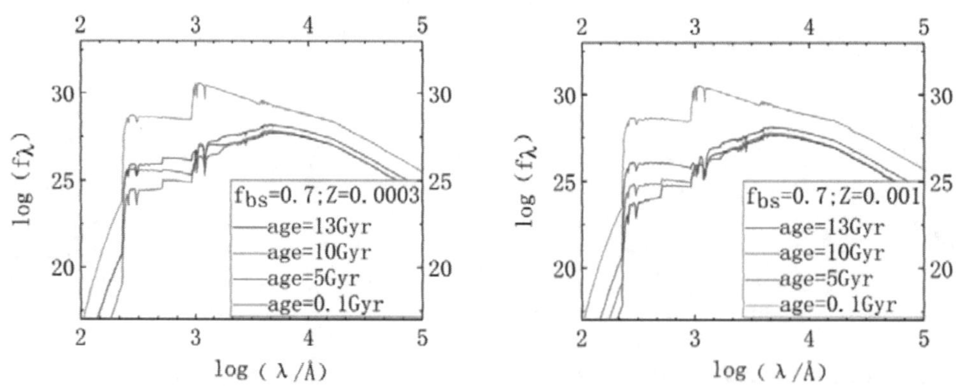

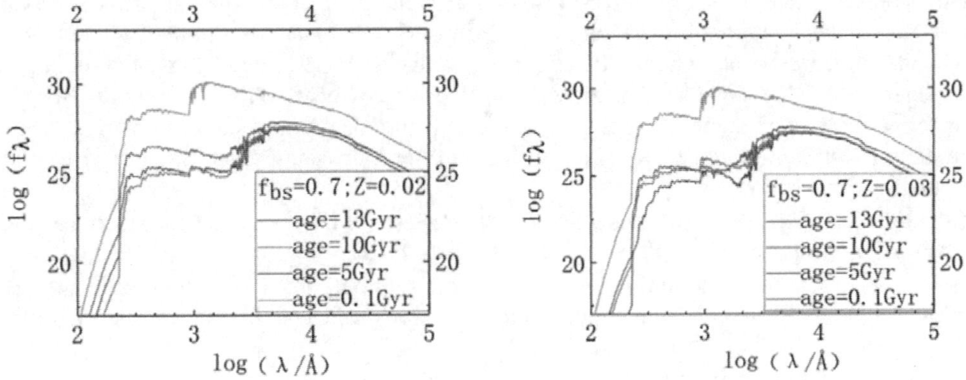

Figure 2. Similar to figure 1, but for a binary fraction of 0.7.

4.3. SSPs with 100% Binary Stars

There are similar trends for the SEDs of SSPs when taking 100 per cent binary stars instead of 30 and 70 per cent, but the UV-upturn phenomenon is more obvious, in particular in the near UV band. Figure 3 shows the SEDs of SSPs with a binary fraction of 1.0.

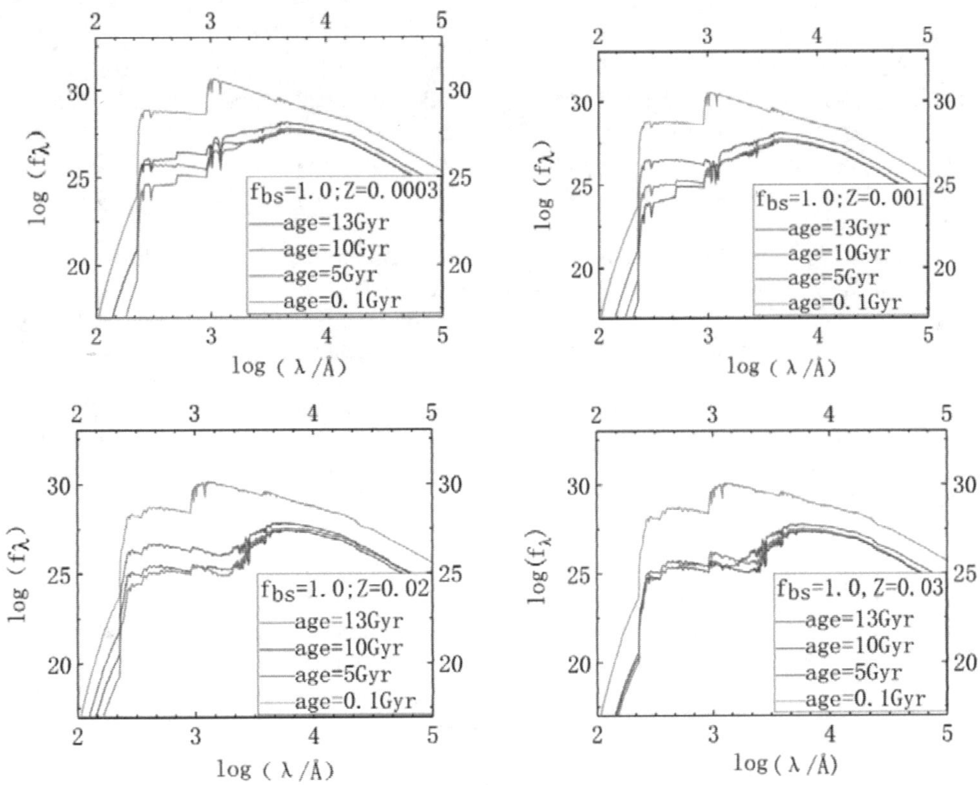

Figure 3. Similar to figure 1, but for a binary fraction of 1.0.

5. Conclusion

This paper presents a modeling of SEDs of SSPs with various binary fractions. The SEDs of SSPs with binary fractions of 0.3, 0.7 and 1.0 have been shown. This is the first time to compare the SEDs of **SSPs with different binary fractions, metallicities and ages**. The results are important for SED studies of star clusters and galaxies.

We find that binary fraction affects the SEDs of SSPs clearly, in particular in the UV band. This means that UV SEDs are possibly useful for determining the binary fraction of star clusters and galaxies. In addition, we are shown that the effects of binary fractions, young stars, and metallicity are relatively degenerate. This suggests that there should be larger uncertainties in previous SED studies of the star formation histories, stellar ages, and extinctions of galaxies and star clusters.
According to our results, we suggest to take into account the following points in future SED studies of stellar populations:

- Stellar populations with different binary fractions should be used, in particular for old (> 5 Gyr) populations.
- The degeneracy between binary fraction and young population should be investigated deeper. This will help to get more accurate binary fractions and star formation histories of stellar populations.
- The effect of binary fraction on optical SED, line indices, and colors remains unclear and should be investigated further by future works.

Acknowledgments
Authors acknowledge that this work has been supported by the Yunnan Academician Workstation of Wang Jingxiu (No. 202005AF150025), National Natural Science Foundation of China (No. 11863002), and Sino-German Cooperation Project (No. GZ 1284).

References
[1] Li Z., Zhang F. and Han Z., A Study of Binary Stellar Population Synthesis of Elliptical Galaxies [J]. The Chinese Journal of Astronomy and Astrophysics, 2006, 6 (06): 669-679.
[2] Li Z. and Han Z., An isochrone data base and a rapid model for stellar population synthesis [J]. Monthly Notices of the Royal Astronomical Society, 2008a, 387 (01): 105-114.
[3] Li Z. and Han Z., How Binary Interactions Affect Spectral Stellar Population Synthesis [J]. The Astrophysical Journal, 2008b, 685 (01): 225-234.
[4] Li Z. and Han Z., The role of binary stars in stellar population synthesis [J]. Proceedings of the International Astronomical Union, 2008, 4 (252): 359-364.
[5] Li Z., Mao C., Chen L., and Zhang Q., Combined Effects of Binaries and Stellar Rotation on the Color-Magnitude Diagrams of Intermediate-age Star Clusters [J]. The Astrophysical Journal Letters, 2012a, 761 (02): L22.
[6] Li Z., Zhang L. and Liu J., Integrated spectral energy distributions of binary star composite stellar populations [J]. Monthly Notices of the Royal Astronomical Society, 2012b, 424 (02): 874-883.
[7] Li Z., Mao C., Chen L., Zhang Q. and Li M., The Potential Importance of Binary Evolution in Ultraviolet-Optical Spectral Fitting of Early-type Galaxies [J]. The Astrophysical Journal, 2013, 776: 37.
[8] Li Z., Mao C., Zhang L., Zhang X. and Chen L., A Systematic Study of Effects of Stellar Rotation, Age Spread, and Binaries on Color-Magnitude Diagrams with Extended Main-sequence Turnoffs [J]. The Astrophysical Journal Supplement Series, 2016, 225 (01): 7.
[9] Li Z. and Mao C., Evolution of Optical Binary Fraction in Sparse Stellar Systems [J]. The Astrophysical Journal, 2018, 859 (01): 36.
[10] Li Z., Spurzem R., Mao C., Ma Y. and Deng Y., A study of stellar population of intermediate-age globular cluster NGC 2249 [J]. Results in Physics, 2020a, 17: 103144.
[11] Li Z., Chen J., Zhang S., Deng Y. and Zhao W., Study of color-magnitude diagram of star cluster SL 506 [J]. Astrophysics and Space Science, 2020b, 365 (08): 134.
[12] Conroy C., Graves G. J., van Dokkum P. G., Early-type Galaxy Archeology: Ages, Abundance Ratios, and Effective Temperatures from Full-spectrum Fitting [J], The Astrophysical Journal, 2014, 780 (01): 33.
[13] Han Z., Podsiadlowski P. and Lynas-Gray A. E., Binary Stars as the Source of the Far-UV Excess in Elliptical Galaxies [J]. Monthly Notices of the Royal Astronomical Society, 2007, 380 (03): 1098-1118.

[14] Eldridge J. J., Stanway E. R., Xiao L., McClelland L. A. S., Taylor G., Ng M., Greis S. M. L. and Bray J. C., Binary Population and Spectral Synthesis Version 2.1: Construction, Observational Verification, and New Results [J]. Publications of the Astronomical Society of Australia, 2017, 34: e058.

[15] Zhang F., Han Z., Li L. and Hurley J. R., Evolutionary population synthesis for binary stellar populations [J]. Astronomy and Astrophysics, 2004, 415: 117-122.

[16] Zhang F., Han Z., Li L. and Hurley J. R., Inclusion of binaries in evolutionary population synthesis [J]. Monthly Notices of the Royal Astronomical Society, 2005, 357 (03): 1088-1103.

[17] Chabrier G., The Galactic Disk Mass Function: Reconciliation of the Hubble Space Telescope and Nearby Determinations [J]. The Astrophysical Journal Letters, 2000, 586 (02): L133-L136.

[18] Hurley J. R., Pols O. R., and Tout C. A., Comprehensive analytic formulae for stellar evolution as a function of mass and metallicity [J]. Monthly Notices of the Royal Astronomical Society, 2000, 315(3): 543-569.

[19] Hurley J. R., Tout C. A. and Pols O. R., Evolution of binary stars and the effect of tides on binary populations [J]. Monthly Notices of the Royal Astronomical Society, 2002, 329 (04), 897-928.

[20] Lejeune T., Cuisinier F. and Buser R., Standard stellar library for evolutionary synthesis. I. Calibration of theoretical spectra [J]. Astronomy and Astrophysics Supplement Series, 1997, 125: 229-246.

[21] Lejeune T., Cuisinier F. and Buser R., A standard stellar library for evolutionary synthesis. II. The M dwarf extension [J]. Astronomy and Astrophysics Supplement Series, 1998, 130: 65-75.

Simulation of Formation of Binary Compact Objects in Globular Cluster

Yangyang Deng[1,2], Zhongmu Li[1,*] and Wen Chen[1,2]

[1]Institute of Astronomy and History of Science and Technology, Dali University, Dali 671003, China
[2]Shanghai Key Lab for Astrophysics, Shanghai Normal University, Shanghai 200234, China

[*]Corresponding author email: zhongmuli@126.com

Abstract. Binary compact objects are the special binary systems, which were composed of compact objects (i.e., white dwarf, neutron star and black hole). They contribute a lot to the sources of gravitational waves. The study of binary compact objects in star clusters and galaxies can provide a theoretical guidance for gravitational wave detection and improve the probability of detection effectively. We simulate the formation and fraction of binary compact objects in a small globular cluster, via NBODY6++GPU, an efficient N-body simulation code. We obtain the fractions of white dwarf binaries and neutron star binaries at different ages. The results show that the dynamic interactions among stars can also result in some binary compact objects besides binary evolution.
Keywords: Binary compact object, N-body simulation, globular cluster.

1. Introduction

Gravitational wave (GW) is not only an important way to verify general relativity, but also a new approach for astronomical researches. It opened up a new direction for exploring the universe. It is well known that binary compact objects (BCOs), such as neutron star binaries (NSBs), black hole binaries (BHBs) and neutron star-black holes (NSBHs), are the major sources of GWs. The studies of BCOs are helpful for understanding the formation and evolution of GWs. Many studies showed that most of the GWs came from BCOs, including BHB, NSB and NSBH merging (Abbott et al. 2019, Abbott et al. 2020[1][2]. However, the GWs of white dwarf binaries (WDBs) are not detected. There are still some limitations in current GW detection techniques, and the theoretical researches of GWs are not comprehensive. The study of BCOs components in star clusters and galaxies can provide a theoretical basis for the detection of GW. In this way, the detected range and efficiency can be improved significantly.

There are two major formation modes of BCOs, evolution of massive binaries (Sigurdsson & Hernquist 1993)[3] and dynamic processes (Di Carlo et al. 2019)[4] among stars. So, the probability of BCOs being produced in a dense stellar environment is very high. Globular clusters are the collections of numerous stars. They have large numbers of stars and high densities. Hence, globular clusters are the ideal sites of BCOs formation. On the one hand, BCOs in globular cluster can be formed through a series of complex stellar evolution processes. On the other hand, BCOs can be produced under the dynamic processes among member stars.

Content from this work may be used under the terms of the Creative Commons Attribution 3.0 licence. Any further distribution of this work must maintain attribution to the author(s) and the title of the work, journal citation and DOI.
Published under licence by IOP Publishing Ltd

This paper aims to simulate the formation and fraction of all kinds of BCOs in a globular cluster by N-body simulation. Section 2 briefly introduces the method. Section 3 presents the input parameters. The result is shown in section4 and we make a conclusion in section 5.

2. Method

We apply a powerful numerical simulation code to investigate the globular cluster, which includes a large number of particles with masses. An updated version of NBODY6++GPU (Wang et al, 2015)[5] is used to calculate BCO fractions in this work. It is widely-used because of its powerful computing capability by using GPU parallelization. Besides the dynamic processes among stars, this code also calculates the evolution of stars in the globular cluster. NBODY6++GPU uses a stellar evolution code, BSE (Hurley et al. 2002)[6], which takes both single and binary stars into account. The code can calculate stellar evolution quickly and accurately.

3. Input Parameters

In order to shorten the computing time, we build a small globular cluster model with a star number of 10000. It is similar to some observed globular clusters. The simulation finished at the age of 5.0 Gyr. We set the metallicity to a typical value, i. e., solar metallicity (Z=0.02). The masses of stars are generated by taking the initial mass function (IMF) of Kroupa (2001)[7]. According to the stellar physics, the stellar masses cover a range of 0.08~100M$_\odot$, and the average mass is 0.55M$_\odot$. At zero age, stars distribute randomly with a Plummer sphere (Plummer 1911)[8] in space.
Binary stars are common objects in the universe. In Milky Way, more than 50% stars are binaries. The binary fractions in young star clusters are extremely high, they can be as large as 80% ~ 100% (Milone et al. 2012)[9], so we define the primordial binary fraction as 50%. For the primordial binaries, the upper and lower limits of masses are set to 58.33M$_\odot$ and 0.08M$_\odot$. The eccentricities of binaries follow a thermal distribution (the square of eccentricities follows a uniform distribution). In addition, the orbital periods are given randomly in the range from 0 to 100000 yr. According to the theory of stellar evolution, the binaries whose orbital periods are larger than 100 yr have no binary material interaction. The fundamental parameters of the simulation are listed in Table 1. Note that, the masses of secondary stars (stars with lower masses) in binaries are lower than 6M$_\odot$, thus it is hard to form NSBs and BHBs in binary evolution. It implies that the resulting NSBs and BHBs in the simulation are probably produced by dynamic processes.

Table 1. Input parameters of simulation. N, f_b and Z are star number, primordial binary fraction and metallicity, respectively. M_{max} and M_{min} indicate the upper and lower limits of stellar masses.

Parameter	Value
N	10000
f_b	50%
Z	0.02
Age	5.0 Gyr
M_{max}	100 M$_\odot$
M_{min}	0.08 M$_\odot$

4. Results

In this work, the cluster is simulated to the age of 5.0 Gyr. The computation spent 16 days. The initial number of stars is 10000, and it then decreases with time, because of the merging and escaping of stars. We focus on the formation of BCOs in the simulation. According to the output files, we can obtain the numbers of BCOs. The fractions of BCOs can be calculated by comparing the numbers of BCOs and all stars in the star cluster. The results are shown in Table 2. We find that WDBs form in a very short time with the evolution of star cluster. As we see, the fraction of WDBs reaches 0.031% in the first 0.5 Gyr, and increases until 5.0 Gyr. The fraction of NSBs changes slightly within this time in the range of 0.0 ~

5.0 Gyr. The fraction of NSBs is 0.012%, the corresponding binary number is only 1. Figure 1 shows the evolution of BCO fractions in this work. We can see that the fractions of WDBs are always much larger than that of NSBs.

Table 2. BCO fractions in N-body simulation.

Age (Gyr)	BCO fraction (%)										
	0.0	0.5	1.0	1.5	2.0	2.5	3.0	3.5	4.0	4.5	5.0
WDBs	0.000	0.031	0.052	0.066	0.111	0.135	0.179	0.195	0.220	0.222	0.235
NSBs	0.000	0.000	0.000	0.000	0.000	0.000	0.000	0.000	0.000	0.012	0.012

The formation processes of the two kinds of BCOs are also studied. There are two WDBs come from primordial binaries within three in total WDBs when WDBs form within the first 0.5 Gyr. We find that dynamic processes usually separate the WDBs evolved from isolated binaries. As a result, all WDBs are formed by dynamic interactions among stars.

The other kind of BCOs, NSB, is also well studied in this simulation. The only one NSB is produced in the process of dynamics. Because NSBs are one of the main sources of GWs, we analyze the formation of the NSB in this simulation in detail. Figure 2 summarizes the entire process of the NSB formation. The evolution processes of two components (Nos. 135 and 10024) that finally form the NSB are shown. Star No. 135 goes through a very complicated formation process. In detail, at the beginning of the evolution, the star is included in a binary system. With the evolution of the binaries, the original binary system was destroyed. Then star No. 135 becomes a single red-giant. Next, it evolves to a single neutron star. After that, the single neutron star encounters a single main sequence star, a new binary system is formed. Meanwhile, the massive star No. 10024 evolves to a single neutron star. Finally, Star No. 10024 collides the new binary system, a NSB and a single star are formed.

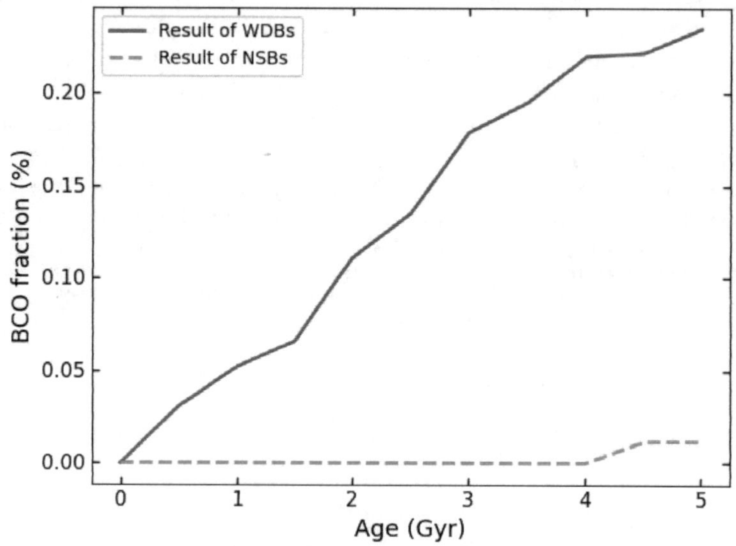

Figure 1. Evolution of fractions of BCO in N-body simulation. The blue solid line is the result of WDBs, the orange thick line is for NSBs.

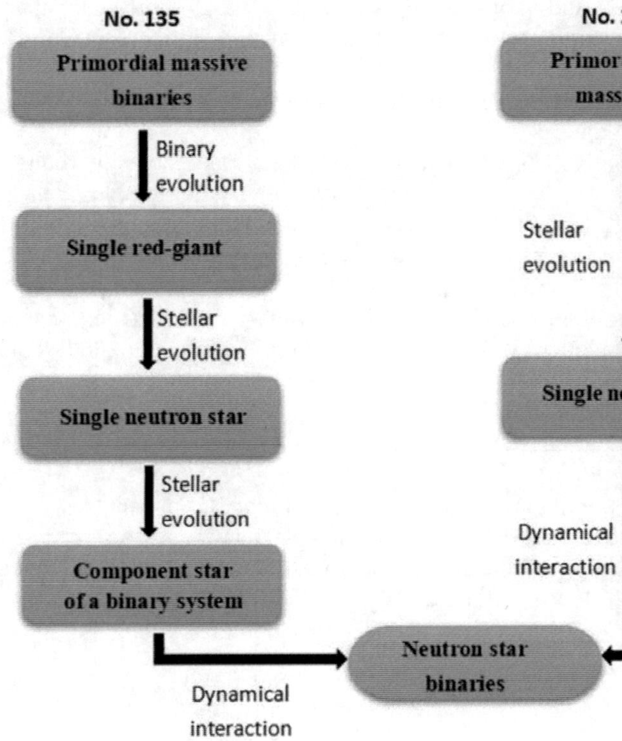

Figure 2. Process of the NSB formation. Star Nos. 135 and 10024 are the component stars of the NSB.

5. Conclusion

In this work, we simulate the formations and fractions of BCOs in a small globular cluster. This paper first introduces the importance of binary stars in the future development, then describes the method used in this paper and the specific simulation process, and finally carries on the further analysis and summary of the simulation results. We use an advanced N-body simulation code, which has comprehensive consideration to simulate a globular cluster. The fraction of WDBs increases significantly in the age range of 0.0 ~ 5.0 Gyr, but the result of NSBs increases very slowly. Therefore, the formation of NSBs is much slower than that of WDBs. This is normally according to the IMF. Both WDBs and NSBs are formed mainly from dynamic interaction among stars.

We do not observe NSBHs and BHBs in our simulation. The reasons are that there are few super massive binary stars according to the IMF, and the small number of stars in the mimic globular cluster. In future works, it is necessary to enlarge the number of stars in the cluster and extend the simulated age.

Acknowledgments

This work is supported by Yunnan Academician Workstation of Wang Jingxiu (No. 202005AF150025), National Natural Science Foundation of China (No. 11863002), and Sino-German Cooperation Project (No. GZ 1284).

References

[1] Abbott B. P., Abbott R. and Abbott T. D., et al., GWTC-1: A Gravitational-Wave Transient Catalog of Compact Binary Mergers Observed by LIGO and Virgo during the First and Second Observing Runs [J]. Physical Review X, 2019, Vol. 9, No. 3, pp. 031040.

[2] Abbott R., Abbott T. D. and Abraham S., et al. GWTC-2: Compact Binary Coalescences Observed by LIGO and Virgo During the First Half of the Third Observing Run [J]. arXiv e-prints, 2020, arXiv:2010.14527.
[3] Sigurdsson S., Hernquist L. Primordial black holes in globular clusters [J]. Nature, 1993, Vol. 364, No. 6436, pp. 423-425.
[4] Di Carlo U. N., Giacobbo N., Mapelli M., et al., Merging black holes in young star clusters [J]. Monthly Notices of the Royal Astronomical Society, 2019, Vol. 487, No. 2, pp. 2947-2960.
[5] Wang L., Spurzem R. and Aarseth S., et al., NBODY6++GPU: ready for the gravitational million-body problem [J]. Monthly Notices of the Royal Astronomical Society, 2015, Vol. 450, No. 4, pp. 4070-4080.
[6] Hurley J. R., Tout Ch.A. and Pols O. R., Evolution of binary stars and the effect of tides on binary populations [J]. Monthly Notices of the Royal Astronomical Society, 2002, Vol. 329, No. 4, pp. 897-928.
[7] Milone A. P., Piotto G. and Bedin, L. R., et al., The ACS survey of Galactic globular clusters. XII. Photometric binaries along the main sequence [J]. Astronomy & Astrophysics, 2012, Vol. 540, pp. A16-A56.
[8] Plummer, H. C., On the problem of distribution in globular star clusters [J]. Monthly Notices of the Royal Astronomical Society, 1911, Vol. 71, pp. 460-470.
[9] Kroupa, P., On the variation of the initial mass function [J]. Monthly Notices of the Royal Astronomical Society, 2001, Vol. 322, No. 231, pp. 231-246.

Modeling Prediction and Research on Leaf Moisturizing Effect of Tobacco Redrying

Keqiang Li[1], Ruidong Li[1,*], Wenyong Huang[1], Zehui Yang[1], Xiaobing Zhang[2] and Jianguo Liu[2]

[1]Yunnan Leaf Tobacco Redrying Co. Ltd, Kunming 650031, China
[2]China Tobacco Zhejiang Industrial Co. Ltd, Huangzhou 310024, China

[*]Corresponding author email: 371856443@qq.com

Abstract. This paper studies the influence of the process parameter setting of the hot-air leaf moisturizer on the quality index of the exit leaf during the secondary leaf conditioning of threshing and redrying, and establishes the corresponding prediction model. According to the characteristics of the characteristic process data of the secondary moisturizing, the prediction models of BP neural network and cyclic neural network are established. Call the current popular neural network writing framework TensorFlow's high-level API interface to build the neural network structure. Gradually optimize the activation function, optimizer, number of hidden layer neurons and other key parameters in the neural network structure to make the prediction results of the test set reach the best state. By inputting the parameters of the front steam nozzle pressure, the front-end water flow rate, the hot air temperature, the return air temperature, the temperature of the feed blade, and the water combination of the feed blade, two key tobacco leaf evaluation indicators are predicted for the outlet leaf moisture and temperature. Based on the mean square error, root mean square error, and average absolute error of the prediction results as quantitative indicators, the three indicators of BP neural network are 7.17, 2.68, 1.83, and the three indicators of cyclic neural network are 4.70, 2.16, 1.74. It is concluded that the cyclic neural network has the best prediction effect. It has certain reference value for the tobacco factory to adjust the process parameters of the secondary leaf conditioning according to different conditions.
Keywords: BP neural network, cyclic neural network, secondary moisturizing, tensorflow, outlet leaf temperature, outlet leaf moisture, mean square error.

1. Introduction

Threshing and redrying is the basis for the development of the tobacco industry[1]. Threshing and redrying is an advanced raw tobacco processing technology, and the heating and humidification technology of tobacco leaves (including the secondary moisturizing stage) is the key link[2]. In the moisturizing stage of tobacco redrying, Liu Zhengqin[3] explored how to improve the traditional method, and proposed a method of heating and humidifying with pure steam. Experiments verified that the changes in various index parameters of the moisturizing leaf affected the tobacco after moisturizing. Quality has an impact. Long Minghai et al.[4] took Hongda B1F and Yunyan 87XZF as the research objects and proved that the steam moisturizing method can well retain the aroma components of middle and high-grade tobacco leaves, and the soda-water mixed moisturizing method is beneficial to the degradation of macromolecular compounds. He Guangyu[5] proposed an adaptive nonlinear PID control method based on the complex system characteristics of the two-run tube equipment with typical large, pure hysteresis, and strong interference. Through a digital simulation

Content from this work may be used under the terms of the Creative Commons Attribution 3.0 licence. Any further distribution of this work must maintain attribution to the author(s) and the title of the work, journal citation and DOI.
Published under licence by IOP Publishing Ltd

platform combined with an improved intelligent optimization algorithm, an adaptive nonlinear PID control method was proposed. Scholars such as Lian Changwei and Wang Fayong[6,7] have proved through different field tests that the sensory quality and combustion performance of the obtained sheet smoke will change when the leaf moistening intensity, hot air humidification amount, and steam pressure at the outlet are different. Some scholars[8-10] proposed adopting new equipment and control methods in the leaf conditioning stage, and using different processing methods and workflows can effectively improve the quality of leaf conditioning. Zhao Chaowen [11] determined that the main processes of threshing and redrying were leaf moisturizing, threshing and destemming and redrying, and constructed an evaluation model of the influence of these processes on the quality of tobacco slices. Through this model, the comprehensive index score was calculated and standardized. Finally, the score Q is obtained, the BP neural network is built with the score Q as the output expectation, and the genetic algorithm is used to optimize the network, and an effective evaluation model is established. Many scholars have conducted research on the leaf moistening stage, but they are all based on experiments, and the conclusions of each experiment are not the same. They did not thoroughly explore the relationships existing in the data and establish general models. Deep learning is currently a hot spot in scientific research, and has been widely used in image recognition, natural language processing, automatic control systems, etc[12-14]. Based on previous researches and the characteristics of the secondary leaf conditioning process data, this paper summarizes it as a black box optimization problem. Based on the adjustable process parameters of the hot-air leaf moisturizer and the characteristics of the tobacco leaves, a BP (Back Propagation) neural network prediction model is proposed to solve the instability and uncertainty of the quality indicators of exported tobacco at this stage. Reasonably adjust the process parameters to provide a theoretical reference.

2. The Establishment of BP Neural Network Model

2.1. Statement of Problem

This paper selects the data collected from the second leaf conditioning experiment conducted by a tobacco factory process laboratory within one year for research[15], the total number of samples is 100 groups, and some sample information is intercepted as shown in Table 1.

Table 1. Featured process parameters for secondary leaf moistening.

No.	Front steam nozzle pressure (Bar)	Front end water flow (L/h)	Hot air temp. (°C)	Return air temp. (°C)	Feeding blade temp. (°C)	Feeding leaf moisture (%)	Outlet blade temp. (°C)	Outlet leaf moisture (%)
1	0.00	250.00	80.00	50.00	29.30	14.47	45.33	22.80
2	0.10	250.00	94.00	60.00	29.60	13.36	53.00	18.63
3	0.20	170.00	106.00	67.00	29.70	13.81	60.00	17.15
4	0.35	220.00	118.00	78.00	28.90	14.32	61.00	20.28
5	0.46	190.00	135.00	82.00	29.70	14.32	70.33	18.01
6	0.00	200.00	80.00	50.00	29.90	14.57	41.67	22.40
7	0.10	200.00	92.00	57.00	29.20	15.12	50.00	20.06
8	0.20	250.00	106.00	67.00	28.60	14.81	58.67	21.86
9	0.33	210.00	117.00	73.00	29.40	19.50	62.33	21.25
10	0.46	195.00	132.00	80.00	29.50	15.31	68.33	19.90

It can be seen from the above table that each group of samples is to conduct process experiments on the quality indicators of export materials under different conditions. The first four parameters are equipment control parameters, the fifth and sixth parameters are inlet process parameters, and the last two are outlet quality indicators. For 100 groups of experiments, based on a large amount of research experience, it can be known that 80 groups of them can be used as the training set, and the remaining 20 groups of samples can be used as the test set as a reasonable allocation plan.

The main purpose of modeling is to form the relationship between the first six parameters and the latter two parameters. The first six parameters can be used to predict whether the tobacco leaf quality indicators meet the corresponding requirements after moistening. To solve the problem of the need to adjust parameters in production and life to deal with different situations. For example, in some areas, tobacco leaves have less moisture, and the corresponding parameters can be adjusted to increase the export leaf moisture to meet production needs.

2.2. Modeling of BP Neural Network

When modeling a BP neural network, we must first clarify a few points. First, there are several indicators (outputs) to be predicted. Secondly: There are several control parameters (inputs) to be adjusted. Finally, select the number of hidden layers and the number of neurons to determine the structure model of the neural network. For the structural design of the hidden layer, many researchers have done relevant research. The number of hidden layers and the number of neurons contained in each layer largely determines the memory capacity, training speed and generalization ability of the BP neural network. Too many nodes in the hidden layer will increase the amount of calculation and reduce the fault tolerance of the model; too few nodes in the hidden layer may cause problems such as increased local optimal conditions and decreased convergence accuracy[16,17]. From a theoretical analysis, the addition of the activation function allows the BP neural network with only one hidden layer to approximate any nonlinear system. Since the sample data is not large, this paper uses a three-layer BP neural network with only one hidden layer, six neuron input layers, and two neuron output layers.

In theory, there is an optimal number of neurons for each neural network model, but it is difficult to draw a conclusion directly through a specific formula. You can use the empirical formula for preliminary calculations, and then verify the accuracy of the model to get Optimal number of neurons.

$$l = \sqrt{n+m} + a \quad (1)$$

In the above formula, n is the number of neurons in the input layer, m is the number of neurons in the output layer, and a is a constant between 1 and 5. In this paper, n is 6 and m is 2, so the value range of l is 5-9. In the process of model training, set different numbers of hidden layer neurons and observe the number of neurons, the model training effect is the best, and the number of neurons is selected accordingly. The structure of the neural network is shown in Figure 1 below.

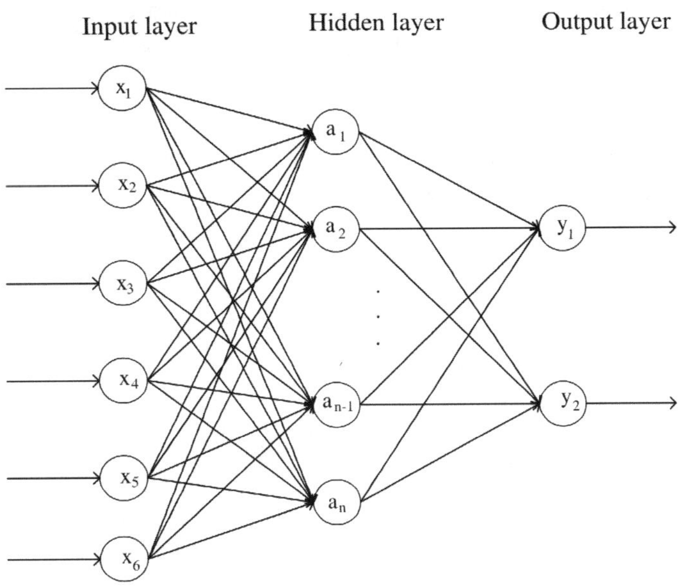

Figure 1. Neural network structure diagram.

After establishing the neural network model and determining the input and output data, write the neural network code and debug it. The first step: Read the data, use the first 80 sample data of the overall sample data as the training set, and the other 20 groups of data as the test set, and normalize the data and disorder the order. There are six adjustable process parameters in the secondary leaf conditioning process, with different dimensions and units. In order to eliminate this influence and speed up the convergence of the data, normalization is required. The purpose of data disorder is to eliminate the correlation between data and enhance the generalization ability of the model. The normalization method used in this article is a linear normalization method, and the calculation formula is as follows.

$$x' = \frac{x - \min(x)}{\max(x) - \min(x)} \tag{2}$$

Step 2: Build a neural network model, call the Sequential() function in Keras to build a three-layer neural network. Step 3: Set the hyperparameters in the model, call the compile() function, set the optimizer, learning rate, loss function, accuracy function, etc. Step 4: Train the neural network model, call the fit() function, specify the input features and corresponding labels of the training set, the number of cycles, and the input features and labels of the test set. Step 5: Save the parameters of the model, make predictions, and draw the image of the predicted value and the actual value, so as to better observe the prediction effect.

2.3. Parameter Analysis of BP Neural Network

In the BP neural network model, the number of neurons in the hidden layer has a significant impact on the accuracy of the entire model. The two factors are analyzed and discussed below to determine a reasonable plan to get the best prediction effect and model parameter.

To study the influence of a single factor on the overall model, the controlled variable method is adopted, and other parameters are kept uniform, and this factor undergoes different levels of changes. The main settings of several parameters are as follows: activation function (Activation='relu'), optimizer (Optimizer='rms- prop'), the size of each bunch of data (batch_size=20), and the data samples are made into a bunch of 20, sent to the neural network, to improve the computer's computing efficiency, the number of loop iterations (epochs=100), that is, the data samples should be fed to the neural network model in a loop, and the model is continuously trained to improve the accuracy.The regularization method chooses L2 regularization (kernel_regularizer=tf.keras.regularizers.l2()), regularization can effectively alleviate the overfitting problem of the model. Explain several important parameters, as described below.

The full name of RMSprop algorithm is Root Mean Square Prop, which is an optimization algorithm proposed by Geoffrey E. Hinton[18,19]. Because the learning rate decay of the Adagrad algorithm is too radical, consider changing the calculation strategy of the second-order momentum, not accumulating all the gradients, and only focusing on the gradients in a certain window in the past. The exponential moving average is approximately the average over a period of time in the past, reflecting "local" parameter information, and using this method to calculate the second-order cumulative momentum.

$$w_{t+1} = w_t - \eta_t \tag{3}$$

$$m_t = g_t \quad V_t = \beta_2 \cdot V_{t-1} + (1-\beta_2) \cdot g_t^2 \tag{4}$$

$$\eta_t = \frac{\alpha}{\sqrt{V_t}} \cdot m_t = \frac{\alpha}{\sqrt{\beta_2 \cdot V_{t-1} + (1-\beta_2) \cdot g_t^2}} \cdot g_t \tag{5}$$

The main idea of regularization is to add an index that can describe the complexity of the model on the basis of the loss function. The loss function (loss) is used to express the accuracy of the model on the training set. In the optimization, the loss function is not directly used, but the complexity of the model R(w) is added. The goal of model optimization is loss(p)+q×R(w). Where q represents the proportion of the model's complex loss in the total loss, and p represents the weight w and bias b in the model.

There are two main functions describing the complexity of the model, one is L1 regularization, and the formula is as follows:

$$R(w) = \|w\|_1 = \sum_i |w_i| \tag{6}$$

The other is L2 regularization, the formula is as follows:

$$R(w) = \|w\|_2^2 = \sum_i |w_i^2| \tag{7}$$

After setting the parameters, in order to observe the quality of the model training results, the corresponding observation indicators are formulated. The main indicators are the loss of the training set and the test set. The smaller the loss, the higher the accuracy of the model. In addition, in order to show the relationship between the prediction result and the true value more clearly, the predict() function is called to predict the sample data of the test set and compare it with the true value. The evaluation indicators are Mean Squared Error (MSE) and mean square Root Mean Squared Error (RMSE), Mean Absolute Error (MAE).

MSE is used to express a measure of the degree of difference between the estimator and the estimated value. The sample size in this article is fixed, and the square function of the distance between the point estimate $\hat{\theta}$ and the actual value θ is used, and the expectation of the function is calculated. The calculation formula is as follows:

$$MSE(\hat{\theta}) = E(\hat{\theta} - \theta)^2 \tag{8}$$

RMSE is the square root of the ratio of the square of the deviation between the predicted value and the actual value to the number of observations n, di^2 Represents the square of the deviation of a set of measured values from the true value, the formula is:

$$RMSE = \sqrt{[\sum \frac{di^2}{n}]} \tag{9}$$

MAE is the average of the absolute values of the deviations of all single predicted values from the arithmetic mean. For a set of data$\{x_1,\dots,x_n\}$, the formula is:

$$MAE = \frac{1}{n}\sum_{i=1}^{n} |x_i - m(x)| \tag{10}$$

Set the number of neurons in the hidden layer to a number from 5 to 9, and the results obtained after BP neural network training are shown in the following figures and table.

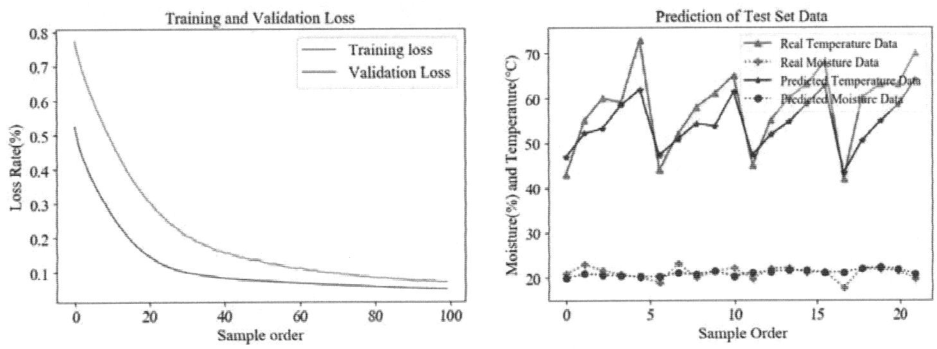

Figure 2. Loss and prediction result graph for 5 neurons.

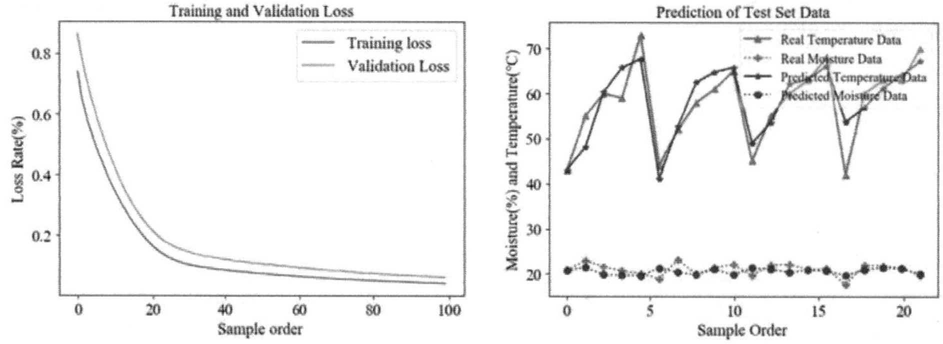

Figure 3. Results of loss and prediction at 6 neurons.

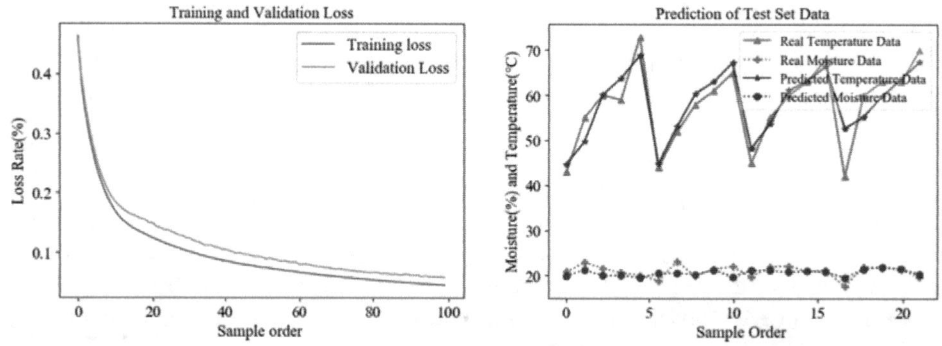

Figure 4. Loss and prediction result graph for 7 neurons.

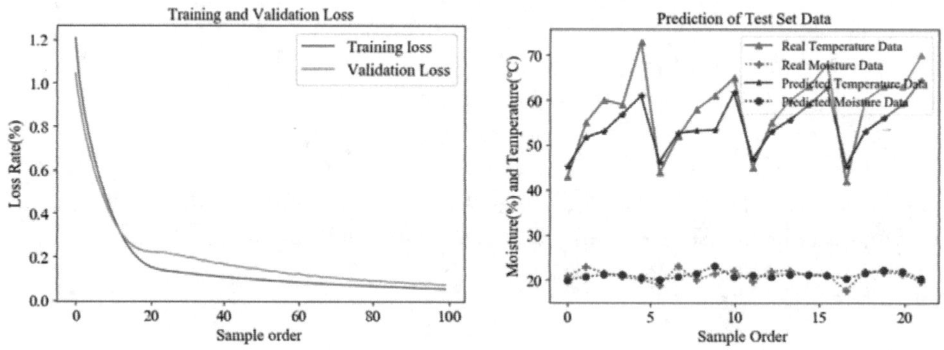

Figure 5. Graph of loss and prediction results at 8 neurons.

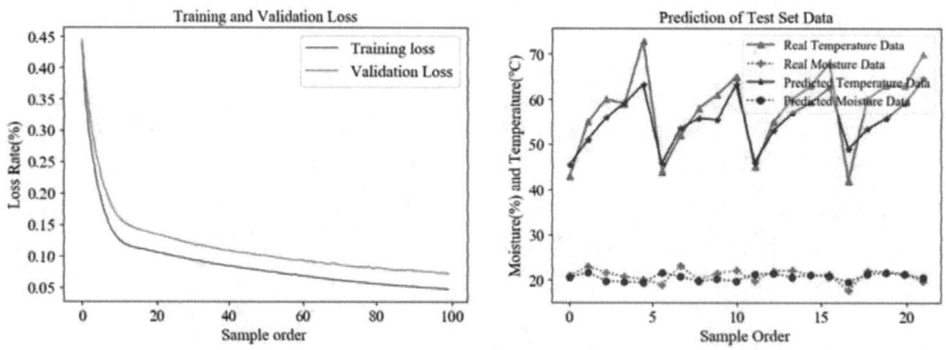

Figure 6. Results of loss and prediction at 9 neurons.

It can be seen from the above five figures that the difference in the number of neurons in the hidden layer has a significant impact on the prediction model. From the loss diagram on the left, it can be seen that the accuracy of the BP neural network prediction model basically meets expectations. The straight line is the loss rate of the training set, and the dashed line is the loss rate of the test set. Regardless of the number of neurons, the loss is reduced to less than 0.1 for the training set or the test set. When there are 5 neurons in the hidden layer, the first half of the loss map has a large difference between the training set and the test set. The main reason is that the number of neurons is small and the characteristics of the input data cannot be trained well. Adjust the weight matrix to an appropriate value. When there are 9 neurons in the hidden layer, the loss of the test set and the training set in the first half of the loss graph is very close, and the gap between the two curves in the second half increases significantly. The reason is that the number of neurons may be too large, which was created during the initial loop iteration. The model has a slight overfitting phenomenon. As a result, the accuracy of the second half of the forecast has decreased. Observe the comparison between the predicted results of the picture on the right and the real results. The upper part of the picture is the temperature forecast, and the lower part is the moisture forecast. The curves with triangles and plus signs are real data, and the other two curves are BP neural network prediction data. It can be seen from Figure 2 that the BP neural network model with 5 neurons in the hidden layer has an average performance in predicting the moisture content of the exported tobacco slices. It does not fluctuate much with the change of real data, and it is basically a straight line. As the number of neurons increases, the prediction effect is significantly improved. However, when the number exceeds 7, it is not possible to predict the values with large changes in the real temperature data. Therefore, it is concluded that when the number of hidden layer neurons is 7, the prediction effect is the best. See Table 2 for specific prediction and evaluation indicators.

Table 2. The prediction results of the BP model with different numbers of neurons.

Number of hidden layer neurons(n)	MSE	RMSE	MAE
5	15.54	3.94	2.82
6	9.65	3.11	2.09
7	7.17	2.68	1.83
8	14.15	3.76	2.76
9	11.49	3.39	2.51
10	9.99	3.16	2.40
11	12.76	3.57	2.68

Observe the above table, because when the number of neurons in the hidden layer is 9, all indicators have decreased. In order to prevent the occurrence of a local optimum, two sets of experiments were carried out downward. It can be seen that when the number of neurons is 7, MSE, RMSE, and MAE all perform better than other cases. It can be determined that when the number of neurons in the hidden layer is 7, it is the optimal BP neural network structure. It is verified from the side that there is an optimal number of neurons in each neural network. Therefore, the final BP neural network structure selected in this article is a three-layer BP neural network structure with 6 neurons in the input layer, 7 neurons in the hidden layer, and 2 neurons in the output layer.

3. Recurrent Neural Network Modeling

The experimental data of the second leaf moisturizing characteristic process has been trained and predicted by the BP neural network and good results have been achieved. It is impossible to improve the accuracy rate by adjusting the parameters, so we should start from the data itself, observe its characteristics and propose a more reasonable and more realistic modeling method. The second set of leaf moisturizing test process in threshing and redrying plant is as follows: adjusting the first set of parameters, taking samples after the hot air moisturizer has worked for a period of time; adjusting the second set of parameters, and then taking samples after hot air moisturizing the leaves for a period of time…Repeating the above steps until all the experiments are completed. The experimental procedures of the threshing and redrying plant are continuous and there are no shutdown, cleaning or other processes. As a result, the previous group of tests will have a certain impact on the latter group of tests.

The main reason is that the tobacco leaves of the last test are left in the hot air moisturizer, and the change process of the first group of parameters and the second group of parameters is not a sudden change process, there will be a small amount of buffer time in the middle. This leads to an unique phenomenon that each adjacent group of experiments influencing each other and the non-adjacent experimental groups basically having no influence. Recurrent neural network is also proposed under a similar background. The main problem it solves is to consider the context when forecasting, and establish the connection between adjacent data. RNN will use the information stored in the memory of the previous step as a part of predicting the result of the next step. More complex cyclic neural networks such as LSTM and GRU will consider long-term dependence issues, not only establishing the connection between adjacent memories, but also will establish connections between memories which are far apart. Then, starting from the data itself and the principle of modeling, RNN is better than BP neural network in predicting the data of the characteristic process of secondary moisturizing. Therefore, based on this point of view, a recurrent neural network is established to verify the correctness of the hypothesis.

3.1. Recurrent Neural Network Design

In essence, the recurrent neural network is not much different from the BP neural network, except that the neurons are replaced by memories. The number of memories will affect the calculation effect of the entire RNN, the number and speed of reading information. According to the empirical formula, the amount of memories are about 1 to 1.5 times the number of training sets. The model training conducted this time includes a training set of 80 samples and a test set of 20 samples, so the number of memories is approximately between 80 and 120. The number of recurrent cores contained in the recurrent calculation layer in the RNN largely determines the training speed, the accuracy of the recurrent neural network and the generalization ability of the model. Too many cores will increase the amount of calculation and reduce the fault tolerance of the model; rarely cores may cause an increase in local optimal conditions and obvious oscillations in the accuracy of prediction results. In the case of a small amount of data, based on previous experience, select 2 loop cores as the main structure of the RNN. In order to reduce the over-fitting of the data, the Dropout layer is added to the RNN structure design[20,21]. In 2012, in order to reduce the phenomenon of over-fitting, Professor Hinton proposed the concept of Dropout. To put it simply, in the training of the neural network, the training unit is removed according to a certain probability, so that the parameters are not updated in this round of training, and the overall structure of the neural network will become as shown in Figure 7.

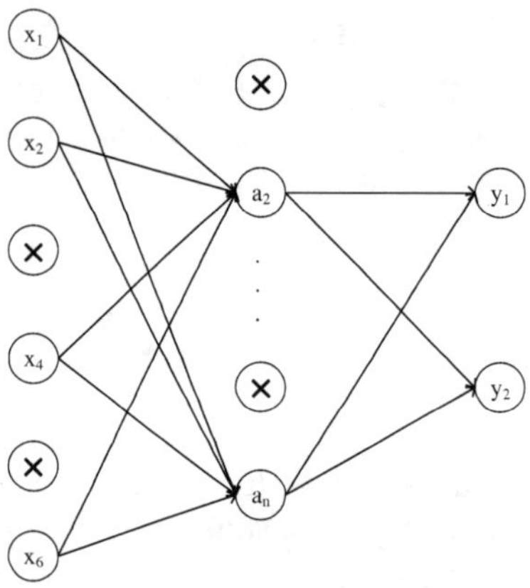

Figure 7. Neural network running diagram with the Dropout layer added.

As shown in the figure above, after adding the Dropout layer, some training units do not update the parameters. Among them, the unit that is not trained is to generate vector 0 and vector 1 with a certain probability through the Bernoulli function, vector 0 is the unit that does not participate in training. Because the final result contains the temperature of the outlet leaves and the moisture of the outlet leaves, a fully connected layer containing two neurons is added at the end. When using Sequential() in Keras to build a network, the structure of a recurrent neural network are mainly composed of a simple

RNN layer with 100 memories, a Dropout layer, a simple RNN layer with 120 memories and a fully connected layer with 2 neurons.

3.2. Parameter Selection and Calculation of Recurrent Neural Network

The recurrent neural network model is based on the context to make predictions, so the division of the sample data test set and training set is slightly different from that of the BP neural network. The BP neural network uses the first 80 sets of samples as the training set, and the last 20 sets of samples as the test set. The RNN test set has 4 samples per interval. One sample is taken into the test set, and the remaining samples are used as the training set, then the test set contains samples numbered 5, 10, 15, 20.... According to the dividing principle, the size of each piece of data should be as small as possible (batch_size=3). Through the experience of adjusting the parameters of the BP neural network, other parameters are set as: optimizer (Optimizer='RMSprop'), loss function (Loss='Mean Squared Error'), callback function monitoring value (monitor='Validation Loss'). Loop iterations required further discussion. Three sets of model training experiments were carried out, with 100, 200, and 500 loop iterations as adjustable parameters. The results are as follows.

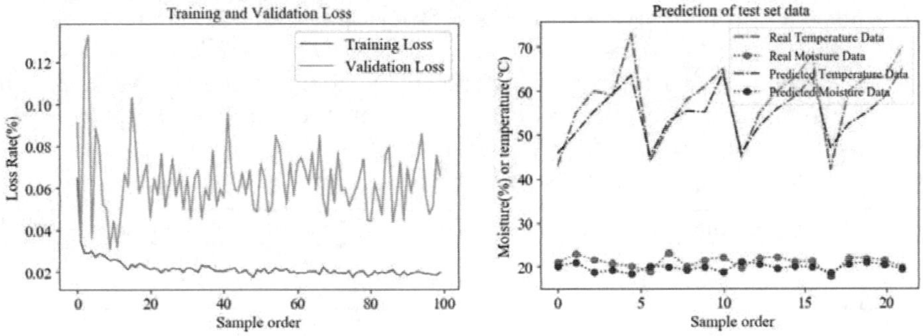

Figure 8. Figure of RNN model loss and prediction results with 100 iterations.

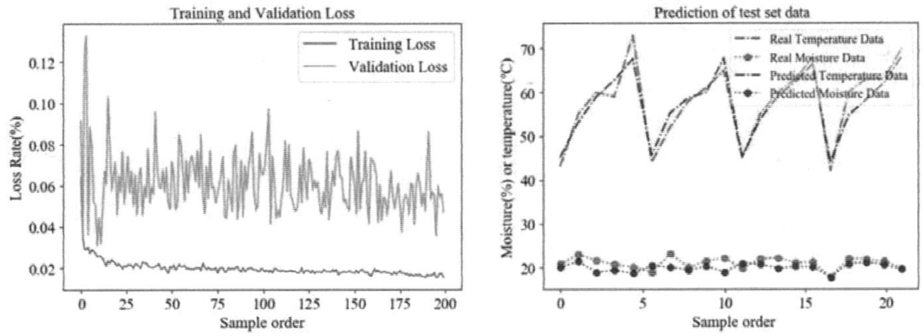

Figure 9. Figure of RNN model loss and prediction results with 200 iterations.

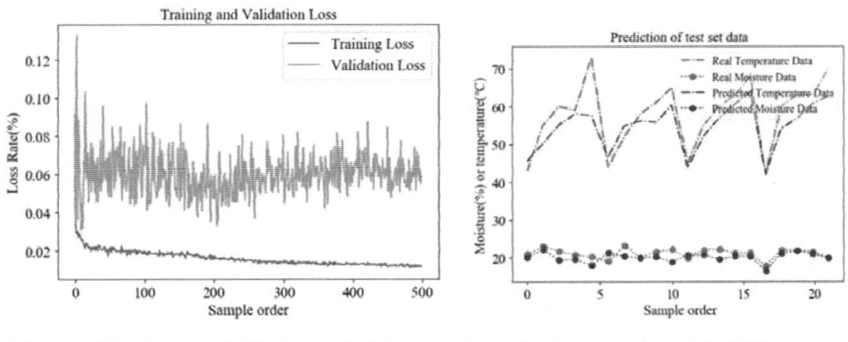

Figure 10. Figure of RNN model loss and prediction results with 500 iterations.

It can be seen from the figure on the left that the increase or decrease of the number of iterations does not significantly affect the loss of the test set. The loss mainly oscillates up and down with the horizontal axis 0.06 as the center line which also shows that the increase in the number of iterations does not significantly improve its accuracy. Observing the figure on the right, it is found that when the number of iterations of the data loop is too many or too few, it will cause the problem of insufficient fitting ability of the model. When the number of loop iterations is 200, the final test indicators MSE, RMSE, and MAE are much smaller than the other two cases. It is reasonable to set the parameter to 200, and the specific prediction results are shown in Table 3.

Table 3. RNN model prediction results for different loop iterations.

Number of loop iterations	MSE	RMSE	MAE	Time consuming (s)
100	12.65	3.56	2.79	12.37
200	4.70	2.16	1.74	19.01
500	14.58	3.82	2.68	42.59

4. Forecasting Models of Comparative Analysis and Discussion

This paper models, analyzes and predicts the characteristic process data of the secondary moisturizing leaves through three methods of BP neural network and recurrent neural network. The predicted values of moisture and temperature of tobacco leaves at the outlet of the last 20 groups of samples by three different methods are shown in Table 4.

Table 4. The prediction results of BP algorithm, RNN algorithm.

Number	Temperature of Outlet Leaves (°C)			Moisture of Outlet Leaves (%)		
	Actual Value	BP	RNN	Actual Value	BP	RNN
1	43	44.78	45.66	20.85	19.94	20.07
2	55	49.81	52.59	22.97	21.31	21.35
3	60	60.13	58.87	21.62	20.02	18.94
4	59	63.80	62.43	20.70	20.10	19.40
5	73	68.82	66.72	20.15	19.56	18.63
6	44	44.84	45.71	18.97	20.54	20.46
7	52	53.16	55.22	23.10	20.62	20.17
8	58	60.40	58.71	20.01	20.27	19.33
9	61	63.01	59.74	21.50	21.31	20.31
10	65	67.22	66.97	22.07	19.82	18.96
11	45	48.28	45.94	19.87	21.19	21.04
12	55	53.68	53.75	22.00	21.20	20.00
13	60	61.14	58.99	22.18	20.85	19.87
14	63	63.12	61.86	21.02	21.12	20.27
15	68	66.52	65.80	21.26	20.85	20.06
16	42	52.71	45.91	17.75	19.47	18.11
17	60	55.26	54.73	21.94	21.45	20.86
18	63	59.85	58.19	21.86	22.03	21.15
19	63	63.61	61.81	21.42	21.55	20.62
20	70	67.44	67.43	19.81	20.35	19.49

In order to observe and compare the results of the three methods more intuitively, subtract the true value from the predicted value of the three methods and take the absolute value, which is recorded as the absolute error. Calculating the error distribution of each method. In the multiple linear regression, the results are two linear equations, but when the error results are calculated, the errors of the outlet leaf temperature and outlet leaf moisture prediction indicators are counted together. The final results are shown in the table 5.

Table 5. Comparison of absolute error of predicted value of BP algorithm, RNN algorithm.

Absolute error range (i)	Multiple linear regression	BP neural network	Recurrent neural network
0<i<0.5	12	9	3
0.5<i<1	6	6	9
1<i<1.5	4	6	9
1.5<i<2	4	5	8
2<i<3	8	8	5
3<i<4	3	2	3
4<i<5	2	3	1
5<i	1	1	2

Combining the above two tables, it can be concluded that the prediction effect of RNN is better than that of BP neural network. The prediction data with absolute error less than 2 can be considered as small error. BP neural network prediction results have 26 prediction data with small errors, and RNN has 29 predictions, indicating that the prediction of recurrent neural network is more stable than the previous method, and the prediction error is smaller.

If the absolute error is greater than 3, it can be considered that the error is relatively large. In this regard, the prediction data of the two methods have little difference. The mean square error (MSE) is the sum of the predicted value minus the square of the true value, and the mean value is calculated. To a certain extent, it can directly reflect the pros and cons of the prediction effect. The mean square error of the two methods is 7.17 and 4.70. It can be seen that the performance of the BP neural network prediction method is poor, mainly because the amount of data is not large and the characteristics of the method cannot be used effectively. The prediction result of the recurrent neural network is the best. The mutual influence of adjacent test groups in the secondary leaves moistening process provides the basis for RNN modeling and prediction. Because the various memories are related to each other, improving the accuracy of the model is not particularly dependent on the amount of data. Accurate predictions can be made with only a small amount of data. Therefore, the method of recurrent neural network should be used when modeling and predicting the characteristic process data of the secondary leaves moisturizing.

5. Conclusion

In this paper, by extracting the characteristics of the secondary leaf moisturizing process data, establishing the BP neural network and the Recurrent neural network prediction model, the influence of the process parameter setting of the hot air moisturizing machine on the quality index of the exported leaves is studied. The results show that when the BP network predicts the secondary leaf conditioning process data, the number of neurons selected affects the training effect of the network, resulting in over-fitting or under-fitting. Secondly, the small data size also limits the prediction results of the BP network. The prediction result of the RNN is better among them. The mutual influence of adjacent test groups in the second leaf moistening process provides the basis for RNN modeling and prediction. Because the various memories are related to each other, it is not particularly dependent on the amount of data. To improve the accuracy of the model, only a small amount of data is needed to make accurate predictions. Therefore, the method of RNN should be used when modeling and predicting the characteristic process data of the secondary moisturizing. In the future, multiple sources and large amounts of data will be introduced for this process link, and the neural network model will be optimized to make the prediction results more accurate, so as to guide the processing technology.

Acknowledgments

This work is supported by China National Tobacco Corporation's major science and technology project "Research and application of "characterized and modular" threshing and redrying characteristics that highlight the tobacco leaf style in the producing area" (No.110201901031 (FK-02)), China National Tobacco Corporation Yunnan Branch Technology The planned key project "Research on the uniformity control of threshing and redrying slices" (No. 2018530000241024).

References

[1] Mingde Hua. Study on some problems in the development of threshing and redrying industry in China[J]. Acta Tab. Sin, 1998(02):3-5.

[2] Yuekun Wang, Yikun Hua, Hongwei Li et al. Overview of the research progress of threshing and redrying[J]. Acta Agric.Jiangxi, 2013,25(12):57-61.

[3] Zhengqin Liu. Research on the new processing method of threshing, redrying and moisturizing leaves[C]. Guangxi: Constructing a new system Stimulate new vitality. 2020:10.

[4] Minghai Long, Xiaolong Zhang, Xianguo Wang et al. The effect of threshing and redrying and moistening methods on the quality of tobacco leaves[J]. Hubei Agricultural Sciences, 2016,55(01):108-111+116.

[5] Guangyu He. Research on Auto Disturbance Rejection Control Algorithm of Threshing and Redrying Hot Air Moisturizer[D]. Hunan:Hunan University, 2018.

[6] Changwei Lian, Tengjian Zhang, Minrui Lu et al. The effect of different leaf moisturizing intensities on the quality and energy consumption of threshing and redrying in-process products[J]. Journal of Anhui Agricultural Sciences, 2020,48(15):194-196.

[7] Fayong Wang, Shaohui Niu, Chao Xu et al. Screening of leaf moisturizing parameter model based on tobacco slice structure and aroma preservation processing[J]. Jiangsu Agric.Sci, 2020,48(09):223-227.

[8] Cheng Chen, Ming Zhou, Xingcheng Yang et al. Discussion on Improving the Feeding Mode of Leaf Conditioner [J]. Modern Agricultural Machinery, 2019(06):45-46.

[9] Chaoming Wang. The design and application of a new type of hot air moisturizing leaf circulation system[J]. Plant Engineering Consultants, 2019(05):48-50.

[10] Yanling Liu, Zeli Wang, Qiang zhou et al. Establishment of a water temperature control model for the moisturizing section of threshing and redrying leaves[J]. Anhui Agricultural Science Bulletin, 2017,23(22):109-111.

[11] Chaowen Zhao. Research on the influence of threshing and redrying technology on the processing quality of tobacco leaves[D]. Kunming University of Science and Technology,2020.

[12] Yujia Duan, Ting Ju. Evaluation of the effectiveness of code review opinions based on deep learning[J]. Electronic Science and Technology, 2020,33(01):39-45..

[13] Junhua Cheng, Guohui Zeng, Jin Liu. Research on classification method of complex background image based on deep learning [J]. Electronic Science and Technology, 2020,33(12):59-66.

[14] Chao Yan, Zhanquan Sun, Engan Tian et al. Research progress of medical image segmentation technology based on deep learning[J]. Electronic Science and Technology, 2021,34(02):7-11..

[15] Yunlan Tang. Cigarette product parameter optimization design method and application research integrating neural network and multi-objective evolutionary algorithm[D]. Hunan:National University of Defense Technology, 2008.

[16] Bing Cai. The design method of BP neural network hidden layer structure[J]. Journal of Tonghua Teachers College, 2007,28(2):18-19.

[17] Jiani Fan, Zhenlei Wang, Feng Qian. Research Progress of BP Artificial Neural Network Hidden Layer Structure Design [J]. Control Engineering of China, 2005 (S1):109-113.

[18] Zou F, Shen L,Jie Z et al. A Sufficient Condition for Convergences of Adam and RMSProp[C].California: Conference on Computer Vision and Pattern Recognition (CVPR). 2019.

[19] Xu D, Zhang S, Zhang H et al. Convergence of the RMSProp deep learning method with penalty for nonconvex optimization[J]. Neural Networks, 2021.vol 139:17-23.

[20] Yuanyuan Chen, Zhang Yi. Adaptive sparse dropout: Learning the certainty and uncertainty in deep neural networks[J]. Neurocomputing, 2021, 450: 354-361.

[21] Nitish Srivastava, Geoffrey Hinton, Alex Krizhevsky, et al. Dropout: a simple way to prevent neural networks from overfitting[J].The Journal of Machine Learning Reseach, 2014,15(1): 1929–1958.

Modelling and Implementation of a Knowledge Question-answering System for Product Quality Problem Based on Knowledge Graph

Hanxu Liu[1], Fangxu Dong[1,*], Meiqing Wang[1] and Qiu Lin[2]

[1]School of Mechanical Engineering and Automation, Beihang University, Beijing 100191, China
[2]Beijing Institute of Control and Electronic Technology, Beijing 100038, China

[*]Corresponding author email: dongfang@buaa.edu.cn

Abstract. Aiming at the problem of difficulty in understanding the semantics of the problem in the traditional quality problem management system, the knowledge retrieval technology of product quality problem based on the knowledge graph is carried out. The process model for knowledge retrieval of quality problem based on semantic templates is constructed. A domain corpus is built, which consisting of thousands of quality problem handling records. The TF-IDF (Term Frequency-inverse Document Frequency) algorithm was used to extracted the vocabulary from the quality problem analysis reports. A natural language question semantic classification process model based on Naive Bayes classifier is established to improve the accuracy of semantic template matching. On the basis of theoretical study, a quality problem knowledge question-answering system-QQ-KQAS based on knowledge graph is developed, and the effectiveness of the proposed method is verified through examples.
Keywords: Quality problem, knowledge graph, natural language retrieval, semantic templates, naive bayes classifier.

1. Introduction

With the continuous deepening of the informatization construction of aerospace enterprises, the aerospace product quality problem knowledge graph, as an important part of the aerospace quality big data management platform, has become the basis for the development of product quality management in the direction of intelligence [1-2]. Question-answering based on the knowledge graph is an important application of the quality problem knowledge graph which provides a natural and intuitive way for the retrieval of quality problem knowledge, and provides an effective way to improve the reuse rate and effectiveness of the quality problem knowledge.

In recent years, question-answering technology based on knowledge graph has received extensive attention from scholars in China and abroad. In addition to its large-scale application in open fields such as the Internet [3], it has also carried out preliminary study and application in vertical fields such as military, medical and manufacturing industries [4-6]. In summary, the question-answering technology based on knowledge graph mainly includes two categories: the method based on semantic analysis and the method based on representation learning. The method based on semantic analysis is to parse the natural language question into the grammatical logic form according to the grammar of the natural language question, and then convert it into the query statement corresponding to the knowledge graph and obtain the answer in the knowledge base [7]. The method based on representation learning regards question-answering as a semantic matching process. The numerical

Content from this work may be used under the terms of the Creative Commons Attribution 3.0 licence. Any further distribution of this work must maintain attribution to the author(s) and the title of the work, journal citation and DOI.
Published under licence by IOP Publishing Ltd

vector of low dimensional space is obtained through the representation learning of knowledge base and problems, and the answer with the largest semantic similarity with question is obtained by numerical calculation [8]. Representation-based learning methods need large-scale corpus for model training. Mourad et al. [9] have developed a fully automatic QA system in the biomedical domain, which can deal with four types of biomedical questions. Shekarpour et al. [10] have proposed an approach that is able to automatically generate explanations for the QA pipeline, which paves the way for further developing the explainable QA systems. Yang et al. [11] have proposed a Two-stage Multi-teacher Knowledge Distillation method for web QA system, and have tackled the challenge that appears in previous model compression methods, which is the information usually loss during the model compression procedure. Phan and Do [12] have developed a Vietnamese QA system through experimenting on three models, and have achieved both high accuracy and time improvement with the combination of deep learning and knowledge graph. Sawant et al. [13] have presented a system that can deal with the full spectrum of query styles between keyword queries using knowledge graph and web corpus.

Compared with the QA system of knowledge graph in open fields such as the Internet, the quality problem knowledge QA system has the characteristics of limited problem forms and limited data resources, and it needs to make deep use of quality problem for knowledge reasoning. As a result, it is difficult to obtain effective language models through representation-based learning. Therefore, based on the constructed quality problem knowledge graph. The remainder of the paper is organized as follows. First, related semantic templates and the process model of knowledge retrieval based on the templates are proposed in Section 2. Then, methods and advantages on constructing the quality problem domain corpus are presented in Section 3. Question classification method for knowledge retrieval of quality problem based on Naive Bayes classifier are introduced in Section 4. The implementation of the QA system is performed in Section 5 where experimental question and the answer for it with the knowledge graph are presented. Conclusion and future work are finally presented in Section 6.

2. Knowledge Retrieval Method of Quality Problem Based on Semantic Template

2.1. Semantic Template Oriented to Knowledge Retrieval of Quality Problem

The semantic template for knowledge retrieval of quality problem is a formal semantic description model created to describe formatted question text. Semantic templates are generally composed of elements such as "question object", "question type", "basic concept", "event" and "constraint condition". These elements can be combined to form various types of semantic templates. For example, in the template "which [product] has the most [scrap]?", "question object" is "product", "question type" is "interrogative sentence", "basic concept" is "scrap", and "constraint condition" is "most". Among them, "scrap" and "product" in square brackets are parts that users can input on their own. Through this variable parameter standard question statement, a standardized formatting template can be formed. Users can improve the efficiency of retrieval by selecting templates according to their requirements. At the same time, since the semantic relationship of templates is clear when they are defined, the use of semantic templates is helpful for the decomposition of questions and the study of semantic association between components, and the accuracy of knowledge retrieval can be improved. However, with the increasing number of semantic templates, the accurate matching of semantic templates becomes more important.

2.2. The Process Model of Knowledge Retrieval of Quality Problem Based on Semantic Template

The process of knowledge retrieval of quality problem based on semantic template is a process of natural language word segmentation, keyword extraction, keyword-based semantic template matching and entity-relation subgraph query. The process model is shown in Figure 1.

(1) Input a section of knowledge retrieval problem text based on semantic template, and use word segmentation tool to segment the problem text. In the process of word segmentation, domain corpus is needed to improve the accuracy of word segmentation;

(2) Analyse and match the keywords obtained after word segmentation with the entities, concepts, and attributes in the quality problem knowledge graph, and obtain knowledge elements that match the retrieval problem—key entities, concepts, or attributes;
(3) Calculate the correlation between the question text and the semantic template using Naive Bayes classifier, then match and select the semantic template according to the level of correlation;
(4) Integrate the key entities and concepts of the retrieval problem with the matching semantic template to construct a retrieval subgraph, and conduct knowledge retrieval oriented to the quality problem knowledge graph;
(5) Return the search results and visualize them.

In the process of knowledge retrieval of quality problem based on semantic template, the construction of domain corpus is the basis for improving the accuracy of natural language problem recognition, and the classification algorithm based on Naive Bayes classifier is the key to achieving accurate template matching.

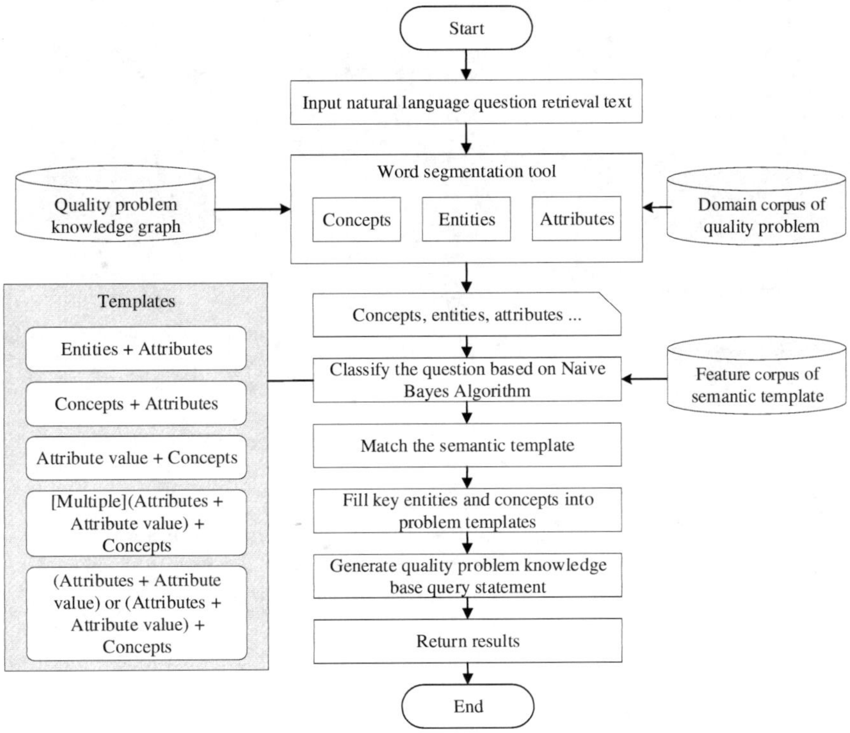

Figure 1. Knowledge retrieval process model of quality problem based on semantic template.

3. Construction of Quality Problem Domain Corpus

The key of quality problem knowledge retrieval based on natural language is the understanding of question semantics, and the key of semantic understanding is the accuracy of the word segmentation of natural language question. In this paper, HanLP(Han Language Process), a Chinese language processing package, is used as the word segmentation tool. Since its corpus is constructed by the corpus training of *People's Daily* in recent ten years, it has a good word segmentation effect in the general domain. But in the domain of aerospace quality problem, the effect of word segmentation is not satisfactory. For example, the domain-specific phrase "quality problem close loop" cannot be identified by the original word segmentation tool. Therefore, in order to make the word segmentation in natural language processing more accurate, it is necessary to construct a domain corpus for knowledge of aerospace quality problems.

3.1. Strategies for the Construction of Quality Problem Domain Corpus

In the process of quality problem solving, quality manager input a large number of quality problem case data into the database. Each quality problem case information includes dictionary data such as product, model, problem occurrence stage and problem cause type et al, and text data such as description of the quality problem phenomenon, corrective measures and cause analysis of the problem et al. For the dictionary data, they can be directly imported into the domain corpus; for text data such as description of the quality problem, the general description form is: a certain failure phenomenon occurs in the function and performance of a product. Most of the core vocabulary are mainly subject-verb phrases and gerunds, which have the characteristics of refined wording, strong independence, language standardization, and a certain degree of repetition. Therefore, this paper uses natural language processing technology to extract keywords from the description text in the quality problem information and import them into the domain corpus to lay the foundation for more accurate word segmentation in natural language processing. Figure 2 shows the model of the construction process of the quality problem domain corpus.

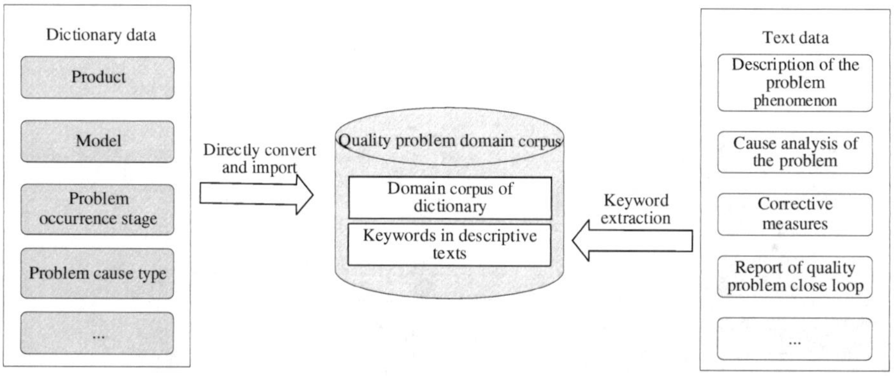

Figure 2. Construction of quality problem domain corpus.

3.2. Domain Keyword Extraction for the Descriptive Text

The quality problem case information contains descriptive text information such as description of the problem phenomenon, cause analysis of the problem, corrective measures, and the report of quality problem close loop, which contains a large amount of domain knowledge. In order to improve the accuracy of natural language problem retrieval, it is necessary to extract keywords that appear frequently in the descriptive text and have a certain degree of representativeness, and use them as a component of the quality problem domain corpus.

In the process of extracting keywords from descriptive texts, since professional domain vocabulary is generally mixed with general words and they usually appear at the same time, the frequency of general vocabulary will be higher than that of professional vocabulary. If keywords are only extracted by counting word frequencies, a few real key domain words may be missed. This paper uses Term Frequency-Inverse Document Frequency (TF-IDF) algorithm to extract key domain vocabulary from descriptive texts [14]. TF-IDF is a commonly used weighting technology for information retrieval and information exploration. The process model of the construction of a keyword dictionary based on the descriptive text is shown in Figure 3. Firstly, the noise data that do not meet the specification are eliminated, and then the descriptive text is segmented and analysed by HanLP to filter out the identified stop words. Then the useless auxiliary words, conjunctions and other words are filtered out based on the part-of-speech filtering. The next step is to use the TF-IDF algorithm for keyword recognition with the assistance of the synonym dictionary to calculate the scores of each word. The keywords are ranked according to the calculated scores. Then the domain experts will identify the keywords based on actual experience. The keywords that have passed the review are added to the domain corpus.

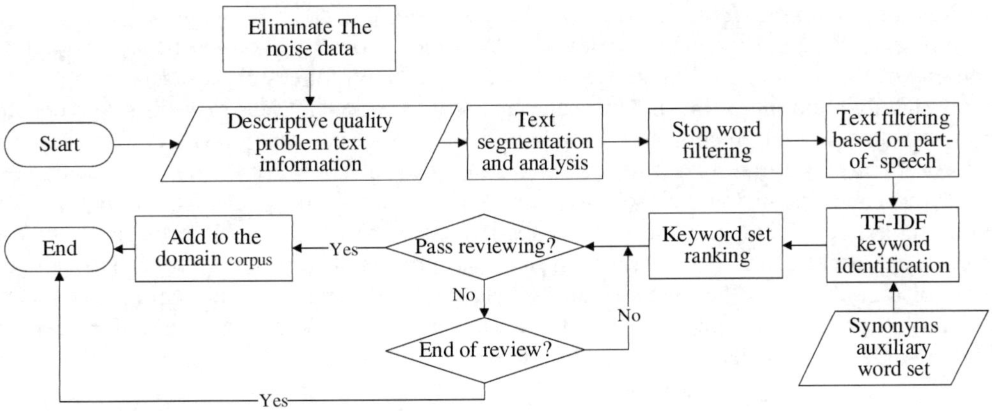

Figure 3. Keyword extraction process model for descriptive text data.

4. Semantic Template Matching Based on Naive Bayes Classifier

4.1. Principle of Naive Bayes Classification

Naive Bayes classification is derived from Bayes theorem. It is a widely used classification model based on probability theory. Bayes theorem is used to describe the numerical relationship between the probability P(B|A), the probability of event B under the premise of event A, and the probability P(A|B), the probability of event A under the premise of event B. The formula is:

$$P(A|B) = \frac{P(B|A) \times P(A)}{P(B)} \tag{1}$$

Where P(A) represents the probability value of event A, P(B) Represents the probability of occurrence of event B.

The Naive Bayes classification algorithm assumes that the feature conditions are independent of each other, and then obtains the classification algorithm of different class probabilities on the basis of the Bayes principle. The category with the highest probability among the classification categories corresponding to the predicted value is the result of the classification. The process is as follows:

(1) Suppose $x = \{a_1, a_2, \dots, a_m\}$ is an item to be classified, where a is an independent characteristic property of x;

(2) $C = \{y_1, y_2, \dots, y_n\}$ is the set of all categories;

(3) Calculate each category's conditional probability $P(y_1|x)$, $P(y_2|x)$, … , $P(y_n|x)$;

(4) Assuming that $P(y_k|x)$ is the maximum value of each conditional probability, then $x \in y_k$.

The calculation method of the conditional probability value in step (3) is as follows:

1) Prepare a set of items to be classified as a training sample set, and the classification of the items to be classified is known;
2) Calculate the conditional probability estimation value $P(a_j|y_i)$ corresponding to each category of each feature attribute under each category (where j=1, 2, …, m; i=1, 2, ..., n);
3) It can be found through the Bayesian formula $P(y_i|x) = \frac{P(x|y_i)*P(y_i)}{P(x)}$ that the probability of occurrence of x is the same. For the item to be classified, the value of the denominator P(*x*) in the Bayes formula is the same. Therefore, when calculating the posterior probability and comparing the magnitudes, it is enough to find the maximum value of the numerator. The value of the numerator can be expressed as follows because the characteristic attributes are independent of each other:

$$P(x|y_i)P(y_i) = P(a_1|y_i)P(a_2|y_i) \dots P(a_m|y_i)P(y_i) = P(y_i)\prod_{j=1}^{m} P(a_j|y_i)$$

4.2. Question Classification Method for Knowledge Retrieval of Quality Problem

The key problem of natural language retrieval in the field of quality problem lies in the computer's understanding of the input natural language questions. The Naive Bayes classifier is used as the basis of the text classification algorithm in this paper. Firstly, a problem classification set is constructed according to business requirements, and a semantic template set corresponding to the classification is manually constructed as a training set for Bayesian classification. Because the quality problem domain is obviously problem-oriented and closely integrated with business requirements, the feature words of the training set are determined manually. After the feature word set number is vectorised, the training set is used as the input of the Naive Bayes classifier to train and form a classifier. Then the natural language question input by the system can be classified and predicted to realize the computer's understanding of the natural language question. The quality problem classification process model based on Naive Bayes classifier is shown in Figure 4.

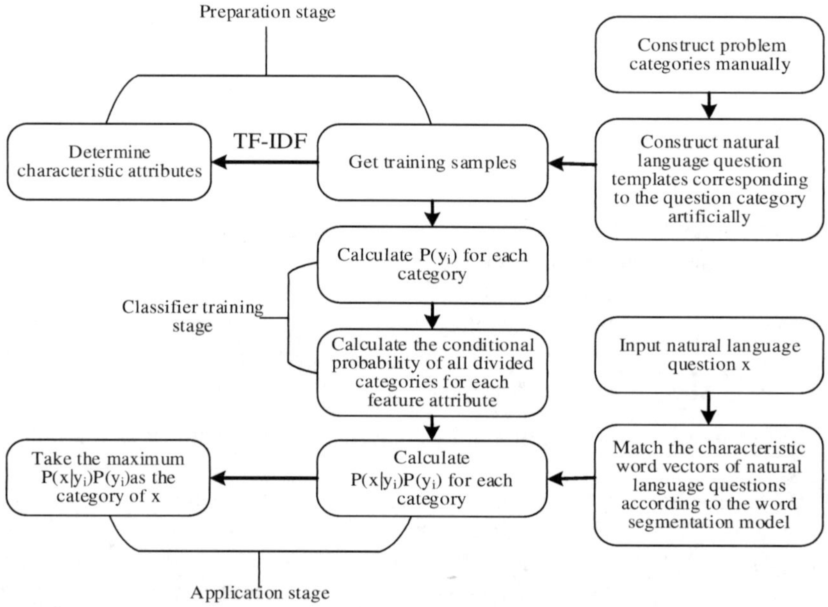

Figure 4. Problem classification process model based on Naive Bayes *classifier.*

There are many ways to express natural questions, and many ways to ask questions for the same question category. Although the questions can be asked in many ways, the intentions of users' inquiries are similar. By classifying natural language questions with consistent query intentions, and manually defining category labels, accurate classification of questions can be achieved. The Naive Bayes classifier realizes the classification of the problem by identifying the characteristic words in the natural language question.

Table 1. Concept entity -- character code.

Conceptual entity	character code
Quality problem	qm
Model	mq
Product	pt
Work stage	we
Problem cause type	pn
Department	dt
Model development stage	de
Standard terms	sd

After defining the problem classification template, the feature words in the problem training set are manually extracted. For the identified key concept entities in the quality problem domain, the feature word codes should be used to represent the key entity features for the computer to understand, as shown in Table 1. The identified key concept entities include quality problem, model, product, work stage, problem cause type, department, model development stage, and standard terms. The corresponding feature word codes are “qm”, “ml”, “pt”, “we”, “pn”, “dt”, “de”, and “sd”.

5. Implementation and Case Study

On the basis of the product quality problem knowledge graph established by the Quality Engineering Laboratory of Beihang University, the "Quick Quality-Knowledge Question-answering System (QQ-KQAS) " was developed. The system is based on B/S architecture, using front-end and back-end separation technology. The back-end uses Springboot integration framework, while the front-end uses Vue.js framework. Neo4j is used as the graph database. And Echarts plug-in is used to realize the visualization of knowledge graph. Besides, the system uses HanLP word segmentation tool to realize the integration of natural language word segmentation, synonyms and domain corpus. The system realizes the definition of semantic templates and the function of quality problem knowledge retrieval based on semantic template on the basis of the constructed quality problem knowledge graph.

Figure 5 shows the query results and visualization of the results based on the natural language problem "what nonconformity quality problems have occurred in the pipeline centrifugal pump" as an example. There are three nonconformity problems in this product, they are “damage of pump body of pipeline centrifugal pump”, “flow is lower than the designed flow” and “mechanical seal failure of centrifugal pump”.

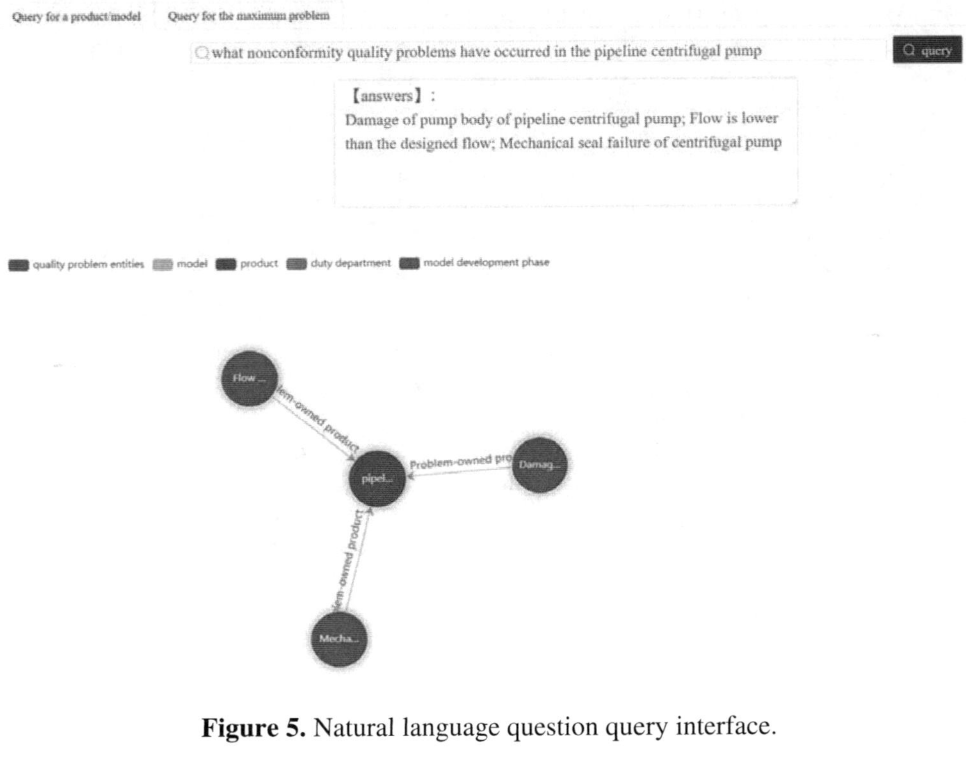

Figure 5. Natural language question query interface.

6. Conclusions and Future Work

This paper studies on the knowledge retrieval technology of model quality problem based on the knowledge graph, and then establishes the process model of quality question knowledge retrieval based on semantic templates. A method for constructing quality problem domain corpuss for dictionary data and descriptive text data is proposed in order to segment natural language questions more accurately. The Naive Bayes classification model is introduced to realize the semantic classification based on natural language questions to solve the problem of accurate matching of semantic templates, which improves the efficiency of knowledge retrieval of quality problem based on semantic templates. The developed "QQ-KQAS for quality problem based on knowledge graph" has been initially applied in an aerospace company and achieved good application results.

Manually constructing semantic templates is time-consuming and laborious. Therefore, in the next step of the thesis, we will focus on the automated construction of semantic templates and the precise matching of semantic templates based on deep learning to further improve the level of intelligence of the question-answering system.

References

[1] Yang B., Mingyan S., Tong Z. Research on the application of aerospace equipment quality knowledge graph [J]. Information Technology and Standardization, 2021(5):27-30.

[2] Yanan C., Pengwu H., Wenbo H., et al. Research on quality management method of aerospace equipment based on knowledge graph[J]. Quality and Reliability, 2019(6):56-60.

[3] Hoy Matthew B. Alexa, Siri, Cortana, and More: An Introduction to Voice Assistants [J]. Medical reference services quarterly, 2018, 37(1): 81-88.

[4] Ben D. A study on automatic generation of template library in Intelligent question and answering system in automobiles domain [D]. Shanghai Jiao Tong University, 2017.

[5] Daiyi L., Jie S., Yunxing L., Zongmin M. Military weapon QA system based on knowledge graph[J]. Command Information System and Technology, 2020, v.11; No.65(05):62-69.

[6] Chenhao M. Design and implementation of automatic question answering system based on thyroid knowledge map [J]. Intelligent Computer and Application, 2018, 008(003): P.102-107.

[7] Berant J., Chou A., Frostig R., et al. Semantic parsing on freebase from question-answer pairs[C]. Proceedings of the 2013 Conference on Empirical Methods in Natural Language Processing. Seattle, 2013:1533-1544.

[8] Cui W., Xiao Y., Wang H., et al. KQBA: learning question answering over QA corpora and knowledge bases [J]. Proceedings of the VLDB Endowment, 2017, 10(5):565-576.

[9] Sarrouti M, Said O. SemBioNLQA: A semantic biomedical question answering system for retrieving exact and ideal answers to natural language questions[J]. Artificial intelligence in medicine, 2020, 102(Jan.):101767.1-101767.16.

[10] Shekarpour S, Nadgeri A, Singh K. QA2Explanation: Generating and Evaluating Explanations for Question Answering Systems over Knowledge Graph[J]. 2020.

[11] Ze Yang, Linjun Shou, Ming Gong, Wutao Lin, Daxin Jiang. Model Compression with Two-stage Multi-teacher Knowledge Distillation for Web Question Answering System[P]. Web Search and Data Mining,2020.

[12] Truong H. V Phan, Phuc Do. BERT+vnKG: Using Deep Learning and Knowledge Graph to Improve Vietnamese Question Answering System[J]. International Journal of Advanced Computer Science and Applications (IJACSA),2020,11(7):479-487.

[13] Uma Sawant, Saurabh Garg, Soumen Chakrabarti, Ganesh Ramakrishnan. Neural architecture for question answering using a knowledge graph and web corpus[J]. Information Retrieval Journal,2019,22(3-4):

[14] Stephen Robertson. Understanding Inverse Document Frequency: On theoretical argument for IDF [J]. Journal of Documentation, 2004, 60(5): 503-520.

Research on System Development and Application of Electrical Equipment Intelligent IoT Platform

Xinsheng Chen[1], Meng Sun[2], Ling Li[2], Xi Chen[3] and Sheng Wang[3,*]

[1]State Grid Shanghai Municipal Electric power Company, Shanghai, China
[2]State Grid Materials Co. Ltd., Beijing, China
[3]State Grid Electric Power Research Institute, Beijing, China

[*]Corresponding author email: ls_0225@163.com

Abstract. With the increasingly development of the new information and communication technologies, China has proposed the integration of new information technology, electrical equipment, and intelligent Internet of Thing (IoT) system to accelerate the innovative development of the manufacturing industry. However, the information isolation and data scattering cause the poor services of electrical equipment. Aiming at the question, the preliminary construction of electrical equipment intelligent IoT Platform (EIP) is carried out in this article through system analysis and system design, which helps enterprises to improve the quality of manufacturing, realizing the transformation and upgrading and healthy development of the enterprise. At the same time, through the accumulation of heavy and complicated IoT big data, the quality evaluation and total life cycle management of electrical equipment are strongly supported to enhance the level of supply chain operations management, which lays a more solid foundation for the electrical equipment manufacturing industry to enhance the core competitiveness and for China to accelerate the realization of high-quality development by the intelligent manufacturing.
Keywords: Electrical equipment, intelligent IoT platform, smart manufacturing, system analysis, system design.

1. Introduction

In recent years, the next-generation of information and communication technologies represented by cloud computing, big data, Internet of Things (IoTs), mobile Internet, and artificial intelligence have come increasingly wide use in the construction of Energy Internet. As the implementation of Energy Internet, the synchronous development of strong smart grid and Ubiquitous Power Internet of Thing (UPIoT) is the new type of business with the deep integration of these new technologies and the energy industry, which is an important cornerstone for promoting the transformation and upgrading of the smart electricity and the real economy [1]. In order to accelerate the innovative development of the manufacturing industry and realize the transformation from a major manufacturer to a strong one, China has proposed the integrated development of next-generation information technology, electrical equipment, and intelligent IoT system [2], which requires the application of advanced technologies to build the intelligent supply chain.

At present, most of China's electric enterprises have realized the whole digital process of material business, and have relevant intelligent application functions. In addition, most of the electrical equipment suppliers have also greatly improved their digital and intelligent manufacturing standard. However, there still exist some defects during the process of equipment supply. On the one hand, due to the lack of effective data exchange channels, the electrical equipment enterprises fail to accurately

Content from this work may be used under the terms of the Creative Commons Attribution 3.0 licence. Any further distribution of this work must maintain attribution to the author(s) and the title of the work, journal citation and DOI.
Published under licence by IOP Publishing Ltd

acquire the process information such as supply capacity, production schedule, manufacturing quality, etc. from the suppliers, so it is necessary to send a large number of experts and staff to track process regularly, resulting in unnecessary cost problems [3]. On the other hand, the suppliers cannot timely obtain the feedback on demand changes, as well as the information of the delivery time, production and operation, fault handling, etc., which means that the suppliers are unable to provide satisfactory services [4]. Therefore, it is of great significance and urgently needed to rely on an information platform to open up the information sharing channel between the supply side and the demand side for improving the energy efficiency of the production of electrical equipment, which is the key to promote the digital development in electric power industry.

By building the electrical equipment intelligence IoT platform (EIP), electrical equipment company and equipment organic connection, the testing data, equipment operation defect data timely feedback to the bidding procurement and production manufacturing, improving the quality of equipment procurement and production, from the source service provider equipment product quality promotion, power electrical equipment capacity upgrade, efficient development.

2. Literature Review

In terms of the architecture research and prototype research and demonstration application of EIP, the United States, Japan, Europe, South Korea and a few other countries started earlier and have a strong overall strength [5]. In 2008, IBM of the United States put forward the "Smart Earth" strategy, and in 2009, the EU put forward the 'Internet of Things Action Plan' [6]. The purpose is to use a variety of information technology means to break through the various restrictions of the Internet, to achieve the UPIoTs. The United States emphasizes the use of sensors and other sensing technologies to build an intelligent infrastructure, while the European Union mainly carries out the wide use of RFID and pays attention to information security [7]. With the increasingly widespread application of RFID technology, its distributed characteristics are combined with the large flow of processing, and the design of the technical architecture of the intelligent Internet of Things platform for electrical equipment, as well as the specification of the architectural requirements, function Settings, integration methods, system deployment and other requirements of each construction component have become the top priority of the research. Li and Wang [8] proposed a new smart city service management and application development platform, based on which different people can collaborate and make contributions, such as demand acquisition and service development. Amine et al. [9] carried out a careful study on the overall framework of the environmental protection Internet of Things and the key technologies of intelligent environmental protection applications, and finally designed a new gateway software architecture of the environmental protection Internet of Things.

For the system development, there are relevant researches. Alvarez et al. [10] theoretically analyzed the building non-central information system, and makes a comprehensive and in-depth analysis and discussion on the feasibility of the system, the design of basic elements, the design of system architecture, the stability of system operation and the effectiveness of control decisions. Wu et al. [11] introduced the research status of electric power information system control from the present situation of electric power dispatching information and work requirement. Hao et al. [12] combined with the ability of SOA architecture of business process reengineering, puts forward a new kind of three-layer model as the core support fast, smooth, reusable application system development framework. Lin et al. [13] discussed the architecture and related technologies of the power dispatching integrated data platform, and proposes that the power dispatching integrated data platform is an information product, which can use modern information technology to calculate and analyze the power data, so as to get the final power consumption ratio.

In spite of the above researches, the majority of them mainly focuses on the design of the platform technology architecture, while there is a lack of research on the architectural requirements, function Settings, integration methods, system deployment and other aspects of the specification of each construction component. Therefore, it is urgent to conduct the in-depth research on the development of EIP to realize the intelligent management for the electrical equipment.

3. System Analysis

3.1. Target and Demand of the Platform

3.1.1. Overall target
The overall target of EIP is to achieve the efficient supervision and management on the 7 major categories and 20+ kinds of materials in the electric power industry by integrating all kinds of big data such as production data, test data, logistics data, etc., and conducting the module construction of IoT gateway, IoT management, and category center, so as to create the 'transparent factory' in the electrical equipment manufacturing industry. Therefore, based on the analysis of the demand positioning of EIP and the necessity and feasibility of its development, this article designs the system architecture and module composition of EIP, sorts out its business process and data process, and integrates multi-source heterogeneous data which is cross-systems through the thinking of "Internet +", so as to expand new business patterns driven by data.

3.1.2. Functional demands
Functional demand refers to the basic requirements that the system must provide users with specific operation in order to accomplish the target contents, which requires the system to possess the corresponding capabilities. According to the overall target, the functional demands of EIP should follow the following points.

(1) Data control of electrical equipment in whole process. Efficient data control is an effective solution to manage equipment in an orderly way, which includes data acquisition, data synchronization, data processing, and data fidelity. This is the fundamental demand directly determining whether EIP can be used to its maximum effect.

(2) Online business. Since the business of electrical equipment enterprises involves equipment manufacturing, logistics and transportation, product operation and maintenance, etc., the online business of EIP should cover the entire process which can be divided into three phases of production monitoring, logistics tracking, and after-sale services.

(3) Combination between industry and finance. Equipment suppliers should be able to choose financing institutions independently and fill in the required information for the financing application by EIP to realize the close combination between industry and finance, which is conducive to the collaborative development of the material supply chain. It needs to be noted that EIP only acts as a role providing the links of financial institutions such as banks, trust institutions, and insurance institutions for suppliers to apply for financing rather than serving as a qualification guarantee institution.

3.1.3. Non-functional demands
Different from the functional demands, the non-functional demands are usually put forward to measure the performance and feature of a system, ensuring the stable and secure running. The concrete non-functional demands of EIP involve the system performance, system reliability, and system manageability, which are listed respectively in Tables 1 to 3.

According to Table 1 which gives the evaluation criteria and corresponding explanations on the system performance of non-functional demands, the multi-point flexible deployment is carried out for the databases by the way of library division and the application services are constructed by the distributed micro-service architecture, to speed up the running efficiency of EIP.

Table 2 shows the criteria and corresponding explanations on the system reliability of non-functional demands. In order to satisfy the above requirements, the distributed architecture ought to be applied to reduce the single point of failure, and the load balancing of services should be provided through the multi-node deployment of application servers. Furthermore, designing the local backup schemes and log monitoring reports will contribute more to the stability of EIP. And EIP must be able to run on all versions of operating systems and browsers which are popular currently.

The criteria and corresponding explanations on the system manageability of non-functional demands are listed in Table 3, which ensures the orderly organization of system data. EIP is designed around the above target and demands.

Table 1. Non-functional demand of EIP on system performance.

Criteria	Demand explanation	Target
System capacity	Number of registered users	500 people during the early implement
	Maximum number of online users	100 people during the early implement
	Size of structured data	500 GB during the early implement
	Size of unstructured data	1 TB during the early implement
	Business throughput	50/minute during normal hours and 200/minute during rush hours
	Growth forecast of business throughput (after three years)	100/minute during normal hours and 500/minute during rush hours
Average response time	Home page browse	≤3s
	System login	≤5s
	Comprehensive business execution	≤8s
	Basic statistical statement query	≤8s
	Monthly statistical statement query	≤20s
	Yearly statistical statement query	≤30s
	Tabular statement processing	≤20s
	Graphic statement processing	≤30s
Transaction failure rate	2 hours of running time with the maximum online users	≤0.1%
	8 hours of running time with 40% of the maximum online users	≤0.1%
Database and server performance	Average CPU utility	≤80%
	Average memory utility	≤80%
	Network throughput per unit	5Mbps during normal hours and 20 Mbps during rush hours

Table 2. Non-functional demand of EIP on system reliability.

Criteria	Demand explanation
Fault tolerance	First, the code logic should be rigorous with all kinds of system exception processing to ensure that every method and process has exception handling statements. Second, the system transaction failure, communication failure, and other cases should be automatically identified and resolved to ensure the availability. Third, when EIP is abnormal, the data being manipulated by servers should be automatically stored into a temporary data table to realize the data protection and recovery.
Robustness	EIP should establish the protection mechanism under the condition of large concurrent or overloaded business to ensure the stable operation. For example, in the case of large concurrency, a single service node should be able to undertake $1/(N-2)$ of the maximum concurrency designed by the system with a service cluster of N nodes that the system can bear.
Self-monitoring	The system should be capable of self-monitoring, which can record, monitor, and alert the important processes and services such as running status, key operations, fault repair, etc., by providing the monitoring interfaces of related components.
Compatibility	A good compatibility of EIP with other systems should be available to support the respective business of different enterprises. For the C/S (Client/Server) architecture, the clients need to be compatible with the current mainstream Windows (Win 7, Win 8, and Win 10). For the B/S (Browser/Server) architecture, the browsers need to be compatible with the current mainstream IE, Chrome, Firefox covering the latest version and the previous versions. For the APP applications, the mobile devices need to be compatible with IOS and Android.

Table 3. Non-functional demand of EIP on system manageability.

Criteria	Demand explanation
Log management	EIP needs to be able to collect and query logs, including user behavior, system exception, and error information.
Monitoring	EIP should have the ability to manage the monitoring data, providing indicator data according to the

management	requirements of relevant interface standards to give the real-time feedback on system operation.
Audit management	EIP should standardize the business files and system files related to audit projects, including at least: business serial number, IP address, user information, operation time, specific operation content, operation result, etc.
Authority management	On the one hand, EIP should realize the authority management of internal employees through corresponding interfaces. On the other hand, EIP should realize the authority management of external users by adding corresponding functions

3.2. Development Necessity of the Platform
EIP is the core of supply chain management system with digital, networked, intelligent, and standardized characteristics, working on the basis of three business chains of intelligent procurement, digital logistics, and panoramic quality control, which is highly consistent with the decision and deployment to build the Energy Internet. As a significant part and the concrete embodiment in material department of the UPIoTs, on the one hand, the construction of EIP gives full play to the hub advantage of material resources coordination to promote the integration and interaction of resources, technology as well as information, and creates an open, equal, collaborative, and sharing supply chain management platform, realizing the improvement of quality and efficiency of material management and the reinforce of supply chain service capacity. On the other hand, EIP is constructed on the center of boosting the quality of purchase equipment with the implementation of smart business scenarios as the starting point, which promotes the construction of strong smart grid and UPIoTs to a certain degree, and consequently achieves the integration of internal business and the cooperation of external industries. In particular, the necessity to develop EIP can be described from the following aspects.

(1) In terms of the construction concept, EIP promotes the internal integration as well as the quality and efficiency of enterprises, drives the process of collaboration and in-depth cooperation for upstream and downstream industries, and realizes the collaboration and sharing of technologies, logistics and data resources, which is consistent with the strategic goal and development direction of the company.

(2) In terms of the construction method, intelligent operations such as intelligent procurement, digital logistics, and panoramic quality control are closely combined with novel information technologies and fresh practical applications to comprehensively improve the quality, efficiency, and benefits of traditional material management from the whole chain aspect, which is in line with the strategic path of the company.

(3) In terms of the construction effect, EIP enables the purchase equipment all interconnected, the supply chain data entirely connected, the business operations carried out online, and the enterprise resources whole shared. Accordingly, the quality of purchase equipment, the efficiency of procurement and supply, the service effect of users, the standard management level, and the value creation ability will all get comprehensively improved, which will strongly support the construction of UPIoTs.

(4) In terms of the existing platform construction, there are a fair number of systems to manage all kinds of information, however, only several of them are suitable for energy material supply with some defects. On the one hand, these systems cover the narrow categories of power materials and are applied to some kind of material as well as the corresponding supplier. On the other hand, the IoT management capability is not perfect enough to support the access of massive data, which consequently fails to realize the remote maintenance and intelligent management of the gateway. Therefore, it is necessary to develop a new platform to make up for the deficiencies of existing platforms by adding corresponding functions to raise the effectiveness and efficiency.

3.3. Development Feasibility of the Platform

3.3.1. Technical support
The information processing technology of intelligent IoT for electrical equipment and the key technology of coordination between edge agent and category center for electrical equipment can provide

technical support for the research on the architecture design and prototype development of EIP. The specific applied technologies can be listed as follows.

(1) Big data analysis technologies. Big data as the popular conception in IT industry, is followed by data warehouse, data fusion, data mining, data security, and so on.
(2) Industrial big data management technologies. This kind of technologies is used for the collection of diverse data, management of multimodal data, and writing of high-throughput data.
(3) Data security technologies. By adopting the edge computing, applications are launched at the source of data to generate faster network service response and to converse some sensitive data through desensitization rules, which meets the basic needs of material industry for real-time business, application intelligence, as well as security and privacy protection.

3.3.2. Methodology basis
The methodology basis will provide more theoretical support for the construction of related models by EIP, laying a solid foundation for the power enterprises to accelerate the high-quality development. The specific methodologies can be listed as follows.

(1) Comprehensive evaluation theory. Aiming at the objective evaluation of electrical equipment suppliers, it is necessary to use multi-attribute comprehensive evaluation theory to filtrate the core evaluation indicators, construct the evaluation system, and establish the data model, so as to provide the decision-making basis for electrical equipment enterprises.
(2) Typical scenario method. In the design of system business process, the method can be adopted to start from a certain process, determine the business process by describing the path of business information, and finally traversal all information flow and business flow to complete the design of the entire system process.
(3) Statistical process control. The method, through statistical analysis to find abnormal fluctuations caused by system factors and accidental factors in the production process for quality control, is a mathematical statistical method used in process control.

3.3.3. Application benefits
The research aims at accelerating the construction of intelligent IoT for electrical equipment and promoting the win-win development of energy industry chain through the development of EIP with which the market prospects are broad and the application benefits are considerable.

(1) Management benefits. EIP integrates the information of the whole lifecycle in which the equipment is from manufacturing to operation then to scrap, scientifically evaluating the high-quality suppliers with strong quality assurance ability, high technological level, and timely service and accurately identifying the high-quality products with advanced technology, exquisite materials, and good environmental adaptability in the bidding evaluation, so as to improve the procurement management level.
(2) Economic benefits. Aiming at the crucial business in the whole process of material procurement, EIP makes full use of advanced information technologies such as big data, artificial intelligence, edge computing, and so on to carry out the innovation and optimization of core business to achieve the more timeliness, better economy, and higher compliance during procurement.
(3) Social benefits. By giving full play to the hub advantage and sharing value, EIP continuously expands the information sharing range of upstream and downstream enterprises in the material supply chain and improves the quality and efficiency of business operation driven by data, accelerating the realization of enterprise goals and promoting the implementation of company strategies.

4. System Design

4.1. Overall Architecture Design
Relying on the basic infrastructures, EIP manages the mass data of electrical equipment for supporting

the business by integrating itself with the related systems. The overall architecture of EIP, which is divided into three parts, is depicted in Figure 1.

The cloud infrastructures are the necessary conditions of data acquisition, data transmission, data procession, and data storage, which requires the hardware support of sensor, transmission network, storage database, and computing device.

On this basis, EIP is designed for a six-layer architecture which reveals the infrastructure service layer, technology service layer, business service layer, business logic layer, application service layer, and application presentation layer from bottom to top.

Further, in order to support the six-layer architecture, relevant systems need to be integrated to complete the docking between functions and applications. These systems involve Supplier Production System (SPS), Supplier Data Acquisition Center (SDAC), Mobile Device Management Platform (MDMP), E-Commerce Platform (ECP), Intelligence Operation Center (IOC), Structured Storage Database (SSDB), Unstructured Storage Database (USDB), e-mail platform, Short Message Service Platform (SMSP), and Identity Security Center (ISC). MDMP, ECP, and IOC are connected through the Data Middle Platform (DMP) to finish the interaction between EIP and the systems, while the others are connected directly with EIP for the integrated applications.

4.2. Function Module Design

On the basis of the architecture design, the system functions of EIP are divided into three aspects of enterprise-oriented, supplier-oriented, and data-oriented. For electrical equipment enterprises, the function modules are designed for real-time online supervision, order monitoring in manufacturing, capacity allocation, and online support service. For suppliers, the function modules are designed for quality evaluation and supplier collaboration. For data management, the function modules are designed for data acquisition in factory side, IoT data integration, category management, and information and business fusion. The 10 function modules can be further subdivided into 37 function submodules, and the overall module design is depicted as Figure 2.

The division of above functional modules aims to improve the efficiency and effectiveness of the intelligence IoT management for electrical equipment to achieve the high-quality development in the industry.

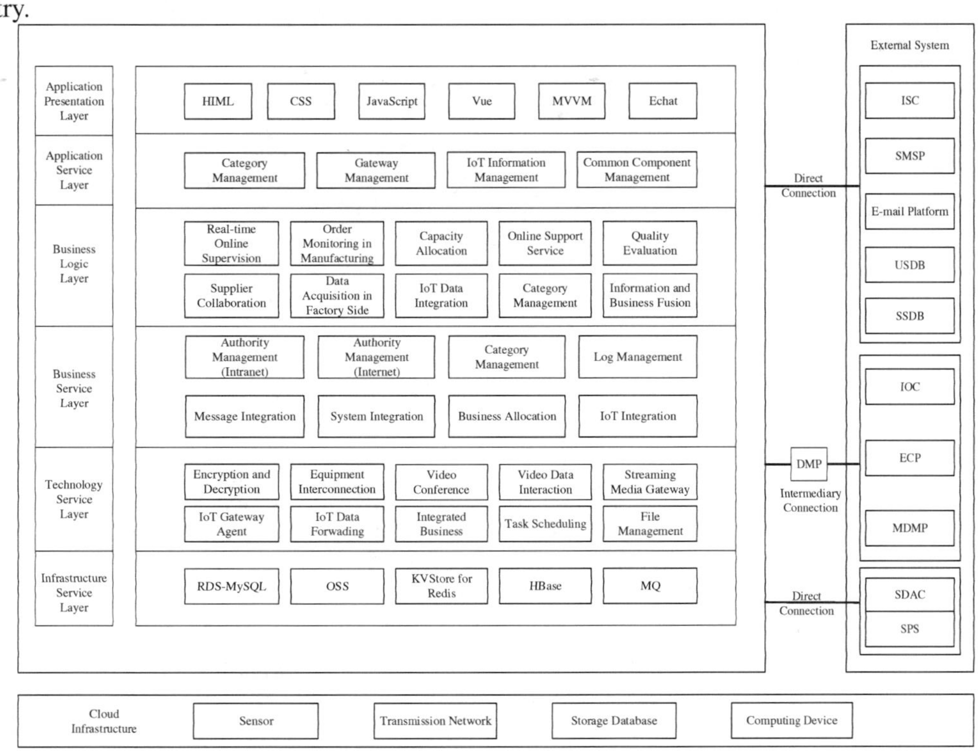

Figure 1. Overall architecture diagram of EIP.

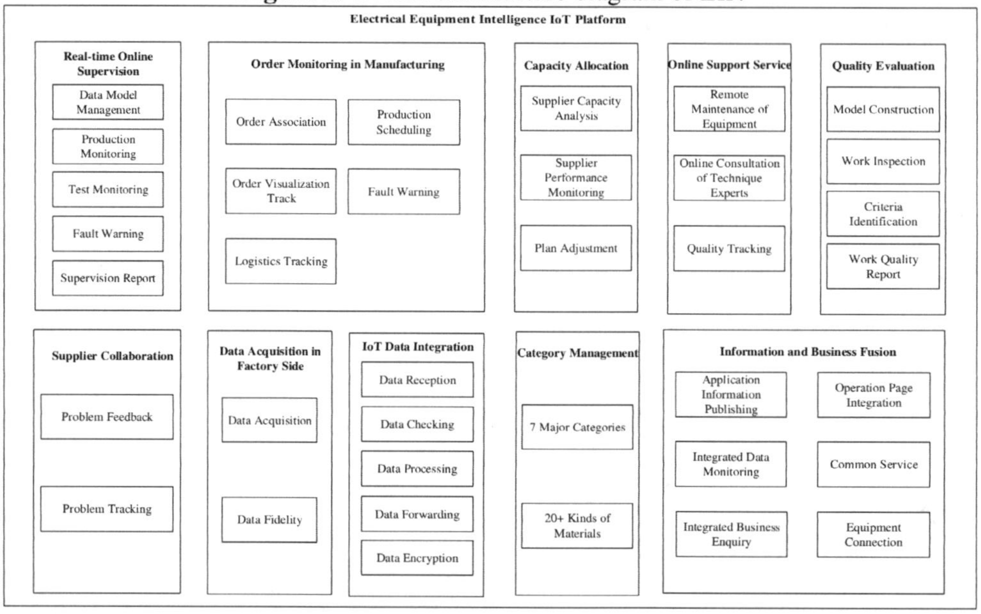

Figure 2. 10 function modules and 37 function submodules of EIP.

4.3. Database Design

4.3.1. Data integration

According to the overall architecture, EIP is integrated with 10 external systems with the data integration process depicted in Figure 3.

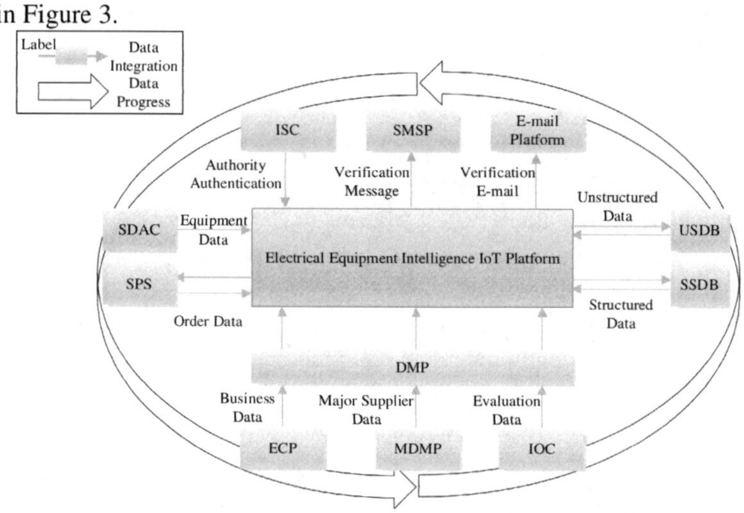

Figure 3. Data integration process between EIP and external systems.

Firstly, EIP is combined with ISC to realize the authority authentication of system users, during the period of which the external users log in EIP with the verification message by SMSP and the internal users log in EIP with the verification e-mail by e-mail platform. Secondly, users having logged in EIP acquires the production data, experiment data, and video data from SDAC. Meanwhile, integrating with SPS, EIP provides the purchase orders to equipment suppliers from which EIP receives the sales orders and production orders in return. Thirdly, through the intermediary agent of DMP, EIP acquires the order data, contract data, project data, etc. from ECP to achieve related functions such as order monitoring and capacity allocation, acquires the major supplier data from MDMP to track the logistics information, and acquires the related evaluation data on equipment quality from IOC to support the improved services. Finally, all the data is stored in SSDB, which includes the structured data such as document, text, table, etc., and USDB, which includes the unstructured data and semi-structured data such as image, audio, video, etc., by classification for the flexible call of EIP.

4.3.2. Data logic model

EIP mainly involves three types of data which are respectively common business data, category management business data, and IoT management data. The logic model of common business data describes the relationship of data entities which are suppliers, orders, schedules, etc., as is shown in Figure 4.

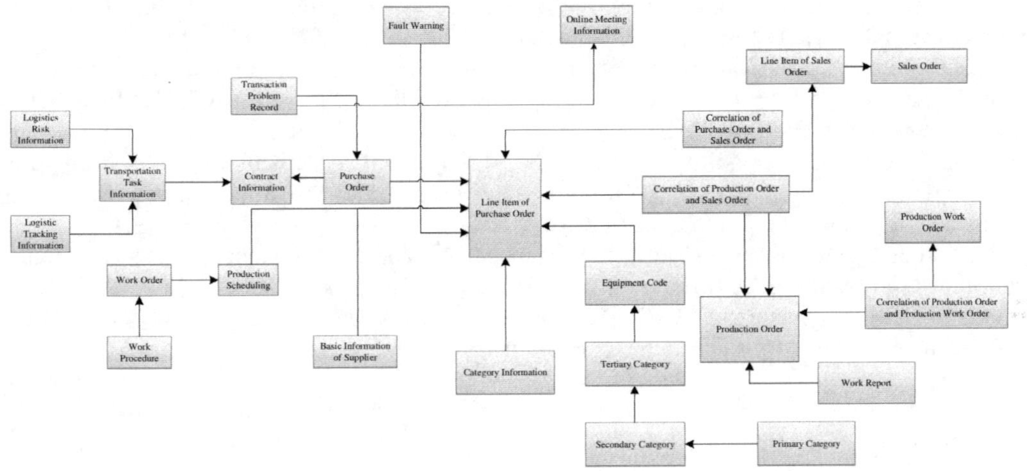

Figure 4. Logic model of common business data.

The logic model of category management business data describes the relationship of data entities which are electrical equipment, orders, production and test data, etc., as is shown in Figure 5.

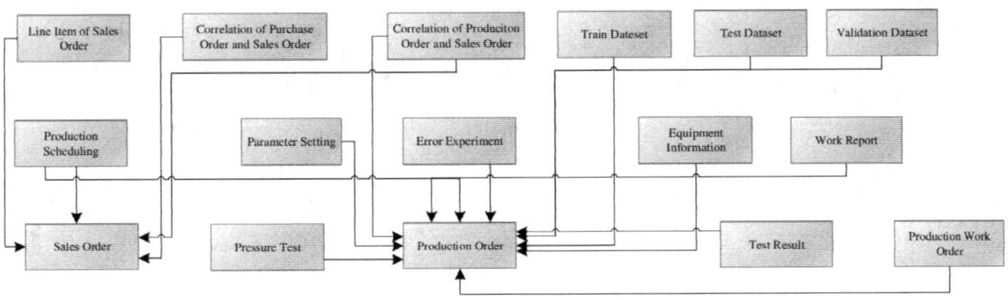

Figure 5. Logic model of category management business data.

5. Conclusion

With the increasingly development of the new information and communication technologies, China has proposed the integration of new information technology, electrical equipment, and intelligent IoT system to accelerate the innovative development of the manufacturing industry. However, the information isolation and data scattering cause the poor services of electrical equipment. Aiming at the question, the preliminary construction of EIP is carried out in this article, through system analysis and system design build platform, help enterprises to improve the quality of manufacturing, realizing the transformation and upgrading and healthy development of the enterprise to enhance the core competitiveness for the electrical equipment manufacturing industry.

The development prospect of EIP is very broad, and its constant upgrade and wide application contribute to the several benefits: (1) It solves the problem on standardized access of differentiated data in equipment production. (2) It realizes the objective, scientific, and systematic quality evaluation of electrical equipment. (3) It builds a shared data security system of information interaction between Internet and Intranet and entity interconnection covering nearly ten thousand equipment suppliers. (4) It boosts the speed of data processing by realizing the calculation of mass data in the edge side. However, the construction of EIP is still in its preliminary stages, thus more technical and methodological breakthroughs are needed to improve the system, which is also the key research content in the future work.

References

[1] Liu, JC; Li, YH; Lu, YY; et al. Research on the influence factors of ubiquitous power Internet of things for promoting consumption of wind power based on fuzzy G1-ISM in China[J]. International Journal of Electrical Power & Energy Systems, 2020, 121.

[2] Lu, YY. China's electrical equipment manufacturing in the global value chain: A GVC income analysis based on World Input-Output Database (WIOD)[J]. INTERNATIONAL REVIEW OF ECONOMICS & FINANCE, 2017, 52:289-301.

[3] Shen, X; Cao, M; Lu, Y; et al. Life Cycle Management System of Power Transmission and Transformation Equipment based on Internet of Things[C]. 2016 China International Conference on Electricity Distribution (CICED), 2016.

[4] Malhotra, R.; Mcleod, E.; Alzahawi, T. Management and Maintenance of Electrical Equipment in Industrial Facilities[J]. IEEE Industry Applications Magazine, 2021, 27(1):48-54.

[5] Quan J. The Architecture of MOOCs Platform Based on Virtual Technology[C]. 6th International Conference on Electronic, Mechanical, Information and Management Society. Atlantis Press, 2016.

[6] Yibai Wang. Architecture Research on Cloud-computing Platform Based on Virtualization Technology[C]. International Information and Engineering Association: Computer Science and Electronic Technology International Society, 2018, 5.

[7] Information Technology - Data Analytics; Study Data from Pukyong National University Update Knowledge of Data Analytics (Developing a big data analytics platform for manufacturing systems: architecture, method, and implementation)[J]. Computers, Networks & Communications,2019.

[8] Li F, Wang J. Citizen fusion service platform for smart cities: architecture, technologies and practice[C]. 2014 IEEE International Conference on Global Software Engineering Workshops, 2014: 51-56.

[9] Amine Ait Si Ali, Hamza Djelouat, Abbes Amira Amine Bermak. Electronic nose system on the Zynq SoC platform[J]. Microprocessors and Microsystems, 2017, 53.

[10] Alvarez, M.; Arquero, A.; Martinez, E.; et al. Domogis: prototype of a interface of a building control system integrated into a GIS[J]. Informes De La Construccion, 2010, 52(518):15-24.

[11] Wu, H; Pang, GKH; Choy, KL; et al. A Scheduling and Control System for Electric Vehicle Charging at Parking Lot[C]. 2017 11th Asian Control Conference (ASCC), 2017:13-18.

[12] Hao, J; Chang Q; Zhang, G; et al. A New Workflow System Architecture and Modeling Platform[J]. Application of Electronic Technique, 2015, 41(2):171-173,176.

[13] Lin, F; Hu, M; Jiang, YC; et al. Architecture and Related Techniques of a Power Dispatching Data Platform[J]. Automation of Electric Power Systems, 2007, 31(1):61-64.

AUTHOR INDEX